Global Cryosphere: Ice Sheets and Glaciers

Global Cryosphere: Ice Sheets and Glaciers

Edited by **Cortez Ford**

Syrawood Publishing House
New York

Published by Syrawood Publishing House,
750 Third Avenue, 9th Floor,
New York, NY 10017, USA
www.syrawoodpublishinghouse.com

Global Cryosphere: Ice Sheets and Glaciers
Edited by Cortez Ford

International Standard Book Number: 978-1-68286-311-4 (Hardback)

Printed in China.

Contents

Preface

The cryosphere is an important part of the global climate system, particularly in relation to atmospheric and oceanic cycles and hydrology. They are also crucial because the environmental changes and climatic variability are most evident in these regions of earth. This book brings forth some of the most significant topics for discussion in this field of study like glacial flow, ice sheet cover and ice thickness, impact of climate change on glaciers, etc. It presents researches and studies performed by experts across the globe. Students and researchers would find this book an invaluable source of knowledge on this topic.

This book is the end result of constructive efforts and intensive research done by experts in this field. The aim of this book is to enlighten the readers with recent information in this area of research. The information provided in this profound book would serve as a valuable reference to students and researchers in this field.

At the end, I would like to thank all the authors for devoting their precious time and providing their valuable contributions to this book. I would also like to express my gratitude to my fellow colleagues who encouraged me throughout the process.

Editor

1

Seabed topography beneath Larsen C Ice Shelf from seismic soundings

A. M. Brisbourne, A. M. Smith, E. C. King, K. W. Nicholls, P. R. Holland, and K. Makinson
British Antarctic Survey, High Cross, Madingley Road, Cambridge, CB3 0ET, UK

Correspondence to: A. M. Brisbourne (aleisb@bas.ac.uk)

Abstract. Seismic reflection soundings of ice thickness and seabed depth were acquired on the Larsen C Ice Shelf in order to test a sub-ice shelf bathymetry model derived from the inversion of IceBridge gravity data. A series of lines was collected, from the Churchill Peninsula in the north to the Joerg Peninsula in the south, and also towards the ice front. Sites were selected using the bathymetry model derived from the inversion of free-air gravity data to indicate key regions where sub-ice shelf oceanic circulation may be affected by ice draft and seabed depth. The seismic velocity profile in the upper 100 m of firn and ice was derived from shallow refraction surveys at a number of locations. Measured temperatures within the ice column and at the ice base were used to define the velocity profile through the remainder of the ice column. Seismic velocities in the water column were derived from previous in situ measurements. Uncertainties in ice and water cavity thickness are in general <10 m. Compared with the seismic measurements, the root-mean-square error in the gravimetrically derived bathymetry at the seismic sites is 162 m. The seismic profiles prove the non-existence of several bathymetric features that are indicated in the gravity inversion model, significantly modifying the expected oceanic circulation beneath the ice shelf. Similar features have previously been shown to be highly significant in affecting basal melt rates predicted by ocean models. The discrepancies between the gravity inversion results and the seismic bathymetry are attributed to the assumption of uniform geology inherent in the gravity inversion process and also the sparsity of IceBridge flight lines. Results indicate that care must be taken when using bathymetry models derived by the inversion of free-air gravity anomalies. The bathymetry results presented here will be used to improve existing sub-ice shelf ocean circulation models.

1 Introduction

Ice-shelf disintegration is of global significance due to two important processes. Firstly, although the loss of the floating portion of an ice shelf following retreat or break-up has only a minor direct effect on sea level, the removal of the buttressing effect of the ice shelf and the subsequent acceleration of the tributary ice streams can result in a significant contribution to sea level rise (e.g., Rignot et al., 2004; Scambos et al., 2004). Secondly, cold and low-salinity water produced by the basal melting of ice shelves influences Antarctic Bottom Water formation and consequently the properties of the global oceans (Jacobs, 2004). Reduced production of Ice Shelf Water following the loss of an ice shelf decreases the production of bottom water formed by ice shelf-ocean interaction. However, Antarctic Bottom Water production potentially increases through enhanced sea ice production as a result of more open water available for atmosphere-ocean exchange. In some sectors of Antarctica, such as in the South Eastern Weddell Sea, ice shelf melting reduces the salinity to the point where bottom water formation is not possible (Fahrbach et al., 1994).

The susceptibility to break-up of eastern Antarctic Peninsula ice shelves has been highlighted over the last two decades by the well-documented collapse of the most northerly sections of Larsen Ice Shelf, Larsen A in 1995 (Rott et al., 1996; Vaughan and Doake, 1996) and Larsen B in 2002 (Glasser and Scambos, 2008). There is some evidence that the freshwater released after these break-ups has contributed to a significant freshening of the Antarctic Bottom Water produced in this region (Jullion et al., 2013). Larsen C Ice Shelf (Fig. 1), the largest ice shelf on the Antarctic Peninsula at around 50 000 km^2 (Cook and Vaughan, 2010)

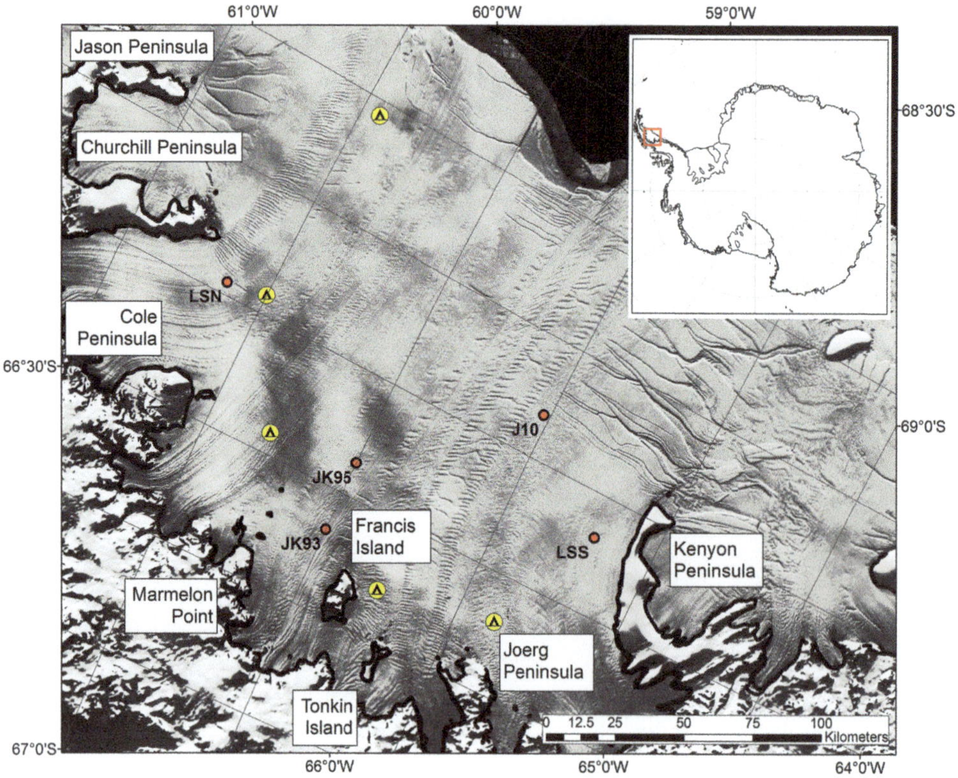

Fig. 1. Larsen C Ice Shelf with MODIS image (Haran et al., 2005) annotated with the main geographical features referred to in the text. Seismic field camps in yellow. Red circles are sites of previously measured ice thickness and water depth: JK93 Seismic (King et al., 1993); JK95 Seismic (Jarvis and King, 1995); LSN/LSS BAS Moorings (Nicholls et al., 2012); J10 Seismic (Kulessa et al., 2010). The grounding line as derived from DInSAR data (Rignot et al., 2011) is represented by the thick black coastline.

lies just to the south of these ice shelves, and is currently just beyond the encroaching −5 °C mean annual air temperature isotherm, the proposed heuristic limit of ice-shelf viability (Vaughan and Doake, 1996). Ice-shelf retreat has been attributed to atmospheric warming in a number of cases (Vaughan and Doake, 1996; Rott et al., 1998; Skvarca et al., 1999). Consequently, with the Antarctic Peninsula exhibiting one of Earth's highest rates of atmospheric warming (Vaughan et al., 2003), eventual break-up of the more southerly Larsen C Ice Shelf is likely.

Ice shelf break-up is preceded by thinning, which results from processes at both the surface and the ice base. However, the relative contribution of surface processes and basal melting to the observed thinning of Larsen C is as yet unresolved. Satellite radar altimeter data indicate that surface lowering of Larsen C has been on-going since at least 1992 (Shepherd et al., 2003; Fricker and Padman, 2012). Several authors suggest that this thinning may be caused by a reduction in firn air, a manifestation of increased surface melting resulting from circumpolar air currents flowing over the peninsula and descending on the eastern side (Holland et al., 2011; Pritchard et al., 2012). Shepherd et al. (2003) argue that a significant component of thinning must result from increased basal melting caused by warmer Weddell Sea Deep Waters observed at depths greater than the ice-shelf draft. However, Nicholls et al. (2012) presented results from two drill sites which demonstrate that the sub-ice shelf cavity is flushed by water at the surface freezing point, indicating that increased basal melting is unlikely to have contributed to surface lowering.

Accurate knowledge of the bathymetry beneath ice shelves is critical to modelling and understanding the contribution of basal melting to ice shelf thinning and subsequent instability (Jenkins et al., 2010; Grosfeld et al., 1997). Due to the difficulty of mapping the cavity directly, sub-ice shelf bathymetry remains relatively poorly known: radar energy is highly attenuated by sea water and therefore precludes airborne radio echo sounding; direct measurements with autonomous underwater vehicles are expensive and complex. Seismic surveys, although time-consuming, are the only geophysical method available to measure accurately ice thickness, water column thickness and seabed properties. Consequently, a number of seismic surveys have been carried out, and these have provided most of the presently available maps of sub-ice shelf bathymetry (e.g., Determann et al., 1988; Johnson and Smith, 1997; Nost, 2004).

The importance of sub-ice shelf geometry has led to a number of indirect techniques being used to derive geometry models from sparse observations. Hemer et al. (2006) and Galton-Fenzi et al. (2008) minimised the misfit between ocean-tide models and tidal elevation observations by adjusting sub-ice shelf cavity geometry models for the Amery Ice Shelf. Roy et al. (2005) inverted airborne gravity data to derive water depth and sediment thickness of Lake Vostok, East Antarctica, constrained by available seismic and radar profiles. Similarly, Muto et al. (2013) constrained their aerogravity inversion for bathymetry and sediment distribution beneath Pine Island Glacier with cavity geometry derived from autonomous underwater vehicle tracks. Tinto and Bell (2011) modelled bathymetry beneath Thwaites Glacier in the Amundsen Sea region using known bathymetry to constrain inhomogeneous geology. Of direct relevance to this study, Cochran and Bell (2012) used Operation IceBridge free-air gravity anomaly (FAA) data over the Larsen Ice Shelf. The Parker-Oldenburg technique (Oldenburg, 1974) used in that study relates the observed free-air anomaly to topographic relief superimposed on a horizontal interface. Two variables are permitted in the inversion technique used. Firstly, mean depth to the seabed is fixed to provide the best-match at a known point, such as a previous seismic or drill site. Secondly, density contrast is specified. As only one density contrast can be used for the entire model, it is equivalent to assuming a uniform geology. Any lateral variation in gravity anomaly is therefore attributed to lateral bathymetry variation. The sensitivity of the resulting bathymetry model to both these variables can be tested. Cochran and Bell (2012) used previous seismic measurements (King et al., 1993; Jarvis and King, 1995) north of Francis Island (Fig. 1) to constrain the mean depth of the bathymetry model beneath Larsen C. A density contrast of $1.67\,\mathrm{g\,cm^{-3}}$ was assumed based on the regional geological framework.

The bathymetry model presented by Cochran and Bell (2012) exhibits over-deepening along the grounding line and also two broad troughs extending from the grounding line to the ice front. These features could provide conduits for oceanic circulation beneath the entire ice shelf. Mueller et al. (2012) modelled ice-ocean interaction for Larsen C Ice Shelf, predicting basal melt rates using a numerical model with a bathymetry that appears similar to that of Cochran and Bell (2012), and also variants of this bathymetry. They demonstrated the sensitivity of the predicted melt rates to cavity geometry: localised narrowing of the cavity results in significantly elevated melt rates because of local intensification of tidal flow.

In order that the Cochran and Bell (2012) bathymetry map beneath Larsen C could be validated for use in oceanographic modelling, a seismic survey was undertaken in 2012/13 to obtain point measurements of ice and sub-ice shelf water cavity thickness (Fig. 2). Here we present the results of this survey and compare them with those obtained by inversion of the Operation IceBridge gravity data.

2 Data and methods

Logistics and time constraints precluded surveying the entire ice shelf in a single field season. Therefore, the bathymetry map of Cochran and Bell (2012) was used to prioritise specific regions of interest in sub-ice shelf circulation models. The Cochran and Bell (2012) cavity geometry (Fig. 2) infers a potential pathway for whole-shelf circulation via overdeepenings along the grounding line. Restrictions to flow along this conduit would significantly alter sub-shelf circulation and therefore basal melt rates. Consequently, a series of point measurements were collected along lines radiating from coastal promontories, where the IceBridge bathymetry model would predict restricted water flow (Fig. 2). Also, a single line was acquired closer to the seaward edge of the ice shelf, between an extreme maximum and an extreme minimum of the cavity thickness as predicted by the gravity inversion. Eighty-seven new seismic measurements of ice thickness and seabed depth were made. Previously, only five in-situ measurements have been available (Figs. 1 and 2) (King et al., 1993; Jarvis and King, 1995; Kulessa et al., 2010; Nicholls et al., 2012).

2.1 Data acquisition and processing

Seismic soundings were acquired along 10 lines at 3, 5 or 10 km spacing (Fig. 2). At each site, a 4 m shot hole was drilled using a hand-operated PICO drill. Charges of 150 to 600 g of Pentolite high explosive were used and the holes backfilled with loose snow. Twenty-four 40 Hz georods (Voigt et al., 2013) were deployed at a 10 m spacing and buried to ensure good coupling and to provide protection from the wind. Two-second records were acquired with a 24 channel Geometrics Geode data logger at 0.125 ms sample interval.

Seismic reflection quality was highly variable (Fig. 3). In general, the seabed reflector was obvious in the raw seismic records. However, the ice base reflection was not always clear. Where necessary a frequency-wavenumber (F-K) filter was applied to remove unwanted noise (groundroll) which masked the ice base reflection where ice thickness was less than around 270 m (e.g., Fig. 3b). In some cases, an automatic gain control (AGC) filter was applied to the records to enhance weaker signals later in the records (Fig. 3b). In many cases, the AGC filter emphasises internal multiple reflections from the ice layer which can be used to determine ice thickness when the primary reflection is unclear (return i2 in Fig. 3c).

2.2 Seismic velocities in ice and water

As the density of the snow and firn increases with depth so does the seismic velocity. In general, the velocity of solid ice is reached at a depth of 80–100 m. Below this depth the temperature gradient down to the ice-ocean interface results

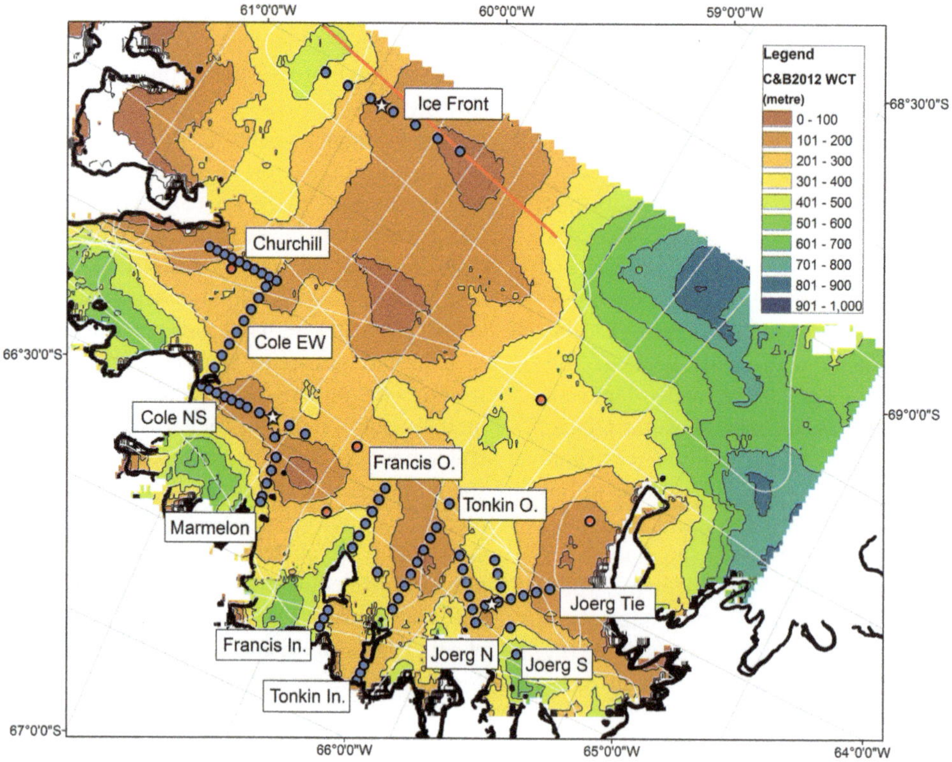

Fig. 2. Seismic reflection acquisition sites presented in Fig. 6 (blue circles) with line names as referred to in the text overlain on the contoured water cavity thickness derived from the inversion of IceBridge free-air gravity anomaly (Cochran and Bell, 2012). White stars indicate the three shallow seismic refraction sites where velocity-depth profiles were derived. Previous sites as per Fig. 1 (red circles). The grounding line as derived from DInSAR data (Rignot et al., 2011) is represented by the thick black coastline. The IceBridge flight line segment towards the ice front referred to in the text is indicated by the red line, all other IceBridge flight lines are in white (an edited version of which was used in the FAA inversion).

in a velocity decrease with depth. From the surface down to the depth of maximum velocity, seismic reflection traveltimes are converted to depth using measured velocity profiles. In order that a velocity-depth profile for the ice column could be established, three shallow seismic refraction surveys were carried out (Fig. 2). A series of surface shots were recorded into georods spaced at 2.5 to 610 m offset and the first arrivals picked (Fig. 4a). Arrival times are converted to a velocity-depth relationship using the method described by Kirchner and Bentley (1990). This method relies on a monotonic increase in velocity with depth. Seasonal melting of the near-surface, as previously observed on Larsen C (Fahnestock et al., 2002), could cause ice layers that potentially invalidate this assumption. However, smooth and increasing traveltime picks (Fig. 4a) observed at all three shallow refraction sites indicate that the assumption is valid in these cases: in general, at the seismic wavelength (> 10 m) velocities increase uniformly with depth. For each seismic survey point, the velocity profile corresponding to the nearest shallow refraction survey was used. Density-depth profiles are derived from the P-wave velocity using the empirical relationship of Kohnen (1972). Below the maximum depth of penetration obtained by seismic refraction surveys, temperature measurements from within the ice column obtained from the boreholes discussed by Nicholls et al. (2012) are used to derive the mean seismic velocity using the temperature-velocity relationship of Kohnen (1974). Downhole measurements indicate an ice temperature of -13 °C at 100 m depth, equivalent to a seismic velocity in ice of $3825\,\mathrm{m\,s^{-1}}$. A shelf-base temperature from conductivity-temperature-depth (CTD) measurements of -2 °C (Nicholls et al., 2012) indicates an ice velocity of $3800\,\mathrm{m\,s^{-1}}$ at the ice base. Therefore, below 100 m depth a mean seismic velocity in the ice of $3812\,\mathrm{m\,s^{-1}}$ is used. Scatter in traveltime picks and insufficient spread-length preclude well-constrained deep velocities from the refraction data. This velocity is therefore also used to constrain velocities at depths below 100 m in the inversion of the seismic refraction traveltimes.

A mean seismic velocity in the water column of $1445 \pm 1\,\mathrm{m\,s^{-1}}$ was derived from CTD measurements made close to the Churchill Peninsula in the north and the Kenyon Peninsula in the south (Nicholls et al., 2012).

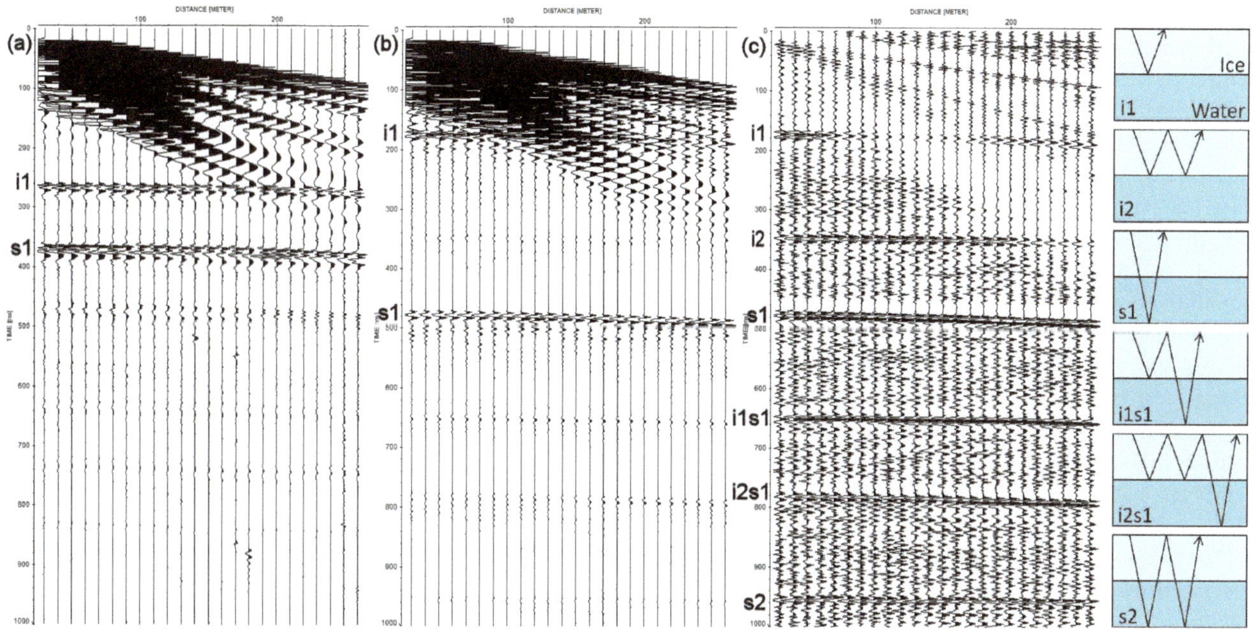

Fig. 3. Example 24-channel seismic records with multiples: **(a)** Raw data from a single Francis Island site with clear ice-base (i1) and seabed (s1) direct arrivals, multiples also present but weak; **(b)** Raw data from a single Churchill Peninsula site with clear s1 arrival but the direct ice-base arrival (i1) is obscured by groundroll noise; **(c)** as b but with a frequency-wavenumber filter to remove groundroll and AGC filter to enhance weaker signals. Direct reflections and multiples annotated as per the schematic. Raypaths on the schematic are for illustrative purposes only and do not represent true raypaths, which are significantly distorted by the velocity gradient of the firn layer and velocity contrast at the ice base.

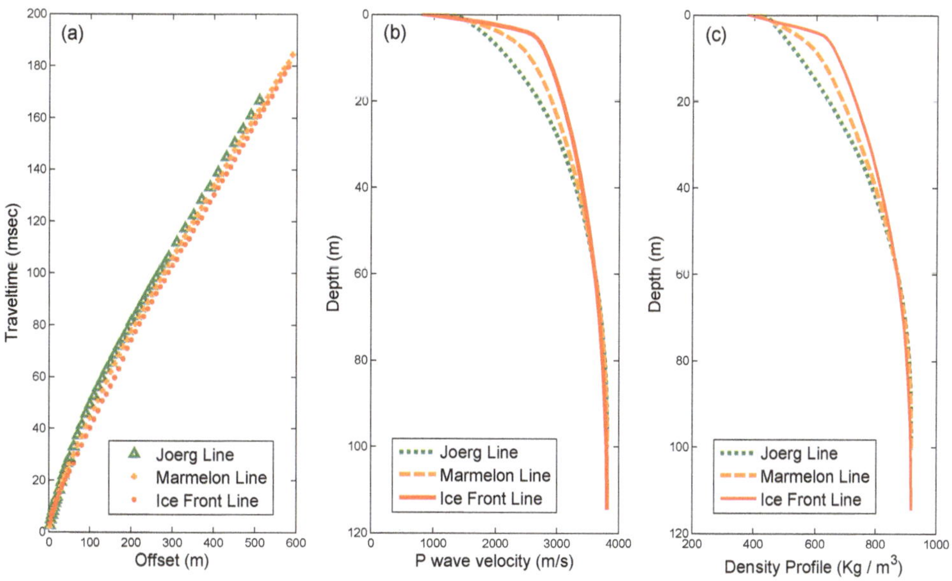

Fig. 4. (a) Traveltime against offset for the 3 seismic refraction experiments indicating a monotonic velocity increase with depth, inferred from the continuous increase in traveltime with offset; **(b)** seismic velocity-depth profiles derived using the method of Kirchner and Bentley (1990) for the three seismic refraction profiles. Velocities at depths beyond 100 m are constrained by temperature profiles discussed by Nicholls et al. (2012) and the velocity-temperature relationship of Kohnen (1974); **(c)** density-depth profiles derived using the method of Kohnen (1972).

The velocity-depth profiles are presented in Fig. 4b and the derived density-depth profiles in Fig. 4c. Densification rates were higher down to pore close-off towards the ice front than close to the grounding line (Ice Front compared to Marmelon, both sites are at a similar latitude). Also, firn densification rates were higher in the north than the south (Marmelon compared to Joerg, both at a similar longitude). Integrating the density profiles produces a total firn-air content of 11.3 ± 1.3 m at the ice-front site, 12.9 ± 1.3 m at the Marmelon site and 14.4 ± 1.3 m at the Joerg site, compared with Holland et al.'s (2011) estimates of 11.0 ± 1.8 m, 10.0 ± 1.8 m, and 12.7 ± 1.8 m respectively, all of which agree within error bars. These results are consistent with a higher degree of melting of firn in the north than the south, as found in the borehole measurements of Nicholls et al. (2012) and the firn air content variation presented by Holland et al. (2011).

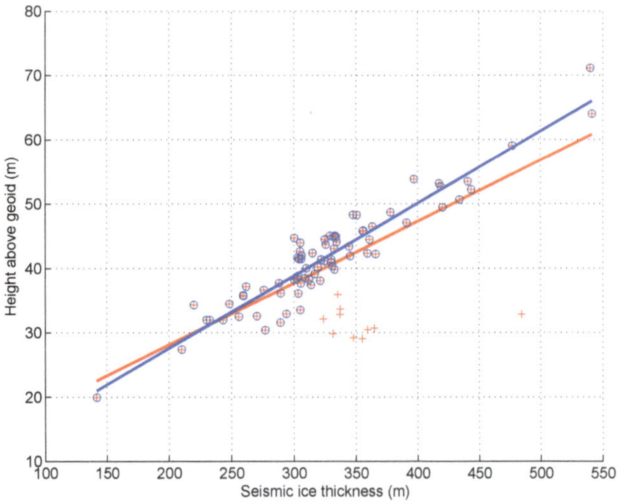

Fig. 5. Geoid-corrected height (h) against seismic ice thickness (H). Red crosses represent all sites. Blue circles highlight those sites deemed unaffected by the proximity of the grounding line, i.e., fully floating excludes sites with negative firn correction calculated from elevation and seismic thickness as per Griggs and Bamber (2009). The red line is the linear fit to all data; the blue line is the linear fit to fully-floating sites (Fit: h = (0.113 ± 0.005) H + (5.003 ± 1.525); $R^2 = 0.89$). No correction for tidal variation in surface elevations was made and this therefore contributes to the scatter observed in the data (± 2 m at the Churchill Camp, Fig. 1).

2.3 Uncertainties

The most significant assumption made in the derivation of depth from ice base and seabed reflections is a flat and planar reflector. Unless the line of georods is directly along strike of any dipping interface the moveout of the returns (variation of arrival time with offset due to ray path geometry) will deviate from that expected. Such discrepancies become clear when picking the traveltimes of the reflections. In the majority of cases the moveout deviation was < 10 m s^{-1} at 260 m offset. During the field campaign, at a site where a 10 m s^{-1} deviation in normal moveout (NMO) at 250 m offset was observed, an orthogonal line was shot: this allows determination of slope and thus allows us to quantify the uncertainty associated with non-standard moveout. Results indicate that although moveout and arrival times varied with shot offset and array orientation, a maximum of ± 8 m variation in seabed depth could be expected.

Nost (2004) presented an analysis of the uncertainties associated with a sloping seabed, indicating a slope of 0.205 (∼ 12°) introduces an uncertainty of ± 60 m for ice and water cavity thicknesses of 500 m each. The maximum large-scale dip of reflectors along the profiles acquired here (∼ 2.5° along the Francis Offshore Line) indicates that assuming a horizontal and planar seabed is reasonable and does not significantly increase uncertainties. Without carrying out a number of shots at each site or at the very least utilising an L-shaped array (Roethlisberger, 1972) it is impossible to determine interface geometry accurately and therefore we made the assumption that interface topography is negligible.

Picking errors in the seismic reflection data are in general < 0.5 ms. Uncertainties in the seismic velocities in the firn layer cannot readily be quantified here due to the statistically small number of measurements made. However, using a similar technique and multiple sites Kirchner and Bentley (1990) estimate a velocity uncertainty of ± 60 m s^{-1} near the surface, reducing to ± 30 m s^{-1} at 10 m depth and ± 15 m s^{-1} at 50 m depth. Our use of in situ temperature measurements provides greater accuracy beyond these depths.

Booth et al. (2010) used semblance analysis of seismic reflection data from Larsen C Ice Shelf (Site J10 in Fig. 1), combined with radar sections constraining layer thickness, to imply a marine ice P-wave velocity of 2162 m s^{-1} (40 % lower than meteoric ice). This was shown to reduce the overall root-mean-square (rms) velocity of the ice base reflector by 10 % which would result in an overestimation of ice thickness of 10 % for that particular location. In this study it is not possible to determine the thickness of any marine ice present. We therefore assume a meteoric ice velocity for the entire ice column. Ice thickness measurements determined here are however consistent with surface elevation measurements (Fig. 5) indicating that this assumption does not significantly bias the results.

At each site a dual-frequency Leica 1200 or GS10 GPS receiver was used to measure ellipsoidal surface heights. The EIGEN-GL04C geoid model (Forste et al., 2008) was used to correct measured ellipsoidal elevation to the height above the geoid, which was then subtracted from the ice thickness to determine ice draft. The shelf ice rises and falls with the ocean tides, which were not corrected for in this study. This will contribute some time-dependent variation in both the surface height measurements and the water cavity thickness. A tidal range of over 4 m was recorded at the Churchill Camp, indicating an implicit variation in cavity thickness of

± 2 m. This is insignificant with respect to other uncertainties and is therefore not corrected for in this study.

This results in an uncertainty in both the ice thickness and water column thickness of 5 to 10 m depending on the reflection quality and moveout, or a seabed depth uncertainty of around 10 m, increasing to 20 m in areas where reflector quality is diminished. Ice thickness could be greatly underestimated where a significant thickness of low-velocity marine ice is present. Importantly, the thickness of the water column derived from the seismic data is independent of the ice velocity-depth profile used.

Cochran and Bell (2012) attribute uncertainties to their gravimetrically derived model of as much as 100 m due to the possible presence of low-density sediments, especially in the deep troughs.

3 Results

3.1 Ice thickness and draft

Figure 5 presents the ice thickness derived from seismic soundings against geoid-corrected surface elevation. As expected, there is a clear linear relationship between surface elevation and thickness for floating ice. However, there are some outliers from this general relation, which we suggest to be points that are not freely floating. Using the derived surface elevation and seismic ice thicknesses, we calculate the firn-air correction required to satisfy the assumption that the ice is floating in hydrostatic equilibrium (Griggs and Bamber, 2009). Where the required firn-air correction is negative, we contend that the ice is not freely floating. A number of outliers are observed where proximity to the grounding line precludes fully-floating ice (Marmelon Line, proximal to Tonkin Island and the Cole Peninsula). Due to the different firn profiles and melting histories between north and south we could expect different elevation-thickness relationships, as reflected in the refraction seismic velocity-depth profiles. The seismically measured ice thickness is in good agreement with the thickness measurements derived from surface elevations and airborne radar soundings (Holland et al., 2009) with an rms difference of 26.7 m for the 76 fully floating sites. Ice thickness measurements are not affected by tidal height.

3.2 Results: Bathymetry and sub-ice shelf cavity thickness

Profiles comparing ice draft and water cavity thickness from the seismic measurements with those derived by the free-air gravity anomaly inversion are presented in Fig. 6. There are some significant discrepancies in the gravity inversion results when compared with the seismically measured depths. The rms error of the gravity inversion at 87 seismic sites is 162 m in bathymetry and 143 m in sub-ice shelf cavity thickness. Errors in the cavity thickness determined by the gravity inversion range from less than 10 m up to 320 m.

3.2.1 Ice front

The gravimetric results for the ice front line indicate a significant shallowing in the bathymetry from north to south (Fig. 6a), resulting in a water cavity which reduces from around 380 m in the north to less than 100 m in the south. However, the seismic results indicate a uniform water cavity thickness of around 220 to 240 m, resulting in an rms error of 103 m for the gravimetrically derived bathymetry along this line. Ice draft is consistent between the methods.

3.2.2 Churchill Peninsula

There is good agreement between the gravimetrically and seismically derived cavity thickness along the Churchill Line, as presented in Fig. 6b, with an rms error of 20 m in bathymetry.

3.2.3 Cole Peninsula – EW and NS

Although there is general agreement between the observed and modelled depths for the two profiles radiating from the Cole Peninsula (Figs. 6c and d), there is no evidence in the seismic data for the thick sub-ice shelf cavity close to the peninsula. Also, on the north-south line the cavity narrows significantly within 10 km of the peninsula, to <100 m, a feature not observed in the gravity inversion results.

3.2.4 Marmelon Point

There is a significant discrepancy between the gravity results and the seismic observations east of Marmelon Point (Fig. 6e). The seismic sites close to the point indicate that the ice is most likely grounded in this area. This result is supported by field observations of strand cracks (Swithinbank, 1955) between the seismic sites. Furthermore, the DInSAR grounding line (Rignot et al., 2011) indicates that islands are buried within the ice shelf in this region (visible in Fig. 1, directly east of Marmelon Point). The size of these sub-ice features is likely below the resolution of the gravity data. However, the difference in the water cavity implied by the two methods (~ 200 m) is significant. The lateral resolution of the Cochran and Bell (2012) gravity inversion is determined by the flight-line spacing (areas > 10 km from data points masked) and flight speed (a low-pass filter results in a resolution scale of ~ 5 km). Anomalies below these dimensions are resolved at reduced amplitude. Features such as the submerged islands off Marmelon Point, which are in the sub-kilometre dimension range, are therefore unlikely to be well resolved in the gravity inversion.

3.2.5 Francis Island – inshore and offshore

The seismic results indicate that although not grounded between the peninsula and Francis Island, the water cavity is in general <100 m thick (Fig. 6g), and relatively thin just

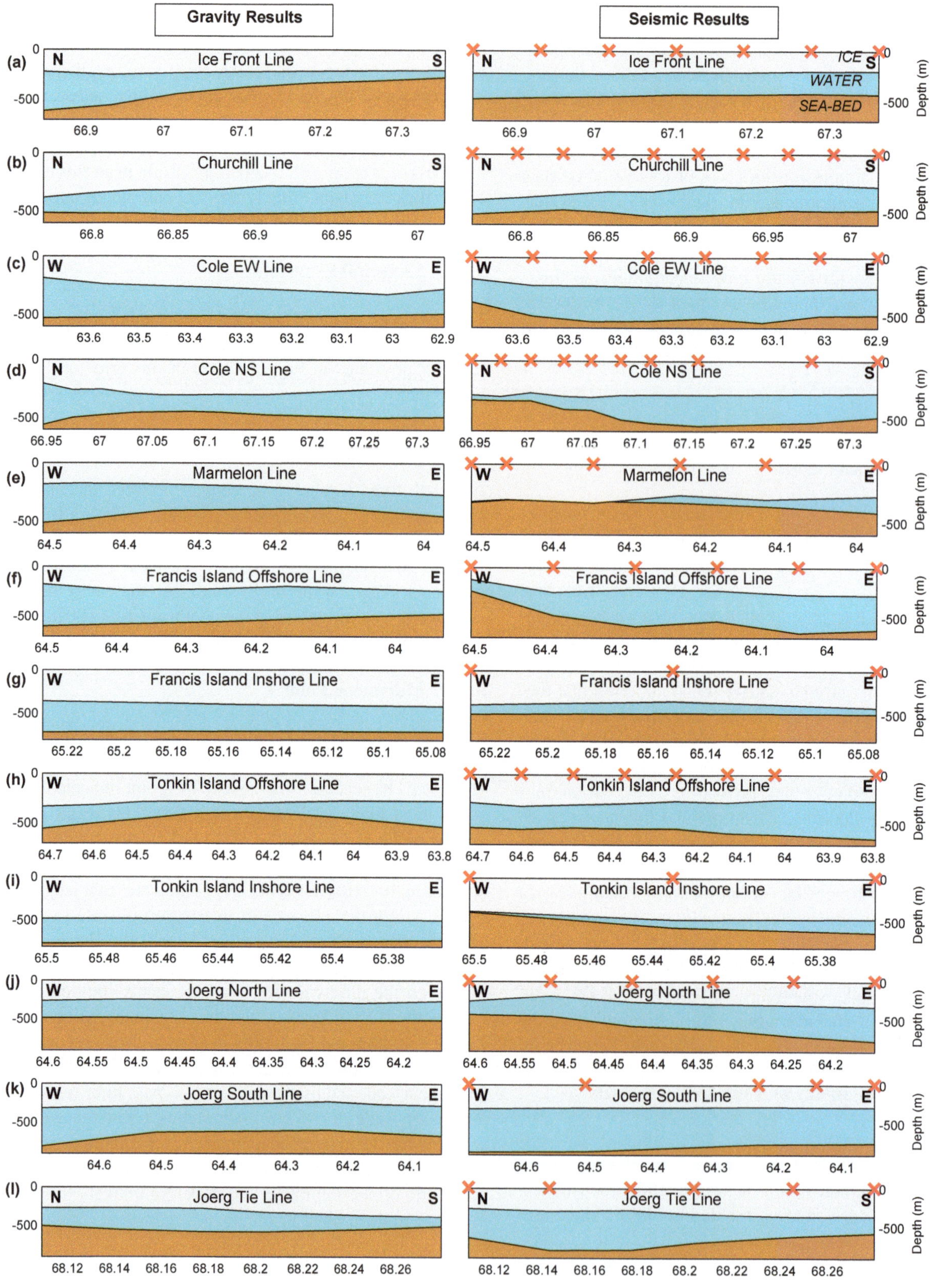

Fig. 6. Ice and sub-ice shelf cavity profile pairs comparing Cochran and Bell (2012) gravity inversion (left) and seismic (right) results. Red crosses indicate locations of seismic measurements. Vertical exaggeration varies between plots. The horizontal coordinate is latitude (S) or longitude (W) as appropriate.

offshore of Francis Island (Fig. 6f). The results of Cochran and Bell (2012) imply a >400 m cavity in both these areas.

3.2.6 Tonkin Island – inshore and offshore

Similar to Francis Island, and again contrary to the gravity inversion, the cavity is <100 m thick between Tonkin Island and the mainland (Fig. 6i) but thickens eastwards offshore of the island (Fig. 6h).

3.2.7 Joerg Peninsula

The gravity inversion results indicate a deep but narrow trough running NE-SW to the south of the Joerg Peninsula. The seismic results confirm that the trough is present but is significantly broader than expected (Figs. 6j–l).

4 Effect of lateral density variations on free-air anomaly inversion

Non-uniqueness in the inversion of gravity data is a result of the trade-off between the magnitude, geometry and depth of a subsurface density contrast. A priori knowledge, such as likely lithology, structure and density contrasts, is therefore essential to constrain uniquely inversions for subsurface structure. With knowledge of water column thickness, and constrained by regional geological structure, gravity data over subglacial lakes or ice shelves have been inverted for density contrasts in the sub-surface (e.g., Roy et al., 2005; Muto et al., 2013). Similarly, a priori knowledge of subsurface density contrasts is necessary to determine uniquely the seabed topography from gravity data acquired over an ice shelf. Due to the lack of knowledge of subsurface structure in the region, Cochran and Bell (2012) assume that the seafloor beneath Larsen C consists of volcanic sequences overlain by a negligibly thin or discontinuous layer of sediments. However, to the west of the Antarctic Peninsula, Bart and Anderson (1995), for example, demonstrate that a series of glacial episodes resulted in a complex offshore stratigraphy. The bathymetry results presented here demonstrate that sub-ice shelf cavity models derived by the inversion of airborne free-air gravity anomalies are subject to significant error and can introduce bathymetric features which are not present. Where the assumption of a uniform geology is incorrect, i.e., a subsurface density contrast is present, spurious bathymetric features will always be introduced. Complex stratigraphy north of the Jason Peninsula, offshore from the former Larsen B Ice Shelf, was presented by Sloan et al. (1995). By extrapolating from geology described onshore to interpret marine seismic sections, volcanics appear to dominate the western margin of the former Larsen B Ice Shelf. However, further offshore significant marine, glacial-marine and till deposits are inferred, including large scale cut-and-fill geometries and deposition sequences consistent with ice stream retreat.

To demonstrate the non-uniqueness of the gravity inversion we modelled a simple 2-D anomaly, i.e., a linear feature striking perpendicular to a survey line (Fig. 2). The method of Talwani et al. (1959) is used to derive the gravity anomaly of arbitrary polygons. For example, a potentially realistic model is sediment filled channels at either end of the ice front flight line, separated by a sill of volcanics. Assuming unconsolidated sedimentary infill of $1.9\,\mathrm{Mg\,m^{-3}}$ over volcanics of $2.7\,\mathrm{Mg\,m^{-3}}$ (Fig. 7), modelling indicates an infilled channel of around 800 m depth is sufficient to produce the 25 mgal anomaly observed along the ice front flight line segment. More consolidated, and hence higher density, sediments require a deeper infill layer. For example, infill of $2.3\,\mathrm{Mg\,m^{-3}}$ would require the sediment to extend to depths of 2500 m beneath the seabed. Such sediment thicknesses are not unrealistic and at the shallower end of the scale are comparable to seabed depths indicated near the grounding line. Potentially, this anomaly is an extension of the north-south gravity high which manifests itself as incoherent hyperbolic reflections in marine seismic reflection data, attributed to Jurassic volcanics by Sloan et al. (1995) and Smith and Anderson (2010). This simple 2-D model illustrates the non-uniqueness of the gravity inversion method for deriving bathymetry where subsurface geology is unconstrained: lateral variation in the subsurface density contrast or variable bathymetry over a uniform density can produce an identical free-air gravity anomaly.

5 Implications for sub-ice shelf ocean circulation and basal melting

The effects of significant topographic features in the sub-ice shelf cavity have previously been demonstrated by Mueller et al. (2012), who showed the differences in modelled basal melt rates that result from using different sub-ice shelf cavity geometries. In the eastern sector of the gravimetrically derived Larsen C bathymetry there is a narrowing of the cavity to <100 m near the ice front south of 67° S (Fig. 2). Although the bathymetry used by Mueller et al. (2012) was derived by extrapolation from offshore bathymetry, it also features a similar narrowing, slightly to the north but in general consistent with Cochran and Bell (2012). Mueller et al. (2012) predict a $\sim 2\,\mathrm{m\,a^{-1}}$ melt rate as a result of rapid tidal flow through this narrow cavity in the northeast region, compared with a background rate of $<0.2\,\mathrm{m\,a^{-1}}$. Mueller et al. (2012) also tested a bathymetry with a minimum 350 m water cavity, more closely emulating the seismic observations towards the ice front. Predicted basal melt rates towards the ice front were significantly reduced in the thicker cavity, by as much as $2.3\,\mathrm{m\,a^{-1}}$. This result is more consistent with observed ice shelf draft, since localised melt rates of $\sim 2\,\mathrm{m\,a^{-1}}$ would be expected to be reflected in thinning, and thus lower ice surface topography, in the northeastern area of the ice shelf.

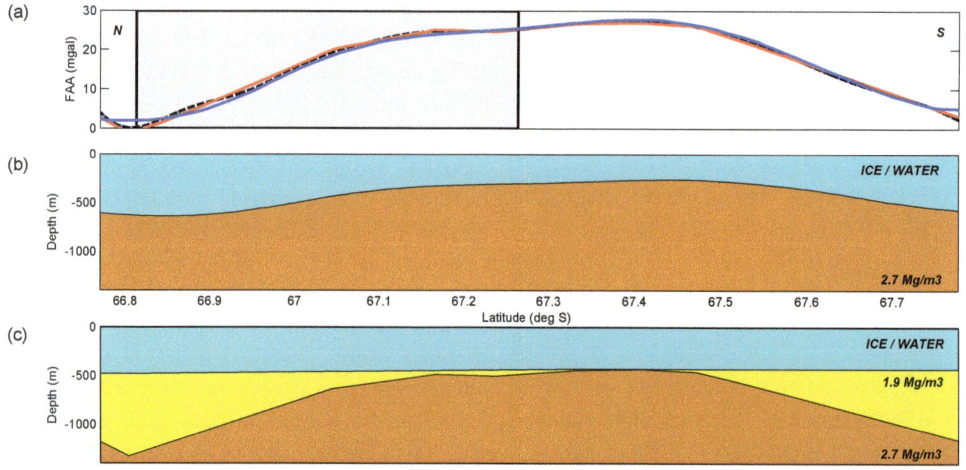

Fig. 7. Modelling of the IceBridge free-air gravity anomaly using the method of Talwani et al. (1959): **(a)** Black dashed line: measured IceBridge FAA; Blue line: modelled FAA with Cochran and Bell (2012) bathymetry model and constant sub-surface density of 2.7 Mg m^{-3}; Red line: modelled FAA assuming 1.9 Mg m^{-3} sediment over 2.7 Mg m^{-3} basement and bathymetry as derived by seismic survey (seabed assumed laterally invariant south of last shot point). The length of the IceBridge line coincident with the seismic line is outlined by the grey box (i.e., where bathymetry is constrained by seismic measurements); **(b)** Cochran and Bell (2012) bathymetry from FAA inversion; **(c)** Simple 2-D model of sediment infill along the IceBridge flight line.

The gravity-derived cavity model also implies a consistently thick water column along the grounding line, potentially presenting a route for whole-shelf oceanic circulation, readily allowing mixing between the northern and southern ends of the cavity. However, the results presented here indicate that there are significant barriers to flow at Marmelon Point and around Francis and Tonkin Islands. This is consistent with the preferred ocean circulation scenario of Nicholls et al. (2012) who propose that the southern and northern parts of the shelf cavity are ventilated from sources crossing beneath the ice front at different locations, rather than whole-shelf circulation via the grounding line cavity.

6 Marine ice beneath Larsen C

A qualitative correspondence between ice base reflector quality and proposed areas of marine ice freeze-on (Holland et al., 2009) was observed (Fig. 8). In the Holland et al. (2009) study the presence of marine ice was inferred from the absence of basal reflections in radar data where a surface return was obtained. Freeze-on results in a mushy transitional interface (Lewis and Perkin, 1986; Craven et al., 2009), rather than a sharp reflector, and would therefore result in reduced seismic reflection amplitude. There are however a number of other factors that can contribute to reflector amplitude: source coupling, receiver coupling, attenuation within the ice column, reflector geometry, the presence of water-filled crevasses near the ice-base and acoustic impedance contrast across the reflector. The correspondence observed here indicates that at the early stages of marine ice freeze-on, where forming close to the grounding line in shallow-draft areas between glacier flow units (Holland et al., 2009), the interface is not a sharp boundary. Further from the grounding line the freeze-on mechanism appears to cease and this transition becomes more abrupt, resulting in more obvious ice-base seismic reflections.

7 Conclusions

Seismic measurements have been acquired at 87 sites across Larsen C Ice Shelf to constrain ice thickness and sub-ice shelf water column thickness. Most of our seismic reflection data from Larsen Ice Shelf are of good quality, reflections from the ice-base and seabed are clear, and the traveltimes of these reflections can be picked precisely. At a few stations, the ice-base reflection is not easy to identify and the main uncertainty in the ice thickness measurement at these locations arises from this. In these cases, standard processing techniques can be used to increase the signal-to-noise ratio and improve the identification of primary reflections and their multiples. Another source of error in the ice thickness measurements comes from uncertainty in the seismic velocity, which must be derived using seismic refraction surveys and a simple temperature-depth model. Variations in the seismic velocities are observed across Larsen Ice Shelf, consistent with varying degrees of melting and firn compaction. Where present, seabed topography cannot be quantified unambiguously without detailed site surveys. The moveout of reflections can however be used to estimate uncertainty.

On Larsen C Ice Shelf, our seismic measurements indicate an rms error of 162 m in the bathymetry derived from airborne gravity inversion, and 143 m in the sub-ice shelf cavity

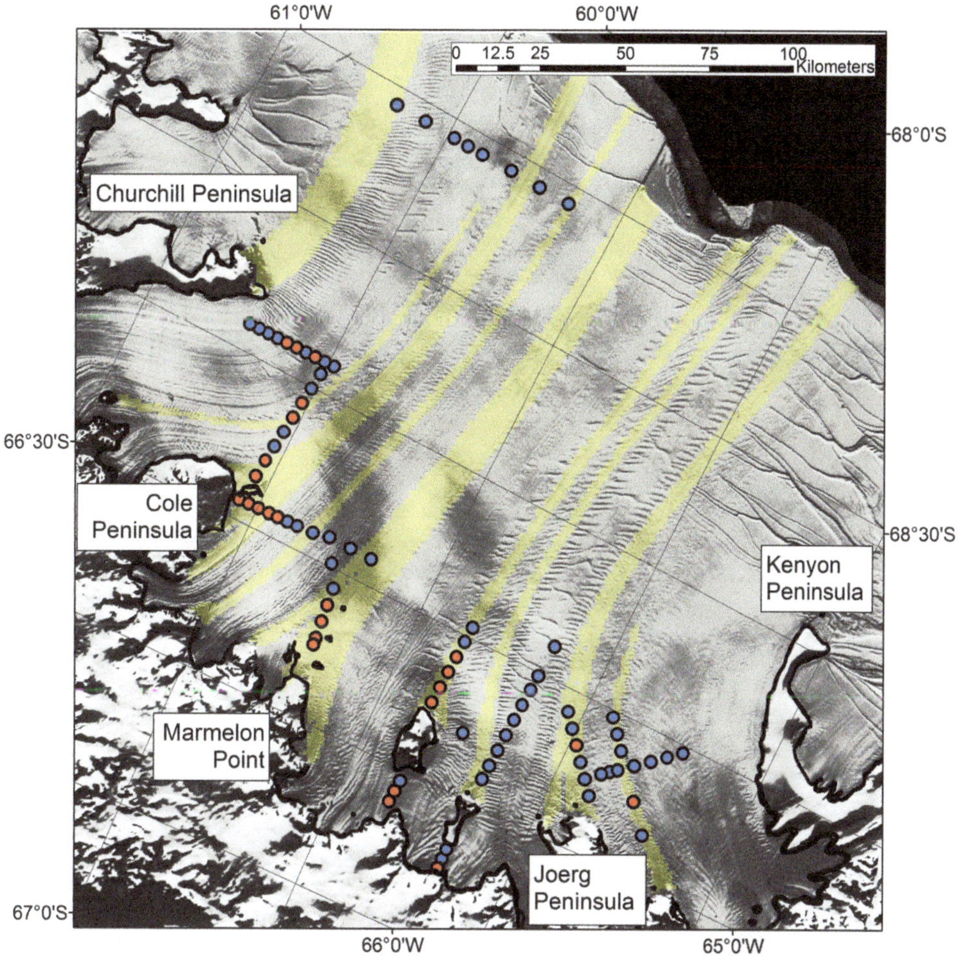

Fig. 8. Proposed marine ice bands in yellow (Holland et al., 2009) over MODIS imagery (Haran et al., 2005). Seismic sites with good ice-base reflections are in blue and poor ice-base reflections in red. The marine ice is formed close to the grounding line just offshore from major promontories. There is strong correspondence between sites with poor ice-base reflections and the proposed areas of marine ice formation.

thickness. Within the seismic survey areas, significant errors exist in the gravity-derived model in both the detail and the regional nature of seabed topography. There is however good agreement with ice draft observations from airborne radio echo sounding, with an rms error of 26.7 m.

The seismic results indicate that beneath most of the ice shelf out to its eastern edge, there are no seabed features that will inhibit or concentrate water circulation. In contrast, closer to the grounding line (western edge of the ice shelf), the seismic results indicate restricted, or even zero, cavity thickness, forming significant barriers to ocean circulation, especially close to the Cole Peninsula, Marmelon Point and both Francis and Tonkin Islands. This is consistent with the preferred circulation hypothesised by Nicholls et al. (2012) based on oceanographic data from two boreholes.

Mueller et al. (2012) demonstrated the oceanographic significance of the bathymetric features introduced by the gravity inversion model. A narrowing of the sub-ice shelf cavity towards the ice front introduced basal melt rates as high as $\sim 2\,\mathrm{m\,a^{-1}}$. However, a model which more closely resembles the seismic results precludes such high basal melt rates. Such high melt rates are not consistent with observations of ice surface topography.

In this paper we do not re-invert the IceBridge gravity data for seabed bathymetry using the true, seismically-derived bathymetry for improved control. That remains for a future study, preferably with more seismic data to give a more complete coverage of the ice shelf. In this case, the gravity data will improve interpolation between the seismic stations, rather than being the sole source of derived bathymetry. It is also possible to model the free-air anomaly using the seismically derived bathymetry to indicate sub-bed geological structure. Using the derived bathymetry along with typical sedimentary and igneous rock densities, we have demonstrated that realistic models of infilled sedimentary channels produce gravity anomalies similar to those observed over Larsen Ice Shelf. It is clear that to obtain a more reliable model of the bathymetry of the Larsen C Ice Shelf region from gravity inversion, a more realistic implementation of the regional geology is required. As such, any future seismic acquisition campaigns could be modified to acquire data which could constrain seabed properties and potentially discriminate areas of significant sedimentary cover.

The inshore ends of proposed marine ice bands (Holland et al., 2009), resulting from freeze-on at the base of the ice shelf are consistent with sites where the ice base seismic reflection is difficult to resolve. This result would be expected where frazil ice is forming a more diffuse ice-water interface (Craven et al., 2009).

Acknowledgements. This work was funded by UK's Natural Environment Research Council (NERC) and the British Antarctic Survey (BAS). Field support was provided by BAS. We thank Dave Routledge for his invaluable assistance in the field and with data acquisition. Data are available from the author. The authors thank Leo Peters and one anonymous reviewer for their constructive comments.

Edited by: R. Bingham

References

Bart, P. J. and Anderson, J. B.: Seismic record of glacial events affecting the Pacific margin of the northwestern Antarctic Peninsula, Ant. Res. Series, 68, 75–95, 1995.

Booth, A. D., Kulessa, B., King, E. C., Clark, R. A., Jansen, D., Sammonds, P., and Luckman, A.: Physical properties of meteoric and marine ice in Larsen C Ice Shelf, Antarctic Peninsula, from Q and AVA analyses of reflection seismic data, EGU General Assembly, Vienna, Austria, 2010, EGU2013-6531, 2013.

Cochran, J. R. and Bell, R. E.: Inversion of IceBridge gravity data for continental shelf bathymetry beneath the Larsen Ice Shelf, Antarctica, J. Glaciol., 58, 540–552, 2012.

Cook, A. J. and Vaughan, D. G.: Overview of areal changes of the ice shelves on the Antarctic Peninsula over the past 50 years, The Cryosphere, 4, 77–98, doi:10.5194/tc-4-77-2010, 2010.

Craven, M., Allison, I., Fricker, H. A., and Warner, R.: Properties of a marine ice layer under the Amery Ice Shelf, East Antarctica, J. Glaciol., 55, 717–728, 2009.

Determann, J., Thyssen, F., and Engelhardt, H.: Ice thickness and sea depth derived from reflection-seismic measurements on the central part of Filchner-Ronne Ice Shelf, Antarctica, Ann. Glaciol., 11, 14–18, 1988.

Fahnestock, M. A., Abdalati, W., and Shuman, C. A.: Long melt seasons on ice shelves of the Antarctic Peninsula: an analysis using satellite-based microwave emission measurements, Ann. Glaciol., 34, 127–133, 2002.

Fahrbach, E., Peterson, R., Rohardt, G., Schlosser, P., and Bayer, R.: Suppression of bottom water formation in the southeastern Weddell Sea, Deep Sea Res. Pt. I, 41, 389–411, 1994.

Forste, C., Schmidt, R., Stubenvoll, R., Flechtner, F., Meyer, U., Konig, R., Neumayer, H., Biancale, R., Lemoine, J. M., Bruinsma, S., Loyer, S., Barthelmes, F., and Esselborn, S.: The GeoForschungsZentrum Potsdam/Groupe de Recherche de Geodesie Spatiale satellite-only and combined gravity field models: EIGEN-GL04S1 and EIGEN-GL04C, J. Geodesy, 82, 331–346, doi:10.1007/s00190-007-0183-8, 2008.

Fricker, H. A. and Padman, L.: Thirty years of elevation change on Antarctic Peninsula ice shelves from multimission satellite radar altimetry, J. Geophys. Res.-Oceans, 117, C02026, doi:10.1029/2011jc007126, 2012.

Galton-Fenzi, B. K., Maraldi, C., Coleman, R., and Hunter, J.: The cavity under the Amery Ice Shelf, East Antarctica, J. Glaciol., 54, 881–887, 2008.

Glasser, N. F. and Scambos, T. A.: A structural glaciological analysis of the 2002 Larsen B ice-shelf collapse, J. Glaciol., 54, 3–16, doi:10.3189/002214308784409017, 2008.

Griggs, J. A. and Bamber, J. L.: Ice shelf thickness over Larsen C, Antarctica, derived from satellite altimetry, Geophys. Res. Let., 36, L19501, doi:10.1029/2009gl039527, 2009.

Grosfeld, K., Gerdes, R., and Determann, J.: Thermohaline circulation and interaction between ice shelf cavities and the adjacent open ocean, J. Geophys. Res.-Oceans, 102, 15595–15610, doi:10.1029/97jc00891, 1997.

Haran, T., Bohlander, J., Scambos, T., Fahnestock, M., and compilers: MODIS mosaic of Antarctica (MOA) image map: Digital media, National Snow and Ice Data Center., Boulder, CO, USA, 2005.

Hemer, M. A., Hunter, J. R., and Coleman, R.: Barotropic tides beneath the Amery Ice Shelf, J. Geophys. Res.-Oceans, 111, C11008, doi:10.1029/2006jc003622, 2006.

Holland, P. R., Corr, H. F. J., Vaughan, D. G., Jenkins, A., and Skvarca, P.: Marine ice in Larsen Ice Shelf, Geophys. Res. Let., 36, L11604, doi:10.1029/2009gl038162, 2009.

Holland, P. R., Corr, H. F. J., Pritchard, H. D., Vaughan, D. G., Arthern, R. J., Jenkins, A., and Tedesco, M.: The air content of Larsen Ice Shelf, Geophys. Res. Let., 38, L10503, doi:10.1029/2011gl047245, 2011.

Jacobs, S. S.: Bottom water production and its links with the thermohaline circulation, Ant. Sci., 16, 427–437, 2004.

Jarvis, E. P. and King, E. C.: Seismic Investigation of the Larsen Ice Shelf, Antarctica – in Search of the Larsen Basin, Ant. Sci., 7, 181–190, 1995.

Jenkins, A., Dutrieux, P., Jacobs, S. S., McPhail, S. D., Perrett, J. R., Webb, A. T., and White, D.: Observations beneath Pine Island Glacier in West Antarctica and implications for its retreat, Nat. Geosci., 3, 468–472, doi:10.1038/ngeo890, 2010.

Johnson, M. R. and Smith, A. M.: Seabed topography under the southern and western Ronne Ice Shelf, derived from seismic surveys, Ant. Sci., 9, 201–208, 1997.

Jullion, L., Naveira Garabato, A. C., Meredith, M. P., Holland, P. R., Courtois, P., and King, B. A.: Decadal freshening of the Antarctic Bottom Water exported from the Weddell Sea, J. Climate, in press, doi:10.1175/jcli-d-12-00765.1, 2013.

King, E. C., Jarvis, E. P., and Mowse, E. A.: Seismic characteristics of an airgun fired over snow, Cold Reg. Sci. Tech., 21, 201–207, 1993.

Kirchner, J. F. and Bentley, C. R.: RIGGS III: Seismic short-refraction studies using an analytical curve-fitting technique, Ant. Res. Series, 42, 109–126, 1990.

Kohnen, H.: Uber die Beziehung zwischen seismischen Geschwindigkeiten und der Dichte in Firn und Eis, Zeitschrift fur Geophysik, 38, 925–935, 1972.

Kohnen, H.: The temperature dependence of seismic waves in ice, J. Glaciol., 13, 144–147, 1974.

Kulessa, B., King, E. C., Barrett, B. E., Jansen, D., Luckman, A. J., and Sammonds, P.: Joint inversion of multi-component seismic and ground-penetrating radar (GPR) data for ice-physical properties, and application to the Larsen C ice shelf, AGU Fall Meeting Abstracts, San Francisco, 2010, C21A-0509, 2010.

Lewis, E. L. and Perkin, R. G.: Ice Pumps and Their Rates, J. Geophys. Res.-Oceans, 91, 1756–1762, doi:10.1029/Jc091ic10p11756, 1986.

Mueller, R. D., Padman, L., Dinniman, M. S., Erofeeva, S. Y., Fricker, H. A., and King, M. A.: Impact of tide-topography interactions on basal melting of Larsen C Ice Shelf, Antarctica, J. Geophys. Res.-Oceans, 117, doi:10.1029/2011JC007263, 2012.

Muto, A., Anandakrishnan, S., and Alley, R.: Subglacial bathymetry and sediment layer distribution beneath the Pine Island Glacier ice shelf, West Antarctica, modeled using aerogravity and autonomous underwater vehicle data, Ann. Glaciol., 54, 27–32, 2013.

Nicholls, K. W., Makinson, K., and Venables, E. J.: Ocean circulation beneath Larsen C Ice Shelf, Antarctica from in situ observations, Geophys. Res. Let., 39, L19608, doi:10.1029/2012gl053187, 2012.

Nost, O. A.: Measurements of ice thickness and seabed topography under the Fimbul Ice Shelf, Dronning Maud Land, Antarctica, J. Geophys. Res.-Oceans, 109, C10010, doi:10.1029/2004jc002277, 2004.

Oldenburg, D. W.: The inversion and interpretation of gravity anomalies, Geophysics, 39, 526–536, 1974.

Pritchard, H. D., Ligtenberg, S. R. M., Fricker, H. A., Vaughan, D. G., van den Broeke, M., and Padman, L.: Antarctic ice loss driven by ice-shelf melt, Nature, 484, 502–505, 2012.

Rignot, E., Casassa, G., Gogineni, P., Krabill, W., Rivera, A., and Thomas, R.: Accelerated ice discharge from the Antarctic Peninsula following the collapse of Larsen B ice shelf, Geophys. Res. Let., 31, L18401, doi:10.1029/2004gl020697, 2004.

Rignot, E., Velicogna, I., van den Broeke, M. R., Monaghan, A., and Lenaerts, J.: Acceleration of the contribution of the Greenland and Antarctic ice sheets to sea level rise, Geophys. Res. Let., 38, 5, L05503, doi:10.1029/2011gl046583, 2011.

Roethlisberger, H.: Seismic exploration in cold regions, Pt II ed., Cold Regions Engineering, Section A2a, U.S. Cold Regions Research and Engineering Laboratory, Hanover, N. H., 1972.

Rott, H., Skvarca, P., and Nagler, T.: Rapid collapse of northern Larsen Ice Shelf, Antarctica, Science, 271, 788–792, doi:10.1126/science.271.5250.788, 1996.

Rott, H., Rack, W., Nagler, T., and Skvarca, P.: Climatically induced retreat and collapse of northern Larsen Ice Shelf, Antarctic Peninsula, Ann. Glaciol., 27, 86–92, 1998.

Roy, L., Sen, M. K., Blankenship, D. D., Stoffa, P. L., and Richter, T. G.: Inversion and uncertainty estimation of gravity data using simulated annealing: An application over Lake Vostok, East Antarctica, Geophysics, 70, J1–J12, doi:10.1190/1.1852777, 2005.

Scambos, T. A., Bohlander, J. A., Shuman, C. A., and Skvarca, P.: Glacier acceleration and thinning after ice shelf collapse in the Larsen B embayment, Antarctica, Geophys. Res. Lett., 31, L18402, L18402, doi:10.1029/2004gl020670, 2004.

Shepherd, A., Wingham, D., Payne, T., and Skvarca, P.: Larsen ice shelf has progressively thinned, Science, 302, 856–859, 2003.

Skvarca, P., Rack, W., Rott, H., and Donangelo, T. I. Y.: Climatic trend and the retreat and disintegration of ice shelves on the Antarctic Peninsula: an overview, Polar Res., 18, 151–157, 1999.

Sloan, B., Lawver, L., and Anderson, J.: Seismic stratigraphy of the Larsen basin, eastern Antarctic Peninsula, Ant. Res. Series, 68, 59–74, 1995.

Smith, R. T. and Anderson, J. B.: Ice-sheet evolution in James Ross Basin, Weddell Sea margin of the Antarctic Peninsula: The seismic stratigraphic record, Geol. Soc. Am. Bull., 122, 830–842, 2010.

Swithinbank, C. W. M.: Ice shelves, Geograph. J., 121, 64–76, 1955.

Talwani, M., Worzel, J. L., and Landisman, M.: Rapid gravity computations for two-dimensional bodies with application to the Mendocino submarine fracture zone, J. Geophys. Res., 64, 49–59, 1959.

Tinto, K. J. and Bell, R. E.: Progressive unpinning of Thwaites Glacier from newly identified offshore ridge: Constraints from aerogravity, Geophys. Res. Let., 38, L20503, doi:10.1029/2011gl049026, 2011.

Vaughan, D. G. and Doake, C. S. M.: Recent atmospheric warming and retreat of ice shelves on the Antarctic Peninsula, Nature, 379, 328–331, doi:10.1038/379328a0, 1996.

Vaughan, D. G., Marshall, G. J., Connolley, W. M., Parkinson, C., Mulvaney, R., Hodgson, D. A., King, J. C., Pudsey, C. J., and Turner, J.: Recent rapid regional climate warming on the Antarctic Peninsula, Clim. Change, 60, 243–274, 2003.

Voigt, D., Peters, L. E., and Anandakrishnan, S.: "Georods": the development of a four-element geophone for improved seismic imaging of glaciers and ice sheets, Ann. Glaciol., 54, 142–148, 2013.

2

Feedbacks and mechanisms affecting the global sensitivity of glaciers to climate change

B. Marzeion[1], A. H. Jarosch[2], and J. M. Gregory[3]

[1]Center of Climate and Cryopshere, Institute of Meteorology and Geophysics, University of Innsbruck, Austria
[2]Institute of Earth Sciences, University of Iceland, Reykjavík, Iceland
[3]NCAS-Climate, University of Reading, Reading, and Met Office Hadley Centre, Exeter, UK

Correspondence to: B. Marzeion (ben.marzeion@uibk.ac.at)

Abstract. Mass loss by glaciers has been an important contributor to sea level rise in the past, and is projected to contribute a substantial fraction of total sea level rise during the 21st century. Here, we use a model of the world's glaciers to quantify equilibrium sensitivities of global glacier mass to climate change, and to investigate the role of changes in glacier hypsometry for long-term mass changes. We find that 21st century glacier-mass loss is largely governed by the glacier's response to 20th century climate change. This limits the influence of 21st century climate change on glacier-mass loss, and explains why there are relatively small differences in glacier-mass loss under greatly different scenarios of climate change. The projected future changes in both temperature and precipitation experienced by glaciers are amplified relative to the global average. The projected increase in precipitation partly compensates for the mass loss caused by warming, but this compensation is negligible at higher temperature anomalies since an increasing fraction of precipitation at the glacier sites is liquid. Loss of low-lying glacier area, and more importantly, eventual complete disappearance of glaciers, strongly limit the projected sea level contribution from glaciers in coming centuries. The adjustment of glacier hypsometry to changes in the forcing strongly reduces the rates of global glacier-mass loss caused by changes in global mean temperature compared to rates of mass loss when hypsometric changes are neglected. This result is a second reason for the relatively weak dependence of glacier-mass loss on future climate scenario, and helps explain why glacier-mass loss in the first half of the 20th century was of the same order of magnitude as in the second half of the 20th century, even though the rate of warming was considerably smaller.

1 Introduction

Glaciers[1] have lost a substantial fraction of their mass during the past century (Cogley, 2009; Marzeion et al., 2012), with the globally averaged mass balance turning negative probably around 1850 (Leclercq et al., 2011). Within the 20th century, mass loss of glaciers was likely the largest single cause of sea level rise, followed by thermal expansion of the ocean, mass loss of the Greenland and Antarctic ice sheets, and changes in terrestrial water storage (Gregory et al., 2013). Even though the rise of global mean air temperature accelerated in the 20th century, the mass loss rate of glaciers during the second half of the 20th century was not higher than during the first half of the century (Leclercq et al., 2011; Marzeion et al., 2012).

Mass loss from glaciers will continue to contribute to sea level rise substantially during the 21st century, even though their total sea level rise potential is limited by their total mass. The total ice mass in glaciers is estimated as 35 ± 7 cm sea level equivalent (SLE, Grinsted, 2013), 43 ± 6 cm SLE (Huss and Farinotti, 2012), 41–52 cm SLE (depending on the assumed fraction of ice caps, Radić et al., 2013), or 49 ± 6 cm SLE (by Marzeion et al., 2012, including the mass estimate for peripheral glaciers in Antarctica from Radić et al., 2013). Driven by climate scenarios obtained from the

[1]In using the term glaciers, we are referring to global glaciers and ice caps with the exclusion of the Antarctic and Greenland ice sheets, but including Greenland peripheral glaciers and ice caps. Antarctic peripheral glaciers are not included because of the lack of long-term climate observations, which we rely on for model calibration and validation.

Coupled Model Intercomparison Project phase 5 (CMIP5) database, and depending on the emission scenario applied (see van Vuuren et al., 2011, for an overview of the different Representative Concentration Pathways, RCPs), the contribution of glaciers to sea level rise during the 21st century is estimated as 16 ± 4 cm SLE (RCP4.5) to 22 ± 4 (RCP8.5) by Radić et al. (2013), or as 15 ± 4 cm SLE (RCP2.6) to 22 ± 5 cm SLE (RCP8.5) by Marzeion et al. (2012). Remarkably, there is considerable overlap of the uncertainty ranges of projected mass loss from glaciers even for very different climate scenarios. For instance, RCP2.6 and RCP8.5, which during the period 2081–2100 arrive at a global mean temperature change of 1.2 ± 0.4 K and 3.8 ± 0.8 K respectively relative to 1986–2005. The overlap of uncertainty ranges in glacier projections for these scenarios is not caused by the uncertainties of the glacier models. It is rather the large spread of the ensemble of climate projections used to drive the glacier models, which gives a spread in glacier-mass loss projections that is relatively large compared to the ensemble mean. Moreover, the rates of mass loss projected for the different RCP scenarios are relatively similar during the 21st century. The latter also reflects the result that over long timescales, there was no simple relation between rates of glacier-mass loss and temperature change during the 20th century (Leclercq et al., 2011; Marzeion et al., 2012).

There are a number of assessments that explain some of the mechanisms responsible for this behavior. For example, Huss (2012) points out that in the Alps during the period 1900–2011, the mass balance year of 2003 had the most negative specific mass balances, but the greatest loss of ice volume occurred in 1947, when the glacier surface area was considerably larger. However, it is not only changes in surface area that change a glacier's response to climate forcing over time: dynamic changes in ice thickness and terminus elevation, reflected together in changes of the ice surface topography, feed back to the mass balance and dampen (in case of terminus elevation) or enhance (in case of thickening/thinning) the glacier's response to climate forcing (Huss et al., 2012). Paul (2010) comes to a similar conclusion that without changes in glacier hypsometry, mass balances in the Alps since 1850 would have been two to three times more negative than observed. This issue has also been discussed in detail by Leclercq et al. (2010) and Huss et al. (2010).

Since dynamic adjustments of a glacier's hypsometry to changed forcing do not happen instantaneously, glaciers may be found out of balance with climate forcing long after a change in the climate forcing occurred. Jóhannesson et al. (1989) and Oerlemans (2001) developed different frameworks that allow the estimation of the response time of a glacier's hypsometry to changes in climate forcing. Zuo and Oerlemans (1997) pointed out that an imbalance between the states of glaciers and climate may be an important factor for sea level rise, and Oerlemans et al. (1998) investigated the dependence of mass balance response of 12 glaciers to hypsometric changes using an ice flow model.

But to date, there are no studies that quantify how mass balances of glaciers are influenced by these mechanisms on a global scale, that is, to what extent present and future sea level rise from glaciers reflects past climate change, and to what extent past and future sea level rise from glaciers is influenced by the feedback of glacier hypsometric changes on glacier mass balance. In order to illuminate these issues, we use the global glacier model of Marzeion et al. (2012) which captures most of the relevant mechanisms required to study the glaciers' response to a number of idealized climate forcings. We briefly describe the glacier model in Sect. 2. In Sect. 3, we present experiments determining equilibrium sensitivities of the world's glaciers, including a distinction of the effects of changes in temperature and precipitation. The effect of glacier hypsometry changes on the response of glaciers to future climate change is investigated in Sect. 4. We discuss the results in Sect. 5 and conclude in Sect. 6.

2 Description of the glacier model

The glacier model is based on calculating the annual specific mass balance B for each of the world's individual glaciers as

$$B = \left[\sum_{i=1}^{12} \left[P_i^{\text{solid}} - \mu^* \cdot \max\left(T_i^{\text{terminus}} - T_{\text{melt}}, 0 \right) \right] \right] - \beta^*, \qquad (1)$$

where P_i^{solid} is the monthly solid precipitation onto the glacier surface per unit area, which depends on the monthly mean total precipitation and the temperature range between the glacier's terminus and highest elevations (i.e., temperature at terminus elevation below a certain threshold implies all precipitation is solid, temperature at the glacier's maximum elevation above the threshold implies all precipitation is liquid, and within that temperature range, the precipitation fraction is interpolated linearly; see Marzeion et al. (2012) for a detailed description), μ^* is the glacier's temperature sensitivity, T_i^{terminus} is the monthly mean air temperature at the glacier's terminus, T_{melt} is the monthly mean air temperature above which ice melt is assumed to occur, and β^* is a bias correction (see below). The model does not attempt to capture the full energy balance at the ice surface, but relies on air temperature as a proxy for the energy available for melt (Ohmura, 2001; Hock, 2003; Sicart et al., 2008). P_i^{solid} and T_i^{terminus} are determined based on gridded climate observations (New et al., 2002; Mitchell and Jones, 2005), to which temperature and precipitation anomaly fields from CMIP5 models are added, depending on the experiment performed (see descriptions of experimental setup in Sects. 3 and 4). Changes affecting the glacier hypsometry (i.e. changes in its volume, surface area, and elevation range) are reflected in the determination of P_i^{solid} and T_i^{terminus}, are modeled based on B, and on linearly adjusting the glacier's surface area and length towards their respective values obtained from volume-area and volume-length scaling (Bahr et al., 1997; Bahr,

1997); that is, the surface area change $\mathrm{d}A$ of a glacier during each mass balance year t is calculated as

$$\mathrm{d}A(t) = \frac{1}{\tau_A(t)}\left(\left(\frac{V(t+1)}{c_A}\right)^{1/\gamma} - A(t)\right), \quad (2)$$

where $\tau_A(t)$ is the area relaxation timescale (see Eq. 5), $V(t+1)$ is the glacier's volume at the end of the mass balance year, $c_A = 0.0340\,\mathrm{km}^{3-2\gamma}$ (for glaciers), $c_A = 0.0538\,\mathrm{km}^{3-2\gamma}$ (for ice caps), $\gamma = 1.375$ (for glaciers), $\gamma = 1.25$ (for ice caps) are scaling parameters (Bahr et al., 1997; Bahr, 1997), and $A(t)$ is the surface area of the glacier at the end of the preceding mass balance year. Similarly, length changes $\mathrm{d}L$ (and terminus elevation changes associated with them) during each mass balance year are estimated as

$$\mathrm{d}L(t) = \frac{1}{\tau_L(t)}\left(\left(\frac{V(t+1)}{c_L}\right)^{1/q} - L(t)\right), \quad (3)$$

where $\tau_L(t)$ is the length relaxation timescale (see Eq. 4), $c_L = 0.0180\,\mathrm{km}^{3-q}$ (for glaciers), $c_L = 0.2252\,\mathrm{km}^{3-q}$ (for ice caps), $q = 2.2$ (for glaciers), $q = 2.5$ (for ice caps) are scaling parameters (Bahr et al., 1997; Bahr, 1997), and $L(t)$ is the glacier's length at the start of the mass balance year. The glacier length response timescale τ_L is estimated following Jóhannesson et al. (1989) as

$$\tau_L(t) = \frac{V(t)}{\sum_{i=1}^{12} \int P_{i,\mathrm{clim}}^{\mathrm{solid}}}, \quad (4)$$

where $\int P_{i,\mathrm{clim}}^{\mathrm{solid}}$ is the monthly climatological solid precipitation integrated over the glacier surface area, calculated over the preceding 30 yr. The glacier area response timescale is estimated as

$$\tau_A(t) = \tau_L(t)\frac{A(t)}{L(t)^2} \quad (5)$$

based on the assumption that area changes caused by glacier width changes occur instantaneously, while area changes caused by glacier length changes occur with the timescale of glacier length response.

The volume change $\mathrm{d}V$ of a glacier in year t is calculated as

$$\mathrm{d}V(t) = B(t) \cdot A(t). \quad (6)$$

The temperature sensitivity μ^* is determined from observed past variations for each of the glaciers with available mass balance in Cogley (2009). In that data set, there is a global total of 255 glaciers that have all the metadata needed for the parameter estimation, are covered by the temperature and precipitation data set we use (see below), are indicated to be reliable by the status flag of the data set, and have at least two annual mass balance measurements. The procedure is as follows. We assume that there exists some 31 yr reference period, centered on year t^*, whose climatology is such that the glacier with its present-day hypsometry would be in equilibrium (i.e., with its mass not changing). For this reference period, by construction

$$B = \sum_{i=1}^{12}\left[P(t^*)_{i,\mathrm{clim}}^{\mathrm{solid}} - \mu(t^*) \cdot \left(\max\left(T(t^*)_{i,\mathrm{clim}}^{\mathrm{terminus}} - T_{\mathrm{melt}}, 0\right)\right)\right] = 0, \quad (7)$$

where $P(t^*)_{i,\mathrm{clim}}^{\mathrm{solid}}$ and $T(t^*)_{i,\mathrm{clim}}^{\mathrm{terminus}}$ are the monthly climatological values of P_i^{solid} and T_i^{terminus}, during the 31 yr period centered around the year t^*. Note that we do not assume t^* to be a time at which the glacier was actually in balance. If the climate has been warming and the glacier retreating, as is generally the case, t^* would be in the past, and the glacier actually would have had a negative mass balance at time t^*. The assumption is that if the climate of time t^* had been maintained, the glacier eventually would have contracted until it reached its present-day hypsometry.

We obtain a total of 109 monthly climatologies of precipitation and temperature (the data set of Mitchell and Jones (2005) provides 109 yr of monthly precipitation and temperature; at the end and beginning of the time series, the climatologies are calculated over shorter time periods), and subsequently obtain an estimate of μ from Eq. (7) for each of the 109 choices of t^*. We then apply the glacier model to all glaciers for which direct mass balance observations are available, for each of the 109 possible values of $\mu(t)$. For each of these glaciers, we identify t^* as that time, for which applying the corresponding temperature sensitivity $\mu^* \equiv \mu(t^*)$ yields the smallest mean error of the modeled mass balances. This minimum difference is denoted by β^*.

For glaciers without observed mass balances (i.e., the vast majority of glaciers), t^* is interpolated from surrounding glaciers with mass balance observations, and μ^* is subsequently determined from solving Eq. (7) for μ^*, using precipitation and temperature obtained from the climatology centered around the interpolated value of t^*.

The bias correction β^* is determined by interpolating the minimized bias obtained during the determination of t^* from surrounding glaciers with mass balance observations. A cross validation of the determination of μ^* shows that the spatial interpolation of t^* leads to substantially smaller errors than the spatial interpolation of μ^* (Marzeion et al., 2012). This can be understood as an effect of neighboring glaciers experiencing a similar history of climate forcing, but having potentially very different temperature sensitivities.

Given any pair of glaciers for which the historical calibration is carried out, we may calculate the temporal correlation between the annual time series for those two glaciers of the errors in the modeled mass balance. Considering all such pairs, we can calculate the correlation of this temporal error correlation with the distance between the two glaciers. This latter correlation is 0.008, indicating that the model errors for the individual glaciers can be treated as independent of each other. Taking the temporal correlation between the annual time series of the mass balances instead of the errors of the modeled mass balances, and calculating the correlation of this temporal mass balance correlation with the distance between the two glaciers, we obtain a value of −0.209. This indicates that correlations of the mass balances of neighboring glaciers do not transfer to correlations in the errors of the modeled mass balances.

Initial values for surface area and elevation distribution of a glacier are obtained by draping ice outlines from the Randolph Glacier Inventory (RGI, Arendt et al., 2012) version 1 over version 2 of the ASTER global digital elevation model (GDEM), applying a suitable watershed algorithm (Ehlschlaeger, 1989) to separate ice complexes into individual glaciers, and extracting glacier elevation statistics (minimum, mean and maximum elevation) from the GDEM. The model accounts for the differing dates of surface area measurement in the RGI by ensuring that the observed glacier extent is reproduced in the year of observation. The initial volume at the start of the model run is determined from this constraint: iteratively, we determine the ice volume (as well as surface area, length and terminus elevation, following the scaling relations mentioned above) at the start of the simulation that yields the observed surface area in the year of observation. Unless mentioned otherwise, all the runs of the glacier model presented here were initialized using the "historical" experiment of the respective CMIP5 model.

A more detailed and complete description of the determination of the model's parameters, both glacier specific and global, and a comprehensive validation of the model can be found in Marzeion et al. (2012). Uncertainties of the modeled results are obtained by propagating the uncertainties of the modeled specific mass balance, which are determined independently during a leave-one-glacier-out cross validation, through the entire model system, also taking into account uncertainties of the representation of the dynamic glacier response to volume changes. The propagated and temporally accumulated uncertainties themselves are also independently validated using geodetically measured volume and surface area changes (see Marzeion et al., 2012).

3 Equilibrium sensitivities

3.1 Experimental setup and forcing

Results from 15 different CMIP5 experiments (see Table 1) were used to force the glacier model in the equilibrium experiments with constant climate over a range of global mean temperature anomalies. For each of the RCP8.5 experiments, monthly anomaly fields of precipitation and near-surface air temperature were determined relative to the monthly climatology of 1961 to 1990. Then, monthly climatologies of the anomalies were determined for each overlapping 30 yr period contained in the combined historical and RCP8.5 experiment (i.e., between 1850 or 1860 and 2100 or 2300, depending on the climate model, see Table 1). From this data set, the climatological anomaly fields of precipitation and temperature were extracted for global mean temperature anomalies of 1 to 10 K. (The number of fields extracted from each CMIP5 experiment therefore depends on the maximum global temperature anomaly reached over a 30 year mean, and since the global mean temperature anomalies do not necessarily contain the required integer anomalies, we took the anomaly field with the smallest difference in global mean temperature anomaly. This difference is smaller than 0.1 K for all cases.) These anomaly fields were added to the observed climatological fields of New et al. (2002) to obtain the climate forcing for the glacier model. Additionally, the glacier model was forced by the observed climatological fields of New et al. (2002) only (i.e., without any modeled anomalies, giving a global mean anomaly of 0 K).

To obtain the equilibrium response of the glaciers to constant climate forcing, the same forcing was repeatedly applied for each glacier until volume changes of the glacier became negligible. This was defined to be the case when the volume change over the last 100 modeled years was smaller than 1 % of the glacier volume. Reaching the equilibrium took up to approximately 700 yr for some glaciers. On the global scale, ice volume changes are small after 200 yr (Fig. 1). Note that in an experimental setup like this, glaciers may reach an equilibrium, while the state of the climate system that was used to drive the glaciers into equilibrium is not itself an equilibrium.

The equilibrium state of the glaciers will be a response both to temperature and precipitation changes. Additionally, the effect of precipitation changes will potentially be modified by temperature through changes in the fraction of solid to liquid precipitation. In order to isolate the effects of both, the experiment was repeated once applying only the temperature anomaly fields (i.e., ignoring any precipitation anomalies resulting from future climate change), and once applying only the precipitation anomaly fields (i.e., ignoring the temperature anomalies).

3.2 Results

Figure 2 shows the resulting equilibrium volume changes of the worlds glaciers as a function of global mean temperature anomaly[2]. When forced with the observed climatology of 1961 to 1990, glaciers would lose mass corresponding to

[2]Peripheral glaciers in Antarctica cannot be modeled by our model because the climate data sets of New et al. (2002) and Mitchell and Jones (2005) do not cover Antarctica. Marzeion et al. (2012) estimated volume change of peripheral glaciers of Antarctica by upscaling. Since snow fall increase may dominate increased melt (Barrand et al., 2013), this may be problematic, and we excluded Antarctic peripheral glaciers here. For a direct comparison

Table 1. CMIP5 models and experiments used for forcing.

Models	Historical	RCP2.6	RCP4.5	RCP6.0	RCP8.5
bcc-csm1-1	1850–2005	2006–2300	2006–2300	2006–2100	2006–2300
CanESM2	1850–2005	2006–2300	2006–2300	–	2006–2100
CCSM4	1850–2005	2006–2100	2006–2100	2006–2100	2006–2100
CNRM-CM5	1850–2005	2006–2100	2006–2300	–	2006–2300
CSIRO-Mk3-6-0	1850–2005	2006–2100	2006–2300	2006–2100	2006–2300
GFDL-CM3	1860–2005	2006–2100	2006–2100	2006–2100	2006–2100
GISS-E2-R	1850–2005	–	2006–2300	2006–2100	2006–2300
HadGEM2-ES	1860–2005	2006–2300	2006–2300	2006–2099	2006–2300
inmcm4	1850–2005	–	2006–2100	–	2006–2100
IPSL-CM5A-LR	1850–2005	2006–2300	2006–2300	2006–2100	2006–2300
MIROC5	1850–2005	2006–2100	1850–2100	2006–2100	2006–2100
MIROC-ESM	1850–2005	2006–2100	2006–2100	2006–2100	2006–2100
MPI-ESM-LR	1850–2005	2006–2300	2006–2300	–	2006–2300
MRI-CGCM3	1850–2005	2006–2100	2006–2100	2006–2100	2006–2100
NorESM1-M	1850–2005	2006–2100	2006–2300	2006–2100	2006–2100

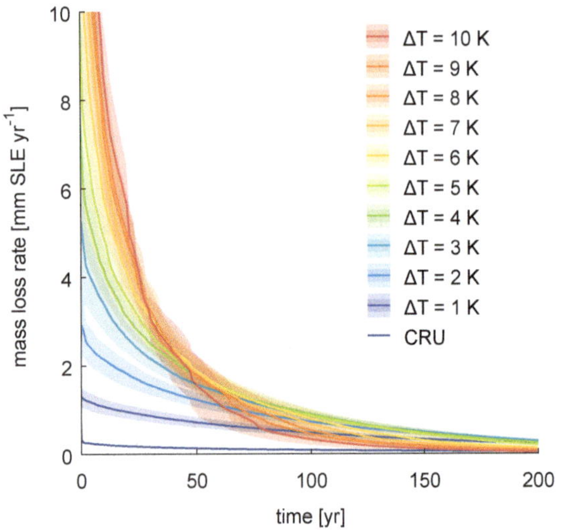

Fig. 1. Rates of global glacier-mass loss during the first 200 yr of the equilibrium experiment (Fig. 2). Solid lines show model mean, shading indicates one standard deviation. Colors indicate global mean temperature anomaly applied. CRU corresponds to the experiment where the observed climatology is applied (i.e., $\Delta T = 0$ K). The volume response within the first 200 yr is generally > 95 % of the total.

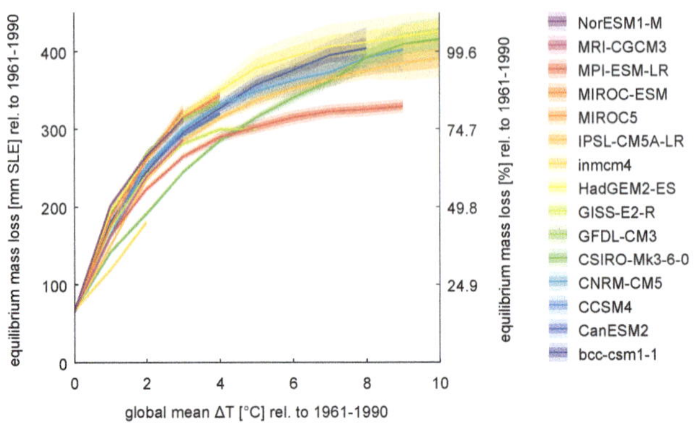

Fig. 2. Equilibrium mass loss of glaciers as a function of global mean temperature anomaly. Dark (light) shading indicates one (two) standard errors. Colors indicate which CMIP5 model was used for forcing. The volume estimates differ slightly between the models. Therefore, the mean of the volume estimates was used for the right axis, and values > 100 % may occur for individual models.

6.6 ± 0.2 cm SLE before reaching a new equilibrium. Note that the uncertainty range given – one standard error – to a large degree depends on the timescale of glacier adjustment in this experimental setup, since errors are accumulated over time. The equilibrium sensitivity to small temperature and associated precipitation changes with respect to the climatology is 10.5 ± 2.3 cm SLE K^{-1} (see Table 2) with considerable differences between different climate models. The sensitivity decreases to zero for progressively larger temperature changes, because of the reduction in the ice mass that remains to be removed.

Generally speaking, global mean precipitation increases with increasing global mean temperatures (e.g., Andrews et al., 2010). It can therefore be expected that the precipitation anomalies associated with the warming dampen the glacier equilibrium response to the warming. This is confirmed in Fig. 3, showing the additional equilibrium mass loss when precipitation anomalies from the climate model are not applied to the glacier model. Increasing precipitation decreases the glacier sensitivity by 1.2 ± 0.3 cm SLE K^{-1} (see Table 2), again with considerable differences between the climate models. But for warming greater than 1 K this

with the results of Marzeion et al. (2012), apply a scaling factor of 1.22 ± 0.01 to the volume change estimates presented here.

Table 2. Equilibrium sensitivities obtained as the difference between the equilibrium states at global mean $\Delta T = 0$ K and global mean $\Delta T = 1$ K, and the anomalies of these sensitivities when either precipitation or temperature anomalies are ignored.

Model	Equilibrium sensitivity [mm SLE K^{-1}]	Δ(equi. sensitivity) [mm SLE K^{-1}] at P = const.	Δ(equi. sensitivity) [mm SLE K^{-1}] at T = const.
bcc-csm1-1	114.4 ± 3.1	15.7 ± 5.4	−132.7 ± 5.5
CanESM2	111.1 ± 3.1	12.6 ± 5.4	−127.3 ± 5.5
CCSM4	127.4 ± 3.1	10.2 ± 5.4	−137.3 ± 5.4
CNRM-CM5	97.9 ± 3.1	14.8 ± 5.5	−125.9 ± 5.4
CSIRO-Mk3-6-0	75.2 ± 3.1	11.9 ± 5.4	−86.1 ± 5.4
GFDL-CM3	122.8 ± 3.1	13.4 ± 5.4	−136.0 ± 5.4
GISS-E2-R	105.9 ± 3.1	13.1 ± 5.4	−121.1 ± 5.4
HadGEM2-ES	128.5 ± 3.1	10.7 ± 5.5	−144.2 ± 5.5
inmcm4	52.2 ± 3.1	14.4 ± 5.4	−75.8 ± 5.4
IPSL-CM5A-LR	82.8 ± 3.1	14.6 ± 5.4	−99.6 ± 5.5
MIROC5	120.9 ± 3.1	6.3 ± 5.4	−135.5 ± 5.4
MIROC-ESM	109.5 ± 3.2	13.3 ± 5.5	−124.3 ± 5.4
MPI-ESM-LR	96.5 ± 3.1	10.2 ± 5.4	−113.6 ± 5.4
MRI-CGCM3	95.5 ± 3.1	9.4 ± 5.5	−119.6 ± 5.6
NorESM1-M	135.3 ± 3.1	7.2 ± 5.4	−151.1 ± 5.4
Mean ± std. dev.	105.1 ± 22.5	11.8 ± 2.8	−122.0 ± 20.9

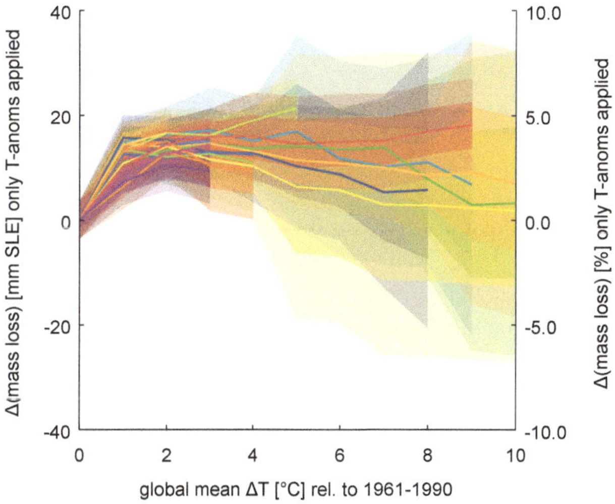

Fig. 3. Additional equilibrium mass loss of glaciers when only temperature anomalies are applied, and precipitation anomalies are ignored. Dark (light) shading indicates one (two) standard errors. Colors as in Fig. 2.

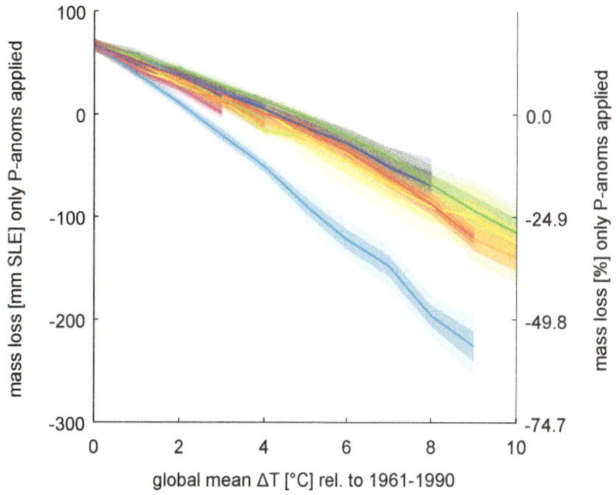

Fig. 4. Equilibrium mass loss (negative numbers imply mass gain) of glaciers when temperature anomalies are ignored, and only precipitation anomalies associated with the temperature anomalies are applied. Dark (light) shading indicates one (two) standard errors. Colors as in Fig. 2.

dampening effect of precipitation becomes smaller, even though precipitation can be expected to increase further. The reason is that with higher temperatures an increasing fraction of precipitation is liquid and does not contribute to glacier mass gain. At very high temperature anomalies, essentially all precipitation falling at the glaciers becomes liquid, such that changes in the amount of precipitation do not affect glacier mass balance in our model, since energy input from liquid precipitation was neglected.

It would therefore be wrong to assume that precipitation anomalies at the glacier sites are small – it is just that much of the increasing precipitation is liquid because of the increased temperatures: Fig. 4 shows the equilibrium mass loss of glaciers when temperature anomalies are ignored (i.e., negative values imply glacier mass gain). If the warming is ignored, precipitation changes associated with 4 K global mean temperature change would roughly compensate for the mass loss that would occur in response to the temperatures of the 1961–1990 reference period, and glaciers would stabilize at their present-day extent.

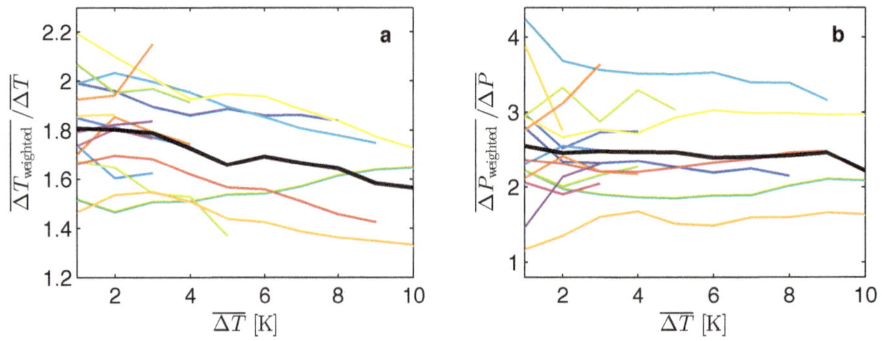

Fig. 5. Temperature and precipitation amplification at the locations of glaciers. **(a)** Annual mean amplification of temperature anomalies at the locations of glaciers as a function of global mean temperature anomaly $\overline{\Delta T}$. **(b)** Annual mean amplification of relative precipitation anomalies at the locations of glaciers as a function of global mean temperature anomaly $\overline{\Delta T}$. Colors as in Fig. 2.

To understand this rather strong effect of precipitation changes, it is helpful to set the global mean changes of temperature and precipitation into relation with the changes experienced by glaciers. Figure 5a shows the ratio of glacier surface area weighted mean temperature anomaly to global mean temperature anomaly for each climate model as a function of global mean temperature change. While there are strong differences between climate models, all of them project greater than average warming at the glacier sites, particularly at low temperature changes. The mean amplification factor at small temperature changes is 1.8 (i.e., at a global mean temperature change of 2 K, glaciers experience a mean temperature change of 3.6 K). This temperature amplification is easily explained by the geographical distribution of glaciers, which all are situated on land, which on average experiences a greater warming than the global mean (e.g., Sutton et al., 2007), and a large fraction of glaciers is located at high northern latitudes, where warming is also greater than at lower latitudes (e.g., Manabe et al., 1991). Since this Arctic temperature amplification to a large degree is caused by albedo reduction through loss of sea ice (Screen and Simmonds, 2010) and seasonal snow cover, it is consistent that the amplification factor experienced by glaciers decreases for high global mean temperature changes, since sea ice and snow cover lost at lower temperatures may no longer influence the local albedo.

Glaciers are projected to experience an even stronger amplification of precipitation (Fig. 5b). The spread between the different climate models is even larger than for temperature, and the amplification is nearly independent of temperature. The positive correlation between temperature and precipitation anomalies is determined by the energy balance of the troposphere (Andrews et al., 2010).

4 Feedbacks of glacier hypsometry and mass balance

4.1 Experimental setup and forcing

All the projections of global glacier mass change during the 21st to 23rd centuries presented in Marzeion et al. (2012) were repeated,

- once ignoring all the effects of glacier-mass loss or gain on glacier hypsometry (i.e., glacier surface area and elevation distribution were held constant in time), as obtained from the RGI and GDEM, and glacier volume was treated as infinite, allowing the glacier to respond to changed climate forcing independent of the history of climate forcing;
- and once including all the effects of glacier-mass loss or gain on glacier hypsometry except for the terminus elevation, which was held constant in time, as obtained from the RGI and GDEM (i.e., surface area and volume evolving with time).

The results of these experiments were then compared to the results of Marzeion et al. (2012) who account for terminus elevation changes, surface area changes, and volume changes, in order to isolate and quantify the feedbacks of glacier hypsometry change with mass balance.

4.2 Results

4.2.1 Constant hypsometry, infinite ice volume

If all aspects of a glacier's hypsometry are held constant in time, and ice volume is treated as infinite, the glacier is prevented from approaching an equilibrium with climate forcing, and subjecting the glacier to climate change essentially becomes equivalent to transplanting a glacier into a future climate, without taking into account the temporal evolution of climate and glacier change that lead there. In this scenario, rates of glacier-mass loss are easier to understand than

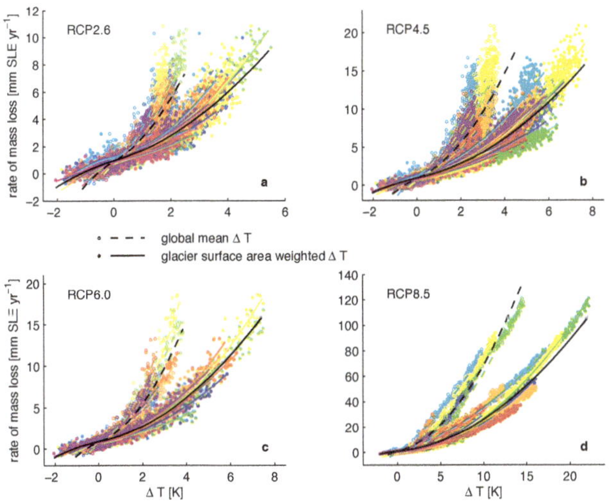

Fig. 6. Global glacier mass balance, expressed in terms of global mean sea level equivalent (SLE), as a function of anomalies in (1) global mean (open circles, dashed lines) and (2) glacier area weighted (filled dots, solid lines) temperature. Results refer to the constant hypsometry, infinite volume scenario. Colored lines show quadratic fits for each individual CMIP5 model, black lines show fits based on mean parameters of colored lines. Colors as in Fig. 2. Different panels refer to different RCPs.

accumulated mass loss, and shown in Fig. 6. Independent of the applied scenario and climate model, rates of mass loss increase (for positive temperature anomalies) or decrease (for negative temperature anomalies) roughly quadratical with temperature. A quadratic relation can be expected from Eq. (1), where both solid fraction of precipitation and monthly ablation, and length of the accumulation and ablation seasons, are related almost linearly to temperature (within certain bounds). We therefore fitted a quadratic function for all scenarios and all models, once using the global mean, and once using glacier surface area weighted temperature anomaly.

Table 3 summarizes the parameters of the function, showing the mean parameter values of the different CMIP5 models, and the standard deviation. We find a contribution of roughly 0.85 mm SLE yr^{-1} mass loss independent of ΔT that corresponds to the adjustment of glaciers to past climate change, as discussed in Sect. 3, and is therefore independent of temperature anomalies. This contribution implies substantial future glacier-mass loss even without further warming. The magnitude here however is substantially higher than the magnitude of the mass loss rates found in the equilibrium experiments (the line corresponding to "CRU" in Fig. 1, on the order of 0.5 mm SLE yr^{-1} at the start of the experiment), indicating that glacier hypsometry changes dampen the glaciers response to climate change.

A further contribution is linearly related to temperature anomalies, and on the order of 0.61 mm SLE yr^{-1} K^{-1} if glacier surface area weighted temperature anomaly is used, and on the order of 1.25 mm SLE yr^{-1} K^{-1} if global mean temperature anomaly is used. This contribution corresponds to the changes in solid precipitation and ablation. The decreased sensitivity of glacier surface area weighted temperature anomaly is consistent with the temperature amplification experienced by the glaciers (Fig. 5). Finally, a quadratic contribution of roughly 0.18 mm SLE yr^{-1} K^{-2} (glacier surface area weighted temperature) or 0.57 mm SLE yr^{-1} K^{-2} (global mean temperature) which – assuming that the length of the ablation season is linearly related to the annual mean temperature – is related to the interaction of changes in the length of ablation and accumulation seasons and changed temperature and solid precipitation.

Instead of applying the model specific parameters for the quadratic fit, it is also possible to apply the mean of the model specific parameters. The last column of Table 3 shows the RMSE of this model mean fitted curve. By this measure, the fit of the curve based on the glacier surface area weighted temperature anomaly is substantially better than the fit based on the global mean temperature anomaly (RMSE about half as big). This indicates that differences in the spatial patterns of the temperature anomaly are responsible for a considerable fraction of the different responses of global glacier mass to climate change. The remaining differences are due to the patterns in precipitation anomaly, history of the climate forcing, and different seasonalities of the temperature changes.

Based on a linear approximation, Gregory and Oerlemans (1998) estimate the linear sensitivity to 0.63 mm SLE yr^{-1} K^{-1}, and Meehl et al. (2007) to 0.80 ± 0.33 mm SLE yr^{-1} K^{-1} based on observations, and to 0.61 ± 0.12 or 0.49 ± 0.13 mm SLE yr^{-1} K^{-1} based on different models (all using global mean temperature). To better understand our strongly enhanced sensitivity, we estimated a linear sensitivity for the case of fixed glacier hypsometry as before, but this time driving the glacier model with past, observed climate (Mitchell and Jones, 2005), essentially reproducing the method applied by Meehl et al. (2007) in our model. This experiment results in a linear sensitivity of 0.41 ± 0.49 mm SLE yr^{-1} K^{-1}, consistent with all the previous results (the error range given is the 95 % confidence interval, and is large because of a relatively weak correlation of mass loss rates with global mean temperature anomalies). This indicates that the reason for the substantially different sensitivities when determined based on either future or past climate change is that the present glacier hypsometries have already responded to past climate change; that is, the actual sensitivity for a fixed reference hypsometry is high, but the sensitivity is strongly dampened by the glaciers' hypsometry response. This response has already had a considerable influence on past glacier mass change.

The temporally accumulated mass loss from the experiment with constant hypsometry and infinite volume is shown in Fig. 7. From this too it is apparent that changes in

Table 3. Model mean parameters of a function of the form $a + b \cdot \Delta T + c \cdot \Delta T^2 \cdot \frac{\Delta T}{|\Delta T|}$ fitted to the modeled global glacier mass balance when glacier hypsometry is held constant, and ice volume treated as infinite (Fig. 6). Standard deviation between CMIP5 models given as uncertainty. Last column: goodness-of-fit (RMSE) from applying the model mean parameters instead of the model-specific parameters.

Scenario		a $\left[\frac{\text{mm SLE}}{\text{yr}}\right]$	b $\left[\frac{\text{mm SLE}}{K\ \text{yr}}\right]$	c $\left[\frac{\text{mm SLE}}{K^2\ \text{yr}}\right]$	RMSE $\left[\frac{\text{mm SLE}}{\text{yr}}\right]$
RCP2.6	global mean ΔT	0.94 ± 0.15	1.19 ± 0.36	0.56 ± 0.59	1.19
	glacier surface area weighted ΔT	0.92 ± 0.14	0.62 ± 0.16	0.16 ± 0.09	0.72
RCP4.5	global mean ΔT	0.86 ± 0.17	1.24 ± 0.45	0.55 ± 0.32	2.96
	glacier surface area weighted ΔT	0.91 ± 0.16	0.61 ± 0.19	0.17 ± 0.07	1.45
RCP6.0	global mean ΔT	0.81 ± 0.10	0.97 ± 0.44	0.67 ± 0.39	1.22
	glacier surface area weighted ΔT	0.86 ± 0.08	0.53 ± 0.16	0.20 ± 0.06	0.77
RCP8.5	global mean ΔT	0.72 ± 0.36	1.65 ± 1.23	0.51 ± 0.40	9.74
	glacier surface area weighted ΔT	0.74 ± 0.47	0.68 ± 0.40	0.19 ± 0.06	4.69

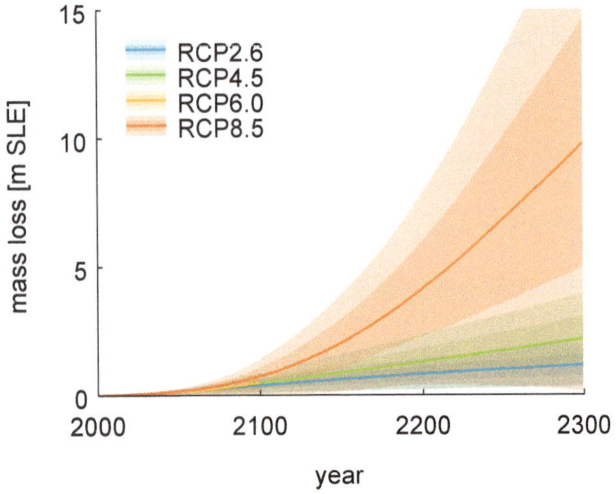

Fig. 7. Global glacier-mass loss when glacier hypsometry is held constant, and ice volume treated as infinite. Solid line is ensemble mean, dark/light shading indicates ensemble spread (one/two standard deviations).

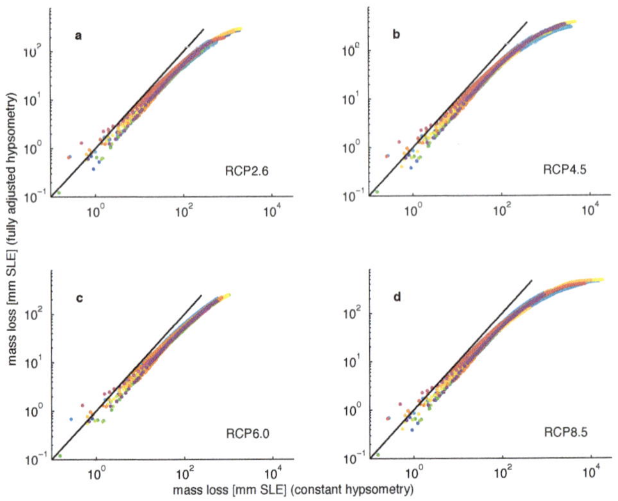

Fig. 8. Global glacier-mass loss of Marzeion et al. (2012) vs. global glacier-mass loss when glacier hypsometry is held constant, and ice volume treated as infinite. Black line is the bisector. Colors as in Fig. 2. Different panels refer to different RCPs.

glacier hypsometry and volume must impose strong limits on glacier-mass loss in reality, since the projected sea level rise far exceeds the (fixed) initial ice volume. Given the excessive projected mass loss of this experiment, it is clear that the ice mass must be the dominant limiting factor of future mass loss in all the scenarios, and for all climate models.

Plotting the temporally integrated glacier-mass loss of this experiment as a function of the temporally integrated mass loss of Marzeion et al. (2012, see Fig. 8) reveals that glacier hypsometry changes dampen the mass loss response of glaciers to climate change even for fairly small mass losses of less than 10 mm SLE, supporting the conclusion that the feedbacks considered here have already played a role in shaping the 20th century mass response of glaciers to climate change.

4.2.2 Constant terminus elevation

Changes in glacier terminus elevation can be expected to provide a negative feedback to the glacier mass balance, since a glacier losing mass eventually retreats to higher elevation, where colder temperatures prevail. Without the possibility of a response of the glacier terminus to mass balance anomalies, glaciers cannot reach an equilibrium with a changed climate. Terminus elevation changes are neglected, for example, in the approach taken by van de Wal and Wild (2001) and Slangen and van de Wal (2011). Figure 9 shows that if this feedback is ignored, glaciers would lose substantially more mass under global warming. Figure 10 shows that the mass loss is not only greater under all scenarios, but also at almost all times the rates of mass loss are higher. The feedback is strongest in the RCP2.6 scenario, and weakest in the

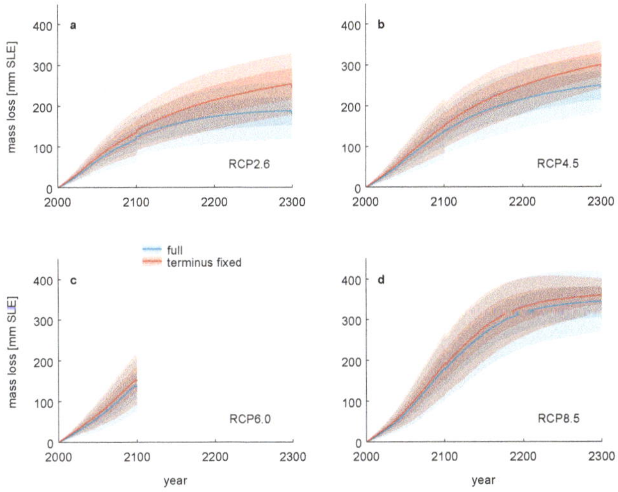

Fig. 9. Effect of fixed glacier terminus altitude on glacier mass change; "full" refers to the results of Marzeion et al. (2012) with fully responsive glacier hypsometry. Solid line is ensemble mean, dark/light shading indicates ensemble spread (one/two standard deviations). Different panels refer to different RCPs.

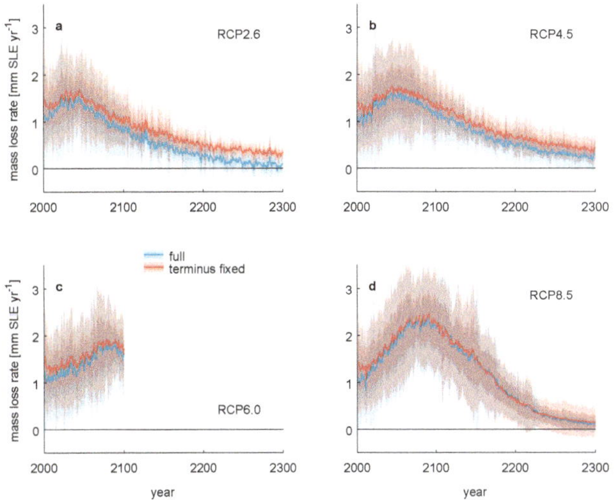

Fig. 10. Effect of fixed glacier terminus altitude on glacier mass change rates; "full" refers to the results of Marzeion et al. (2012) with fully responsive glacier hypsometry. Solid line is ensemble mean, dark/light shading indicates ensemble spread (one/two standard deviations). Different panels refer to different RCPs.

RCP8.5 scenario. The reason for this difference is the presence of glaciers that may be affected: in the RCP2.6 scenario, mass loss is relatively small and relatively few glaciers disappear completely, implying that many glaciers are affected by the feedback. But under a scenario of greater warming, many glaciers lose all their mass, implying that terminus elevation does not matter any longer for their mass balance. Figure 11, showing the difference in mass loss in 2300 (i.e., at a time when the fully responsive model is close to equilibrium) confirms that the effect of changing terminus elevations is largely a function of past mass loss, independent of climate model and RCP scenario. During the transient period, the relation between past mass loss and the terminus elevation feedback is more complex. The glacier terminus may be both higher or lower than its equilibrium value. Which is the case depends both on how strong climate variability is compared to climate change, and what the response time of the glacier is. Since the response times are different for each glacier, Fig. 11 becomes noisier at periods of strong mass loss rates.

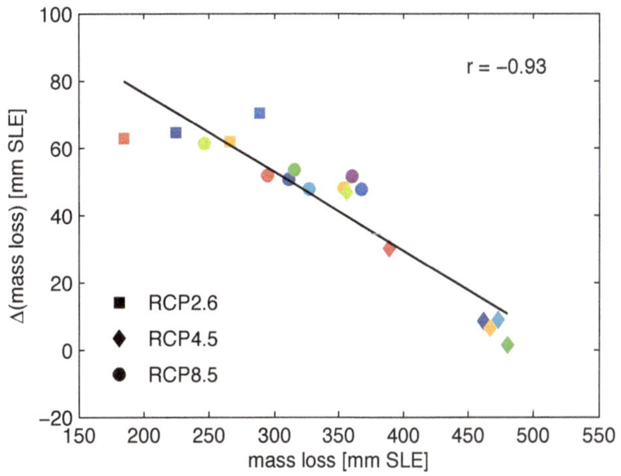

Fig. 11. Summary of the effect of glacier terminus retreat to higher altitude on glacier-mass loss. Plotted are the anomalies in the year 2300 as a function of respective mass loss projected in Marzeion et al. (2012) at the same time. Colors as in Fig. 2. Different markers refer to different RCPs.

5 Discussion

In the global view, glaciers are considerably out of balance with the climate of the recent past (Jacob et al., 2012; Gardner et al., 2013), and glacier-mass loss would continue even if climate changed no further. Using observed accumulation-area ratios, Mernild et al. (2013) come to the conclusion that glaciers are already committed to a mass loss of 16.3 ± 7.3 cm SLE. While this estimate is higher than ours of 6.6 ± 0.2 cm SLE, a direct comparison with our results is not possible because of different reference times. But considering that they included observations up to 2010 which includes more negative mass balances than our reference period of 1961–1990, our estimate might be consistent with theirs. Glaciers' mass loss during the next decades is therefore to a large extent predetermined from the past. This helps explain why there is great similarity in glaciers response to widely differing climate scenarios (Marzeion et al., 2012; Radić et al., 2013).

Our result that the wide ensemble spread within each RCP scenario to a large extent is attributable to differences in temperature anomaly patterns is in line with the results of Giesen and Oerlemans (2013), even though not directly comparable since their model involves – apart from precipitation – atmospheric transmissivity as a forcing parameter. Giesen and Oerlemans (2013) also report that applying changes in temperature only increased the glaciers' response to future warming significantly, but the compensation by increasing precipitation in our case is not as strong as in theirs.

When comparing glacier surface area weighted to global mean temperature changes (Fig. 5), we neglected seasonality. This is problematic especially concerning the arctic amplification of global warming, which is stronger in winter than in summer, thereby having a lesser influence on glacier mass balance than a comparable temperature change of the annual mean (Gregory and Oerlemans, 1998). That said, Table 3 shows that the RMSE of fitting mass loss with area weighted temperature changes is about half as big as the RMSE when using global mean temperature change. This implies that differences in the temperature change patterns of the difference GCMs are responsible for about half of the differences in the mass losses – or in other words, area-weighted temperatures are a better predictor for glacier-mass loss than global mean temperatures, even if seasonality is neglected. The fit would probably improve further if seasonality was accounted for in the glacier surface area weighted temperature, but doing so is non-trivial, as the length of ablation and accumulation seasons is glacier specific, and variable in time.

Observed sea level rise during the 20th century can only be explained if the glaciers' contribution did not develop in parallel to global mean temperature, but was high already in the first half of the 20th century (Gregory et al., 2013). Our result that changes in glacier hypsometry play a significant role in shaping the glaciers' response to climate change, in particular that loss of low-lying surface area (i.e., terminus retreat to higher elevations) decreases the sensitivity is critical for explaining the strong glacier-mass losses during the first half of the 20th century. Being able to explain and account for this mechanism is also important for the ability of process-based projections of sea level rise (Church et al., 2013). Knowledge of the equilibrium response of glaciers to climate change is furthermore necessary in order to develop scenarios of long-term sea level rise (Levermann et al., 2013).

By providing a positive feedback on the mass balance, ice thickness changes may be more important than terminus retreat in determining the mass balance response to climate change for some glaciers (Raymond et al., 2005), particularly if the response of the glaciers is not dynamic (i.e., ice is melting over a large fraction or all of the surface), ice thickness change may become important (Paul and Haeberli, 2008). Similarly, for glaciers with long, flat tongues, even a dynamic response to volume change may not necessarily lead to strong changes in terminus elevation. The surface topography may change considerably nevertheless (Larsen et al., 2007; Bolch et al., 2008). Our glacier model does not explicitly account for ice thickness change and associated changes in the surface topography that are unrelated to surface area change and terminus elevation. It is therefore not possible to isolate this effect here. However, we argue that while ice thickness change is not considered explicitly, it is contained in the model implicitly – more specifically, in the glaciers' temperature sensitivity μ^*: the ice thickness–mass balance feedback increases a glacier's sensitivity to temperature change, by amplifying the temperature anomalies at the ice surface. Since μ^* is essentially calibrated by determining a value that results in the best fit to observations (which of course include the feedback), and since the cross validation of the model in Marzeion et al. (2012) does not indicate a significant bias of modeled mass balances, independent of the length of the observation time series, we can conclude that not explicitly including ice thickness changes in the model does not affect the reliability of the model's results. It would be feasible to explicitly include this mechanism, and it seems desirable to do so in order to be able to quantify it globally. We deliberately chose not to do this here, since it would require a complete recalibration of the model, which would inhibit the direct comparison with the results of (Marzeion et al., 2012). Moreover, the results of Paul (2010) and Huss et al. (2012) indicate that the combined effect of glacier surface topography changes is dampening the glaciers' response to climate forcing over long timescales, even though this may be arguably different for individual glaciers (Harrison et al., 2009). The increase of the sensitivity by a factor of two to three if glacier hypsometry is held constant reported by Paul (2010) for the European Alps is consistent with our results for the global scale.

Other potentially important feedbacks not included in the model but meriting thorough evaluation include, for example, changes of ice albedo due to accumulation of dust on melting glaciers (Oerlemans et al., 2009), and changes to the long-wave radiation energy budget from newly exposed rock faces surrounding the glacier.

6 Conclusions

We have used a model of glacier response to climate change to quantify the equilibrium sensitivity of glaciers, and to distinguish the respective contributions of temperature and precipitation anomalies. Because of the geographic distribution of glaciers, the temperature and precipitation change experienced by glaciers is far greater than the global mean. Precipitation anomalies projected for the future dampen the mass loss of glaciers, but their effect is strongly limited by the increasing temperatures, which increases the liquid fraction of precipitation on the glaciers.

We find that glacier-mass loss during the 21st century is to a significant degree a response to 20th century climate change. This partly explains the relatively weak dependence of 21st century mass loss on future greenhouse gas emissions. A second reason is that the complete loss of individual glaciers imposes a strong restriction to the rates of mass loss in a warming climate. Results from methods not accounting for finiteness of ice mass available for melting, for example, by extrapolating current rates of mass loss, or even increases of rates of mass loss (Meier et al., 2007), will therefore yield substantial overestimates.

Thirdly, changes of glacier hypsometry reduce the response of glaciers to warming, and need to be considered in explaining the observed rates of mass loss during the 20th century and the projected rates for the 21st century. Figure 9 summarizes our quantification of this effect. We find that the retreat of glacier termini to higher altitudes is a strong negative feedback, which is becoming weaker as more glaciers disappear completely.

Acknowledgements. This work was funded by the Austrian Science Fund (FWF): P22443-N21 and PP25362-N26, and supported by the Austrian Ministry of Science BMWF as part of the UniInfrastrukturprogramm of the Research Platform Scientific Computing at the University of Innsbruck. We would like to thank D. Farinotti, A. Gardner, and P. Leclercq for their constructive comments on an earlier version that helped us greatly to improve the manuscript.

Edited by: E. Larour

References

Andrews, T., Forster, P. M., Boucher, O., Bellouin, N., and Jones, A.: Precipitation, radiative forcing and global temperature change, Geophys. Res. Lett., 37, L14701, doi:10.1029/2010GL043991, 2010.

Arendt, A., Bolch, T., Cogley, G., Gardner, A., Hagen, J. O., Hock, R., Kaser, G., Paul, F., Radic, V., Bliss, A., Fountain, A., Mercer, A., Negrete, A., Giffen, B., Menounos, B., Kienholz, C., Mayer, C., Nuth, C., Burgess, D., Hall, D., Kriegel, D., Berthier, E., Burgess, E., Cawkwell, F., Wyatt, F., Hartmann, G., Wolken, G., Frey, H., Brown, I., Howat, I., Lund, J., Rich, J., Filbert, K., Andreassen, L., Copland, L., Beedle, M., Koenig, M., Sharp, M., Moelg, N., Sigurdsson, O., Rastner, P., Forester, R., LeBris, R., Pettersson, R., Wheate, R., Herreid, S., Vorogushin, S., Winsvold, S., Chinn, T., Hagg, W., and Manley, W.: Randolph Glacier Inventory 1.0: A Dataset of Global Glacier Outlines, global Land Ice Measurements from Space, Boulder Colorado, USA, Digital Media, 2012.

Bahr, D.: Global distributions of glacier properties: a stochastic scaling paradigm, Water Resour. Res., 33, 1669–1679, 1997.

Bahr, D., Meier, M., and Peckham, S.: The physical basis of glacier volume-area scaling, J. Geophys. Res., 102, 355–362, 1997.

Barrand, N., Hindmarsh, R., Arthern, R., Williams, C. R., Mouginot, J., Scheuchl, B., Rignot, E., Ligtenberg, S., van den Broeke, M., Edwards, T., Cook, A. J., and Simonsen, S. B.: Computing the volume response of the Antarctic Peninsula ice sheet to warming scenarios to 2200, J. Glaciol., 59, 397–409, 2013.

Bolch, T., Buchroithner, M., Pieczonka, T., and Kunert, A.: Planimetric and volumetric glacier changes in the Khumbu Himal, Nepal, since 1962 using Corona, Landsat TM and ASTER data, J. Glaciol., 54, 592–600, 2008.

Church, J. A., Monselesan, D., Gregory, J. M., and Marzeion, B.: Evaluating the ability of process based models to project sea-level change, Environ. Res. Lett., 8, 014051, doi:10.1088/1748-9326/8/1/014051, 2013.

Cogley, J. G.: Geodetic and direct mass-balance measurements: comparison and joint analysis, Ann. Glaciol., 50, 96–100, doi:10.3189/172756409787769744, 2009.

Ehlschlaeger, C.: Using the A^T search algorithm to develop hydrologic models from digital elevation data, in: International Geographic Information Systems (IGIS) Symposium, Vol. 89, 275–281, 1989.

Gardner, A. S., Moholdt, G., Cogley, J. G., Wouters, B., Arendt, A. A., Wahr, J., Berthier, E., Hock, R., Pfeffer, W. T., Kaser, G., Ligtenberg, S. R. M., Bolch, T., Sharp, M. J., Hagen, J. O., van den Broeke, M. R., and Paul, F.: A reconciled estimate of glacier contributions to sea level rise: 2003 to 2009, Science, 340, 852–857, 2013.

Giesen, R. H. and Oerlemans, J.: Climate-model induced differences in the 21st century global and regional glacier contributions to sea-level rise, Clim. Dynam., 41, 3283–3300, doi:10.1007/s00382-013-1743-7 2013.

Gregory, J. M. and Oerlemans, J.: Simulated future sea-level rise due to glacier melt based on regionally and seasonally resolved temperature changes, Nature, 391, 474–476, 1998.

Gregory, J. M., Church, J. A., White, N. J., Bierkens, M. F. P., Box, J. E., van den Broeke, M., G., C. J., Fettweis, X., Hanna, E., Leclercq, P. W., Marzeion, B., Oerlemans, J., Wada, Y., Wake, L. M., and van de Wal, R. S. W.: Twentieth-century global-mean sea-level rise: is the whole greater than the sum of the parts?, J. Climate, 26, 4476–4499, doi:10.1175/JCLI-D-12-00319.1, 2013.

Grinsted, A.: An estimate of global glacier volume, The Cryosphere, 7, 141–151, doi:10.5194/tc-7-141-2013, 2013.

Harrison, W. D., Cox, L. H., Hock, R., March, R. S., and Pettit, E. C.: Implications for the dynamic health of a glacier from comparison of conventional and reference-surface balances, Ann. Glaciol., 50, 25–30, 2009.

Hock, R.: Temperature index melt modelling in mountain areas, J. Hydrol., 282, 104–115, doi:10.1016/S0022-1694(03)00257-9, 2003.

Huss, M.: Extrapolating glacier mass balance to the mountain-range scale: the European Alps 1900–2100, The Cryosphere, 6, 713–727, doi:10.5194/tc-6-713-2012, 2012.

Huss, M. and Farinotti, D.: Distributed ice thickness and volume of all glaciers around the globe, J. Geophys. Res., 117, F04010, doi:10.1029/2012JF002523, 2012.

Huss, M., Hock, R., Bauder, A., and Funk, M.: Reply to the Comment of Leclercq et al. on "100-year mass changes in the Swiss Alps linked to the Atlantic Multidecadal Oscillation", The Cryosphere Discuss., 4, 2587–2592, doi:10.5194/tcd-4-2587-2010, 2010.

Huss, M., Hock, R., Bauder, A., and Funk, M.: Conventional versus reference-surface mass balance, J. Glaciol., 58, 278–286, 2012.

Jacob, T., Wahr, J., Pfeffer, W. T., and Swenson, S.: Recent contributions of glaciers and ice caps to sea level rise, Nature, 482, 514–518, 2012.

Jóhannesson, T., Raymond, C., and Waddington, E.: Time-scale for adjustment of glaciers to changes in mass balance, J. Glaciol., 35, 355–369, 1989.

Larsen, C. F., Motyka, R. J., Arendt, A. A., Echelmeyer, K. A., and Geissler, P. E.: Glacier changes in southeast Alaska and northwest British Columbia and contribution to sea level rise, J. Geophys. Res., 112, F01007, doi:10.1029/2006JF000586, 2007.

Leclercq, P. W., van de Wal, R. S. W., and Oerlemans, J.: Comment on "100-year mass changes in the Swiss Alps linked to the Atlantic Multidecadal Oscillation" by Matthias Huss et al. (2010), The Cryosphere Discuss., 4, 2475–2481, doi:10.5194/tcd-4-2475-2010, 2010.

Leclercq, P. W., Oerlemans, J., and Cogley, J. G.: Estimating the Glacier Contribution to Sea-Level Rise for the Period 1800–2005, Surv. Geophys., 32, 519–535, doi:10.1007/s10712-011-9121-7, 2011.

Levermann, A., Clark, P. U., Marzeion, B., Milne, G. A., Pollard, D., Radić, V., and Robinson, A.: The multi-millennial sea-level commitment of global warming, P. Natl. A. Sci. USA, 110, 13745–13750, doi:10.1073/pnas.1219414110, 2013.

Manabe, S., Stouffer, R., Spelman, M., and Bryan, K.: Transient responses of a coupled ocean-atmosphere model to gradual changes of atmospheric CO_2. Part I: Annual mean response, J. Climate, 4, 785–818, 1991.

Marzeion, B., Jarosch, A. H., and Hofer, M.: Past and future sea-level change from the surface mass balance of glaciers, The Cryosphere, 6, 1295–1322, doi:10.5194/tc-6-1295-2012, 2012.

Meehl, G. A., Stocker, T. F., Collins, W. D., Friedlingstein, P., Gaye, A. T., Gregory, J. M., Kitoh, A., Knutti, R., Murphy, J. M., Noda, A., Raper, S. C. B., Watterson, I. G., Weaver, A. J., Zhao, Z.-C., et al.: Global Climate Projections, in: Climate Change 2007: The Physical Science Basis. Contribution of Working Group I to the Fourth Assessment Report of the Intergovernmental Panel on Climate Change, edited by: Solomon, S., Qin, D., Manning, M., Chen, Z., Marquis, M., Averyt, K. B., M., T., and Miller, H. L., Cambridge University Press, Cambridge, UK and New York, NY, USA, 2007.

Meier, M. F., Dyurgerov, M. B., Rick, U. K., O'Neel, S., Pfeffer, W. T., Anderson, R. S., Anderson, S. P., and Glazovsky, A. F.: Glaciers dominate eustatic sea-level rise in the 21st century, Science, 317, 1064–1067, doi:10.1126/science, 2007.

Mernild, S. H., Lipscomb, W. H., Bahr, D. B., Radić, V., and Zemp, M.: Global glacier retreat: a revised assessment of committed mass losses and sampling uncertainties, The Cryosphere Discuss., 7, 1987–2005, doi:10.5194/tcd-7-1987-2013, 2013.

Mitchell, T. D. and Jones, P. D.: An improved method of constructing a database of monthly climate observations and associated high-resolution grids, Int. J. Climatol., 25, 693–712, doi:10.1002/joc.1181, 2005.

New, M., Lister, D., Hulme, M., and Makin, I.: A high-resolution data set of surface climate over global land areas, Clim. Res., 21, 1–25, 2002.

Oerlemans, J.: Glaciers and Climate Change, A. A. Balkema Publishers, Lisse, Abingdon, Exton, Tokyo, 2001.

Oerlemans, J., Anderson, B., Hubbard, A., Huybrechts, P., Johannesson, T., Knap, W. H., Schmeits, M., Stroeven, A. P., van de Wal, R. S. W., Wallinga, J., and Zuo, Z.: Modelling the response of glaciers to climate warming, Clim. Dynam., 14, 267–274, 1998.

Oerlemans, J., Giesen, R., and Van Den Broeke, M.: Retreating alpine glaciers: increased melt rates due to accumulation of dust (Vadret da Morteratsch, Switzerland), J. Glaciol., 55, 729–736, 2009.

Ohmura, A.: Physical Basis for the Temperature-Based Melt-Index Method, J. Appl. Meteorol., 40, 753–761, 2001.

Paul, F.: The influence of changes in glacier extent and surface elevation on modeled mass balance, The Cryosphere, 4, 569–581, doi:10.5194/tc-4-569-2010, 2010.

Paul, F. and Haeberli, W.: Spatial variability of glacier elevation changes in the Swiss Alps obtained from two digital elevation models, Geophys. Res. Lett., 35, L21502, doi:10.1029/2008GL034718, 2008.

Radić, V., Bliss, A., Beedlow, A. C., Hock, R., Miles, E., and Cogley, J. G.: Regional and global projections of 21st century glacier mass changes in response to climate scenarios from global climate models, Clim. Dynam., 42, 37–58, 2013.

Raymond, C., Neumann, T. A., Rignot, E., Echelmeyer, K., Rivera, A., and Casassa, G.: Retreat of Glaciar Tyndall, Patagonia, over the last half-century, J. Glaciol., 51, 239–247, 2005.

Screen, J. A. and Simmonds, I.: The central role of diminishing sea ice in recent Arctic temperature amplification, Nature, 464, 1334–1337, 2010.

Sicart, J., Hock, R., and Six, D.: Glacier melt, air temperature, and energy balance in different climates: The Bolivian Tropics, the French Alps, and northern Sweden, J. Geophys. Res., 113, D24113, doi:10.1029/2008JD010406, 2008.

Slangen, A. B. A. and van de Wal, R. S. W.: An assessment of uncertainties in using volume-area modelling for computing the twenty-first century glacier contribution to sea-level change, The Cryosphere, 5, 673–686, doi:10.5194/tc-5-673-2011, 2011.

Sutton, R. T., Dong, B., and Gregory, J. M.: Land/sea warming ratio in response to climate change: IPCC AR4 model results and comparison with observations, Geophys. Res. Lett., 34, L02701, doi:10.1029/2006GL028164, 2007.

van de Wal, R. and Wild, M.: Modelling the response of glaciers to climate change by applying volume-area scaling in combination with a high resolution GCM, Clim. Dynam., 18, 359–366, doi:10.1007/s003820100184, 2001.

van Vuuren, D. P., Edmonds, J., Kainuma, M., Riahi, K., Thomson, A., Hibbard, K., Hurtt, G. C., Kram, T., Krey, V., Lamarque, J.-F., Masui, T., Meinhausen, M., Nakicenovic, N., Smith. S. J., and Rose, S. K.: The representative concentration pathways: an overview, Clim. Change, 109, 5–31, 2011.

Zuo, Z. and Oerlemans, J.: Contribution of glacier melt to sea-level rise since AD 1865: a regionally differentiated calculation, Clim. Dynam., 13, 835–845, 1997.

3

Brief Communication: Sudden drainage of a subglacial lake beneath the Greenland Ice Sheet

I. M. Howat[1], C. Porter[2], M. J. Noh[1], B. E. Smith[3], and S. Jeong[1]

[1]School of Earth Sciences and Byrd Polar Research Center, Ohio State University, Columbus, Ohio, USA
[2]Polar Geospatial Center, University of Minnesota, St. Paul, Minnesota, USA
[3]Polar Science Center, Applied Physics Lab, University of Washington, Seattle, Washington, USA

Correspondence to: I. M. Howat (ihowat@gmail.com)

Abstract. We report on the appearance of a 2 km wide, 70 m deep circular depression located 50 km inland of the southwestern margin of the Greenland Ice Sheet that provides the first direct evidence for concentrated, long-term storage, and sudden release, of meltwater at the bed. Drainage of the lake may have been triggered by the recent increase in meltwater runoff. The abundance of such lakes and their potential importance to the ice sheet's hydrologic system and flow regime remain unknown.

1 Introduction

Recent observations of the Greenland Ice Sheet have revealed a complex hydrological system in which hundreds of gigatonnes of surface meltwater drain toward the margin each summer through both supraglacial and subglacial pathways. The supraglacial drainage system is typified by dendritic river systems and abundant meltwater lakes, ranging up to several kilometers wide. In areas of high accumulation, infiltrating meltwater can also be stored within perennial aquifers deep within the firn (Forster et al., 2014). The supraglacial and subglacial drainage systems are connected through englacial pathways (e.g., moulins) created by hydraulically driven fracturing through 1 km or more of ice, often facilitated by extensional flow and surface crevassing (Joughin et al., 2013). Meltwater entering the subglacial system disperses rapidly, indicating an efficient drainage system comprised of tunnels extending tens of kilometers into the interior (Chandler et al., 2013; Fitzpatrick et al., 2013).

Supraglacial lakes and firn aquifers store a substantial amount of meltwater, providing a buffer between melting and mass loss to the ocean (Forster et al., 2014; Fitzpatrick et al., 2014). The quantity of perennial subglacial meltwater water storage, however, remains highly uncertain. Thus far, subglacial lakes have only been detected in the far north of the ice sheet (Palmer et al., 2013), where lower snowfall promotes greater geothermal heating of the base and long-term subglacial storage of meltwater. It is not clear whether this water melted from basal ice or reached the bed from the surface. Unlike basally derived meltwater, subglacial storage of surface meltwater would be influenced directly by climate warming and, therefore, is more relevant to understanding ice sheet response to anthropogenic climate change. The high efficiency of Greenland's subglacial drainage system reduces the likelihood that lakes of surface meltwater can form (Palmer et al., 2013). For a hydraulic connection to penetrate to the bed its discharge will likely be high enough to immediately form drainage tunnels and drain toward the margin. While seasonal reductions in drainage system efficiency may lead to slowing or temporary meltwater storage through the winter, this water would likely be flushed from the system the next summer. Finally, the high surface slope of the margin implies a strong hydraulic gradient toward the ice edge that would also prevent lake development.

The most likely location for surface meltwater to be stored subglacially, therefore, would be high in the ablation zone at the uppermost elevation where hydraulic connections to the bed exist. Here, the total supply of melt to the bed would be at a minimum so that the drainage system will be least developed and may not receive input every year. The greater ice thickness and lower surface slope would increase the likelihood that bed relief would cause areas of convergence in

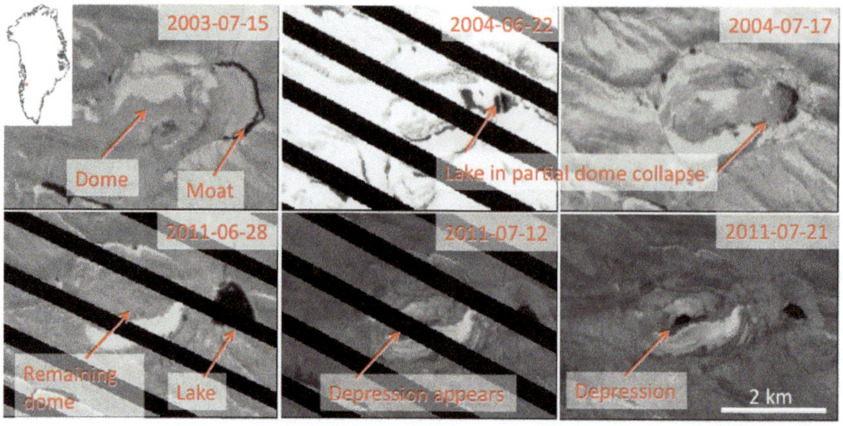

Figure 1. Series of Landsat 7 Enhanced Thematic Mapper Plus panchromatic satellite imagery showing two episodes of the collapse. The 2003 and 2004 images show the formation of a supraglacial lake within a depression on the eastern side of what previously appears to have been a dome edged by a water-filled moat. The 2014 images show the formation of the larger depression within the center of the dome. The black stripes are due to failure of the sensor's scan line corrector system.

the hydraulic potential field. It is in this setting, near the southwest margin of the ice sheet (67.61° N, 48.7° W), where we identified a prominent depression in the surface, first observable in a Landsat 7 Enhanced Thematic Mapper-Plus (ETM+) satellite image acquired on 12 July 2011 (Fig. 1; see Appendix A for detailed data set descriptions). The feature is not visible on the next earlier image, dated 28 June, indicating that the depression formed in that period. Sub-meter resolution panchromatic imagery and stereoscopic digital elevation models (DEMs) from the WorldView-1 (WV1) satellite of the area were acquired for the first time on 28 October 2011 (Fig. 2). The DEM and accompanying orthoimage reveal an approximately 70 m deep depression with a diameter of between 1.5 and 2 km and a total area of 1.6 km^2. The walls of the depression have slopes of 10 to 15°. The bottom of the depression is filled with meltwater covering an area of 0.08 km^2 that, as the Landsat imagery shows, drained into the depression from a nearby lake. Concentric rings of crevasses that indicate collapse of the surface surround the depression. The northern edge of the depression is bounded by a crevasse over 50 m wide and more than 30 m deep.

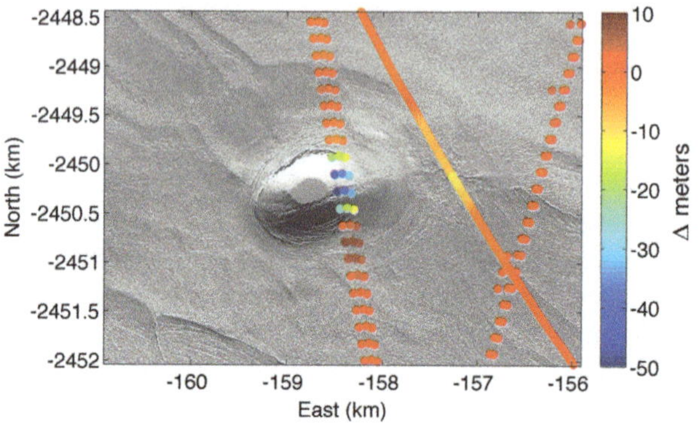

Figure 2. WorldView-1 panchromatic satellite image from 28 October 2011, with (dots) 2005 ICESat-1 and (curve) 1993 Airborne Topographic Mapper lidar tracks overlain. Colors are the difference between the 2011 WorldView DEM and earlier lidar elevations. The November 2005 ICESat track is the westernmost track, running closest to the center of the depression. The map coordinates are in polar stereographic projection with a meridian at 70° W (Worlview image ©2011 DigitalGlobe, Inc.).

2 Observations

Prior to the 28 October 2011 WV1 acquisition, available elevation data for the area of the depression include (1) an April 1993 overflight by the NASA Airborne Topographic Mapper (ATM) lidar 1 km east (up-glacier) of the depression, (2) three Ice and Cloud Elevation Satellite (ICESat) laser altimetry passes over the eastern side of the depression acquired in March, June and November 2005 (Fig. 2) and (3) a 40 m resolution stereoscopic DEM produced from SPOT-5 satellite imagery acquired on 7 July 2008 and distributed by the SPOT 5 stereoscopic survey of Polar Ice: Reference Images and Topographies (SPIRIT) project (Korona et al., 2009). An additional WV1 DEM was acquired on 5 July 2013. Elevation profiles along the 15 November 2005 ICESat track (Fig. 3) reveal that the depression formed through collapse of what was previously a 5 m tall dome on the surface fringed by a moat, which was typically filled with snow in the late-summer imagery. The dome lay on a reversed surface slope (i.e., rising to the east in the direction of flow), with a gradient of 0.012, on the stoss side of a rise (Fig. 3). The down-glacier side of the rise drops twice as steeply to another plateau. Up-glacier of the depression, a small supraglacial lake occupies the base of another

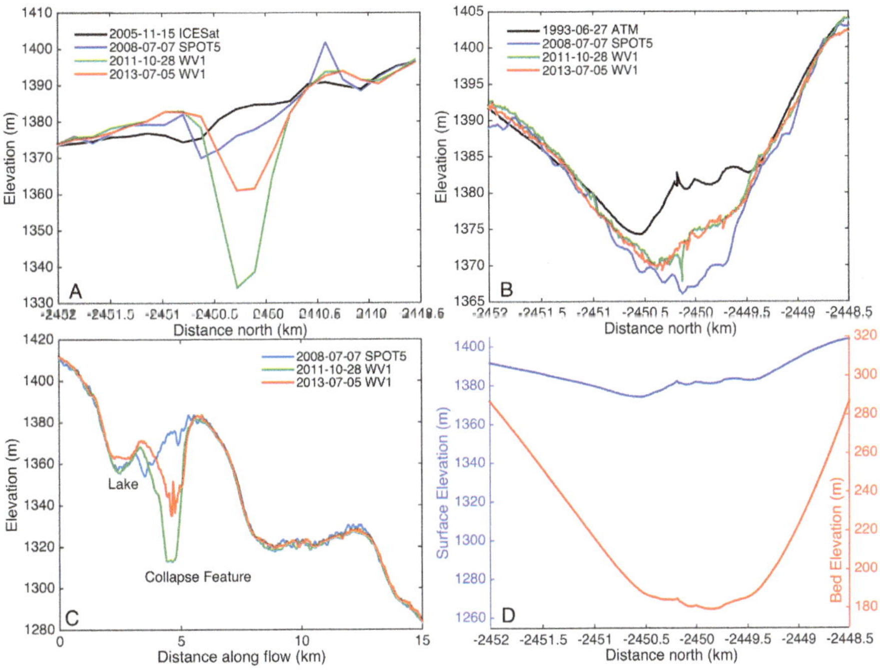

Figure 3. Surface elevations from laser altimetry and stereoscopic digital elevation models along the **(a)** 15 November 2005 ICESat-1 and **(b)** 26 June 1993 ATM tracks shown in Fig. 2. **(c)** Along-flow (east to west) surface elevation profiles through the center of the depression from three stereoscopic DEMs. **(d)** Surface (blue) and bed heights (red) for the 1993 airborne survey over the eastern portion of the depression (see Fig. 2). Both vertical axes have the same scale.

steep slope. Such undulating topography is common over the southwestern margin.

Comparison of elevation profiles along the 1993 ATM track (Fig. 3) indicates an earlier lowering of the surface of up to 20 m prior to the 2008 SPOT DEM over the area immediately upstream of the current depression, at the location where a supraglacial lake has formed intermittently since 2004. An apparent collapse of that surface occurred between the summers of 2003 and 2004 (Fig. 1). Prior to summer 2004, water occupied a narrow moat on the eastern margin of what appeared to be a small dome. In the 15 July 2004 Landsat image, the lake appears in a depression. The surface then rose 5 to 8 m between 2008 and 2011 (Fig. 3). Inspection of DEM profiles along the flow direction (Fig. 3) reveals that this rising of the surface was due to horizontal advection of a small topographic bump and not infilling.

3 Discussion

Based on these data, we conclude that the depression formed due to rapid drainage of a subglacial lake. Drainage occurred in two episodes: a smaller event in 2004 and a larger one in 2011. Prior to the 2004 event, the surface appeared unchanged in imagery extending back to 1972. This, and the raised surface detected in the 1993 ATM survey, indicates that water storage persisted for at least several decades before drainage. Differencing of the 2008 and 2011 DEMs over the area of the depression gives a minimum total volume change of 2.5×10^{-2} km^3. Assuming that vertical displacement of the surface is equal to the depth of the former lake, and that all drainage occurred in the 14 days between the 28 June and 12 July images, gives an average discharge of 20 m^3 s^{-1}. This discharge is similar to that estimated for a supraglacial lake that drained and quickly dispersed into the subglacial drainage system (Das et al., 2008).

Storage of water at the bed requires a flat hydraulic potential surface. This is achieved when the slope of the bed is approximately 11 times steeper than the surface and of opposite sign, so that the force of gravity pulling water along the bed slope balances the gradient in ice overburden. As shown in Fig. 3, the depression occurred on a reversed slope on the stoss side of a rise, with a steeper drop on the lee. A downward step in the bed, such as the down-glacier slope of a subglacial roche moutonnée, would provide the conditions needed for a zero, or reversed, hydraulic gradient. The only available thickness data are from the Kansas University ice-penetrating radar acquired during the same survey as the 1993 ATM flight. These data reveal that the location of the surface depression lies within a shallow bathymetric low, with a maximum ice thickness of 1200 m, into which subglacial water may be routed (Fig. 3). The single transect, however, is not enough to constrain the hydraulic gradient field over the depression.

Unlike Antarctic subglacial lakes, we assume that this water reached the bed from the surface through moulins and was routed into the lake through the subglacial drainage system. This is suggested by numerous draining surface lakes and moulins visible in the satellite imagery in the vicinity of the depression. Ice flow speeds in this area are only 50 to 60 m yr^{-1} (Joughin et al., 2010) indicating little or no generation of basal meltwater due to frictional heating. On the other hand, basal temperatures in the upper ablation zone are likely close to the pressure melting temperature. This, and the continuing influx of heat with surface meltwater, would prevent the subglacial lake from freezing.

The drainage of this subglacial lake, however, may have a similar trigger as the draining and filling of those in Antarctica (Gray et al., 2005; Wingham et al., 2006; Clarke, 2006). Filling of the lake would increase the hydraulic gradient, eventually overcoming the gradient in ice pressure. Drainage would melt channel walls through viscous heat dissipation, enlarging subglacial tunnels and leading to continued drainage despite reduced water pressures. In this case an additional mechanism for triggering drainage may be the transition of the subglacial drainage system from inefficient to efficient modes in the vicinity of the lake. Inefficient systems maintain high water pressures at low subglacial discharges through a network of small cavities. An increase in discharge can cause these cavities to coalesce into efficiently draining tunnels with reduced water pressures (Schoof, 2010). The gradient in water pressure between these modes causes efficient drainage to propagate upstream with increasing water input (Bartholomew et al., 2011). Chemical tracer and ice motion observations at the southwestern Greenland margin suggest efficient drainage up to 50 km inland, which is also the distance of this lake to the ice sheet margin (Chandler et al., 2013; Bartholomew et al., 2011). Runoff increased over the past decade, with an anomalously high melting in 2010 (Van As et al., 2012), which would likely increase subglacial discharge and promote expansion of efficient drainage, potentially triggering drainage of the reservoir.

Without detailed ice thickness and bed information, or maximum rates of surface lowering, it is not clear whether collapse of the lake and formation of the depression occurred through brittle or ductile deformation of the surrounding ice. Sudden drainage of a supraglacial lake resulted in uplift of a 750 m wide column of ice through faulting of the full $\sim$ 1 km ice thickness (Das et al., 2008). Rapid drainage of a subglacial lake may provide the inverse situation by suddenly removing the hydraulic support for the floating ice column above, resulting in shear stresses exceeding the strength of the ice and causing full-thickness faulting. Slower drainage, however, may have been accommodated through viscous creep. A constant surface lowering rate over the 14 days gives a vertical shear strain rate at the lake perimeter on the order of 10^{-7} s^{-1}, implying shear stresses between 300 kPa, for ice near the freezing temperature, and 950 kPa for ice at -30 °C (Cuffey and Paterson, 2010). Distributed around the perimeter of the lake, these shear stresses would only support between 7 and 20 % of the weight of the ice above the lake, with water pressure required to support the remainder. This surface-lowering rate is therefore also consistent with a catastrophic drainage, in which the water pressure dropped substantially as soon as outflow began.

While the 2013 DEM shows shallowing of the depression (Fig. 3), it is uncertain whether this is due to subglacial refilling of the lake or infilling of the surface depression by meltwater, blowing snow and ice talus from the collapse of the surrounding crevasses. If overfilling and pressurization of the reservoir forced rapid drainage into a surrounding, inefficient drainage system we would expect the reservoir to refill as the drainage tunnels collapse. If, however, drainage was triggered by the inland propagation of an efficient drainage system, the persistence of that regime may prevent refilling of the lake. Thus, failure of the lake to refill may signal a more permanent shift in the ice sheet's subglacial drainage system.

4 Conclusions

We have provided the first direct evidence of long-term storage and sudden release of meltwater within a lake beneath the Greenland Ice Sheet. The lake existed in the upper ablation zone where it likely had received relatively low and intermittent discharges of surface meltwater through englacial conduits, preventing efficient drainage, in an area of reversed surface slope that likely created a reversal in the hydraulic gradient. Drainage of the lake may have been triggered by increasing efficiency of the subglacial drainage system with increased meltwater inputs under recent warming. By itself, the volume of the lake is insignificant to the hydrologic budget of the ice margin, equivalent to 1 year of runoff over a 20 km^2 area at the elevation of the depression. However, there is no indication that this feature should be unique. Undulating surface topography with similar slope reversals is common over the upper ablation zone of the margin. Before drainage there is very little surface expression to identify the presence of a lake. Its small size and the abundance of water at the bed reduces the likelihood of detection of undrained lakes with airborne ice-penetrating radar. Nothing anomalous exists in the radar echogram from the 1993 survey over the eastern portion of the lake. Even after drainage and surface collapse, the abundance of drained supraglacial lake basins and other features makes the depression difficult to identify in commonly available satellite imagery. Finally, the depression infills with snow quickly after the collapse, limiting the time in which it could be detected to a few years. Collection of detailed bed topography and ice thickness in the vicinity of the depression would provide insight into the conditions that caused the lake to form and, potentially, drain. We would then be able to assess the likelihood that other lakes exist and where they may be found.

Appendix A: Data set descriptions

Orthorectified Landsat imagery was obtained from the United States Geological Survey Earth Explorer archive (http://earthexplorer.usgs.gov/). These data are radiometrically calibrated and corrected for terrain distortions using a DEM prior to distribution (http://landsat.usgs.gov/Landsat_Processing_Details.php).

WorldView stereopair images were obtained from the Polar Geospatial Center at the University of Minnesota. We used the Ohio State University DEM extraction software Surface Extraction through TIN-Based Minimization (SETSM, http://www.pgc.umn.edu/system/files/SETSM_Product_Sheet_v1.1.pdf) to construct the DEM and generate the orthoimages. The DEMs were coregistered to lidar data collected by NASA's Operation IceBridge in March and April of 2011 and 2013. Following coregistration, vertical errors in the DEM are less than 0.5 m.

We obtained the ATM Level 1B and Kansas University Multichannel Coherent Radar Depth Sounder ice-penetrating radar data from the National Snow and Ice Data Center (http://nsidc.org/data/icebridge/). We use the ICESat 633 products of the GLA12 release corrected for time-varying elevation biases and filtered as described in Shepherd et al. (2012). Elevations were corrected for detector saturation, and the time-varying bias correction should remove offsets associated with campaign-to-campaign variations in the shape of the transmitted pulse (Borsa et al., 2014). Elevations calculated in this way should be accurate to better than 0.1 m.

Author contributions. I. M. Howat led the data compilation, analysis and writing of the paper. C. Porter first identified the surface depression and preprocessed and provided the WV imagery. M. J. Noh constructed the WV DEMs. B. E. Smith provided the ICESat data and assisted with the analysis. S. Jeong aided with Landsat imagery processing. All authors contributed to manuscript preparation.

Acknowledgements. This work was supported by grant NNX10AN61G to I. M. Howat from the US National Aeronautics and Space Administration. The authors thank the numerous instrument teams and data providers that collected and supplied the data used in this study.

Edited by: O. Gagliardini

References

Bartholomew, I., Nienow, P., Sole, A., Mair, D., Cowton, T., Palmer, S., and Wadham, J.: Supraglacial forcing of subglacial drainage in the ablation zone of the Greenland ice sheet, Geophys. Res. Lett., 38, L08502, doi:10.1029/2011GL047063, 2011.

Borsa, A. A., Moholdt, G., Fricker, H. A., and Brunt, K. M.: A range correction for ICESat and its potential impact on ice-sheet mass balance studies, The Cryosphere, 8, 345–357, doi:10.5194/tc-8-345-2014, 2014.

Chandler, D. M., Wadham, J. L., Lis, G. P., Cowton, T., Sole, A., Bartholomew, I., Telling, J., Nienow, P., Bagshaw, E. B., Mair, D., Vinen, S., and Hubbard, A.: Evolution of the subglacial drainage system beneath the Greenland Ice Sheet revealed by tracers, Nature Geosci., 6, 195–198, 2013.

Clarke, G. C.: Glaciology: Ice-sheet Plumbing in Antarctica, Nature, 440, 1000–1001, doi:10.1038/4401000a, 2006.

Cuffey, K. M. and Paterson, W. S. B.: The Physics of Glaciers, 4, Butterworth-Heinemann, Amsterdam, 72–73, 2010.

Das, S. B., Joughin, I., Behn, M. D., Howat, I. M., King, M. A., Lizarralde, D., and Bhatia, M. P.: Fracture propagation to the base of the Greenland Ice Sheet during supraglacial lake drainage, Science, 320, 778–781, doi:10.1126/science.1153360, 2008.

Fitzpatrick, A. A. W., Hubbard, A., Joughin, I., Quincey, D. J., Van As, D., Mikkelsen, A. P. B., Doyle, S. H., Hasholt, B., and Jones, G. A.: Ice flow dynamics and surface meltwater flux at a land-terminating sector of the Greenland ice sheet, J. Glaciol., 59, 687–696, doi:10.3189/2013JoG12J143, 2013.

Fitzpatrick, A. A. W., Hubbard, A. L., Box, J. E., Quincey, D. J., van As, D., Mikkelsen, A. P. B., Doyle, S. H., Dow, C. F., Hasholt, B., and Jones, G. A.: A decade (2002–2012) of supraglacial lake volume estimates across Russell Glacier, West Greenland, The Cryosphere, 8, 107–121, doi:10.5194/tc-8-107-2014, 2014.

Forster, R. R., Box, J. E., van den Broeke, M. R., Miege, C., Burgess, E. W., van Angelen, J. H., Lenaerts, J. T. M., Koenig, L. S., Paden, J., Lewis, C., Gogineni, S. P., Leuschen, C., and McConnell, J. R.: Extensive liquid meltwater storage in firn within the Greenland ice sheet, Nature Geosci., 7, 95–98, doi:10.1038/NGEO2043, 2014.

Gray, L., Joughin, I., Tulaczyk, S., Spikes, V.B., Bindschadler, R., and Jezek, K.: Evidence for subglacial water transport in the West Antarctic Ice Sheet through three-dimensional satellite radar interferometry, Geophys. Res. Lett., 32, L03501, doi:10.1029/2004GL021387, 2005.

Joughin, I., Smith, B. E., Howat, I. M., Scambos, T., and Moon, T.: Greenland flow variability from ice-sheet-wide velocity mapping, J. Glaciol., 56, 415–430, 2010.

Joughin, I., Das, S. B., Flowers, G. E., Behn, M. D., Alley, R. B., King, M. A., Smith, B. E., Bamber, J. L., van den Broeke, M. R., and van Angelen, J. H.: Influence of ice-sheet geometry and supraglacial lakes on seasonal ice-flow variability, The Cryosphere, 7, 1185–1192, doi:10.5194/tc-7-1185-2013, 2013.

Korona, J., Berthier, E., Bernard, M., Rémy, F., and Thouvenot, E.: SPIRIT. SPOT 5 stereoscopic survey of Polar Ice: Reference Images and Topographies during the fourth International Polar Year (2007–2009), ISPRS J. Photogramm., 64, 204–212, doi:10.1016/j.isprsjprs.2008.10.005, 2009.

Palmer, S. J., Dowdeswell, J. A., Christoffersen, P., Young, D. A., Blankenship, D. D., Greenbaum, J. S., Benham, T., Bamber, J., and Siegert, M. J.: Greenland subglacial lakes detected by radar, Geophys. Res. Lett., 40, 6154–6159, doi:10.1002/2013GL058383, 2013.

Schoof, C.: Ice-sheet acceleration driven by melt supply variability, Nature, 468, 803–806, doi:10.1038/nature09618, 2010.

Shepherd, A., Ivins, E. R., Geruo, A., Barletta, V. R., Bentley, M. J., Bettadpur, S., Briggs, K. H., Bromwich, D. H., Forsberg, R., Galin, N., Horwath, M., Jacobs, S., Joughin, I., King, M. A., Lenaerts, J. T. M., Li, J. L., Ligtenberg, S. R. M., Luckman, A., Luthcke, S. B., McMillan, M., Meister, R., Milne, G., Mouginot, J., Muir, A., Nicolas, J. P., Paden, J., Payne, A. J., Pritchard, H., Rignot, E., Rott, H., Sorensen, L. S., Scambos, T. A., Scheuchl, B., Schrama, E. J. O., Smith, B., Sundal, A. V., van Angelen, J. H., van de Berg, W. J., van den Broeke, M. R., Vaughan, D. G., Velicogna, I., Wahr, J., Whitehouse, P. L., Wingham, D. J., Yi, D. H., Young, D., and Zwally, H. J.: A Reconciled Estimate of Ice-Sheet Mass Balance, Science, 338, 1183–1189, doi:10.1126/science.1228102, 2012.

van As, D., Hubbard, A. L., Hasholt, B., Mikkelsen, A. B., van den Broeke, M. R., and Fausto, R. S.: Large surface meltwater discharge from the Kangerlussuaq sector of the Greenland ice sheet during the record-warm year 2010 explained by detailed energy balance observations, The Cryosphere, 6, 199–209, doi:10.5194/tc-6-199-2012, 2012.

Wingham, D. J., Siegert, M. J., Shepherd, A., and Muir, A. S.: Rapid discharge connects Antarctic subglacial lakes, Nature, 440, 1033–1036, 2006.

4

Effect of uncertainty in surface mass balance–elevation feedback on projections of the future sea level contribution of the Greenland ice sheet

T. L. Edwards[1], X. Fettweis[2], O. Gagliardini[3,4], F. Gillet-Chaulet[3], H. Goelzer[5], J. M. Gregory[6,7], M. Hoffman[8], P. Huybrechts[5], A. J. Payne[1], M. Perego[9], S. Price[8], A. Quiquet[3], and C. Ritz[3]

[1]Department of Geographical Sciences, University of Bristol, Bristol BS8 1SS, UK
[2]Department of Geography, University of Liege, Laboratory of Climatology (Bat. B11), Allée du 6 Août, 2, 4000 Liège, Belgium
[3]Laboratoire de Glaciologie et Géophysique de l'Environnement, UJF – Grenoble 1/CNRS, 54, rue Molière BP 96, 38402 Saint-Martin-d'Hères Cedex, France
[4]Institut Universitaire de France, Paris, France
[5]Earth System Sciences & Departement Geografie, Vrije Universiteit Brussel, Pleinlaan 2, 1050 Brussel, Belgium
[6]NCAS-Climate, Department of Meteorology, University of Reading, Reading, UK
[7]Met Office Hadley Centre, Exeter, UK
[8]Fluid Dynamics and Solid Mechanics Group, Los Alamos National Laboratory, T3 MS B216, Los Alamos, NM 87545, USA
[9]Department of Scientific Computing, Florida State University, 400 Dirac Science Library, Tallahassee, FL 32306, USA

Correspondence to: T. L. Edwards (tamsin.edwards@bristol.ac.uk)

Abstract. We apply a new parameterisation of the Greenland ice sheet (GrIS) feedback between surface mass balance (SMB: the sum of surface accumulation and surface ablation) and surface elevation in the MAR regional climate model (Edwards et al., 2014) to projections of future climate change using five ice sheet models (ISMs). The MAR (Modèle Atmosphérique Régional: Fettweis, 2007) climate projections are for 2000–2199, forced by the ECHAM5 and HadCM3 global climate models (GCMs) under the SRES A1B emissions scenario.

The additional sea level contribution due to the SMB–elevation feedback averaged over five ISM projections for ECHAM5 and three for HadCM3 is 4.3 % (best estimate; 95 % credibility interval 1.8–6.9 %) at 2100, and 9.6 % (best estimate; 95 % credibility interval 3.6–16.0 %) at 2200. In all results the elevation feedback is significantly positive, amplifying the GrIS sea level contribution relative to the MAR projections in which the ice sheet topography is fixed: the lower bounds of our 95 % credibility intervals (CIs) for sea level contributions are larger than the "no feedback" case for all ISMs and GCMs.

Our method is novel in sea level projections because we propagate three types of modelling uncertainty – GCM and ISM structural uncertainties, and elevation feedback parameterisation uncertainty – along the causal chain, from SRES scenario to sea level, within a coherent experimental design and statistical framework. The relative contributions to uncertainty depend on the timescale of interest. At 2100, the GCM uncertainty is largest, but by 2200 both the ISM and parameterisation uncertainties are larger. We also perform a perturbed parameter ensemble with one ISM to estimate the shape of the projected sea level probability distribution; our results indicate that the probability density is slightly skewed towards higher sea level contributions.

1 Introduction

The Greenland ice sheet (GrIS) response to climate change has two parts: surface mass balance (SMB), which is the sum of surface accumulation and surface ablation (broadly speaking, the balance of snowfall versus meltwater runoff); and dynamic, the changes in ice flow and discharge from the ice sheet. Various approaches to simulating these have been taken for making projections of the GrIS contribution to sea level. As with all simulation problems, there is a trade-off between representing more processes, with the aim of increasing the physical realism of the simulations, and technical and computational resources, which limit implementation and the number of simulations.

The dynamic response is simulated with ice sheet models (ISMs), which solve the Stokes equations in complete or approximate form. SMB can be simulated with sophisticated, physically based energy balance schemes in regional climate models (RCMs) such as MAR (Modèle Atmosphérique Régional: Fettweis, 2007) and RACMO2/GR (e.g. Ettema et al., 2009). The two aspects of the GrIS response can thus be combined by forcing an ISM with SMB simulated by an RCM. But RCMs usually use a fixed surface topography, neglecting the important effects of ice sheet surface elevation changes on the atmosphere (Edwards et al., 2014) by omitting the SMB–elevation feedback. An alternative to using SMB from an RCM is to simulate it within the ISM, so that the evolving ice sheet topography can dynamically alter the SMB. The most common approach to this is with a positive degree day (PDD) scheme, which parameterises SMB as a function of temperature and precipitation (supplied from observations for the present day, or a regional or global climate model for future projections) and possibly applies a simple snowpack model (e.g. Janssens and Huybrechts, 2000). PDD models incorporate the temperature aspect of the SMB–elevation feedback through a lapse rate correction, but not the precipitation aspect (except, in some cases, through a scaling factor for temperature), and also represent SMB much more simply than RCMs such as MAR and RACMO2/GR (Edwards et al., 2014).

To make the best use of physically based simulations of the GrIS response to climate change – simulating ice flow with an ISM and SMB with an RCM, while also including the SMB–elevation feedback – one can couple an ISM to an RCM. However, this is technically challenging and computationally expensive, effectively precluding exploration of climate and ice sheet modelling uncertainties.

The only way, therefore, to incorporate physical modelling of ice flow, SMB processes, and the SMB–elevation feedback while also exploring model uncertainties is with a parameterisation such as the one we present in a companion paper (Edwards et al., 2014), where we characterise the SMB response to elevation in MAR using a suite of simulations in which the MAR GrIS surface height is altered. The parameterisation is a set of four gradients, or “SMB lapse rates”, that relate SMB changes to height changes below and above the equilibrium line altitude (ELA) and for regions north and south of 77° N. Here we apply the parameterisation in five ISMs to adjust MAR projections of SMB under the SRES A1B scenario (Nakićenović et al., 2000) as the ice sheet geometry evolves.

The climate community have been attempting to quantify uncertainty in global climate model (GCM) predictions for some time, focusing on uncertainty in their parameter values with perturbed parameter ensembles (PPEs) but also attempting to estimate structural uncertainty, which is uncertainty about the remaining discrepancy between a model and reality at the model's most successful parameter values (e.g. Sexton et al., 2011; Sexton and Murphy, 2011). But probabilistic quantification of uncertainties in ice sheet model predictions has hardly yet been attempted. There is also an urgent need to propagate uncertainties along the causal chain from greenhouse gas forcing scenarios to the impacts of climate change. ISM projections are only beginning to tackle these challenges. Multi-model comparisons such as MISMIP (Pattyn et al., 2012) and SeaRISE (Bindschadler et al., 2013) compare ISMs with each other using standardised experiments, but the ice2sea project (http://www.ice2sea.eu) is among the first to explore systematically emissions scenarios and climate and ice sheet model uncertainties within a coherent framework (e.g. Rae et al., 2012; Fettweis et al., 2013).

In Edwards et al. (2014), we make uncertainty assessment an integral part of the parameterisation by estimating probability distributions for the four parameters (elevation feedback gradients). Here we propagate these uncertainties to future projections by sampling values from the four distributions. We also use five different ISMs to assess the effect of ISM structural uncertainty that arises from different representations of ice flow and initialisation procedures, and explore GCM structural uncertainty by forcing MAR with two different GCMs. Our coherent experimental design and statistical framework, unusual in sea level projections, allow us to propagate the three types of model uncertainty along the causal chain from SRES scenario to sea level contribution; assess the relative importance of these through time; and present probabilistic assessments of the effect of elevation feedback parameterisation uncertainty on the projected GrIS sea level contribution under the A1B scenario.

2 Method

2.1 Climate projections

The regional climate model MAR (Fettweis, 2007) has been adapted for simulating the climate over Greenland, with full coupling to a complex snow/ice energy balance model and relatively high horizontal resolution (25 km). Unlike most RCMs, MAR includes the positive feedback between ice surface albedo and melting (Franco et al., 2013). We discuss the processes and responses of MAR in more detail in Edwards et al. (2014).

We use five climate simulations performed for the ice2sea project. The first is a twenty year simulation of 1989–2008 in which MAR is forced at the boundaries by the ERA-INTERIM reanalysis (Dee et al., 2011), which we use as a baseline for initialisation and projections. The next two are twenty year simulations of 1980–1999 in which MAR is forced by two GCMs, ECHAM5 (Roeckner et al., 2003) and HadCM3 (Gordon et al., 2000), under 20th century climate forcings, which we use to calculate projection anomalies. The final two are one hundred year projections (2000–2099) forced by the two GCMs under the SRES A1B emissions scenario (Nakićenović et al., 2000). The A1B scenario is only defined until 2100, but we wish to make GrIS projections for two hundred years (2000–2199). We did not have resources to simulate the second century (the climate responding to the A1B scenario fixed at 2100 values), so we extend the MAR simulations by repeating the final decade (2090–2099) ten times. This method is likely to underestimate the longer-term climate response, as discussed later.

MAR simulates SMB values only over grid cells categorised as permanent ice, so the low spatial resolution (relative to ISMs) leads to missing values for parts of the GrIS margin. As part of the data processing, we extrapolate the SMB field outwards from the ice sheet using a simpler representation of the SMB–elevation relationship than in Edwards et al. (2014). For each land grid cell, we use the ice grid cells within 100 km to fit a linear model of SMB versus surface height, and use this relationship with the land grid cell elevation to assign an SMB value. This was performed before the parameterisation of Edwards et al. (2014) was developed.

2.2 Ice sheet models

We implement the parameterisation in five ISMs with varying structure and complexity: Elmer/Ice, which solves the full Stokes equations; GISM, MPAS and CISM, which use a higher order approximation that reduces computational expense (HO: e.g. Pattyn, 2003); and GRISLI, which uses a hybrid of the first order shallow ice and shallow shelf approximations (SIA and SSA: Bueler and Brown, 2009) to further reduce computational expense. We also use a SIA version of GISM, and call the two versions GISM-HO and GISM-SIA. Elmer/Ice and MPAS use finite element numerical methods on unstructured grids, while the others use finite difference methods on regular grids at 5 km resolution.

Elmer/Ice builds on Elmer, the open-source finite element, partial differential equation solving parallel code mainly developed by the CSC-IT Center for Science Ltd in Finland. The unstructured mesh allows a variable grid resolution to focus computational resources at the ice sheet margin; here we use a minimum horizontal grid size of less than 1 km. GISM-SIA is a thermomechanical ISM (Huybrechts and de Wolde, 1999) that has been modified and extended for projections on centennial timescales using a new higher-order approximation of the force balance (GISM-HO: Fürst et al., 2011, 2013). The MPAS-Land Ice model is based on the MPAS (Model for Prediction Across Scales) climate modelling framework of Ringler et al. (2008); here we use a regular 5 km resolution hexagonal mesh. GRISLI (GRenoble Ice Shelves and Land Ice model) is a thermomechanically coupled ISM; it is a hybrid model that for grounded ice uses the SIA for vertical shearing and the SSA as a sliding law (Ritz et al., 2001; Bueler and Brown, 2009). The Community Ice Sheet Model (CISM) version 2.0 includes improvements to all components of the Glimmer-CISM SIA model (Rutt et al., 2009). The models are summarised in Table 1; more detailed information is given elsewhere (Elmer/Ice: Gillet-Chaulet et al., 2012; GISM-HO: Goelzer et al., 2013; MPAS: Perego et al., 2012; CISM: Price et al., 2011; Lemieux et al., 2011; Evans et al., 2012; GRISLI: Quiquet et al., 2012).

2.3 Initialisation

Determining initial conditions for ISMs involves finding a balance between observations of the present-day ice sheet, reconstructions of past climate changes (to which the ice sheet is still responding), and the physical laws and parameterised processes incorporated in the ISM, while accounting for uncertainties and limitations in all of these. ISM initialisation mostly uses ad hoc tuning methods rather than formal data assimilation as in numerical weather forecasting. Our initialisation procedures use observations of present-day ice sheet geometry (surface elevation, bedrock elevation and ice thickness; e.g. Bamber et al., 2013), ice velocities (Joughin et al., 2010; Bamber et al., 2000), and geothermal heat fluxes (Shapiro and Ritzwoller, 2004).

Methods vary between ISMs but in general there are three stages. First, we estimate the 3-D ice-temperature field by solving the heat equation, in some cases accounting for the response to atmospheric temperature changes over one or more glacial–interglacial cycles. Second, we infer the spatial pattern of the basal drag coefficient that leads to the best agreement with observed ice velocities, given observational uncertainties and model limitations. During these first two stages, the geometry of the ice sheet is held constant to the observed values (Bamber et al., 2013). For the third stage, we finally allow the ice sheet geometry to evolve or "relax" so that it is internally consistent with the ice temperature and flow fields.

We derive ice temperatures for Elmer/Ice using the computationally cheaper SIA model SICOPOLIS (Greve, 1997). We derive them for CISM, MPAS and GRISLI from a quasi-steady-state CISM simulation forced with a constant surface temperature field simulated by the RACMO/GR regional climate model (Ettema et al., 2009, as specified in the SeaRISE project protocol: http://websrv.cs.umt.edu/isis/index.php/Present_Day_Greenland). We obtain ice temperatures for the two GISM models from a GISM-SIA simulation

Table 1. Summary of ice sheet model (top) treatment of Stokes equations and (bottom) derivation of ice temperature (IT) boundary conditions. HO, SIA and SSA are higher order, shallow ice and shallow shelf approximations; SS is steady state; SAT is surface air temperature data used for the spin-up simulation; g–ig is glacial–interglacial. All models use bedrock elevation data from Bamber et al. (2013) and geothermal heat flux from Shapiro and Ritzwoller (2004). References: B13: Bamber et al. (2013); D10: Dukowicz et al. (2010); E09: Ettema et al. (2009); E12: Evans et al. (2012); F11, F13: Fürst et al. (2011) and Fürst et al. (2013); G97: Greve (1997); L11: Lemieux et al. (2011); P11: Price et al. (2011).

	CISM	Elmer/Ice	GISM-HO	GISM-SIA	GRISLI	MPAS
PHYSICS						
Stokes equations	HO (P11, L11, E12)	Full	HO (F11, F13)	SIA	Hybrid SIA-SSA	HO (D10)
DERIVATION OF ICE TEMPERATURE						
Spin-up simulation	Quasi-SS; fixed geom. (B13)	One g–ig cycle with SICOPOLIS (G97)	Two g–ig cycles; IT rescaled to obs. thick.	Two g–ig cycles; IT rescaled to obs. thick.	Quasi-SS; fixed geom. (B13)	Quasi-SS; fixed geom. (B13)
Spin-up SAT	E09, constant	E09, constant	Two g–ig cycles, evolving	Two g–ig cycles, evolving	E09, constant	E09, constant

of several glacial–interglacial cycles, rescaling the ice temperature field to the observed ice thickness (Goelzer et al., 2013). Geothermal heat flux fields in all spin-up simulations are from Shapiro and Ritzwoller (2004). Ice temperature spin-up simulations are summarised in Table 1.

We infer the basal drag coefficient from observed velocities using the control method (Elmer/Ice: Morlighem et al., 2010), an iterative inverse method (GRISLI), or tuning (CISM, MPAS); the methods and velocity targets are also summarised in Table 1.

We obtain the initial ice sheet geometry by forcing the ISM with a present-day mean SMB field. For most models this is the 1989–2008 mean (the reference period for the ice2sea project) of the ERA-INTERIM forced MAR simulation; for GISM we use the 1960–1990 mean SMB calculated with the GISM PDD scheme from ERA-40 data (Uppala et al., 2005) and observed surface elevation (similar to Hanna et al., 2011). We obtain the initial geometry for Elmer/Ice by allowing the upper surface to evolve for 55 yr under the present-day SMB forcing, and use the same initial geometry for GRISLI, allowing it to evolve for a further 145 yr to obtain a coherent initial state for this model. We initialise the GISM models by allowing the geometry to evolve for 1000 yr while the ice sheet margin is fixed to observed values under present-day SMB forcing and ice thickness changes are limited to $0.2\,\mathrm{m\,a^{-1}}$. These relaxation periods are determined arbitrarily or by available computational resource, though most of the flux divergence anomalies that arise from observational and modelling uncertainties die away after a few years (e.g. Gillet-Chaulet et al., 2012). We do not relax the geometry for CISM and MPAS. However, these simulations are already closer to the present-day state than for the other models because they use a quasi-steady-state spin-up forced with modern-day surface temperatures, rather than glacial–interglacial cycles, and calibrate the basal drag with balance velocities (which have complete coverage and are in equilibrium with the geometry), rather than surface velocities.

We use two methods to correct any remaining drift that arises from model imbalances. For Elmer/Ice and GRISLI, we perform a control simulation and subtract this from the results (Sect. 2.5). For the others we diagnose and apply a "synthetic SMB" correction, which is the additional SMB required to keep the ice sheet close to the present-day observed geometry under present-day SMB forcing. The synthetic SMB correction is applied unaltered in perturbed simulations (i.e. the parameterisation adjusts only the MAR SMB forcing and not the synthetic SMB correction). This is similar, in principle, to the flux corrections that were formerly in common use in atmosphere–ocean GCMs. In CISM and MPAS there is no relaxation in the initialisation procedure, so the synthetic SMB accounts for the entire initial mass imbalance generated by the initial geometry. The mean synthetic SMB correction for both GISM models is $9\,\mathrm{cm\,a^{-1}}$, with less than 0.3 % of the total ice sheet area having a correction of greater than $10\,\mathrm{m\,a^{-1}}$, mainly at marine-terminating ice margins (Goelzer et al., 2013). For MPAS the mean synthetic SMB correction is $-7.3\,\mathrm{cm\,a^{-1}}$, with 1.6 % of the total ice sheet area having a correction of greater than $10\,\mathrm{m\,a^{-1}}$, and

Table 2. Summary of ice sheet model basal drag calibration and initialisation. PDD is positive degree day.

	CISM	Elmer/Ice	GISM-HO	GISM-SIA	GRISLI	MPAS
BASAL DRAG CALIBRATION						
Method	Tuning	Control	n/a	n/a	Iterative inverse	Tuning
Target velocities	Balance	Surface	n/a	n/a	Surface	Balance
INITIALISATION						
Relaxation	n/a	55 years	1000 years (restrictions)	1000 years (restrictions)	200 years	n/a
Drift	Synthetic	Control	Synthetic	Synthetic	Control	Synthetic
Climate	ERA-I	ERA-I	ERA-40	ERA-40	ERA-I	ERA-I
SMB	MAR	MAR	PDD	PDD	MAR	MAR
Dates	1989–2008	1989–2008	1960–1990	1960–1990	1989–2008	1989–2008

for CISM it is 2.4 cm a^{-1}, with less than 2 % of the total ice sheet area having a correction of greater than 10 m a^{-1}.

We exclude Greenland's glaciers and ice caps from the projection results using a mask produced for ice2sea (Rastner et al., 2012).

The basal drag calibration and initialisation methods are summarised in Table 2. More details can be found in the references given in the previous section and the table.

2.4 Boundary conditions

The boundary conditions are geothermal heat flux, ice temperature, bedrock elevation, and projections of SMB. Geothermal heat fluxes are from Shapiro and Ritzwoller (2004) and held fixed in the simulations. We also keep the ice temperature fields (determined during initialisation; Sect. 2.3 and Table 1) held fixed because there is negligible ice temperature response to changing atmospheric forcing over two centuries. We use bedrock elevations from the ice2sea data set (Bamber et al., 2013) and hold these fixed because there is negligible isostatic adjustment on this timescale (e.g. Goelzer et al., 2013).

The SMB forcings are therefore the only time-varying boundary conditions. We force the ISMs with "anomaly-corrected" SMB projections, $S^{\mathrm{RCM}'}$, to remove the mean discrepancy in MAR between the GCM-forced and ERA-INTERIM reanalysis-forced simulations (Fettweis et al., 2013). For this we calculate SMB anomalies with respect to the present day by subtracting the 1989–2008 mean SMB (1989–1999: 20th century climate forcings; 2000–2008: A1B scenario) from the A1B projections. We add these anomalies to the present-day SMB used for initialisation, to give

$$S^{\mathrm{RCM}'}_{2000,\ldots,2199} = S^{\mathrm{A1B}}_{2000,\ldots,2199} - S^{\mathrm{20C,A1B}}_{1989-2008} + S^{\mathrm{init}},$$

where S^{init} is generally $S^{\mathrm{ERAI}}_{1989-2008}$, the 1989–2008 mean of the ERA-INTERIM forced simulation (except GISM, $S^{\mathrm{ERA40}}_{1960-1990}$). If used, the synthetic SMB field is also applied. Our initialisation and anomaly-correction methods use the approximation that SMB trends during the period 1989–2008 are small relative to the A1B projections (Rae et al., 2012).

2.5 Parameterisation

We use the SMB–elevation feedback parameterisation to adjust the anomaly-corrected SMB forcing as the GrIS surface height in each ISM evolves. The adjustment is made each year, for each grid cell, using one of four parameters (gradients) selected according to the current "reference" SMB and the region of the grid cell (Edwards et al., 2014).

For a given ISM grid cell in a given year (t), a gradient (b_t) (kg m^{-3} a^{-1}) is used to adjust the SMB forcing (kg m^{-2} a^{-1}) using the height difference (m) between the previous year and the start of the simulation:

$$S^{\mathrm{adj}}_t = S^{\mathrm{RCM}'}_t + b_t(h^{\mathrm{ISM}}_{t-1} - h^{\mathrm{ISM}}_0),$$

where S^{adj}_t is the adjusted SMB, $S^{\mathrm{RCM}'}_t$ is the anomaly-corrected RCM SMB, and h^{ISM}_t is the ISM height. The gradient b_t is selected from one of four values according to the reference SMB ($S^{\mathrm{ref}} < 0$ or $S^{\mathrm{ref}} \geq 0$) and region (north or south of 77° N) of the grid cell, where S^{ref}_t is the mean S^{adj} of the previous decade or, for the first decade, all available years.

In Edwards et al. (2014), we estimate probability distributions for each of the four parameters (gradients). Our best estimate values of the parameters are the modes of these four distributions, and our credibility intervals (CIs) encompass the central 95 % of each. We propagate the parameter uncertainties to sea level projections by performing simulations using different values sampled from the distributions: (a) the four modes, i.e. best estimates; (b) the four 2.5 % quantiles, i.e. lower bounds of the CIs; and (c) the four 97.5 % quantiles, i.e. upper bounds. The parameter estimates are given in Table 3. They differ slightly to those in Edwards et al. (2014) due to changes made to the parameter estimation code for an update to that study: a small error was corrected, unnecessary rounding was removed, and the code was restructured (thus changing the random seed for bootstrapping). One difference is 0.03 ($SMB \geq 0$, 97.5 % bound), and the other differences are 0.02 or less.

With one model, GISM-SIA, we also sample the four distributions more thoroughly by using 99 percentiles: i.e. the four 1 % quantiles in one simulation; the four 2 % quantiles in the next, and so on up to the four 99 % quantiles.

We therefore generate five simulations for each ISM:

- Control: forced with S^{init} (Elmer/Ice, GRISLI) or $S^{\text{init}} + S^{\text{syn}}$ (other models), to check for or subtract model drift from the projections;
- No feedback: forced with $S^{\text{RCM}'}$ with no adjustment, to estimate the response of the GrIS without elevation feedback;
- Best estimate: forced with S^{adj}, using the "best estimate" values of the four parameters (Table 3);
- 2.5 % and 97.5 % quantiles: as for best estimate, but using the parameter values that correspond to the bounds of the 95 % CI (Table 3).

We perform these for the two MAR simulations forced by ECHAM5 (all ISMs) and HadCM3 (Elmer/Ice, GISM-HO, GISM-SIA and GRISLI) under the A1B scenario over 2000–2200. As described above we also use GISM-SIA to perform a simulation with each of the 99 percentile estimates of the four parameters, forming a 99 member PPE; we do this only for the ECHAM5 projection.

The drift in the control simulations is very small (0.03 %, or less, of the cumulative projected sea level contribution at 2200) for all models except Elmer/Ice, which has a drift of 2–2.5 % (−4 mm). For this model, drift is not constrained during the initialisation procedure and the free-surface elevation has been allowed to diverge from observations for a relaxation period of only 55 yr. The applied SMB is not corrected by a synthetic SMB, and the remaining drift shows the drift of the model when directly applying the 1989–2008 mean SMB given by MAR forced under ERA-INTERIM. This drift is corrected by subtracting the control simulation from the projections.

Table 3. The 2.5 % quantile, best estimate, and 97.5 % quantile estimates of the SMB–elevation gradients in $\text{kg m}^{-3}\,\text{a}^{-1}$, below ($SMB < 0$) and above ($SMB \geq 0$) the ELA, for regions north and south of 77° N.

	Region	2.5 %	Best estimate	97.5 %
$SMB < 0$	North	−0.22	0.54	1.34
	South	1.03	1.89	2.61
$SMB \geq 0$	North	−0.03	0.09	0.22
	South	−0.07	0.06	0.56

3 Results

The additional cumulative sea level contribution due to the SMB–elevation feedback, under the A1B scenario and averaged over five ISM projections for ECHAM5 and three for HadCM3, is 4.3 % (best estimate; 95 % credibility interval 1.8–6.9 %) at 2100, and 9.6 % (best estimate; 95 % credibility interval 3.6–16.0 %) at 2200 (Figs. 1 and 2; Tables 4 and 5). We exclude GISM-SIA from summary statistics because it is so similar in structure, and identical in set-up, to GISM-HO that it would effectively give double weighting to GISM.

In all ISM predictions the elevation feedback is positive: the lower bounds of the 95 % CIs (the 2.5 % quantile simulations) are always larger than the "no feedback" case. This amplifying effect is expected, because mean GrIS SMB becomes negative at around 2050 (Rae et al., 2012) and the gradient estimates for negative SMB are generally positive (Edwards et al., 2014) thus adjusting the SMB in the same direction: more negative, giving greater sea level contribution. However, this was not guaranteed, because there are both positive and negative values in the mean SMB forcing, in the SMB changes from 2000–2199, and in the gradient sets (Edwards et al., 2014).

The larger feedback adjustment at 2200 arises from the negative SMB forcing at the end of the first century which is sustained (by repetition of the decade 2090–2099 from the MAR simulation: Sect. 2.1) during the second century.

The Elmer/Ice and GRISLI total sea level contributions are virtually identical, but this is a coincidence: the changes in ice discharge and SMB are different, and the differences just compensate each other so that the sea level contributions are the same. The two models do have similar initialisation procedures, but the differing dynamic and SMB responses lead us to believe this is not the reason for the similar results. CISM seems to be an outlier in the ISM ensemble, particularly at 2200 (Fig. 2; Table 5), in both the no feedback result (low) and the magnitude of the feedback (small), though the latter may be a result of the former. The GISM-SIA results are very similar to those of GISM-HO. This can be explained by the identical model set-up and their similarity in response to perturbations; the only difference between the two models is in the dynamic response, which is a much smaller contribution than the SMB response (Fürst et al., 2013).

Table 4. Projected sea level contribution (mm) at 2100 for the ECHAM5 (top) and HadCM3 (bottom) A1B projections: no feedback, 2.5 % quantile, best, and 97.5 % quantile estimates.

Model	No feedback	2.5 %	Best estimate	97.5 %
ECHAM5				
CISM	53.7	53.8	54.7	56.9
Elmer/Ice*	57.0	58.2	59.8	60.9
GISM-HO*	55.7	56.8	58.3	59.8
GISM-SIA	56.6	57.6	59.2	60.6
GRISLI*	57.1	58.1	59.6	61.0
MPAS	57.4	58.4	60.0	61.4
Mean (*)	56.6	57.7	59.2	60.5
HadCM3				
Elmer/Ice*	64.8	66.1	67.9	69.2
GISM-HO*	62.9	64.3	66.0	67.7
GISM-SIA	63.5	64.9	66.7	68.3
GRISLI *	64.7	66.0	67.7	69.3
Mean (*)	64.1	65.4	67.2	68.7

Table 5. As for Table 4, but at 2200.

Model	No feedback	2.5 %	Best estimate	97.5 %
ECHAM5				
CISM	150.8	151.3	158.1	173.1
Elmer/Ice*	164.9	171.3	181.0	189.5
GISM-HO*	160.3	166.9	176.7	186.6
GISM-SIA	160.4	166.9	176.8	186.6
GRISLI*	165.7	171.6	181.2	190.8
MPAS	170.5	177.7	188.7	199.6
Mean (*)	163.6	169.9	179.6	189.0
HadCM3				
Elmer/Ice*	177.5	184.7	195.8	205.7
GISM-HO*	171.9	179.6	191.1	202.3
GISM-SIA	171.7	179.3	190.7	201.9
GRISLI*	177.2	184.1	195.0	206.0
Mean (*)	175.5	182.8	194.0	204.6

We can compare ECHAM5 and HadCM3 using the three ISMs that used both projections (Elmer/Ice, GISM-HO and GRISLI, from here on called the "starred" models). The HadCM3-forced projection gives consistently higher sea level contributions than the ECHAM5-forced, because the HadCM3-forced MAR simulation projects greater meltwater runoff (Rae et al., 2012). Without elevation feedback, the mean GrIS sea level contributions at 2100 are 57 mm for ECHAM5 and 64 mm for HadCM3; with feedback, these increase by 3 mm (5 %). At 2200, the mean contributions without feedback are 164 mm for ECHAM5 and 176 mm for HadCM3; the feedback contributes an additional 16 mm (10 %) and 19 mm (11 %), respectively.

Not only does the contribution due to the feedback increase with time, but so does the uncertainty. Our experimental design allows us to inspect the changes and relative importance of the different modelling uncertainties. At 2100 (Fig. 1; Table 4), the GCM differences are, respectively, 2.6 and 1.7 times those of the elevation feedback parameterisation and ISM uncertainties: the mean difference between the two GCMs (averaged over the no feedback and three feedback estimates) for the three starred models is 7.8 mm; the mean spread of all five ISMs for ECHAM5 is 4.5 mm; and the mean 95 % CI (over all ISMs and GCMs) is 3.0 mm. But by 2200 (Fig. 2; Table 5), the relative importance has changed. The ISM spread and elevation feedback parameterisation uncertainty overtake the difference between the two GCMs: the mean difference between the two GCMs is 13.7 mm; the mean spread of ISMs for ECHAM5 is 25.8 mm; and the mean 95 % CI (over all ISMs and GCMs) is 20.8 mm.

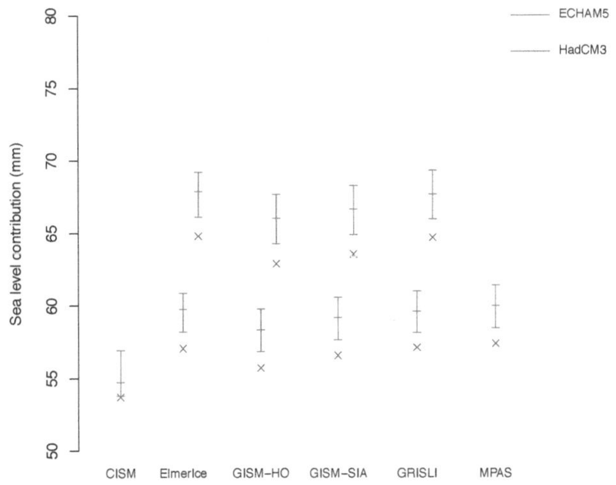

Fig. 1. Projected cumulative GrIS sea level contributions for the ECHAM5 and HadCM3 A 1B projections at 2100. Crosses mark simulations with no elevation feedback. Bars mark the estimated 95 % credibility intervals and (central tick) best estimates of the elevation feedback parameterisation.

In other words, the parameterisation and ISM uncertainties increase by a factor of seven and six, respectively, between 2100 and 2200, while the difference between the two GCMs barely doubles. The relative contributions to our uncertainty thus depend on the timescale of interest.

This can be seen clearly in Fig. 3, which shows the fractional uncertainties as a function of time with a 15 yr running mean applied. Although we have a small number of ice sheet and climate models, it is useful to examine the relative importance of different uncertainties in our study to inform

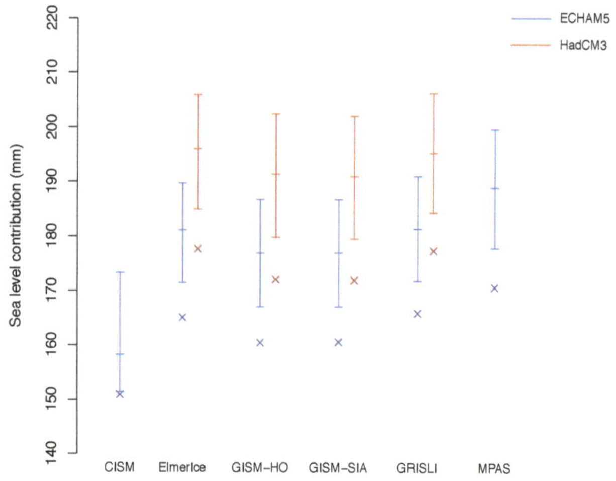

Fig. 2. As for Fig. 1, but at 2200.

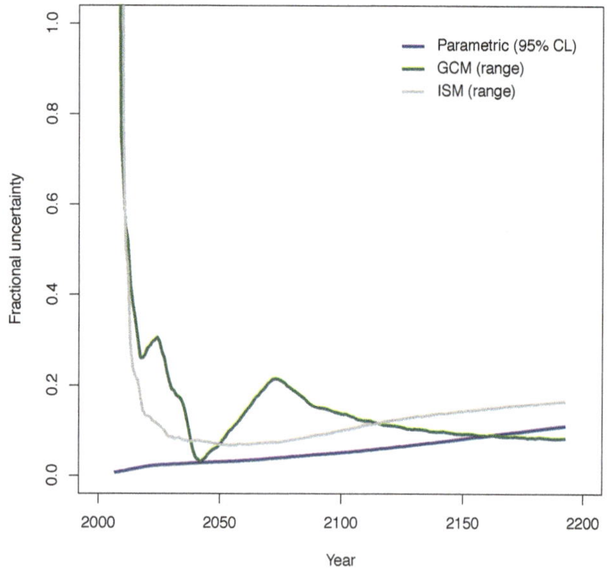

Fig. 3. Fractional uncertainties in projected cumulative GrIS sea level contributions for elevation feedback parameterisation (95 % credibility interval divided by best estimate, averaged over ISMs and GCMs; GCMs (difference between ECHAM5 and HadCM3 divided by their mean, averaged over ISM best estimates; and ISMs (range of ISM best estimates divided by their mean, for ECHAM5 projections). A 15 yr running mean is applied.

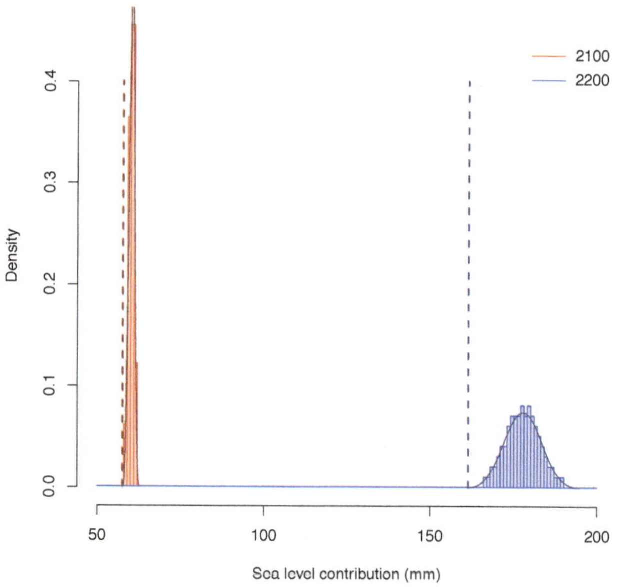

Fig. 4. GISM-SIA projections for sea level contribution in 2100 (red) and 2200 (blue): no elevation feedback (dashed vertical line) and with feedback (histograms and density estimates).

future work: for example, determining which uncertainties are largest in our experimental design, and how others may therefore choose to prioritise their computational resources. At the start of the projections, the ISM structural uncertainty explored (the range of five ISM best estimates divided by their mean, for the ECHAM5 projections) is a large fraction because the signal is small, but the fraction decreases (the signal increases faster than the range) during the first half of the twentieth century. After this the fraction increases again (the range increases faster than the signal). The time dependence of the GCM structural uncertainty explored (the difference between ECHAM5 and HadCM3 results divided by their mean, averaged over the best estimates for the three starred ISMs) appears at first glance to be similar, but the underlying reason is different: it rapidly falls to a minimum when the two GCMs projections happen to coincide for a short period (see also Rae et al., 2012). After a short increase due to the GCMs diverging again, the fractional uncertainty slowly falls because the signal increases faster than the range. The fractional uncertainty due to the parameterisation (95 % CI divided by the best estimate, averaged over all ISMs and GCMs) is initially very small, but increases fairly linearly with time: the width of the CI scales with the signal. In the latter part of the second century the parameterisation uncertainty overtakes that of the GCMs. These results are discussed further in the next section.

The 99 member GISM-SIA PPE gives the entire sea level probability distribution (Fig. 4) because it jointly samples the full probability distributions of each of the four gradient parameters rather than just the bounds of the 95 % CIs. We can inspect any other interval, such as the 90 % CI (177.3–194.7 mm) or 50 % CI (182.3–190.6 mm). We can also test whether the projected distribution of sea level contributions is symmetric about the best estimate of 186.1 mm. The 95 % CI is symmetric (±10.4 mm), and the 90 % CI nearly so (−8.8 mm, +8.6 mm), but the 50 % interval (−3.8 mm, +3.5 mm) indicates the probability density leans slightly towards the higher sea level contributions.

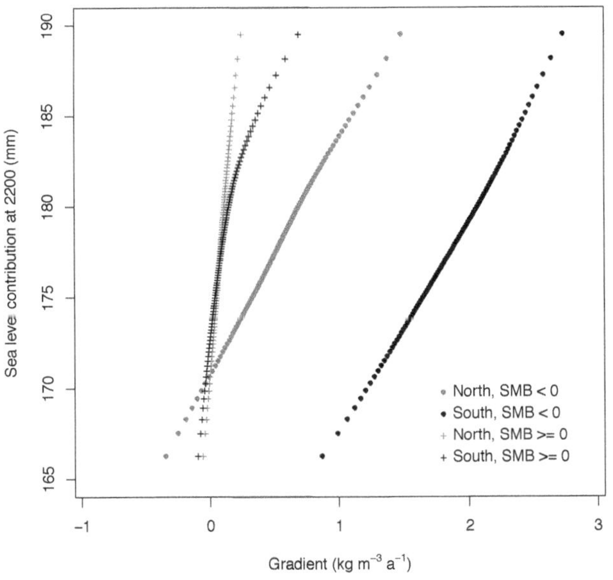

Fig. 5. GISM-SIA projections for sea level contribution in 2200 as a function of elevation feedback gradient values.

The unimodal and (broadly) symmetric shape of the PPE sea level distribution is probably a result of the unimodal and (in most cases) symmetric shapes of the four posterior probability distributions of the gradient values (Edwards et al., 2014), combined with the linear model of adjustment. Figure 5 shows the dependence of the sea level contribution on the gradient sets. However, this cannot show which of the four gradients is most important because they are varied jointly. But it illustrates some characteristics of the parameterisation structure and uncertainties: first, the relationship between sea level contribution and SMB lapse rates is fairly linear, save for a clear change in slope for the south SMB ≥ 0 lapse rate. This is due to the linear adjustment model, but is perhaps surprisingly robust considering the SMB ≥ 0 lapse rates range from negative to positive values, and considering there is a positive feedback in adjusting SMB. In fact the latter effect can be seen in the SMB < 0 slopes, which become slightly steeper at larger SMB lapse rates. Second, the magnitude of the slopes illustrate the characteristics of the SMB lapse rate probability distributions (Edwards et al., 2014): shallower slopes for the two SMB < 0 lapse rates are a result of their broader probability distributions, i.e. larger absolute uncertainties, relative to the SMB ≥ 0 lapse rates. Similarly, the south SMB ≥ 0 slope change is due to the asymmetric probability distribution.

4 Discussion

4.1 Elevation feedback

It is important to quantify the SMB–elevation feedback, relative to projections without it (such as Rae et al., 2012), to improve our assessments of GrIS sensitivity to climate change. Using RCMs to simulate SMB represents the processes more fully than with ISMs; using ISMs to evaluate the feedback itself gives a more complete picture than with RCMs alone (by modifying their topography), because they simulate the entire GrIS response rather than only the SMB aspect. Parameterising the SMB–elevation feedback gives the best of both worlds, allowing ISMs to be forced with SMB from RCMs and include the feedback, while being computationally cheap enough to explore model uncertainties. The response of SMB to changes in ice surface elevation is complex and nonlinear, and even with a probabilistic uncertainty assessment, a linear parameterisation cannot fully represent the SMB changes that would be modelled by a fully coupled RCM-ISM. However, the advantages of our particular parameterisation over other studies are described in the companion paper (Edwards et al., 2014).

We present the first application of an SMB–elevation parameterisation to RCM projections of future sea level rise. The feedback always increases sea level rise, even for the lower bounds of our 95 % CIs (despite the negative gradient values: Table 3). But the additional contribution is relatively small, with best estimates 4.3 % at 2100 and 9.6 % at 2200.

We can compare these results with estimates from the RCM alone. When MAR is forced by the GCM MIROC5 under the Representative Concentration Pathway RCP8.5 (Moss et al., 2010), changing the ice sheet height by an amount equivalent to cumulative projected SMB changes from 2000–2079 gives a much greater feedback contribution than our result: 5–15 % over just one century (Fettweis et al., 2013). At first glance this might appear to be due to the use of a high-end scenario (relative to the medium scenario A1B), giving larger SMB changes and therefore a larger feedback.

However, the greater feedback contribution in the RCM most likely arises from the absence of ice dynamics. Perhaps counter-intuitively, sea level rise from the total GrIS response (SMB and dynamics) is smaller than from the SMB response alone. Recent ice2sea studies (e.g. Gillet-Chaulet et al., 2012; Goelzer et al., 2013) have highlighted the complexity of the interactions between SMB and dynamic responses. Dynamic mass loss and SMB-related mass loss are not additive: mass loss by one process decreases loss by the other, so we cannot neatly divide sea level contributions into, for example, 60 % SMB and 40 % dynamic. SMB removes ice (e.g. through melting) before it can reach the ice sheet margin to be discharged. Conversely, dynamics remove ice before it can melt and also partly compensate surface height decreases (from increased melting) by redistributing mass gained at the top of the ice sheet, thickening the ablation zone and therefore reducing melting. Further discussion can be found in Gillet-Chaulet et al. (2012) and Goelzer et al. (2013), though this is not a recent finding: Huybrechts and de Wolde (1999), for example, find a 24 % lower sea level contribution by 2100 from Greenland when dynamics are included. In our study the time integrated SMB changes for the ECHAM5-forced MAR simulation contribute 62.7 mm to sea level by 2100 and 211.4 mm by 2200 (Goelzer et al., 2013), while the total responses from the ice sheet models are substantially smaller: 53.7–57.4 mm at 2100 and 150.8–170.5 mm at 2200 (Tables 4 and 5).

Therefore, it is not surprising that SMB-only projections, without coupling to an ice sheet model, show larger SMB–elevation feedback contributions. The results from Fettweis et al. (2013) are likely overestimates. Only a coupled RCM-ISM would allow full evaluation of this response, but our results suggest that the feedback effect is small enough that using such models for one- or two-century simulations may not be justified given the additional computational expense, technical difficulty, and severe restrictions on exploring modelling uncertainties.

We can also compare these results with estimates from an ISM alone. Some have previously suggested that the simpler PDD descriptions of ice sheet response are too sensitive to climate change (van de Wal, 1996; van de Berg, 2011), but recent studies find they are less sensitive than MAR and RACMO/GR (Goelzer et al., 2013; Helsen et al., 2011; Vernon et al., 2013; Hanna et al., 2011). Goelzer et al. (2013) compare projections from our parameterisation versus the GISM PDD scheme, using the same ice sheet extent and RCM, and Helsen et al. (2011) compare their parameterisation of the RACMO2/GR SMB–elevation feedback with PDD results. In both cases the PDD schemes predict a smaller sea level contribution in a warmer climate than the RCMs: for example, Goelzer et al. (2013) find the PDD model gives a 31 % lower sea level contribution than MAR. There are many differences between physically based energy balance RCM schemes and empirical PDD schemes, including the presence of albedo feedback in the RCMs, higher spatial resolution in the PDD schemes, and different treatment of refreezing (Goelzer et al., 2013). Our parameterisation of the SMB–elevation feedback is more complex than a PDD scheme, in that it estimates probability distributions rather than a fixed lapse rate, includes nonlinear precipitation aspects, and incorporates variation of the feedback with climate, topography and region (through the use of four probability distributions).

This parameterisation could be used to correct low resolution MAR simulations onto a high resolution digital elevation map; the difference in elevation between the two topographies is analogous to the geometry evolution in ISM simulations. This correction might improve the realism of MAR SMB simulations, which would be useful not only for forcing ISMs but also comparing with observations.

4.2 Modelling uncertainties

It is a research priority to improve quantification of ice sheet model uncertainties, so as to make meaningful and complete statements about the difference between our projections and the real world. Research in this area is sparse, but the ice sheet modelling community can make rapid improvements by applying established uncertainty quantification techniques from climate and other environmental modelling research areas, as we have done here.

Our method is novel because we propagate three types of model uncertainty – GCM structural uncertainty, ISM structural uncertainty, and elevation feedback parameterisation uncertainty – along the causal chain from SRES scenario to sea level within a coherent experimental design and statistical framework. In particular, parameter uncertainty is estimated with a probabilistic method, which gives well-defined credibility intervals (CIs) rather than simple sensitivity analysis; all ISMs use the same parameterisation and parameter sampling, and are forced with the same GCMs and RCM, enabling comparisons across ISMs; and MAR is forced with two GCM forcings under the same SRES scenario, enabling comparisons between ECHAM5 and HadCM3. A similar approach is taken by Shannon et al. (2013) for a new parameterisation of GrIS basal lubrication. This experimental design gives us the opportunity to assess the relative importance of various modelling uncertainties and how these vary with time. We discuss parametric uncertainties, structural uncertainties, time dependence, and future directions.

The feedback parameter uncertainties translate to a unimodal and fairly symmetric probability distribution for sea level contribution (Fig. 4), because of the linear SMB adjustment and the sampling from unimodal and (in most cases) symmetric probability distributions for the gradients (Edwards et al., 2014). If the gradient distributions derived from MAR were strongly asymmetric or multimodal, or the SMB adjustment nonlinear, the sea level distribution might be a different shape. Here is the real value of Bayesian parameter estimation (Edwards et al., 2014) and sampling these uncertainties with perturbed parameter ensembles, because such things are generally not known in advance. The slight asymmetry in the sea level probability distribution described in the previous section might become more pronounced for timescales longer than two centuries.

If substantially greater computational and time resources were available, a natural next step might be a more thorough exploration of parametric uncertainties, by simultaneously perturbing additional parameters of the ISMs and, if possible, parameters of the climate models. Climate model PPEs are used extensively (e.g. Murphy et al., 2009); RCM PPEs are relatively rare due to their computational expense (e.g. Sexton and Murphy, 2011). Two ISM PPEs are presented by Stone et al. (2010) and Applegate et al. (2011), who vary multiple parameters simultaneously in order to explore their interactions. For example, surface lowering driven by parameterised dynamic processes (such as basal lubrication: Shannon et al., 2013) could interact differently with the elevation feedback depending on the values of the relevant parameters. Performing multiple parameter perturbations might affect the shape of the distribution of projected sea level contributions.

Structural uncertainties are much more challenging to quantify. Using multi-model ensembles is a necessary first step. But there is no well-defined space from which to sample (as there is for parameters). Models have common components and errors that are likely to reduce the ensemble range relative to our uncertainty about the climate and GrIS. It is extremely difficult to express this probabilistically – to quantify the degree to which the spread of a model ensemble encompasses our true uncertainty about the system – and there is no consensus about how to do so. Given these difficulties, and our small ensemble sizes, we do not claim to quantify ISM and GCM structural uncertainties, only to sample from them. It is better to sample from this space, to explore it partially, than to ignore it. Coherent experimental design such as this – systematic sampling of models, parameterisation, and parameter values – is essential to better understanding of structural uncertainties.

The structural uncertainty we explore most is in the ISMs. This arises from structural differences in ice sheet modelling, including representation of the Stokes equations, numerical methods, grid resolution, initialisation procedures, and drift correction. Given these differences, it might seem surprising that the results from different ice sheet models are so similar. Goelzer et al. (2013) compare GrIS sea level projections with different plausible modelling choices for a single ice sheet model (GISM-HO) and the same RCM (MAR) and GCM (ECHAM5) as here, and obtain a wider range of projections. But our results confirm the importance of SMB versus dynamic mass loss. The large spread in results in Goelzer et al. (2013) arises from differing SMB projections: an RCM, a PDD model forced with an RCM, and a PDD model forced with a GCM. In our study all models (by definition) use the same projected SMB changes, and also use the same mask to define ice sheet extent (Sect. 2.3). The dynamical response of the GrIS is smaller than the SMB response, so differences in model physics, numerical methods, grid and even initialisation cannot substantially alter the sea level contribution: SMB is the dominant driver of sea level rise. This confirms results by others such as Gillet-Chaulet et al. (2012) and Goelzer et al. (2013).

We use only two GCMs and do not explore RCM structural uncertainty at all. We do incorporate uncertainty in the structure of our parameterisation of MAR, by including a discrepancy variance term when estimating the gradients (Edwards et al., 2014), but this refers to our parameterisation of the MAR elevation feedback, not the MAR simulation of the feedback in the real world; the latter is the RCM structural uncertainty. We could sample RCM structural uncertainty using multiple RCMs, but this is extremely challenging due to their computational expense. An ensemble of RCMs would also require a new parameterisation of the SMB–elevation feedback for each. For further exploration of MAR and GCM uncertainties, we refer to Fettweis et al. (2013) who evaluate the effects of changing the MAR tundra/ice mask and ice sheet topography, and using a range of GCMs for forcing, on simulations of current climate and projections of future climate change.

We therefore do not claim to make a full assessment of modelling uncertainty in sea level projections. Such an assessment would require perturbing other parameters of the ice sheet models, perturbing parameters of the climate models, and quantifying ice sheet model and climate model structural uncertainties through, for example, large multi-model ensembles of ISMs, RCMs and GCMs. This is beyond present capabilities, though we believe it should be our aim. These additional uncertainties have potentially large effects on predictions of future mass loss, which would result in wider credibility intervals than those presented here. We reiterate that our uncertainty assessment pertains mainly to uncertainties from the elevation feedback parameterisation, with additional exploration of the effects of ice sheet and global climate model uncertainties.

Our quantification of the relative magnitude of modelling uncertainties through time (Fig. 3) is specific to our particular choices and opportunities: the RCM, GCMs and ISMs; the emissions scenario; our method of extending the scenario from 2100 to 2199; and (trivially) our chosen CI width. If we used a scenario in which the SMB projections were more negative than for A1B, such as RCP8.5, or less negative, such as E1 (Rae et al., 2012), the elevation feedback would be correspondingly larger or smaller and so would the uncertainty. Similarly, if we were to use 200 yr GCM and RCM simulations, that is, drive MAR with GCM simulations that were forced in the second century with the A1B scenario at 2100, rather than repeating the decade 2090–2099 from the MAR simulations ten times, the SMB forcing would likely continue to decrease rather than remain fixed. This would lead to a more rapid increase in the sea level projections with elevation feedback. The ISM fractional uncertainty might then not increase in the second century, or not by as much (Fig. 3); this would depend on the relative responses of the ISMs. The GCM range might increase more quickly than the mean, giving a time-dependent fractional uncertainty that is more similar in shape to the ISM fractional uncertainty (Fig. 3).

Such choices and limitations are unavoidable given the substantial computational expense of a three stage model chain in which two stages are climate models. Nevertheless, this work begins the process of determining where to focus computational resources and model development for different decision-making timescales.

5 Summary

We present the first application of a parameterisation of the GrIS SMB–elevation feedback to projections of future climate change. This is currently the only practicable method for making projections for the GrIS contribution to sea level that can incorporate physical modelling of ice flow, SMB processes, and the SMB–elevation feedback while also exploring modelling uncertainties.

We use a coherent experimental design and Bayesian framework to make probabilistic statements (credibility intervals) about the effect of feedback parameterisation uncertainty on sea level projections and explore GCM and ISM structural uncertainties. Probabilistic assessments are more meaningful than single projections or sensitivity analyses: they are easier to interpret, more straightforward to compare across studies, and provide more robust and complete information for decision-making.

We make projections for the SRES A1B emissions scenario with the MAR RCM, ECHAM5 and HadCM3 GCMs, and five ISMs. The experimental design allows us to estimate the relative importance of these model uncertainties, and how they vary over time, for the GrIS contribution to sea level over the next two centuries. The additional sea level contribution due to the SMB–elevation feedback, averaged over five ISM projections for ECHAM5 and three for HadCM3 is 4.4 % (best estimate; 95 % credibility interval 1.8–6.9 %) at 2100, and 9.6 % (best estimate; 95 % credibility interval 3.6–16.0 %) at 2200. In all results, the lower bounds of our 95 % credibility intervals for sea level contributions with the SMB–elevation feedback are larger than for the "no feedback" case.

The relative contributions to uncertainty in our study vary with time. In 2100, the GCM differences are larger than the ISM range and the parameterisation uncertainty, but by 2200, the ISM and parameterisation uncertainties are greater. This work indicates that the areas where future research should be directed – such as understanding climate or ice sheet model differences, or performing perturbed parameter ensembles – depend on the timescale of interest. For sufficiently large ensemble sizes, this method can be used to give an indication where future sea level research should be directed.

The relatively small sea level rise contributed by the elevation feedback suggests that century-long simulations with coupled RCM-ISM models may not be justified for the additional computational expense and technical difficulty of coupling, at least for medium to low emissions scenarios.

Acknowledgements. This work was supported by funding from the ice2sea programme from the European Union 7th Framework Programme, grant number 226375. Ice2sea contribution number 151. Elmer/Ice simulations were performed using HPC resources from GENCI-CINES (grant 2011016066) and from the Service Commun de Calcul Intensif de l'Observatoire de Grenoble (SCCI). S. Price, M. Hoffman, and M. Perego were supported by the US Department of Energy (DOE) Office of Science, Advanced Scientific Computing Research and Biological and Environmental Research programs. M. Hoffman was partially supported by the Center for Remote Sensing of Ice Sheets at the University of Kansas through US National Science Foundation grant ANT-0424589. CISM and MPAS simulations were conducted at the National Energy Research Scientific Computing Center and at the Oak Ridge National Laboratory (supported by DOE's Office of Science under Contracts DE-AC02-05CH11231 and DE-AC05-00OR22725, respectively). We thank Jesse Johnson and an anonymous referee for their helpful reviews.

Edited by: E. Hanna

References

Applegate, P. J., Kirchner, N., Stone, E. J., Keller, K., and Greve, R.: An assessment of key model parametric uncertainties in projections of Greenland Ice Sheet behavior, The Cryosphere, 6, 589–606, doi:10.5194/tc-6-589-2012, 2012.

Bamber, J. L., Hardy, R. J., and Joughin, I.: An analysis of balance velocities over the Greenland ice sheet and comparison with synthetic aperture radar interferometry, J. Glaciol., 46, 67–74, 2000.

Bamber, J. L., Griggs, J. A., Hurkmans, R. T. W. L., Dowdeswell, J. A., Gogineni, S. P., Howat, I., Mouginot, J., Paden, J., Palmer, S., Rignot, E., and Steinhage, D.: A new bed elevation dataset for Greenland, The Cryosphere, 7, 499–510, doi:10.5194/tc-7-499-2013, 2013.

Bindschadler, R. A., Nowicki, S., Abe-Ouchi, A., Aschwanden, A., Choi, H., Fastook, J., Granzow, G., Greve, R., Gutowski, G., Herzfeld, U., Jackson, C., Johnson, J., Khroulev, C., Levermann, A., Lipscomb, W. H., Martin, M. A., Morlighem, M., Parizek, B. R., Pollard, D., Price, S. F., Ren, D., Saito, F., Sato, T., Seddik, H., Seroussi, H., Takahashi, K., Walker, R., and Wang, W. L.: Ice-sheet model sensitivities to environmental forcing and their use in projecting future sealevel (the SeaRISE project), J. Glaciol., 59, 195–224, 2013.

Bueler, E. and Brown, J.: Shallow shelf approximation as a "sliding law" in a thermomechanically coupled ice sheet model, J. Geophys. Res., 114, F03008, doi:10.1029/2008JF001179, 2009.

Dee, D. P., Uppala, S. M., Simmons, A. J., Berrisford, P., Poli, P., Kobayashi, S., Andrae, U., Balmaseda, M. A., Balsamo, G., Bauer, P., Bechtold, P., Beljaars, A. C. M., van de Berg, L., Bidlot, J., Bormann, N., Delsol, C., Dragani, R., Fuentes, M., Geer, A. J., Haimberger, L., Healy, S. B., Hersbach, H., Hólm, E. V., Isaksen, L., Kållberg, P., Köhler, M., Matricardi, M., McNally, A. P., Monge-Sanz, B. M., Morcrette, J.-J., Park, B.-K., Peubey, C., de Rosnay, P., Tavolato, C., Thépaut, J.-N., and Vitart, F.: The ERA-Interim reanalysis: configuration and performance of the data assimilation system, Q. J. Roy. Meteor. Soc., 137, 553–597, 2011.

Dukowicz, J. K., Price, S. F., and Lipscomb, W. H.: Consistent approximations and boundary conditions for ice-sheet dynamics from a principle of least action, J. Glaciol., 56, 480–496, 2010.

Edwards, T. L., Fettweis, X., Gagliardini, O., Gillet-Chaulet, F., Goelzer, H., Gregory, J. M., Hoffman, M., Huybrechts, P., Payne, A. J., Perego, M., Price, S., Quiquet, A., and Ritz, C.: Probabilistic parameterisation of the surface mass balance-elevation feedback in regional climate model simulations of the Greenland ice sheet, The Cryosphere, 8, 181–194, doi:10.5194/tc-8-181-2014, 2014.

Ettema, J., van den Broeke, M. R., van Meijgaard, E., van de Berg, W. J., Bamber, J. L., Box, J. E., and Bales, R. C.: Higher surface mass balance of the Greenland ice sheet revealed by high-resolution climate modeling, Geophys. Res. Lett., 36, L12501, doi:10.1029/2009GL038110, 2009.

Evans, K. J., Salinger, A. G., Worley, P. H., Price, S. F., Lipscomb, W. H., Nicols, J. A., White III, J. B., Perego, M., Vertenstein, M., Edwards, J., and Lemieux, J. F.: A modern solver interface to manage solution algorithms in the Community Earth System Model, Int. J. High Perform. C, 26, 54–62, 2012.

Fettweis, X.: Reconstruction of the 1979–2006 Greenland ice sheet surface mass balance using the regional climate model MAR, The Cryosphere, 1, 21–40, doi:10.5194/tc-1-21-2007, 2007.

Fettweis, X., Franco, B., Tedesco, M., van Angelen, J. H., Lenaerts, J. T. M., van den Broeke, M. R., and Gallée, H.: Estimating the Greenland ice sheet surface mass balance contribution to future sea level rise using the regional atmospheric climate model MAR, The Cryosphere, 7, 469–489, doi:10.5194/tc-7-469-2013, 2013.

Franco, B., Fettweis, X., and Erpicum, M.: Future projections of the Greenland ice sheet energy balance driving the surface melt, The Cryosphere, 7, 1–18, doi:10.5194/tc-7-1-2013, 2013.

Fürst, J. J., Rybak, O., Goelzer, H., De Smedt, B., de Groen, P., and Huybrechts, P.: Improved convergence and stability properties in a three-dimensional higher-order ice sheet model, Geosci. Model Dev., 4, 1133–1149, doi:10.5194/gmd-4-1133-2011, 2011.

Fürst, J. J., Goelzer, H., and Huybrechts, P.: Effect of higher-order stress gradients on the centennial mass evolution of the Greenland ice sheet, The Cryosphere, 7, 183–199, doi:10.5194/tc-7-183-2013, 2013.

Gillet-Chaulet, F., Gagliardini, O., Seddik, H., Nodet, M., Durand, G., Ritz, C., Zwinger, T., Greve, R., and Vaughan, D. G.: Greenland ice sheet contribution to sea-level rise from a new-generation ice-sheet model, The Cryosphere, 6, 1561–1576, doi:10.5194/tc-6-1561-2012, 2012.

Goelzer, H., Huybrechts, P., Fürst, J. J., Nick, F. M., Andersen, M. L., Edwards, T. L., Fettweis, X., Payne, A. J., and Shannon, S. R.: Sensitivity of Greenland ice sheet projections to model formulations, J. Glaciol., 59, 733–749, doi:10.3189/2013JoG12J182, 2013.

Gordon, C., Cooper, C., Senior, C. A., Banks, H., Gregory, J. M., Johns, T. C., Mitchell, J., and Wood, R. A.: The simulation of SST, sea ice extents and ocean heat transports in a version of the Hadley Centre coupled model without flux adjustments, Clim. Dynam., 16, 147–168, 2000.

Greve, R.: A continuum-mechanical formulation for shallow polythermal ice sheets, Philos. T. Roy. Soc. A, 355, 217–242, 1997.

Hanna, E., Huybrechts, P., Cappelen, J., Steffen, K., Bales, R. C., Burgess, E., McConnell, J. R., Peder Steffensen, J., van den Broeke, M., Wake, L., Bigg, G., Griffiths, M., and Deniz Savas, D.: Greenland Ice Sheet surface mass balance 1870 to 2010 based on Twentieth Century Reanalysis, and links with global climate forcing, J. Geophys. Res., 116, D24121, doi:10.1029/2011JD016387, 2011.

Helsen, M. M., van de Wal, R. S. W., van den Broeke, M. R., van de Berg, W. J., and Oerlemans, J.: Coupling of climate models and ice sheet models by surface mass balance gradients: application to the Greenland Ice Sheet, The Cryosphere, 6, 255–272, doi:10.5194/tc-6-255-2012, 2012.

Huybrechts, P. and de Wolde, J.: The dynamic response of the Greenland and Antarctic ice sheets to multiple-century climatic warming, J. Climate, 12, 2169–2188, 1999.

Janssens, I. and Huybrechts, P.: The treatment of meltwater retention in mass-balance parameterizations of the Greenland ice sheet, Ann. Glaciol., 31, 133–140, 2000.

Joughin, I., Smith, B. E., Howat, I. M., Scambos, T., and Moon, T.: Greenland flow variability from ice-sheet-wide velocity mapping, J. Glaciol., 56, 415–430, 2010.

Lemieux, J. F., Price, S. F., Evans, K. J., Knoll, D., Salinger, A. G., Holland, D. M., and Payne, A. J.: Implementation of the Jacobian-free Newton–Krylov method for solving the first-order ice sheet momentum balance, J. Comput. Phys., 230, 6531–6545, 2011.

Morlighem, M., Rignot, E., Seroussi, H., Larour, E., Dhia, H. B., and Aubry, D.: Spatial patterns of basal drag inferred using control methods from a full-Stokes and simpler models for Pine Island Glacier, West Antarctica, Geophys. Res. Lett., 37, L14502, doi:10.1029/2010GL043853, 2010.

Moss, R. H., Edmonds, J. A., Hibbard, K., Manning, M., Rose, S. K., van Vuuren, D. P., Carter, T. R., Emori, S., Kainuma, M., Kram, T., Meehl, G., Mitchell, J., Nakicenovic, N., Riahi, K., Smith, S., Stouffer, R. J., Thomson, A., Weyant, J., and Wilbanks, T.: The next generation of scenarios for climate change research and assessment, Nature, 463, 747–756, 2010.

Murphy, J. M., Sexton, D. M. H., Jenkins, G. J., Booth, B. B. B., Brown, C. C., Clark, R. T., Collins, M., Harris, G. R., Kendon, E. J., Betts, R. A., Brown, S. J., Humphrey, K. A., McCarthy, M. P., McDonald, R. E., Stephens, A., Wallace, C., Warren, R., Wilby, R., and Wood, R. A.: Climate change projections, in: UK Climate Projections, no. 2, Met Office Hadley Centre, Exeter, UK, 1–194, 2009.

Nakićenović, N., Alcamo, J., Davis, G., and de Vries, B.: Special report on emissions scenarios: a special report of Working Group III of the Intergovernmental Panel on Climate Change, Intergovernmental Panel on Climate Change, 2000.

Pattyn, F.: A new three-dimensional higher-order thermomechanical ice sheet model: basic sensitivity, ice stream development, and ice flow across subglacial lakes, J. Geophys. Res., 108, 2382, doi:10.1029/2002JB002329, 2003.

Pattyn, F., Schoof, C., Perichon, L., Hindmarsh, R. C. A., Bueler, E., de Fleurian, B., Durand, G., Gagliardini, O., Gladstone, R., Goldberg, D., Gudmundsson, G. H., Huybrechts, P., Lee, V., Nick, F. M., Payne, A. J., Pollard, D., Rybak, O., Saito, F., and Vieli, A.: Results of the Marine Ice Sheet Model Intercomparison Project, MISMIP, The Cryosphere, 6, 573–588, doi:10.5194/tc-6-573-2012, 2012.

Perego, M., Gunzburger, M., and Burkardt, J.: Parallel finite element implementation for higher-order ice-sheet models, J. Glaciol., 58, 76–88, 2012.

Price, S. F., Payne, A. J., Howat, I. M., and Smith, B. E.: Committed sea-level rise for the next century from Greenland ice sheet dynamics during the past decade, P. Natl. Acad. Sci. USA, 108, 8978–8983, 2011.

Quiquet, A., Punge, H. J., Ritz, C., Fettweis, X., Gallée, H., Kageyama, M., Krinner, G., Salas y Mélia, D., and Sjolte, J.: Sensitivity of a Greenland ice sheet model to atmospheric forcing fields, The Cryosphere, 6, 999–1018, doi:10.5194/tc-6-999-2012, 2012.

Rastner, P., Bolch, T., Mölg, N., Machguth, H., Le Bris, R., and Paul, F.: The first complete inventory of the local glaciers and ice caps on Greenland, The Cryosphere, 6, 1483–1495, doi:10.5194/tc-6-1483-2012, 2012.

Ringler, T., Ju, L., and Gunzburger, M.: A multiresolution method for climate system modeling: application of spherical centroidal Voronoi tessellations, Ocean Dynam., 58, 475–498, 2008.

Ritz, C., Rommeleare, V., and Dumas, C.: Modeling the evolution of Antarctic ice sheet over the last 420 000 yr: Implications for altitude changes in the Vostok region, J. Geophys. Res., 106, 31943–31964, 2001.

Rutt, I., Hagdorn, M., Hulton, N., and Payne, A.: The "Glimmer" community ice sheet model, J. Geophys. Res., 114, F02004, doi:10.1029/2008JF001015, 2009.

Rae, J. G. L., Aðalgeirsdóttir, G., Edwards, T. L., Fettweis, X., Gregory, J. M., Hewitt, H. T., Lowe, J. A., Lucas-Picher, P., Mottram, R. H., Payne, A. J., Ridley, J. K., Shannon, S. R., van de Berg, W. J., van de Wal, R. S. W., and van den Broeke, M. R.: Greenland ice sheet surface mass balance: evaluating simulations and making projections with regional climate models, The Cryosphere, 6, 1275–1294, doi:10.5194/tc-6-1275-2012, 2012.

Roeckner, E., Bäuml, G., Bonaventura, L., Brokopf, R., Esch, M., Giorgetta, M., Hagemann, S., Kirchner, I., Kornblueh, L., Manzini, E., Rhodin, A., Schlese, U., Schulzweida, U., and Tompkins, A.: The atmospheric general circulation model ECHAM 5. Part I: Model description, Max-Planck-Institut für Meteorologie, Hamburg, Germany, 126 pp., 2003.

Sexton, D. and Murphy, J.: Multivariate probabilistic projections using imperfect climate models. Part II: Robustness of methodological choices and consequences for climate sensitivity, Clim. Dynam., 38, 2543–2558, 2011.

Sexton D. M. H., Murphy, J., Collins, M., and Webb, M.: Multivariate probabilistic projections using imperfect climate models. Part I: Outline of methodology, Clim. Dynam., 38, 2513–2542, 2011.

Shannon, S. R., Payne, A. J., Bartholomew, I. D., van den Broeke, M. R., Edwards, T. L., Fettweis, X., Gagliardini, O., Gillet-Chaulet, F., Goelzer, H., Hoffman, M., Huybrechts, P., Mair, D., Nienow, P., Perego, M., Price, S. F., Smeets, C. J. P. P., Sole, A. J., van de Wal, R. S. W., and Zwinger, T.: Enhanced basal lubrication and the contribution of the Greenland ice sheet to future sea level rise, P. Natl. Acad. Sci. USA, 110, 14156–14161, doi:10.1073/pnas.1212647110, 2013.

Shapiro, N. M. and Ritzwoller, M. H.: Inferring surface heat flux distributions guided by a global seismic model: particular application to Antarctica, Earth Planet. Sc. Lett., 223, 213–224, 2004.

Stone, E. J., Lunt, D. J., Rutt, I. C., and Hanna, E.: Investigating the sensitivity of numerical model simulations of the modern state of the Greenland ice-sheet and its future response to climate change, The Cryosphere, 4, 397–417, doi:10.5194/tc-4-397-2010, 2010.

Uppala, S. M., Køallberg, P. W., Simmons, A. J., Andrae, U., Da Costa Bechtold, V., Fiorino, M., Gibson, J. K., Haseler, J., Hernandez, A., Kelly, G. A., Li, X., Onogi, K., Saarinen, S., Sokka, N., Allan, R. P., Andersson, E., Arpe, K., Balmaseda, Beljaars, A. C. M., Van De Berg, L., Bidlot, J., Bormann, N., Caires, S., Chevallier, F., Dethof, A., Dragosavac, M., Fisher, M., Fuentes, M., Hagemann, S., Hólm, E., Hoskins, B. J., Isaksen L., Janssen, P. A. E. M., Jenne, R., Mcnally, A. P., Mahfouf, J.-F., Morcrette, J.-J., Rayner, N. A., Saunders, R. W., Simon, P., Sterl, A., Trenberth, K. E., Untch, A., Vasiljevic, D., Viterbo, P., Woollen, J.: The ERA-40 re-analysis, Q. J. Roy. Meteor. Soc., 131, 2961–3012, 2005.

van de Berg, W. J., van den Broeke, M., Ettema, J., van Meijgaard, E., and Kaspar, F.: Significant contribution of insolation to Eemian melting of the Greenland ice sheet, Nat. Geosci., 4, 679–683, 2011.

van de Wal, R. S. W.: Mass-balance modelling of the Greenland ice sheet: a comparison of an energy-balance and a degree-day model, Ann. Glaciol., 23, 36–45, 1996.

Vernon, C. L., Bamber, J. L., Box, J. E., van den Broeke, M. R., Fettweis, X., Hanna, E., and Huybrechts, P.: Surface mass balance model intercomparison for the Greenland ice sheet, The Cryosphere, 7, 599–614, doi:10.5194/tc-7-599-2013, 2013.

5

Impact of physical properties and accumulation rate on pore close-off in layered firn

S. A. Gregory, M. R. Albert, and I. Baker

Thayer School of Engineering, Dartmouth College, Hanover, NH 03766, USA

Correspondence to: M. R. Albert (mary.r.albert@dartmouth.edu)

Abstract. Investigations into the physical characteristics of deep firn near the lock-in zone through pore close-off are needed to improve understanding of ice core records of past atmospheric composition. Specifically, the permeability and microstructure profiles of the firn through the diffusive column influence the entrapment of air into bubbles and thus the ice age–gas age difference. The purpose of this study is to examine the nature of pore closure processes at two polar sites with very different local temperatures and accumulation rates. Density, permeability, and microstructure measurements were made on firn cores from the West Antarctic Ice Sheet (WAIS) Divide, a site that has moderate accumulation rates with a seasonal climate archive, and Megadunes in East Antarctica, a site that is a natural laboratory for accumulation rate effects in the cold low-accumulation desert. We found that the open pore structure plays a more important role than density in predicting gas transport properties, throughout the porous firn matrix. For firn below 50 m depth at both WAIS Divide and Megadunes, finer-grained layers experience close-off shallower in the firn column than do coarser-grained layers, regardless of which grain size layer is the denser layer at depth. Pore close-off occurs at a critical open porosity that is accumulation rate dependent. Defining pore close-off at a critical open porosity for a given accumulation rate as opposed to a critical total porosity accounts for the pore space available for gas transport. Below the critical open porosity, the firn becomes impermeable despite having small amounts of interconnected pore space. The low-accumulation sites, with generally coarse grains, close off at lower open porosities ($\sim < 10\,\%$) than the open porosity ($\sim > 10\,\%$) of high-accumulation sites that have generally finer grains. The microstructure and permeability even near the bottom of the firn column are relic indicators of the nature of accumulation when that firn was at the surface. The physical structure and layering are the primary controlling factors on pore close-off. In contrast to current assumptions for polar firn, the depth and length of the lock-in zone is primarily dependent upon accumulation rate and microstructural variability due to differences in grain size and pore structure, rather than the density variability of the layers.

1 Introduction

As an archive for past atmospheric composition, the polar ice sheets play an important role in understanding climate change, both natural and anthropogenic in origin. The surface of the Greenland and Antarctic ice sheets is covered in a 60–120 m-thick layer of firn, multiyear snow that undergoes further metamorphism with depth until it becomes solid ice at the firn–ice transition. The firn layer acts as a filter on atmospheric signals that are eventually captured within bubbles in the underlying ice. Because air within the firn column can exchange with the atmosphere until deep in the firn at the start of the lock-in zone, or at pore close-off at sites with no lock-in zone, where horizontal layers of firn become impermeable, impeding vertical gas transport, the air entrapped in bubbles of the ice matrix is always younger than the ice (Schwander and Stauffer, 1984). Knowing the depth at which air can no longer exchange with the atmosphere is pivotal for determining the gas age–ice age difference. At present-day polar sites, most studies have focused on the bulk properties of the firn, and not until the recent past have studies started to include the layered nature of the firn (Albert et al., 2004; Courville et al., 2010; Freitag et al., 2004; Fujita et al., 2009; Horhold et al., 2011).

A typical firn column can be divided into three main zones based on the dominating gas transport mechanisms operating at a given depth (Sowers et al., 1992). Gas movement in the top of the firn column, the convective zone, is dominated by convection due to highly permeable firn and/or wind pumping. Below the convective zone is the diffusive zone, where the main gas transport mechanism is molecular diffusion. Near the base of the diffusive zone is the non-diffusive or lock-in zone (LIZ) characterized by layers of firn, some of which are permeable and some are not. At the top of the LIZ is the lock-in depth (LID), the depth at which the first impermeable horizontal layer of firn impedes vertical gas exchange with the atmosphere. The bottom of the LIZ is at the close-off depth beyond which all pores are closed off from one another both vertically and horizontally and no gas transport exists.

A useful technique to examine firn microstructure is X-ray computed tomography. While X-ray computed tomography is well established in soil science (Taina et al., 2008), initial application to polar firn has only happened in the last 10–15 yr (e.g., Coleou et al., 2001; Freitag et al., 2004; Schneebeli and Sokratov, 2004). Several studies have been done using X-ray computed tomography to examine the microstructure of polar firn. Lomonaco et al. (2011) examined the evolution of fine-grained layers from the surface through pore close-off. Freitag et al. (2004) and Fujita et al. (2009) focused primarily on understanding how microstructure differences between fine-grained and coarse-grained firn influences the densification rate of polar firn. Other studies include the influence of microstructure, either from thick sections or X-ray computed tomography on the permeability of firn at the top of the firn column (Albert et al., 2004; Courville et al., 2007; Rick and Albert, 2004). Courville et al. (2010) and Freitag et al. (2002) used 3-D reconstructions of firn cube microstructure to model centimeter-scale permeability of firn using lattice Boltzmann techniques. The specific influence of microstructure on bulk properties affecting gas transport in deep firn has not been well investigated.

Pore close-off has been studied through three approaches: (1) density predicted from densification models, (2) in situ through firn air measurements and modeling, and (3) laboratory-based permeability measurements. An inherent issue in understanding the processes controlling pore close-off is an integration between microscale processes and macroscale processes. For example, in situ firn air measurements are taken on a meter scale over many layers of firn, while laboratory-based permeability measurements are taken on the centimeter scale and often incorporate a single homogenous firn layer. Several studies have been done in an attempt to understand the percolation nature of pore close-off (Entig, 1993; Freitag, 2008). While macroscale processes involved with gas transport through firn are important, understanding the origins of pore close-off on a much smaller scale through the microstructural properties of the firn should provide insight for understanding gas transport through firn and for enabling improvements in firn air modeling.

The traditional assumption is that pore close-off density is site specific, and the first layers of firn to reach this density will close-off and no longer exchange air with the atmosphere. Martinerie et al. (1992) used sixteen different sites to parameterize the mean density of pore close-off based on the temperature of the sites. Parameterizing close-off with a mean density ignores the layered nature of firn and the presence of a lock-in zone. To account for the length of the lock-in zone in firn densification modeling, Goujon et al. (2003) used a closed porosity of 37 % to indicate the close-off depth (bottom of the lock-in zone), a closed porosity of 21 % for Vostok, Antarctica, and 13 % for Summit, Greenland, to indicate the lock-in depth. The thickness of present-day lock-in zones tends to increase with an increase in accumulation rate (Landais et al., 2006). An increase in lock-in zone thickness at high-accumulation sites has been linked by Horhold et al. (2011) to a second period of high density variability using high-resolution density measurements. In this paper, using laboratory measurements on firn cores retrieved from Antarctica, we explore the role of microstructure including pore structure on the nature of layers in deep firn and the range of depths where pore close-off occurs.

The second approach for assessing pore close-off in polar firn is through in situ bulk firn air measurements accompanied by firn modeling. Firn air campaigns have been conducted at a number of Arctic and Antarctic sites where trace gas and isotopic fractionation are used to locate the lock-in depth. The deepest location where air can be sampled from the firn column is typically considered the close-off depth and the bottom of the lock-in zone (Clark et al., 2007; Buizert et al., 2012a; Kawamura et al., 2006; Schwander et al., 1993; Witrant et al., 2012; Etheridge et al., 1996; Trudinger, 2001; Severinghaus et al., 2001, 2010; Fabre et al., 2000; Battle et al., 2011). Firn air models have been developed to construct effective gas diffusivity profiles through a combination of forward and inverse modeling (e.g., Rommelaere et al., 1997; Trudinger et al., 1997, 2002; Buizert et al., 2012a). These models use density profiles derived from densification models to estimate an open porosity profile of an individual site. The open porosity profile is then used to predict effective diffusivity profiles for various tracer gases at the site. The effective diffusivity profile is then tuned based on measurements until a good fit is found. While this method apparently works well for present-day sites where firn air measurements are available, better parameterization of polar firn based on temperature and accumulation rate is needed in order to use firn air models accurately for firn columns existing in past environments (Brook, 2013). Schwander et al. (1988) demonstrated a relationship for tortuosity as a function of porosity at Siple Station, Antarctica, but this relationship does not account for microstructural differences in polar firn from site to site. In an attempt to better characterize gas transport and open porosity, Freitag et al. (2002) fitted both gas diffusivity and permeability to open porosity using two power law functions with exponents 2.1 and 3.4, respectively, for firn from Summit, Greenland. The results from Schwander et al. (1988), Fabre et al. (2000) and Freitag et al. (2002) agree reasonably well for open porosities greater than 0.2 but diverge at low porosities, and none account for the layered nature of firn or density variability of the lock-in zone.

This study aims to better understand the role firn microstructure plays in controlling gas transport in deep firn and to investigate the validity of using density (bulk or local) to predict gas transport in polar firn, including pore close-off, by comparing firn structure from two Antarctic sites, WAIS Divide and Megadunes, which are both sites that do not experience melt but that have very different local climates. The WAIS Divide site in West Antarctica has a mean annual temperature of −31 °C and gets significant annual accumulation. The very cold Megadunes site in the polar desert of East Antarctica is a natural laboratory for the study of investigation of accumulation rate on firn properties (Courville et al., 2007). Our microstructural analysis, density and permeability measurements show the importance of grain and pore structure over local density and open porosity in controlling gas transport and the pore close-off process. We investigate links between accumulation rate, pore structure and gas transport properties affecting permeability, gas diffusivity and pore close-off in deep firn.

2 Methods

2.1 Visual stratigraphy

Stratigraphic layering of the firn was observed and recorded using a backlit light table in a cold room at the Cold Regions Research and Engineering Laboratory in Hanover, NH. Meter-long core sections were placed on the light table where grain size and wind crusts were recorded to the 1 mm scale. Relative grain size was qualitatively described using a five-tier scale from coarse to fine for each meter length section. Coarse grains and fine grains were defined relative to other layers within the meter-long section. The most coarse-grained layers meter by meter are identified as coarse, while the most fine-grained layers are identified as fine grain. All other firn grain layers are classified into four categories between these two identifiers. In this way, the quantitative measurements of a fine-grain layer at the surface and a fine-grain layer at depth are not the same; the qualitative scale was done in order to describe the layering. Comparison between the WAIS Divide site and the Megadunes site was also done visually with a backlit light table in which the coarse-grain layers at WAIS Divide below 55m were observed to be smaller relative to the fine-grain layers at Megadunes below 55 m. Samples of 5–10 cm of similar grain size were cut for further analysis. Emphasis was put on obtaining a single homogenous layer of firn per sample whenever possible, in order to facilitate comparisons of the measured transport properties with quantitative microscopy results.

2.2 Bulk density

Bulk density measurements were made on the same 5–10 cm-resolution samples using volumetric measurements and the mass of the sample. Error in density measurements was small, with less than 0.5 % standard deviation for 10 repeat measurements on a single sample. The mass of the sample from scale measurement is accurate to 1 %, and the volume of the sample calculated from caliper measurements is accurate to within approximately 2 %. To increase the accuracy of bulk property measurements, broken and chipped samples, not of an ideal cylindrical shape, were not included in bulk density or permeability measurements and are not reported in this paper.

2.3 Permeability

Permeability measurements were made using the methods developed by Albert et al. (2000), and used for example in Rick and Albert (2004), and Courville et al. (2007). Using a custom apparatus that had been verified on glass beads, air was drawn through a firn sample, and the associated flow rate and pressure drop across the sample were measured, along with temperature and barometric pressure. A variety of flow rates was employed. Each measured flow rate and associated measured pressure drop was used with Darcy's law to calculate the permeability. Flow rates of air through the samples were kept within the laminar flow regime where Darcy's Law holds true:

$$v = \frac{k}{\mu} \cdot \frac{\mathrm{d}P}{\mathrm{d}x}, \quad (1)$$

where k is the permeability constant, μ is the air viscosity, $\mathrm{d}P$ is the pressure differential, $\mathrm{d}x$ is the height of the sample and v is the flow velocity. Before measurements were taken the permeameter was calibrated using glass bead samples. Calibration was complete when measured glass bead permeability fell within the accepted literature values. Ten measurements with ten different flow rates were done on each sample and usually fell within 5–10 % of one another. At very low permeabilities fewer but at least five measurements were made on each sample due to the low flow rates needed to stay within the laminar flow regime. The lowest permeability measured was $1 \times 10^{-12}\,\mathrm{m}^2$. All samples whose measurements fell below the range of the sensitive pressure transducer were considered impermeable and assigned permeability values of $0\,\mathrm{m}^2$.

2.4 Microcomputed tomography

Firn microstructure properties were obtained using X4-ray microtomography. A Skyscan 1172 model microCT was used in a cold room. Scans were run at 40 kV, a 250 μA current intensity, and a rotation step of 0.7° completing 180° rotation for each run. A total of 275 shadow images were obtained and reconstructed using Skyscan's NRecon software for two-dimensional slice reconstruction. The resolution of the images obtained is 14.8 μm in which each voxel obtained represents a three-dimensional cube with 14.8 μm side length. Firn samples were cut to 1 cm width × 1 cm width × 1.5 cm height segments from the center of the 5–10 cm samples used in density and permeability measurements. The volume of interest analyzed for microstructure properties was 538 × 538 × 673 voxels, or 8 mm × 8 mm × 10 mm in size. For thresholding 256 grey levels were used and a thresholding value of 89 was set between the air and the ice phase. The large difference in linear attenuation coefficients (X-ray absorption) between ice and air enables a simple thresholding limit to be used to binarize the images. To reduce noise and minimize falsely counting incorrectly binarized voxels as pores or small ice clusters, all white specs less than 25 voxels surrounded by black voxels were removed and vice versa. Image analysis on both the ice phase and pore phase was done to obtain the microstructure properties of both the ice structure and pore structure of the firn sample.

The microstructural properties, derived using Skyscan's CTan software, include total porosity, open porosity, closed porosity, structural model index, surface-to-volume (S/V) ratio, and anisotropy. The total porosity of the sample is determined by counting the total number of voxels present for the phase being analyzed and dividing that by the total number of voxels within the region of interest. Open porosity is defined as any pore that intersects with the edge of the region of interest at least once. From total and open porosity values, closed porosity of a sample can quickly be determined through subtraction. To account for closed pores that were cut during sample preparation, pore size distribution was determined for all samples where the majority of closed pores had volumes of 1 mm^3 or less. Through individual object analysis all pores with a center within 0.62 mm (the radius of a 1 mm^3 spherical pore) of the edge of the region of interest were considered closed, and open porosity and closed porosity were adjusted accordingly for each sample.

Structure model index (SMI) gives an estimation for the type of shape present in the analyzed phase. SMI values of 0, 3 and 4 correspond to an ideal plate, cylinder, and sphere, respectively. Calculation of SMI follows that developed by Hildebrand and Rüegsegger (1997) in which one voxel thickness is added to the surface of the phase analyzed. SMI is then calculated as follows:

$$\mathrm{SMI} = 6\frac{S' \cdot V}{S^2}, \tag{2}$$

where S' is the change in surface area between the initial surface area and the artificially increased surface area, S is the original surface area, and V is the initial volume of the analyzed phase. For convex shapes, SMI is positive while concave structures have negative SMI values. The surface-to-volume ratio gives the ratio of the analyzed phase surface area to the volume of the phase in three dimensions. It gives estimation of how tortuous an object is where a low S/V indicates less complexity in shape than a high S/V (morphometric parameters measured by SkyscanTM CT-analyser software, Bruker-MicroCT CT-Analyser, http://www.skyscan.be).

Anisotropy is a measure of alignment or three-dimensional symmetry within the region of interest. The value of anisotropy is determined using the mean intercept length: the length of a line traveling through an object divided by the number of times the line crosses the analyzed phase. The higher the mean intercept length the more the object is aligned in a single direction, indicating a high degree of anisotropy. The mean intercept length is found at many 3-D angles within the region of interest, and each MIL is plotted to create an ellipsoid. The ellipsoid is described using an orthogonal tensor whose maximum and minimum eigenvalues are used to determine the degree of anisotropy as follows:

$$\mathrm{DA} = 1 - \left(\frac{\mathrm{min}_{\mathrm{eigenvalue}}}{\mathrm{max}_{\mathrm{eigenvalue}}}\right). \tag{3}$$

Using this definition, the degree of anisotropy (DA) ranges from a value of 1, totally isotropic, to infinity, totally anisotropic (morphometric parameters measured by SkyscanTM CT-analyser software, Bruker-MicroCT CT-Analyser, http://www.skyscan.be).

3 Site characteristics

The Megadunes and WAIS Divide sites were chosen to provide two contrasting local climates to understand how climate influences firn microstructure, gas transport and pore close-off. WAIS Divide, which was chosen for drilling of a deep ice core for climate science, is the Antarctic counterpart to Summit, Greenland, and it has been described by Battle et al. (2011) as an intermediate site in terms of temperature and accumulation rate. The Megadunes site in the extremely cold polar desert of East Antarctica was chosen for comparison because it essentially constitutes a natural laboratory for the study of effect of accumulation rate. Courville et al. (2007) provide a complete description of the manner in which two different accumulation rates exist under the same temperature and other climatic conditions at Megadunes. The windward faces of the low-amplitude, long-wavelength antidunes experience accumulation, while the leeward faces undergo zero accumulation or ablation. As the dunes slowly migrate upwind over decades, past dunes are slowly buried, and the resulting layering and firn structure provide enduring evidence of alternation between accumulation and hiatus firn layers as evidence of the layers persist down the firn column (Courville et al., 2007).

3.1 Megadunes

Measurements of firn air and extraction of a firn core from a megadunes site in East Antarctica were accomplished at an undisturbed site in December 2004–January 2005 at 80.77914° S, 124.48796° E. Courville et al. (2007) thoroughly describe the site details and near-surface measurements, and Severinghaus et al. (2010) describe in situ firn air measurement results. The average temperature at the site is −49 °C with an accumulation rate less than 4 cm weq a^{-1}. In situ firn air measurements at Megadunes done by Severinghaus et al. (2010) indicate a LIZ from 64.5 m to 68.5 m. The convective zone at Megadunes is notably very thick, reaching a depth of ~23 m (Severinghaus et al., 2010).

3.2 WAIS divide

The WDC05C firn core was retrieved from a previously undisturbed site at 79.46300° S, 112.12317° W near the WAIS Divide ice coring site for firn air measurements in December 2005–January 2006 (Battle et al., 2011). The mean annual temperature at the site is −31 °C and the average accumulation rate is 21 cm weq a^{-1} (Banta et al., 2008). The accumulation rate and average temperature at the site are both significantly higher than those at the Megadunes location. From in situ firn air measurements done by Battle et al. (2011) a LIZ exists from approximately 66 m to 76.54 m. Similar to other high-accumulation sites, the convective zone is small, likely ranging between 1.4 m and 5.2 m in depth (Battle et al., 2011).

4 Results

This paper primarily discusses the nature of firn at two depth ranges: one at 30–36 m, which is within the diffusive zone, and a deeper range encompassing just above the lock-in zone through pore close-off. Measurements of density, permeability, and microstructure in very deep firn are investigated starting at 55 m, approximately ten meters above the start of the LIZ at both sites that had been determined through in situ firn air measurements, to a few meters below the close-off depth at each site. The deepest measurements at WAIS Divide were 78 m, while the deepest measurements done at Megadunes were 75 m. Including measurements of gas transport and microstructure above the LIZ enabled examination of trends and how they relate to pore close-off in both bulk and microstructural properties. To relate relationships observed in deep firn to firn at shallower depths, a series of permeability and density measurements were done on each core from 30 to 36 m, and are reported here, along with the complete bulk density profile of WAIS Divide. Analysis at all depths for both cores includes both fine-grained and coarse-grained firn layers.

Comparison of bulk properties taken from 5–10 cm-long samples to microstructural properties taken from 8 mm × 8 mm × 10 mm was done on visually homogenous 5–10 cm-long samples. By avoiding layered samples, the small microCT samples were representative of the larger single firn layered samples used for density and permeability measurements, enabling a valid comparison of properties between the two sample sizes. While firn air campaigns typically retrieve gas samples coming from several layers of firn on the decimeter to meter scale, gathering a more macroscale interpretation that inherently confounds effects of properties and processes across multiple scales, a detailed inspection of firn microstructure (grain size, open porosity, closed porosity, and pore structure) and its evolution with depth provides insight into the microscale processes involved. Efforts to relate microstructure and firn permeability to the percolation process have been discussed by Entig (1993) and Freitag (2008), and are beyond the scope of this paper.

4.1 Permeability and density

Density and permeability measurements made between 30 and 36 m and between 55 and 80 m at Megadunes and WAIS Divide are displayed in Fig. 1. For both sites, the 30–36 m depths are many meters above the lock-in zone, and the lock-in zone falls within the 55–80 m depths. The firn at the initiation of the lock-in zone at Megadunes originated at an accumulation site when that firn was on the surface. It can be seen from Fig 1a and b that in the 30–36 m depths above the lock-in zone, the Megadunes site is much more permeable than the WAIS Divide site, but their density profiles at this depth range are similar. Figure 1c shows that even for a given density in the 30–36 m depth range, Megadunes firn is more permeable than WAIS Divide firn, above the lock-in zone.

The relationships are different within the lock-in zone. For depths between 55 and 80 m just above and within the lock-in zone, it can be seen from Fig. 1d and e that Megadunes firn is denser than WAIS Divide but the permeability profiles are similar. Very near and within the lock-in zone, the relationship existing between density and permeability (Fig. 1f) shows that Megadunes firn is more permeable for any given density than is the WAIS Divide firn. The LID, defined as the depth where multiple sequential samples are impermeable, is at 64.4 m at Megadunes and 63.8 m at WAIS Divide, indicated by a red and black line, respectively.

4.2 Microstructure

Microstructure imaging helps to understand the reasons for different density and permeability profiles at Megadunes and WAIS Divide. Figures 2 and 3 display three-dimensional reconstructions obtained using microcomputed tomography. The width and thickness of the reconstructed firn cubes is 8 mm while the height is 10 mm. Visually Megadunes firn (Fig. 2) has larger but less total pore space at an equivalent depth than WAIS Divide (Fig. 3). The pore structure at WAIS Divide also looks more complex and divided. For both sites, total pore space decreases and the occurrence of bubbles increases with depth.

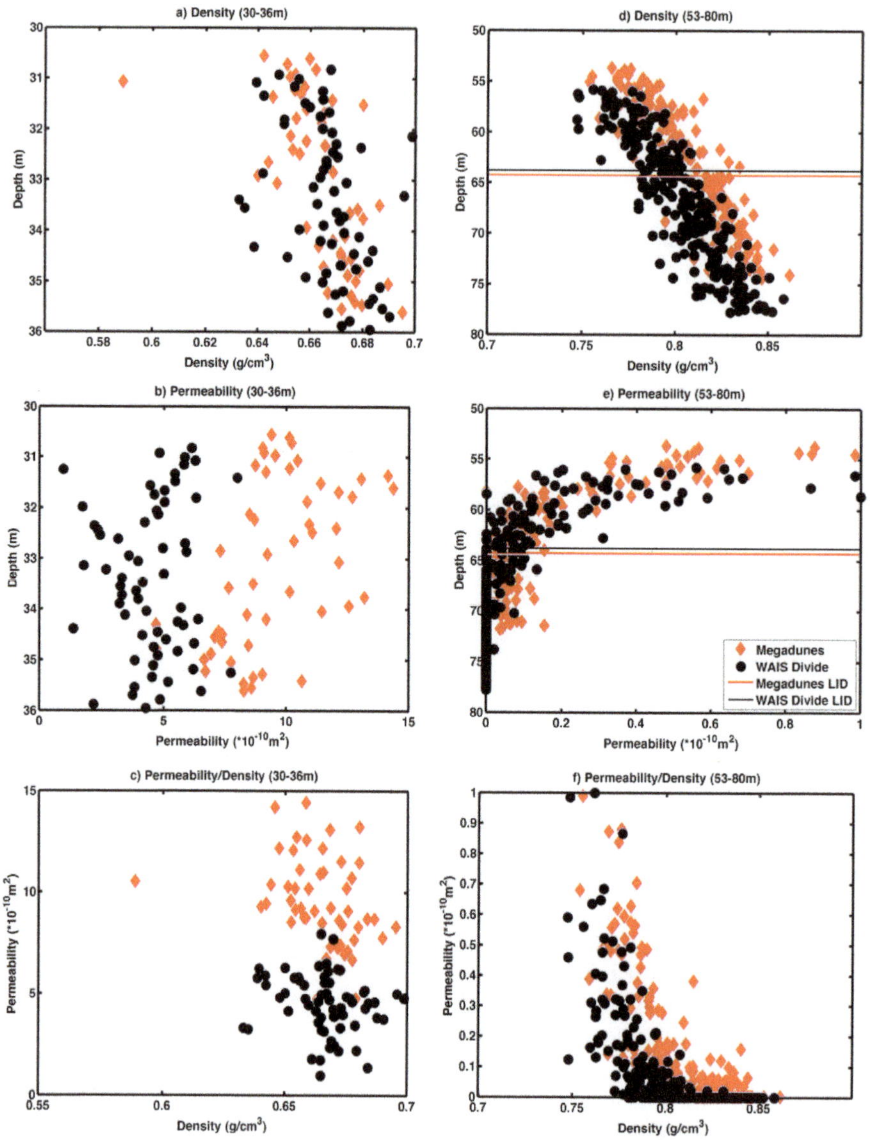

Fig. 1. (a) Density 30–36 m, **(b)** permeability 30–36 m, **(c)** density–permeability relationship 30–36 m, **(d)** density 55–80 m, **(e)** permeability 55–80 m, **(f)** density–permeability relationship 55–80 m. In all graphs, black circles indicate data from WAIS Divide and red diamonds indicate data from Megadunes.

Microstructural analysis enables quantification and validation of differences in visual observations seen between the two sites. The SMI profile, an indicator of pore shape, is similar for Megadunes and WAIS Divide with depth. The pore structure at Megadunes and WAIS Divide (Fig. 4) ranges from semi-cylindrical, an average value of 2 at 56 m with increasing SMI, to average values of 2.7 at 75 m, indicative of an almost perfectly cylindrical pore structure. Toward the end of the LIZ (70–75 m) the pore structure is very close to cylindrical with some high end values reaching 3.5. The data from 56 to 75 m show a trend in increasing SMI with depth that would likely continue to increase to a value of 4 (perfect spheres) around the firn–ice transition. S/V characterizes the complexity of a structure and can serve as a proxy for tortuosity (Spaulding et al., 2011). Megadunes firn has consistently lower values of S/V of the pore space than WAIS Divide, indicating a less tortuous pore structure at a given depth. S/V at Megadunes is lower than at WAIS Divide at equivalent densities as well (Fig. 4). Anisotropy profiles of both sites overlap and are almost completely isotropic. Using these microstructure parameters to describe pore structure, Megadunes and WAIS Divide pores are isotropic and evolving from semi-cylindrical type shapes at mid-depths toward cylinders in the LIZ. Occasionally, within the LIZ, firn layers consist of a pore structure closer to spheres, and the frequency of these layers increases with depth. The difference

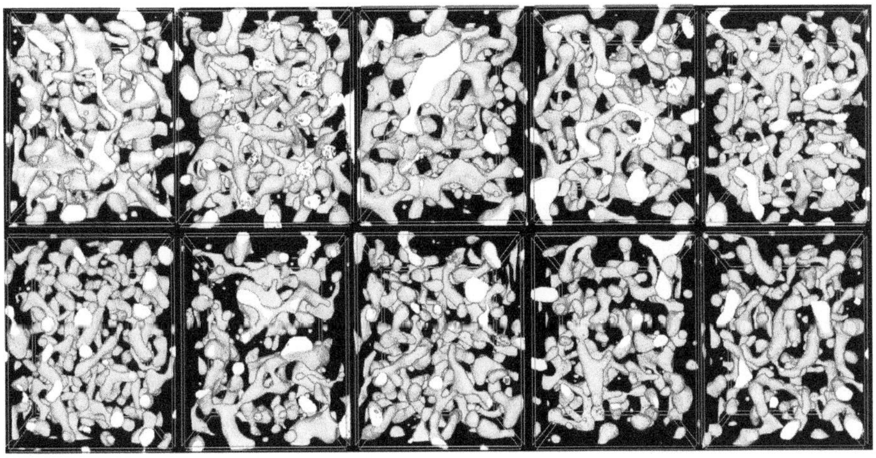

Fig. 2. Megadunes 3-D pore space reconstruction where white is the pore phase. Top row (from left to right): 56.25 m, 58.28 m, 60.43 m, 62.75 m, 64.98 m. Bottom row (from left to right): 66.79 m, 68.17 m, 70.30 m, 72.34 m, 73.27 m.

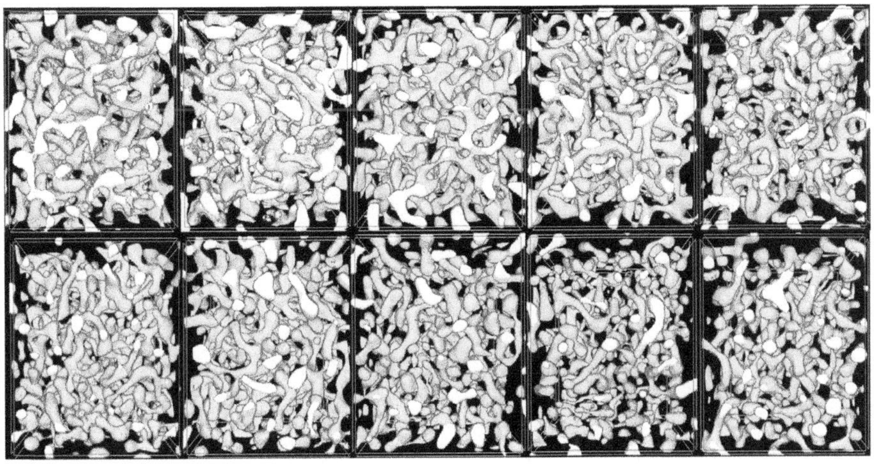

Fig. 3. WAIS Divide 3-D pore space reconstruction where white is the pore phase. Top row (from left to right): 56.36 m, 58.52 m, 60.45 m, 62.44 m, 64.61 m. Bottom row (from left to right): 66.56 m, 67.38 m, 70.56 m, 73.51 m, 74.50 m.

occurs in tortuosity of the pore matrix, indicated by S/V, where WAIS Divide is more tortuous than Megadunes at a given depth and at a given density.

4.3 Porosity

Porosity of polar firn decreases with depth as firn densifies and reflects the amount of space that is not filled by ice for a given volume at a given depth in the firn column. Total porosity of polar firn is the ratio of the air space to the ice matrix within a given volume. Total porosity is a direct reflection of density in porous media, and is generally calculated directly from the measured density of bulk samples as

$$P = 1 - \left(\frac{\rho}{\rho_{\text{ice}}}\right), \tag{4}$$

where P is the porosity of the sample, ρ is the density of the sample, and ρ_{ice} is the density of pure ice ($0.917\,\text{g}\,\text{cm}^{-3}$). While total porosity is derived from the density of a sample, open porosity reflects the amount of interconnected interstitial air space available for gas transport. All pores categorized as open are assumed to connect via the complex pore network to the atmosphere. In contrast, closed porosity indicates the amount of air volume completely surrounded by ice, where interstitial gas transport cannot occur. All porosity measurements in this study were derived from microcomputed tomography analysis on individual firn cubes.

Figure 5a shows that both total porosity and open porosity of Megadunes firn is consistently lower than WAIS Divide firn at a given depth near pore close-off. Despite having lower open porosity, the less complex and less tortuous pore matrix causes Megadunes firn to have larger and straighter air flow channels, which leads Megadunes firn to be more permeable

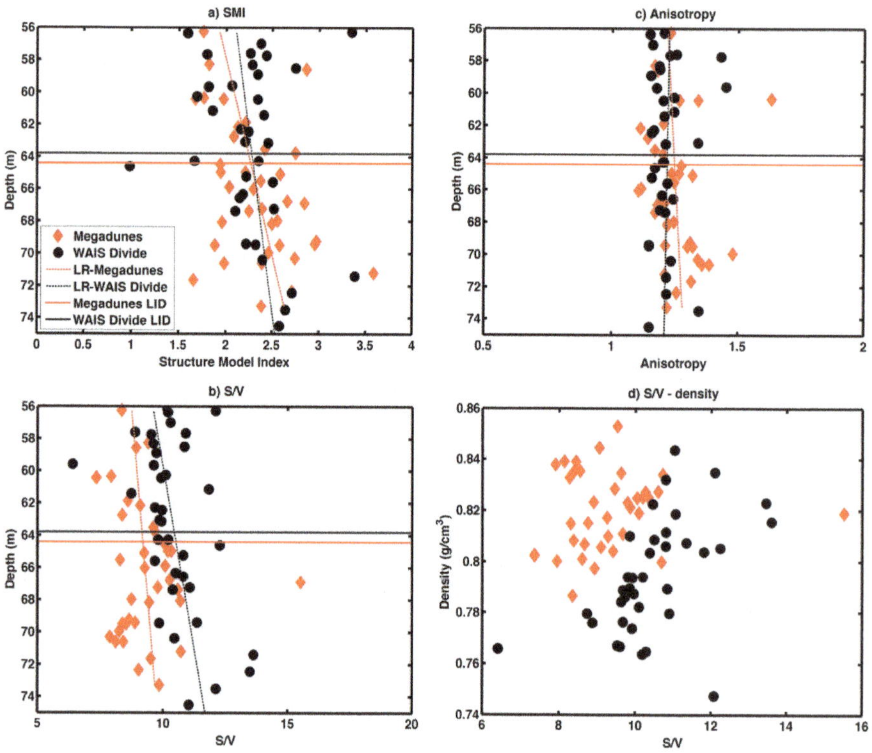

Fig. 4. (a) Structure model index, **(b)** surface-to-volume ratio, **(c)** anisotropy, **(d)** surface-to-volume ratio vs. density for deep firn at Megadunes (red diamonds) and WAIS Divide (black circles).

than firn at WAIS Divide, for a given density. A less tortuous pore structure allows for greater air movement than if the pore structure has many twists and turns. While total and open porosity gradually decrease with depth, closed porosity dramatically increases below the LID as seen in Fig. 5c. Closed porosity at Megadunes and WAIS Divide is less than 8 % above the LID (Fig. 5). Below the LID, the rate at which closed porosity increases accelerates.

To understand the acceleration in closed porosity with no change seen in the rate of decrease in total porosity, the number and size of pores were examined. Figure 6 shows a decrease in pore size below the LID with a large increase in the total number of pores below the LID. As pore close-off progresses, WAIS Divide has more variability in the total number of pores than Megadunes within the LIZ (Fig. 6c). This is likely due to the small-scale layering of firn at high-accumulation sites in comparison to firn at low-accumulation sites (e.g., Landais et al., 2006). For pores less than 10 mm^3, the pores at WAIS Divide tend to be smaller than those at Megadunes (Fig. 6b). For both sites, the abrupt increase in closed porosity at the LID corresponds with dissection of larger open pores in multiple locations to form many smaller closed or almost closed pores. The dissection of large open pores begins at an open porosity of ~ 8 % at Megadunes and ~ 11 % at WAIS Divide.

Another method used by Lomonaco et al. (2011) to observe the pore close-off process is the closed pore fraction (CPF), defined as the number of pores divided by the total pore volume of a given sample. Plotting CPF with depth for Megadunes and WAIS Divide (Fig. 7) shows a steady CPF of ~ 1, slightly lower for WAIS Divide, until a depth of 63 m. Below 63 m, the CPF increases with depth at a constant slope through the LIZ. While the magnitude of CPF is slightly lower in the present study, the same trend in the rate of CPF increase was observed through the LIZ at Summit, Greenland, by Lomonaco et al. (2011). The increase in CPF in the LIZ of polar firn is the result of large open pores that are abruptly dissected by the ice matrix initiating the LIZ. Because firn is a layered material, variability in CPF is due to microstructural differences between fine-grain and coarse-grain firn. Thus the microstructural differences in layering between coarse-grained and fine-grained firn dictate the thickness and size of the lock-in zone.

While the Megadunes site is within the cold desert of the East Antarctic plateau, the presence of the megadune features make it different from Vostok or Dome C, for example, which are also very cold, low-accumulation sites in East Antarctica. The presence of the megadune features make the Megadunes site a natural laboratory for the impacts of accumulation rate on firnification processes. At the Megadunes site, the enduring evidence of the relic accumulation provides a contrast to

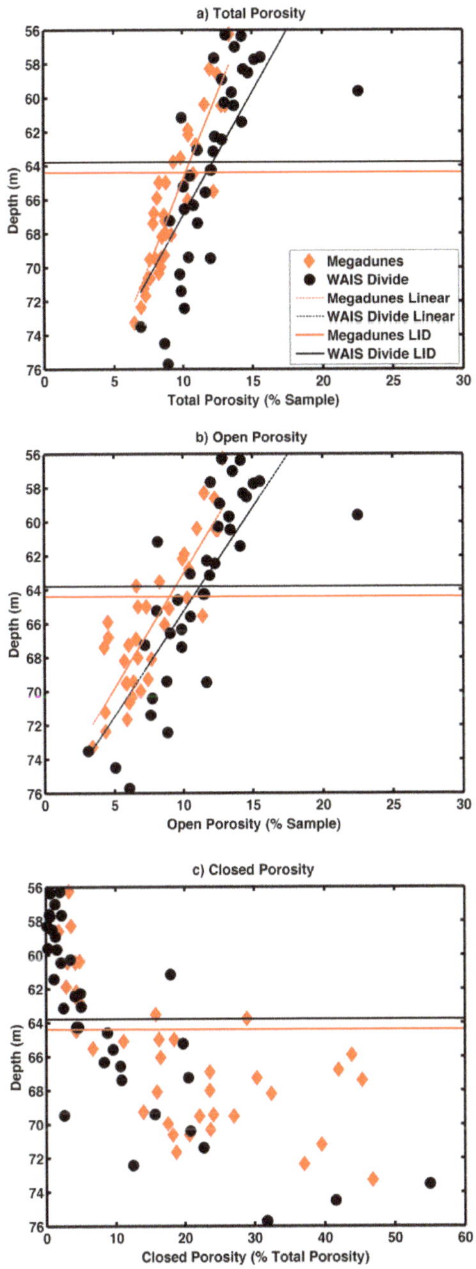

Fig. 5. **(a)** Total porosity % of sample, **(b)** open porosity % of sample and **(c)** closed porosity % of total porosity for deep firn at Megadunes (red diamonds) and WAIS Divide (black circles).

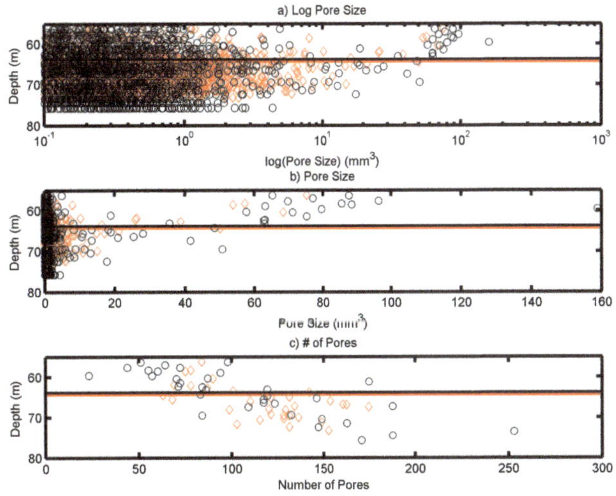

Fig. 6. **(a)** Profiles of log pore size, **(b)** pore size and **(c)** number of pores for pores greater than 0.001 mm^3 in deep firn, 55 m to 80 m, for Megadunes (red diamonds) and WAIS Divide (black circles).

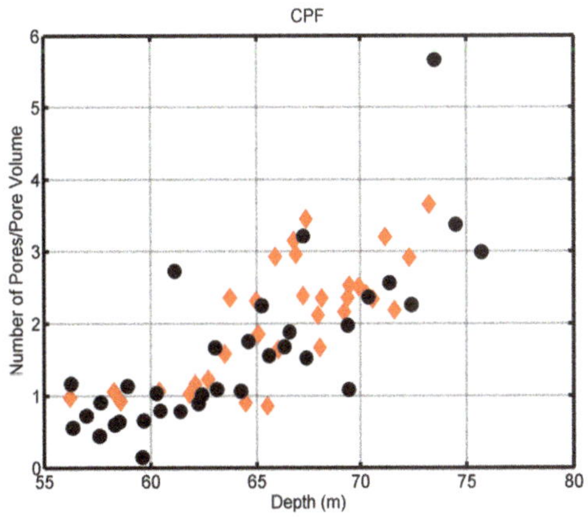

Fig. 7. Closed pore fraction (CPF) for Megadunes (red diamonds) and WAIS Divide (black circles).

the rest of the firn, which had experienced accumulation hiatus, deep in the firn column. The propagation of the slightly higher accumulation rate bands of firn through the column has essentially created a firn column that is shallower than it would have been without the accumulation effect. Hence the pore close-off depth at Megadunes is shallower than at either Vostok or Dome C. Indeed, the presence of the accumulation band at Megadunes created a lock-in zone that very likely would not have existed at that site without that accumulation effect.

5 Discussion

5.1 Application to densification models and LIZ characteristics

Pore close-off is traditionally defined in layered firn as the process in which individual layers of firn become impermeable and proceeds until all layers are impermeable or closed off. In the current literature on pore close-off, density controls the depth at which pore close-off occurs in an individual layer, and denser layers are thought to close off shallower than lower-density layers (Martinerie et al., 1992). Combining density-driven pore close-off and noting the density cross-over observed in polar firn (Hörhold et al., 2011), the first firn layers in deep firn to reach pore close-off should be the higher-density, coarser-grained layers. Hörhold et al. (2012) postulate that the firn layers with the highest impurity concentration should reach the pore close-off density first. In contrast, our observations of close-off including microstructure and density between Megadunes and WAIS Divide suggest that a combination of microstructure and grain size, rather than density, is the main driving force for an impermeable pore structure and pore close-off. Below 45 m, Megadunes firn is consistently denser than WAIS Divide firn at a given depth, yet the permeability profiles above the LID are almost identical. In addition, the LID for both sites is also very similar: 64.4 m at Megadunes and 63.8 m at WAIS Divide. To explain the lack of a direct relationship between density (or porosity) and permeability between Megadunes and WAIS Divide, the tortuosity of the pore structure must be taken into account. Megadunes has larger grains than WAIS Divide at a given density, leading to a less tortuous pore structure and causing higher permeabilities at a given density in Megadunes firn than those observed at WAIS Divide.

It should be noted that previous estimates based on density or porosity, such as those developed by Martinerie et al. (1992) of mean pore close-off density, are close to the peak of a Gaussian type density distribution for all impermeable layers within the lock-in zone at each site (Fig. 8). Megadunes, a cold site with an average temperature of −49 °C, has a predicted mean pore close-off density of 0.830 g cm^{-3}, which is higher than 0.821 g cm^{-3} at WAIS Divide, a warmer site with an average temperature of −31 °C . Our results show that the density range of the distribution and thickness of the LIZ are dependent upon the microstructure variability between firn layers even though the firn layers are not annual layers but were deposited over much longer intervals of time (Albert et al., 2004). This is consistent with Landais et al. (2006), who postulated that low-accumulation sites will have smaller lock-in zones from an homogenous firn column and lack of annual layering.

Our comparison between two sites of significantly different local climates reveals that microstructure and layering, which were originally established at the surface due to the impact of accumulation rate, play a significant role in dictating deep-firn permeability and pore close-off. Continued work to identify the signatures deep in the firn of original accumulation rate impacts at additional sites should further strengthen our finding. As commonly seen in polar sites, WAIS Divide firn exhibits a density cross-over in which initially higher-density layers consisting of finer-grained firn are less dense than corresponding initially lower-density layers consisting of coarser-grained firn. While not highly pronounced on the 5–10 cm scale in Fig. 9, the density cross-over at WAIS Divide has been noted in high-resolution density measurements on firn core WDC06A done by Breton (2011) with a minima in density variability occurring around 30 m. In Fig. 9b, finer-grained layers become impermeable

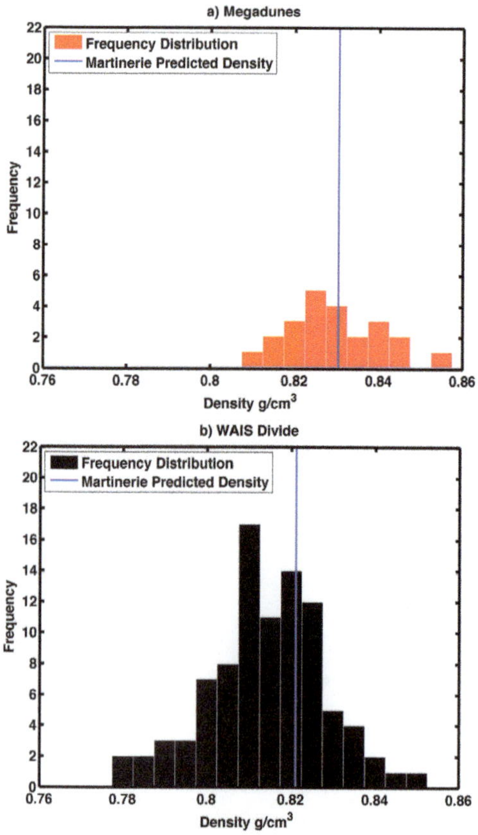

Fig. 8. Density distribution of impermeable firn samples from the LID to the end of the LIZ where all samples are impermeable at **(a)** Megadunes and **(b)** WAIS Divide.

shallower than coarser-grained layers, and a coarser-grained layer is the last permeable layer of the LIZ. Despite having a lower density and higher porosity at depth at WAIS Divide, finer-grained firn reaches pore close-off first there. Accumulation and near-surface metamorphism result in either fine-grained or coarse-grained layers and remain (independent of temperature) as identifiable features of the firn layering as they propagate down the column, leading to shallower pore close-off in fine-grain layers at WAIS Divide. In addition, the propagation of fine-grained accumulation results in shallower pore close-off at Megadunes than at other sites high on the East Antarctic plateau. In contrast, Megadunes does not exhibit a density cross-over, and initially dense fine-grained firn at surface accumulation sites on the upwind dune face remains denser than coarse-grained firn from surface hiatus sites on the leeward dune face through pore close-off (Gregory, 2013). These findings show a direct link between surface deposition microstructure and firn microstructure within the lock-in zone. It has been suggested by Hörhold et al. (2012) that impurity content is the controlling parameter on firn densification as opposed to surface microstructural characteristics. We suggest that the impurity content of the firn is simply a co-located aspect of the firn

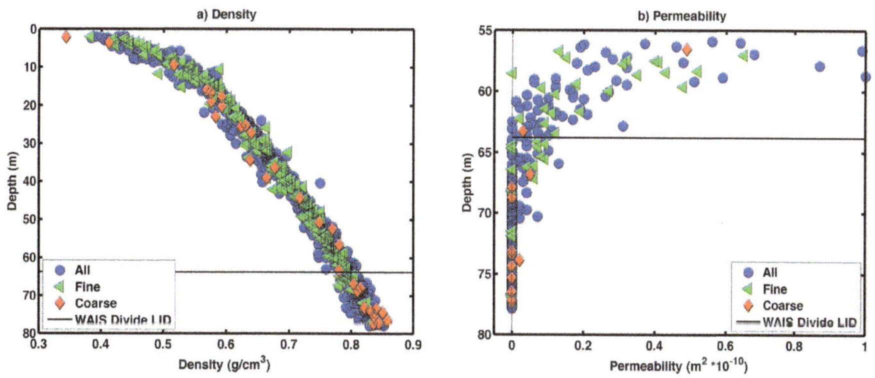

Fig. 9. (a) Density cross-over of fine- and coarse-gain firn at WAIS Divide and **(b)** deep firn permeability profile at WAIS Divide both show fine firn (green triangles), coarse firn (red diamonds), and intermediate firn (blue circles).

due to the depositional event, and that the impurities are not the cause of the densification rate. The fine-grained, dense firn originating from a past accumulation site at Megadunes becomes the first impermeable layers initiating the start of the LIZ. The ability of fine-grained firn to become impermeable before coarse-grained firn in this study at both WAIS Divide and Megadunes as well as in previous studies (Freitag et al., 2004), regardless of whether or not it was the predominately dense layer at depth, emphasizes the importance of grain size and pore structure when predicting gas transport and pore close-off in polar firn. A recent study by Horhold et al. (2011) correlated an increase in accumulation rate to an increase in density variability in deep firn leading to longer LIZs. While we agree that density variability due to layering does indeed increase for high-accumulation sites, the density cross-over seen in most polar firn sites leads to an inverse relationship of density and microstructure in deep firn where fine-grained layers are less dense than coarse-grained layers. Because they are both based on the layering and are originally due to the nature of accumulation rate when that firn was on the surface, microstructure will likely have high layer-to-layer variability in places where layering also makes the density variable. At a given site, the microstructure variability between fine-grained and coarse-grained layers, likely of the same magnitude of the density variability observed by Horhold et al. (2011), controls the thickness of the LIZ. For a given polar site, fine-grained layers will reach pore close-off at shallower depths than coarse-grained layers at that site. To capture this behavior, firn densification models should include two pore close-off densities, one corresponding to the finest-grained layers that will predict the LID and one corresponding to the coarsest-grained layers that will predict the COD. In this way, the LID, COD, and LIZ thickness estimates can be improved, based on the accumulation rate of a site and the resulting microstructure variability. The very distinct layering contrast between the relic surface accumulation and hiatus zones at Megadunes provide a stark indicator of the importance of accumulation rate on the microstructure-dominated lock-in and close-off depths. While Megadunes is unique due to the presence of buried antidunes, very low accumulation will exhibit less microstructure variability in deep firn in conjunction with less density variability observed by Horhold et al. (2011) for low-accumulation sites. High-accumulation sites should, on the other hand, exhibit a higher degree of microstructure variability resulting in a greater depth difference between fine-grain close-off (LID) and coarse-grain close-off (COD). Quantifying the magnitude of microstructure variability within the lock-in zone for a large number of polar sites would be a first step in developing a firn densification model that incorporates two pore close-off depths, one for the finest-grain layers of a site and one for the coarsest-grain layers of a site. A relationship could be developed between the magnitude of grain size variability, lock-in zone length, accumulation rate, and temperature at a given polar site. We will pursue these ideas in future work.

For a physical description of why fine-grain firn reaches pore close-off first, we hypothesize that the grain size of the ice and neck size of the pore structure in firn leads to a threshold in open porosity corresponding to pore close-off in a single layer of firn. Freitag et al. (2004) found that coarse-grained layers densify at a quicker rate than fine-grained layers but that the pore size of both layers decreased at the same rate. Our findings show that whichever layer has smaller pores at the surface (the fine-grained layer) will close off first. Rick and Albert (2004) and Courville et al. (2007) both found that buried layers retain evidence of their character when they were in the near-surface firn in proportion to the amount of time that crystals remain in the near surface due to the local accumulation rate. The larger grains at Megadunes and subsequent larger pore necks enable the firn at Megadunes to reach a higher density and lower open porosity of $\sim 8\,\%$, beneath which the open pore space is quickly dissected into many smaller closed pores. Due to the unique depositional process at Megadunes and the presence of hiatus sites the $\sim 8\,\%$ bound on open porosity is likely

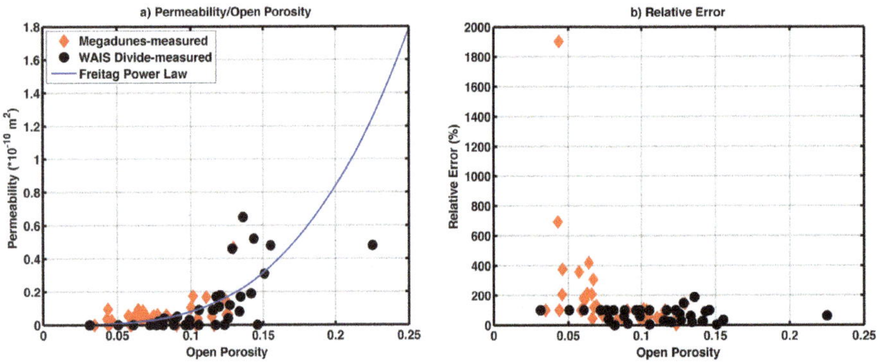

Fig. 10. **(a)** Measured and predicted (blue line; Freitag et al., 2002) permeability values and **(b)** relative error for Megadunes (red diamonds) and WAIS Divide (black circles).

lower than a more typical low-accumulation site. At WAIS Divide, the open porosity threshold occurs higher at ~ 11 % open porosity, due to the smaller ice grains and smaller necks in the pore structure. For a given accumulation rate, fine-grained firn will have the highest open porosity at pore close-off, coarse-grained firn layers will have the lowest open porosity at pore close-off and intermediate grain size layers should have open porosity thresholds for close-off in between the two extremes. Expanding this relationship across many accumulation rates, high-accumulation rates typically lead to generally smaller grains and should have higher open porosity (~ > 10 %) and lower density at pore close-off, while very low accumulation sites with larger grains should have lower open porosity (~ < 10 %) and higher density at pore close-off. The exact physics controlling the sudden pore dissection seen at the open porosity thresholds we observed will be explored further in future studies.

5.2 Application to firn air models

Increasing the understanding of the impact of microstructure on gas diffusivity in deep firn through the LIZ should decrease the necessity for inverse modeling based on present-day firn air measurements within firn gas transport models. Initial gas diffusivity profiles within polar firn are typically modeled off the Schwander et al. (1988) parameterization of firn tortuosity as a function of porosity, despite recognition by Fabre et al. (2000) that tortuosity profiles are site dependent. An attempt to model permeability and diffusivity on the millimeter-to-centimeter scale as a function of open porosity was done by Freitag et al. (2002) on 3-D reconstructed firn cubes. The permeability constant k was fitted to open porosity following the power law commonly used in porous media with exponents experimentally determined:

$$k = 10^{-7.7} n_{\mathrm{op}}^{3.4}, \tag{5}$$

where n_{op} is the open porosity of the sample (Freitag et al., 2002). The power law fit allowed the firn to be permeable below 0.12 open porosity, the threshold for permeable firn in previous studies done by Fabre et al. (2000) and Schwander et al. (1988). Adolph and Albert (2013) show that the Freitag et al. (2002) relationship is site-specific.

Plotting permeability data for Megadunes and WAIS Divide in Fig. 10 shows high scatter around the power law fit from measured permeability for both sites, though the general curve of the power law can be seen. The power law function fails most at low open porosity values (5–10 %) at Megadunes, where the firn is 200–1900 % more permeable than predicted with the power law function. At low open porosities, the Megadunes firn microstructure lends itself to large pores with low complexity resulting in more permeable firn than the Freitag et al. (2002) model could predict. Comparison of the Megadunes data with the Freitag et al. (2002) Greenland study shows that the relationships developed for Greenland do not apply to the Megadunes site in East Antarctica. Pore structure, which is dependent on grain size and accumulation rate, influences gas transport in the diffusive column and also controls pore close-off. The low-accumulation rate, large grain size, and larger less tortuous pore structure at Megadunes enables the firn to be much more permeable than predicted by Freitag et al. (2002). As WAIS Divide has a local climate very similar to Greenland, the error between modeled and measured permeability is not as large. Our results show that at both WAIS Divide and Megadunes, firn layers at a given open porosity display a range of permeabilities due to the open pore structure and its dependence on grain size variability. This span in permeability values at a constant open porosity would likely be seen in other polar firn profiles as well.

In deep firn it has been shown that diffusivity and permeability are linearly related (Adolph and Albert, 2013), and thus the connection between permeability and grain size/pore structure can be extrapolated to diffusivity. Firn air models that use density profiles derived from firn densification models to determine the open porosity which is then used to calculate the effective diffusivity profile of a specific gas (Buizert et al., 2012b) will fall short of capturing the true nature of gas transport in deep firn. A given open porosity at a high-accumulation site will lead to lower observed gas diffusivity than the diffusivity observed at a low-accumulation site. The microstructure must be accounted for to improve physics-based models of diffusivity in polar firn. The dependence of microstructure on post-depositional effects of accumulation rate and near-surface residence time observed by Albert et al. (2004), Rick and Albert (2004), Courville et al. (2007), and Fujita et al. (2009), along with an increase in gas transport due to an increase in grain size shown in this study, will in the future be used to create site-specific effective gas diffusivity profiles. A gas diffusivity profile derived from local climate parameters, specifically accumulation rate, which captures gas transport dependence on pore structure, could provide constraints for forward firn air modeling. If the physics of forward firn air modeling can be accurately captured through climate parameters, it should reduce the need for inverse modeling and enhance the ability of forward modeling to describe past firn columns where inverse modeling is not plausible.

6 Conclusions

Microstructure, specifically open pore structure, and not density, is the main parameter controlling the permeability and gas transport in polar firn. Pore structure is the direct result of microstructure resulting from the initial depositional processes and post-depositional metamorphism controlled by accumulation rate when the firn was near the surface of the ice sheet. Pore close-off will occur when the pore structure restricts air movement, causing the firn layer to become impermeable independent of the density of the layer. At typical firn sites, of both high and low accumulation, where a density inversion of the firn column exists, the finer, less dense firn layers will reach pore close-off shallower than coarse-grained, denser layers below the density inversion. Therefore, the shallowest fine-grained layers to close at a polar site with a density cross-over will contain more air than the coarse-grain layers, which close off deeper and at higher densities. At unique sites such as Megadunes, where no inversion is seen, pore close-off will occur shallower in the denser, finer-grained firn layers, causing the first layers to close and contain less air than the less dense, coarser-grained layers that reach pore close-off deeper in the firn column. Therefore, an important aspect for determining which firn layers, coarse grain or fine grain, have the higher air content is the presence or absence of a density inversion. Because both the density variability and the microstructure variability are features of the firn layering, microstructure will have high variability at sites where high density variability also exists. Thus the microstructure has high variability in the deep firn of high-accumulation sites and low variability in deep firn of low-accumulation sites where megadune features are absent.

In consideration of pore close-off, a threshold is seen in open porosity below which pore segregation increases dramatically and permeability reduces to zero. The exact threshold varies as a result of accumulation rate impacts on near-surface grain size propagating down the firn column into deep firn. At porosities lower than the threshold ($\geq 10\,\%$ for high-accumulation sites similar to WAIS Divide, $\sim 21\,\text{cm weq a}^{-1}$, and $\leq 10\,\%$ for low-accumulation sites similar to Megadunes, $4\,\text{cm weq a}^{-1}$ or less), closed porosity increases rapidly, the total number of pores increases at a greater rate, and the size of pores dramatically decreases, resulting in many small pores opposed to a few large open pores.

These results could be incorporated into next-generation firn air models. First, two pore close-off densities should be incorporated, one for the finest-grained layers at an individual site which would define the top of the LIZ (the LID), and one for coarsest-grained layers at that same site which would define the bottom of the LIZ (the COD). Parameterizing pore close-off with both grain size and density, along with the layered nature of firn in firn densification models, would improve their ability to accurately predict the LID and LIZ thickness. Future studies examining relationships between surface temperature, accumulation rate and density of both coarse-grain and fine-grain layers at the surface, and when each layer type becomes impermeable at a multitude of polar sites, are necessary to create a parameterization based on both grain size and density.

For a given density, coarse-grain firn is more permeable than fine-grain firn due to a less complex and less tortuous pore structure. Next-generation firn air models should include an effective open porosity that accounts for differences in pore structure between high- and low-accumulation-rate sites. An effective open porosity that reflects the pore structure of the site, not simply how much pore space is present, and that correctly captures permeability in deep firn may provide better modeling to contribute to improved interpretation of gas records in ice cores. An increase in the physical understanding of pore close-off in polar firn and its dependence upon local climate enables next-generation firn air modeling under conditions with no present-day analogue, thus constraining the gas age–ice age difference. In ongoing work we will examine data from sites with a variety of local climates in order to further develop the dependence of gas transport and pore close-off on pore structure, grain size, and climatic condition.

Acknowledgements. This research was funded by NSF grants ANT-0944078 and IGERT-0801490. The authors thank Jeff Severinghaus, Zoe Courville and colleagues from the WAIS Divide program for useful discussions, and we thank undergraduate students Emily Harwell, Noah Pfister, and Melina Bartels for assistance with the cold room measurements.

Edited by: M. van den Broeke

References

Albert, M. R., Schultz, E. F., and Perron, F. E. J.: Snow and firn permeability at siple dome, Antarctica, Ann. Glaciol., 31, 353–356, 2000.

Adolph, A. C. and Albert, M. R.: The physical basis for gas transport through polar firn: a case study at Summit, Greenland, The Cryosphere Discuss., 7, 2455–2487, doi:10.5194/tcd-7-2455-2013, 2013.

Albert, M. R., Shuman, C., Courville, Z., Bauer, R., Fahnestock, M., and Scambos,T.: Extreme firn metamorphism: Impact of decades of vapor transport on near-surface firn at a low-accumulation glazed site on the East Antarctic Plateau, Ann. Glaciol., 39, 73–78, 2004.

Banta, J., McConnell, J., Frey, M., Bales, R., and Taylor, K.: Spatial and temporal variability in snow accumulation at the West Antarctic Ice Sheet Divide over recent centuries, J. Geophys. Res., 113, D23102, doi:10.1029/2008JD010235, 2008.

Battle, M. O., Severinghaus, J. P., Sofen, E. D., Plotkin, D., Orsi, A. J., Aydin, M., Montzka, S. A., Sowers, T., and Tans, P. P.: Controls on the movement and composition of firn air at the West Antarctic Ice Sheet Divide, Atmos. Chem. Phys., 11, 11007–11021, doi:10.5194/acp-11-11007-2011, 2011.

Breton, D.: Photonic non-destructive measurement methods for investigating the evolution of polar firn and ice, PhD Thesis, University of Maine, 2011.

Brook, E. J.: Leads and lags at the end of the last Ice Age, Science, 339, 1042–1043, 2013.

Buizert, C., Martinerie, P., Petrenko, V. V., Severinghaus, J. P., Trudinger, C. M., Witrant, E., Rosen, J. L., Orsi, A. J., Rubino, M., Etheridge, D. M., Steele, L. P., Hogan, C., Laube, J. C., Sturges, W. T., Levchenko, V. A., Smith, A. M., Levin, I., Conway, T. J., Dlugokencky, E. J., Lang, P. M., Kawamura, K., Jenk, T. M., White, J. W. C., Sowers, T., Schwander, J., and Blunier, T.: Gas transport in firn: multiple-tracer characterisation and model intercomparison for NEEM, Northern Greenland, Atmos. Chem. Phys., 12, 4259–4277, doi:10.5194/acp-12-4259-2012, 2012a.

Buizert, C., Martinerie, P., Petrenko, V. V., Severinghaus, J. P., Trudinger, C. M., Witrant, E., Rosen, J. L., Orsi, A. J., Rubino, M., Etheridge, D. M., Steele, L. P., Hogan, C., Laube, J. C., Sturges, W. T., Levchenko, V. A., Smith, A. M., Levin, I., Conway, T. J., Dlugokencky, E. J., Lang, P. M., Kawamura, K., Jenk, T. M., White, J. W. C., Sowers, T., Schwander, J., and Blunier, T.: Gas transport in firn: multiple-tracer characterisation and model intercomparison for NEEM, Northern Greenland, Atmos. Chem. Phys., 12, 4259–4277, doi:10.5194/acp-12-4259-2012, 2012b.

Clark, I. D., Henderson, L., Chappellaz, J., Fisher, D., Koerner, R., Worthy, D. E. J., Kotzer, T., Norman, A.-L., and Barnola, J.-M.: CO_2 isotopes as tracers of firn air diffusion and age in an arctic ice cap with summer melting, devon island, canada, J. Geophys. Res., 112, D01301, doi:10.1029/2006JD007471, 2007.

Coleou, C., Lesaffre, B., Brzoska, J. B., Ludwig, W., and Boller, E.: Three-dimensional snow images by X-ray microtomography, Ann. Glaciol., 32, 75–81, 2001.

Courville, Z. R., Albert, M. R., Fahnestock, M. A., Cathles IV, L. M., and Shuman, C. A.: Impacts of an accumulation hiatus on the physical properties of firn at a low-accumulation polar site, J. Geophys. Res., 112, F02030, doi:10.1029/2005JF000429, 2007.

Courville, Z., Hörhold, M., Hopkins, M., and Albert, M.: Lattice-boltzmann modeling of the air permeability of polar firn, J. Geophys. Res., 115, F04032, doi:10.1029/2009JF001549, 2010.

Entig, I. G.: Statistics of firn closure: a simulation study, J. Glaciol., 39, 133–142, 1993.

Etheridge, D. M., Steele, L. P., Langenfelds, R. L., Francey, R. J., Barnola, J. M., and Morgan, V. I.: Natural and anthropogenic changes in atmospheric co2 over the last 1000 years from air in antarctic ice and firn, J. Geophys. Res., 101, 4115–4128, 1996.

Fabre, A., Barnola, J. M., Arnaud, L., and Chappellaz, J.: Determination of gas diffusivity in polar firn: Comparison between experimental measurements and inverse modeling, Geophys. Res. Lett., 27, 557–560, 2000.

Freitag, J.: The layering of polar firn and its linkage to air trapping, Workshop on the mircostructure and properties of firn, 10–1 March 2008, Dartmouth College, Hanover, USA, 2008.

Freitag, J., Dorbrindt, U., and Kipfstuhl, J.: A new method for predicting transport properties of polar firn with respect to gases on the pore-space scale, Ann. Glaciol., 35, 538–544, doi:10.3189/172756402781816582, 2002.

Freitag, J., Wilhelms, F., and Kipfstuhl, S.: Microstructure-dependent densification of polar firn derived from X-ray microtomography, J. Glaciol., 50, 243–250, 2004.

Fujita, S., Okuyama, J., Hori, A., and Hondoh, T.: Metamorphism of stratified firn at dome fuji, antarctica: A mechanism for local insolation modulation of gas transport conditions during bubble close off, J. Geophys. Res., 114, F03023, doi:10.1029/2008JF001143, 2009.

Gregory, S. A.: The impact of microstructure and physical properties on pore close-off at WAIS Divide and Megadunes, MS thesis, Dartmouth College, USA, 2013.

Goujon, C., Barnola, J. M., and Ritz, C.: Modeling the densification of polar firn including heat diffusion: Application to close-off characteristics and gas isotopic fractionation for Antarctica and Greenland sites, J. Geophys. Res.-Atmos, 108, 4792, doi:10.1029/2002JD003319, 2003.

Hildebrand, T. and Rüegsegger, P.: Quantification of bone microarchitecture with the structure model index, Comput. Meth. Biomech. Biomed. Eng., 1, 15–23, 1997.

Hörhold, M. W., Kipfstuhl, S., Wilhelms, F., Freitag, J., and Frenzel, A.: The densification of layered polar firn, J. Geophys. Res.-Earth Surf., 116, F01001, doi:10.1029/2009JF001630, 2011.

Hörhold, M. W., Laepple, T., Freitag, J., Bigler, M., Fischer, H., and Kipfstuhl, S.: On the impact of impurities on the densification of polar firn, Earth Planet Sc. Lett., 325–326, 93–99, 2012.

Kawamura, K., Severinghaus, J. P., Ishidoya, S., Sugawara, S., Hashida, G., Motoyama, H., Fujii, Y., Aoki, S., and Nakazawa, T.: Convective mixing of air in firn at four polar sites, Earth Planet Sc. Lett., 244, 672–682, 2006.

Landais, A., Barnola, J. M., Kawamura, K., Caillon, N., Delmotte, M., Van Ommen, T., Dreyfus, G., Jouzel, J., Masson-Delmotte, V., Minster, B., Freitag, J., Leuenberger, M., Schwander, J., Huber, C., Etheridge, D., and Morgan, V.: Firn-air delta n-15 in modern polar sites and glacial-interglacial ice: a model-data mismatch during glacial periods in antarctica?, Quaternary Sci. Rev., 25, 49–62, 2006.

Lomonaco, R., Albert, M., and Baker, I.: Microstructural evolution of fine-grained layers through the firn column at Summit, Greenland, J. Glaciol., 57, 755–762, 2011.

Martinerie, P., Raynaud, D., Etheridge, D. M., Barnola, J. M., and Mazaudier, D.: Physical and climatic parameters which influence the air content in polar ice, Earth Planet Sc. Lett., 112, 1–13, 1992.

Rick, U. and Albert, M. R.: Firn microstructure impacts on air permeability, ERDC-CRREL Tech. Rep. TR-96, 79 pp., Cold Reg. Res. and Eng. Lab., Hanover, NH, 2004.

Rommelaere, V., Arnaud, L., and Barnola, J. M.: Reconstructing recent atmospheric trace gas concentrations from polar firn and bubbly ice data by inverse methods, J. Geophys. Res.-Atmos., 102, 30069–30083, 1997.

Schneebeli, M. and Sokratov, S. A.: Tomography of temperature gradient metamorphism of snow and associated changes in heat conductivity, Hydrol. Process., 18, 3655–3665, 2004.

Schwander, J. and Stauffer, B.: Age difference between polar ice and the air trapped in its bubbles, Nature, 311, 45–47, 1984.

Schwander, J., Stauffer, B., and Sigg, A.: Air mixing in firn and the age of the air at pore close-off, Ann. Glaciol., 10, 141–145, 1988.

Schwander, J., Barnola, J. M., Andrie, C., Leuenberger, M., Ludin, A., Raynaud, D., and Stauffer, B.: The age of the air in the firn and the ice at summit, greenland, J. Geophys. Res.-Atmos., 98, 2831–2838, 1993.

Severinghaus, J. P., Grachev, A., and Battle, M.: Thermal fractionation of air in polar firn by seasonal temperature gradients, Geochem. Geophy. Geosy., 2, 1048, doi:10.1029/2000GC000146, 2001.

Severinghaus, J. P., Albert, M. R., Courville, Z. R., Fahnestock, M. A., Kawamura, K., Montzka, S. A., Muhle, J., Scambos, T. A., Shields, E., Shuman, C. A., Suwa, M., Tans, P., and Weiss, R. F.: Deep air convection in the firn at a zero-accumulation site, central antarctica, Earth Planet Sc. Lett., 293, 359–367, 2010.

Sowers, T., Bender, M., Raynaud, D., and Korotkevich, Y. S.: Delta-n-15 of n2 in air trapped in polar ice – a tracer of gas-transport in the firn and a possible constraint on ice age-gas age-differences, J. Geophys. Res.-Atmos., 97, 15683–15697, 1992.

Spaulding, N. E., Meese, D. A., and Baker, I.: Advanced microstructural characterization of four East Antarctic firn/ice cores, J. Glaciol., 57, 796–810, 2011.

Taina, I. A., Heck, R. J., and Elliot, T. R.: Application of X-ray computed tomography to soil science: a literature review, Can. J. Soil Sci., 88, 1–20, 2008.

Trudinger, C. M.: The carbon cycle over the last 1000 years inferred from inversion of ice core data, PhD thesis, Monash University, Australia, 2001.

Trudinger, C. M., Enting, I. G., Etheridge, D. M., Francey, R. J., Levchenko, V. A., Steele, L. P., Raynaud, D., and Arnaud, L.: Modeling air movement and bubble trapping in firn, J. Geophys. Res.-Atmos., 102, 6747–6763, 1997.

Trudinger, C. M., Etheridge, D. M., Rayner, P. J., Enting, I. G., Sturrock, G. A., and Langenfelds, R. L.: Reconstructing atmospheric histories from measurements of air composition in firn. J. Geophys. Res.-Atmos., 107, 4780, doi:10.1029/2002JD002545, 2002.

Witrant, E., Martinerie, P., Hogan, C., Laube, J. C., Kawamura, K., Capron, E., Montzka, S. A., Dlugokencky, E. J., Etheridge, D., Blunier, T., and Sturges, W. T.: A new multi-gas constrained model of trace gas non-homogeneous transport in firn: evaluation and behaviour at eleven polar sites, Atmos. Chem. Phys., 12, 11465–11483, doi:10.5194/acp-12-11465-2012, 2012.

6

Measurements of light-absorbing particles on the glaciers in the Cordillera Blanca, Peru

C. G. Schmitt[1,2], J. D. All[3,2], J. P. Schwarz[4,5], W. P. Arnott[6], R. J. Cole[7,2,*], E. Lapham[2], and A. Celestian[3]

[1]National Center for Atmospheric Research, Boulder, Colorado, USA
[2]American Climber Science Program, Eldora, Colorado, USA
[3]Department of Geography and Geology, Western Kentucky University, Bowling Green, Kentucky, USA
[4]Chemical Sciences Division, Earth System Research Laboratory, National Oceanic and Atmospheric Administration, Boulder, Colorado, USA
[5]Cooperative Institute for Research in Environmental Sciences, University of Colorado, Boulder, Colorado, USA
[6]Department of Physics, University of Nevada, Reno, Nevada, USA
[7]Institute of Arctic and Alpine Research, University of Colorado, Boulder, Colorado, USA
[*]now at: Department of Natural Resources and Environmental Management, University of Hawaii, Manoa, Hawaii, USA

Correspondence to: C. G. Schmitt (schmittc@ucar.edu)

Abstract. Glaciers in the tropical Andes have been rapidly losing mass since the 1970s. In addition to the documented increase in temperature, increases in light-absorbing particles deposited on glaciers could be contributing to the observed glacier loss. Here we report on measurements of light-absorbing particles sampled from glaciers during three surveys in the Cordillera Blanca Mountains in Peru. During three research expeditions in the dry seasons (May–August) of 2011, 2012 and 2013, 240 snow samples were collected from 15 mountain peaks over altitudes ranging from 4800 to nearly 6800 m. Several mountains were sampled each of the 3 years and some mountains were sampled multiple times during the same year. Collected snow samples were melted and filtered in the field then later analyzed using the Light Absorption Heating Method (LAHM), a new technique that measures the ability of particles on filters to absorb visible light. LAHM results have been calibrated using filters with known amounts of fullerene soot, a common industrial surrogate for black carbon (BC). As sample filters often contain dust in addition to BC, results are presented in terms of effective black carbon (eBC). During the 2013 survey, snow samples were collected and kept frozen for analysis with a Single Particle Soot Photometer (SP2). Calculated eBC mass from the LAHM analysis and the SP2 refractory black carbon (rBC) results were well correlated ($r^2 = 0.92$). These results indicate that a substantial portion of the light-absorbing particles in the more polluted regions were likely BC. The 3 years of data show that glaciers in the Cordillera Blanca Mountains close to human population centers have substantially higher levels of eBC (as high as 70 ng g^{-1}) than remote glaciers (as low as 2.0 ng g^{-1} eBC), indicating that population centers can influence local glaciers by sourcing BC.

1 Introduction

A substantial portion of the population of South America lives on the west coast, where the water supply is supported by the runoff from glacier melt (Barnett et al., 2005). When a region with glaciers is experiencing warming, the water availability for the region will change even if precipitation patterns remain constant. Initially, runoff from glaciers will increase as melting increases, but eventually the glacier area will be sufficiently reduced so that the increased melting rate will be offset by the decreased glacier area available for melting and increased water loss due to evaporation, sublimation, or absorption into ground surfaces. Thus, over time the typical water discharge increases as warming starts, reaches a peak, then reduces until a new steady state is reached, presumably after the climate has stopped changing or after the glaciers have completely melted (Baraer et al., 2012). Post warming water flow rates are expected to be significantly below pre-warming flow rates. If glacier melt becomes insignificant, surface water flow rates are then driven more by precipitation, and as a result, dry season water flows can be substantially reduced while wet season flows may not be as significantly impacted. Baraer et al. (2012) estimated the evolution of water flow rates for the Cordillera Blanca region. Their results show that most valleys are past their highest levels of runoff, implying diminishing water supplies into the future if melting persists. The effect of increased BC and dust on this glacial "life cycle" is to accelerate the process. Increased BC and dust will lead to faster glacier recession, though it is unclear what the effect on runoff would be during the transition phase.

In our warming climate, glaciers are melting at a fast rate and tropical glaciers are being substantially impacted. Glaciers in Peru account for more than 70 % of the world's tropical glacial area (Kaser, 1999). Rabatel et al. (2013) showed that glaciers in the tropical Andes have receded in area approximately 30 % since the 1970s and that areal loss rates have increased substantially in the first decade of the new millennium. Vuille et al. (2008) showed that surface air temperature has increased by 0.10 °C per decade over the last 70 years in the tropical Andes based on measurements from 279 stations between 1° N and 23° S. Similar trends have been shown to exist in studies focusing on temperature evolution in the last several decades from National Centers for Environmental Prediction/National Center for Atmospheric Research (NCEP/NCAR) climate reanalysis data (Kalnay et al., 1996; Bradley et al., 2009). Mark and Seltzer (2003) used NCEP/NCAR reanalysis data to show that the freezing elevation in the Cordillera Blanca, the location of this study, has risen approximately 200 m between 1955 and 2003. Using climate models, Bradley et al. (2006) have shown that tropical glacial regions (above 5000 m) could experience 4 to 5 °C of temperature increase between 2000 and 2100.

In addition to air temperature changes, dust and BC deposited on snow could potentially be a major factor leading to glacial mass loss. BC and dust reduce snow's albedo by absorbing solar radiation, which leads to increased melt and sublimation rates (Warren and Wiscombe, 1980). It has recently been shown by Painter et al. (2013) that BC likely led to the end of the little ice age in the European Alps in the mid-1800s. Dust on snow in Colorado has led to substantially increased melt rates (Painter et al., 2012) and has decreased total runoff (Painter et al., 2010). The effects of BC on snow in the Northern Hemisphere has been subject to intensive study, however, despite the fact that glaciers are a critical water source for large populations in South America, no studies to date have measured BC and dust in this region.

The Cordillera Blanca range is located in north central Peru. The mountain range is home to 17 peaks over 6000 m and hundreds over 5000 m. The bulk of precipitation comes from the Amazon basin in the wet season, which generally runs from October until May. During the dry season, the atmospheric flows generally come from the west and include more dry air. The dry season in the Cordillera Blanca is not completely dry as it is possible for small storms to pass through the range typically on a weekly basis. These dry season storms can produce snow from a few centimeters to a few tens of centimeters which can sometimes obscure the expected variability of light-absorbing particles by depth. Schauwecker et al. (2014) estimated that for the Cordillera Blanca region, there has been a modest temperature increase as well as a modest (60 mm decade^{-1}) increase in precipitation since the early 1980s. They state that recent changes in temperature and precipitation alone may not explain the glacial recession observed in the 30 years ending in 2012.

Sources of BC and dust in the Cordillera Blanca are numerous. Obvious anthropogenic sources of BC include industry and transportation (especially from diesel-fueled vehicles). Industrial sources are mostly centered in and around Huaraz, the largest city (population: 100 000, altitude: 3052 m altitude) abutting the Cordillera Blanca. Regional emission inventories for BC are not available, and emission inventories developed for global models are generally based on national emission statistics and do not contain regional specifics. There is substantial agriculture and BC from agricultural burning in the Cordillera Blanca region. Dust from land clearing and livestock grazing could impact the Cordillera Blanca cryosphere as well. Another potential source of BC is forest clearing and biomass burning in the Amazon basin. Amazonian burning has been estimated to account for approximately 50 % of the carbonaceous aerosols in the amazon region (Lamarque et al., 2010; Kaiser et al., 2012). Hydroelectric power is one of the most common forms of energy production, and coal burning is not significant in South America. Increases in dust can be caused by agriculture, construction, mining, and increased traffic on dirt roads.

The measurements described in this publication were collected by scientists and volunteers participating in American Climber Science Program (ACSP) expeditions (www.climberscience.com). The ACSP is a research organization with the goal of facilitating research and educational opportunities for scientists and students in regions that are challenging to access. To achieve these goals, experienced climbers are often included in research expeditions in order to increase safety in the mountains. The ACSP research expeditions are also designed to provide opportunities for non-scientists to learn about scientific practices as well as to instruct future scientists on safety in mountain regions. More than 50 international and Peruvian scientists, students, teachers and volunteers have participated in the three ACSP Peru expeditions.

Here, we report on 3 years of measurements of light-absorbing particles sampled on glaciers in the Cordillera Blanca. Section 2 describes the field campaigns and the sample collection methods. Section 3 describes a newly developed technique for the quantification of light-absorbing particles collected on filters. Section 4 presents the results from the 3 years of measurements, and the final section discusses the main conclusions of the project.

2 Field measurements

The ACSP has conducted three research expeditions during the dry season (austral winter) to the Cordillera Blanca region of Peru beginning in 2011. The sampling of snow and ice for light-absorbing particles has been a primary research project during the three expeditions. During the 3 expeditions, 48, 100, and 90 samples were collected respectively. Samples were taken from glacial surfaces from a minimum altitude of 4800 m and then at regular altitude intervals to the mountain top. Several mountains were sampled during all three expeditions, and some mountains have been sampled more than once during a single expedition. On one occasion multiple samples were taken at different depths by collecting ice from the walls of a crevasse; these data are used to identify long-term trends in particle loadings.

When working in remote and demanding conditions it is important to use simple techniques that provide robust scientific results. Several techniques for snow particle sampling were considered before a filtering technique was chosen. The technique for sampling particles in snow is similar to the technique used by the University of Washington's "Soot in Arctic Snow" group (as described in Doherty et al., 2010).

During research expeditions, volunteer scientists collect snow samples from predetermined locations on each mountain. Samples are collected from approximately 100 m above the snow line to the summit at 200 to 500 vertical meter intervals, the exact interval depends on the height of the mountain and the complexity of the terrain. Volunteers are instructed to avoid areas near exposed rock as well as areas that may be near avalanche paths or could be affected by debris from avalanches. Because mountain climbing is a physically demanding and potentially dangerous activity, the ultimate decisions on how high and where to collect samples is left to the discretion of the climbing team with safety being the highest priority. Global Positioning System waypoints are taken and, if possible, samples are collected from the same location as on previous climbs. Snow samples were collected by scooping snow into 4-liter ziplock plastic bags. Bags are labeled on the outside with a permanent marker after the bag has been sealed. Typically, ACSP volunteers collected samples with hands that had been "contaminated" with local snow. This is done by washing hands with snow so that any contaminants on the hands are similar to those in the snow. Since snow is collected on the descent usually well after sunrise, the snow is soft enough that it is easy to fill a bag contacting the snow only minimally. For each sample approximately 1 kg of snow is collected at each site from both the surface (defined as the top 2.5 cm) and the subsurface (deeper than 2.5 cm). The idea was that the surface sample should give an indication of any dry deposition that has occurred since the last snow storm as well as any accumulation of contaminants on the surface due to melting and sublimation while the subsurface samples should contain any contaminants that came with the most recent snowstorm either as ice nuclei or as having been scavenged by falling snow. In reality, it was commonly found that fresh snowfall during the dry season amounted to less than 10 cm. The result was that the subsurface sample often contained snow that had been subject to dry deposition and surface accumulation before the most recent snowfall. Once samples were collected and labeled, the snow samples are packed together into backpacks and returned to base camp for processing.

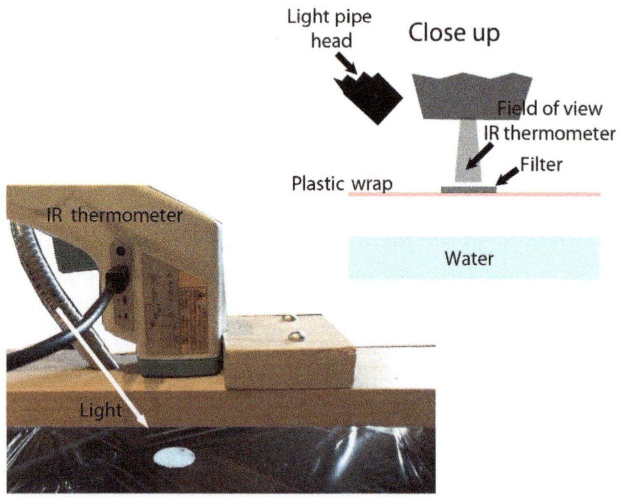

Figure 1. Diagram of the LAHM analysis setup. A light source is aimed at a 45° angle at the filter which rests on plastic wrap approximately 10 cm above the water surface. The infrared thermometer directly views the center of the filter. The field of view of the thermometer is approximately 1.5 cm at the distance of the filter and the contaminant spot on the filter is generally 2 cm in diameter.

In camp, snow samples are melted one at a time by placing the ziplock bags in warm water (usually around 30 °C). Once melted, snow-sourced water is drawn into 60 mL syringes and pumped slowly through 0.7 micron "Pallflex Tissuquartz" type 25 mm quartz fiber filters. A total of 600 mL of snow-water is filtered per sample unless the filter becomes too clogged, in which case less water is filtered, and the amount of filtered water is recorded. The samples are filtered immediately after melting to minimize adhesion of particles to the surface of the bag. The typical time between the starting of melting and the completion of filtering is 15 to 20 min. Filters are removed from filter holders, placed into plastic capsules designed for coin collections, and held in place with a thin foam ring. After each field excursion, filters are returned to Huaraz where they are dried in the sun, then stored in a freezer until the end of the expedition when they are returned to the US for analysis.

3 Filter analysis

A new technique has been developed for the analyzing the light absorption properties of particles collected on filters. The Light Absorption Heating Method (LAHM) is a cost effective technique that can be used to accurately quantify the impact of light-absorbing particles on snow. The premise behind conducting measurements of light-absorbing particles in snow is to estimate the amount of light energy the particles will absorb leading to increased melting or sublimation. Because snow is completely absorptive in the thermal infrared wavelength range, the more critical wavelengths to consider are in the visible range. Understanding the absorption of solar radiation in the visible wavelength range by particles in snow is the goal of the research. The LAHM analysis technique described here approaches the problem directly by measuring the temperature increase of a particle load on a filter when visible light is applied. A diagram of the LAHM setup is shown in Fig. 1. Each filter is suspended on plastic wrap a few centimeters above a pool of cool water. Water is used because it provided a thermally stable background compared to a solid surface which could absorb visible radiation. Plastic wrap does not absorb visible light, and as such it provides a neutral background and is of low mass per unit area thus having a negligible effect on measurements. An infrared thermometer (IR thermometer: Omega OS1327D) which can record data every second is mounted above the filter to measure the temperature of the filter without touching it. To determine the light absorption ability of the materials, a laboratory grade light source (Cole-Parmer Fiber-Lite Fiber Optical Illuminator model 9745-00) with a fiber optic light pipe to direct the light is mounted close to the filter and shines light at approximately a 45° angle to the filter (thus reducing the effects of multiple reflections in the system). The light transmitted through the light pipe only spans the visible wavelengths from 300–800 nm. The particles on the filter absorb light causing the filter to warm. The IR thermometer records the temperature of the filter every second, and data are saved to a computer for later analysis. After an initial 10 s to establish a base temperature, the light source is turned on for 30 s, then extinguished while the temperature of the filter is recorded for an additional 50 s. Temperature increases are normalized by the average temperature recorded during the first 10 s. The maximum temperature increase of a filter ranges from less than 1 °C for a clean filter, up to 10 °C for the most heavily contaminated filters. During analysis, after every 10 filters, 2 control filters were tested to assure consistency in the setup. The control filters included an unused clean filter and a calibration filter with a heavy load of BC. Sample temperature profiles from seven filters collected during the 2011 expedition are shown in Fig. 2.

The LAHM analysis technique was calibrated by sampling a BC standard: water with a known concentration of BC (Schwarz et al., 2012). The BC standard has 10 % uncertainty. Filters with low and high BC loads were made by filtering different amounts of BC standard. The BC standard was made with fullerene soot, a reference material used in the single-particle BC detection community (Baumgardner et al., 2012). The filters used for calibration (Millipore mixed cellulose ester filter membrane 0.22 micron) are different from the filters that have been used in the field (Pallflex Membrane Tissuquartz 0.7 micron). The BC standard was created with BC particles in the size range generally found in the atmosphere (0.2 to 0.8 microns) which were not captured well by the Tissuquartz filters. Torres et al. (2014) found that Tissuquartz filters captured less than 38 % of the BC mass when filtering rainwater. The mass size distribution for their measurements peaked between 210 and 240 nm, similar to atmospheric BC sizes but smaller than the peak sizes expected in snow (Schwarz et al., 2012). The Tissuquartz filters are used in the field because water flow through the filters is much faster making them much easier to use in the field and the filter material enables additional types of analysis that are not possible with the Millipore filters. Temperature profiles recorded for unused filters of both types were nearly identical (well within the SD of measurements), indicating that the filter type did not bias the temperature measurement technique significantly. Three calibration filters at each of four different fullerene soot mass amounts (5–20 μg) were created as well as one calibration filter with 30.0 μg. The temperature increase for each calibration filter was determined at least 4 times, and the overall standard deviation in the temperature increase at each carbon level was used in the uncertainty determination. Figure 3 shows plots of 12 temperature increase profiles measured for the three calibration filters loaded with 10 μg of fullerene soot. The mean values and standard deviations are also shown, along with a calibration curve showing the relationship between BC mass and heating after 10 s. Standard deviation values were generally less than 0.5 °C and remained constant throughout the heating phase. Note that

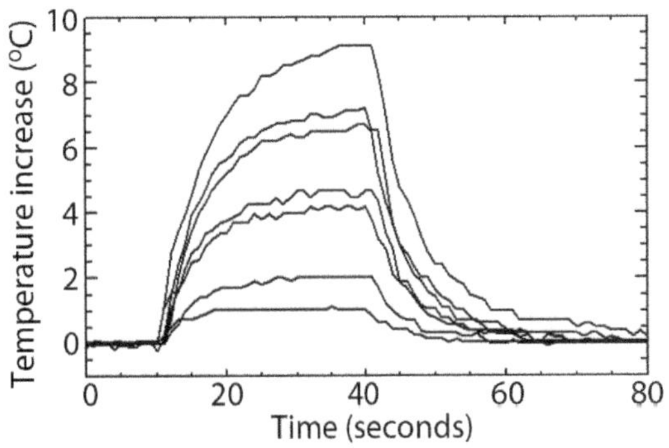

Figure 2. Measured temperature profiles for seven filters with a variety of light-absorbing particle loads. Temperature recording started 10 s before the lamp was illuminated, and the lamp was extinguished 30 s later. Temperatures are normalized to the average of first 10 s of data and the temperature usually returned to the start temperature within a minute of the lamp being extinguished.

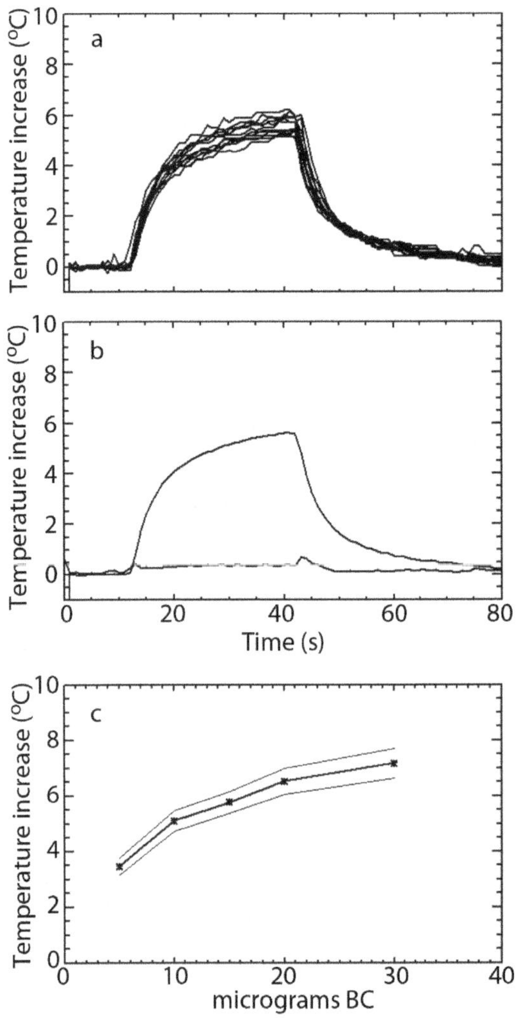

Figure 3. **(a)** Twelve temperature profiles measured for filters with 10 μg of fullerene soot. Profiles are from three different filters measured 4 times each. **(b)** Mean temperature profile and standard deviation for 12 temperature profiles shown in **(a)**. **(c)** Temperature increases after 10 s for five masses of fullerene soot. Error bars are standard deviations for each measurement and bold line is fit to data.

the heating for blank filters was approximately 0.6 °C on average, just above the SD for the shown measurements. Several of the cleanest filters from mountain samples from each year had temperature increases of less than 1.0 °C.

Laboratory measurements of post-filtered samples showed that the Tissuquartz filters did not efficiently collect BC particles smaller than about 0.6 microns. In the atmosphere most BC mass is distributed across particles smaller than this, but BC in snow has been shown to exist, in some samples, at sizes up to a few microns (Schwarz et al., 2012). While it is clear that the Tissuquartz filters may not be collecting all of the BC, it is unclear what portion is being missed. Results of separate mineralogy tests suggest that there were high concentrations of dust which can clog the pores and anecdotal evidence (volunteers noting that it had become very difficult to push the water through the filters) suggests that the effective pore size substantially decreases as dusty samples are being filtered. Given that BC particles in this size range can be missed, the BC on the sample filters is likely an underestimate of the true BC in the snow. Since this under-catch was noticed, tests have been conducted using snow from tropical glaciers that indicate that the missed particles do not account for much absorption. Snow sourced water that had already been passed through a Tissuquartz filter was collected and passed through a 0.22 micron filter. While the Tissuquartz filters in these tests captured high concentrations of particles, the 0.22 micron filters appeared clean after filtering and results from the LAHM technique were below the detection threshold. Of the approximately 20 tests filters, the eBC (effective black carbon) value estimated for the 0.22 micron filter was never more than 20 % of the value determined for the Tissuquartz filter and the average was around 5 %, well within the noise level of the LAHM technique at these levels.

The LAHM technique does not discriminate between BC and dust. Therefore, the values derived from the LAHM technique should be treated as eBC, meaning that the visible light absorptive capacity of the particles on the sample filter are equivalent to the given amount of BC (Grenfell et al., 2011).

The increase in temperature of the calibration filters after 10 s of exposure to light was used to derive a fit equation to predict the mass of BC on the calibration filters. The r^2 value of the fit equation was 0.98 with an uncertainty of approximately ±20 % (10 % from the BC standard and 10 % for the typical SD calculated and shown in Fig. 3b). This uncertainty likely increases with higher filter loads given that the temperature response to BC curve flattens (Fig. 3c). On heavily loaded filters, this flattening could be due to some of the BC being hidden under other layers of particles and therefore not absorbing as much light energy as other BC on the filter. It was found, however, that filter samples collected in the field often had different slopes to their temperature increase, particularly after 20 s of illumination. It is suspected that this is due to dust mixed in with the BC changing the overall heat capacity. Figure 4 shows two temperature profiles, one from a fullerene soot calibration filter and one from a sample filter suspected to contain significant dust as well as a substantial amount of BC. Note that the temperature curves are nearly identical for approximately 10 s after illuminating the lamp before they begin to diverge. It is thought that dust mixed in with BC on a filter could be causing this divergence. Filters collected in other regions with substantially higher dust concentrations and low BC concentrations showed much steeper temperature increases later in the heating phase. This would suggest that the effects of BC and dust could be separated with the LAHM technique, but quantification of this is beyond the scope of this publication.

Data from 2011 were collected with a slightly different filtering setup than subsequent expeditions. The change was made because the type of filter housing used in 2011 occa-

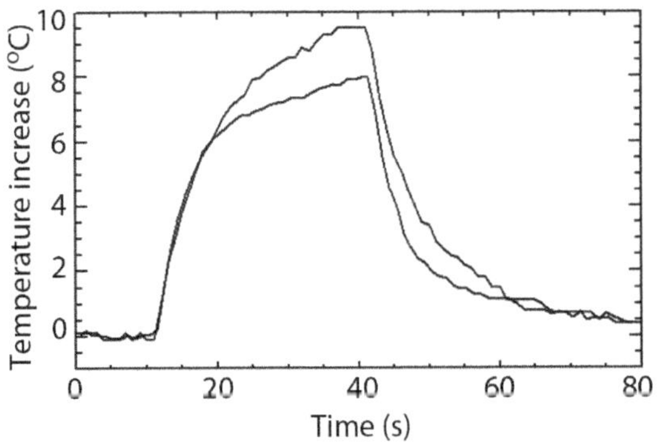

Figure 4. Temperature profiles measured for a calibration filter with fullerene soot (lower curve), and a sample filter with dust and black carbon (upper curve). The curves are nearly identical until 20 s (10 s of heating) after which they diverge. The continued more rapid increase in temperature of the dust – black carbon mixture is thought to be due to the dust.

sionally came into contact with the surface of the filter leading to the removal of some of the particles when the filter housing was opened. We estimate that this may have reduced the filter loading by up to 25 %. This estimate is based on the approximate area of the filter that appeared to be scraped clean as well as the fact that the filter holder had obvious particle loading that corresponded to the cleaned area. While it is not possible to determine the exact amount of material lost, the eBC values estimated for the affected filters were multiplied by 1.25.

During the 2013 expedition, 12 unmelted snow samples were collected in acid washed glass vials and returned to the US for analysis with a Single Particle Soot Photometer (SP2) instrument (see Schwarz et al., 2012). These samples were collected from two mountains: one in a region suspected to be highly polluted due to its proximity to Huaraz and another mountain in a more remote region. Additional snow samples were collected for filtering from the same locations for comparison. Results from the LAHM technique are compared to SP2 refractory black carbon (rBC; Baumgardner et al., 2012) mass concentration measurements in Fig. 5. Note that while there is some variability between the techniques, the results are reasonably correlated ($r^2 = 0.92$). Microscale variability could have accounted for some of the differences. Figure 5 shows that for some of the comparisons, the SP2 rBC values were higher than the LAHM eBC values. This could be due to non-uniformity in snow as the LAHM technique uses much more snow than the SP2 (600 g for LAHM vs. ~ 10 g for SP2). Additionally, the Tissuquartz filter could have missed a fair amount of the BC mass due to the large pore size.

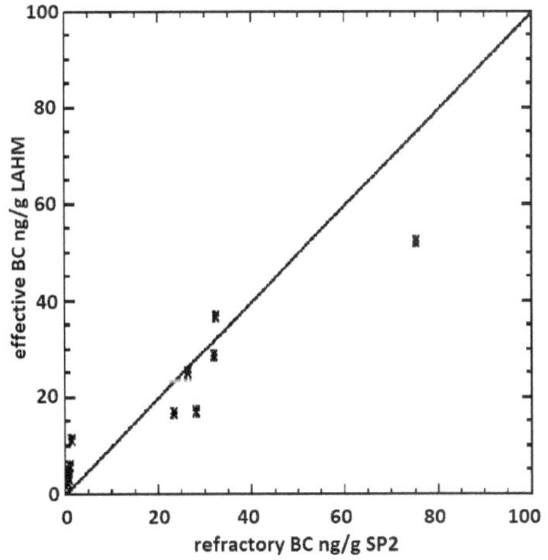

Figure 5. Plot showing the relationship between eBC as determined from the LAHM analysis and refractory black carbon as determined by the SP2 instrument ($r^2 = 0.92$). The values near 0.0 refractory BC are from Pisco mountain in region 2 while the higher values are from Vallunaraju in region 4. The 1 to 1 line is plotted to guide the eye.

4 Results

The eBC values determined by the LAHM technique for all measurements collected during the three expeditions are shown in Fig. 6 plotted by altitude. Values of eBC range from the quite low: 2.0 ng-eBC g-H2O^{-1} up to almost 80 ng g^{-1}. The highest altitudes have substantially less eBC while lower altitude values were larger; although numerous measurements still show values of less than 10.0 ng g^{-1} eBC at low altitudes. The largest values are similar to those reported by Ye et al. (2012) who reported values between 20 and 70 ng g^{-1} of BC in northwestern China seasonal snow at altitudes up to 3500 m. Huang et al. (2011) measured much larger values (over 1000 ng g^{-1}) of eBC in northeastern China, where dust was common as well as high levels of BC. The Cordillera Blanca measurements of eBC compare similarly in value and variability to those observed by Doherty et al. (2013) in Alaska and Greenland and are similar to those reported by Grenfell et al. (1981) for the Washington Cascade mountains (22–59 ng g^{-1}). The values are somewhat lower than those reported by Sergent et al. (1993) for the French Alps (80–280 ng g^{-1}).

4.1 Surface accumulation

As stated in the field measurements section, it was rare to collect samples where the surface sample contained long term dry deposition and surface accumulation and the subsurface sample contained only pristine snow. Often, the snow from the previous storm had compacted enough that the subsurface sample contained snow that had been previously exposed as a surface. This led to the eBC values for the surface and sub-

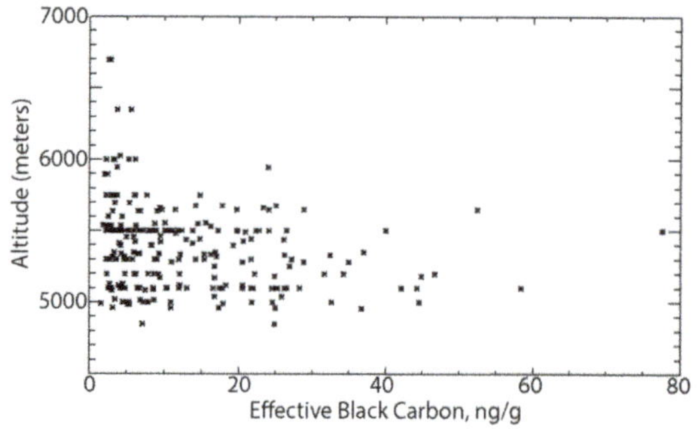

Figure 6. complete data set of eBC values determined from the LAHM analysis plotted versus altitude.

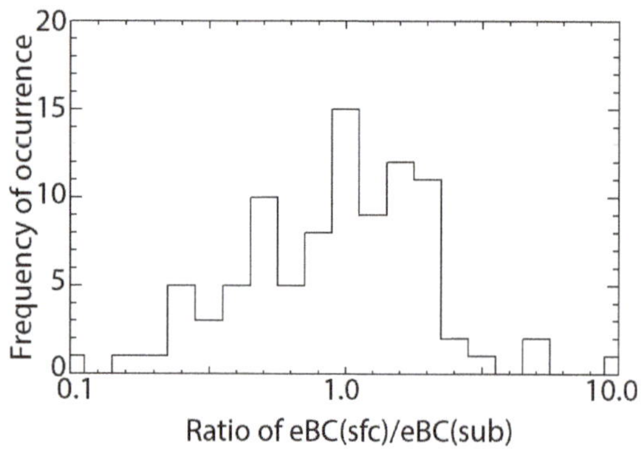

Figure 7. Probability distribution function of the ratio of the eBC value measured for surface samples divided by the eBC value measured for subsurface samples presented. x axis is on a log scale and bins are logarithmically spaced.

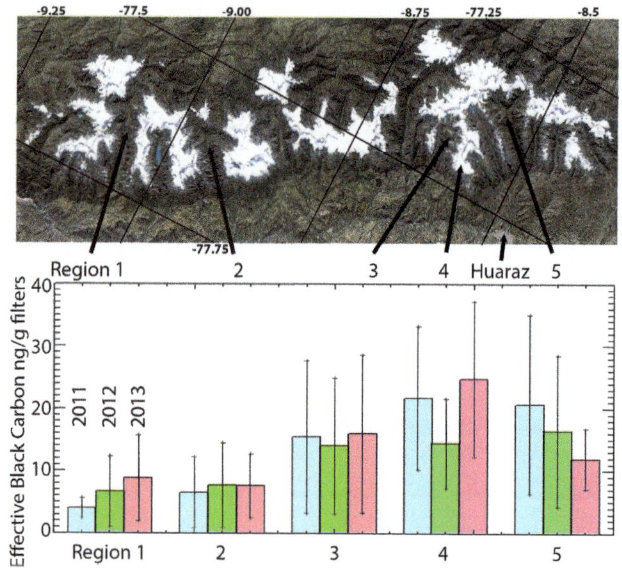

Figure 8. Google map image of the Cordillera Blanca mountain range. The five regions as well as the city of Huaraz are indicated on the map by black arrows. Average eBC values as determined from the LAHM analysis are shown in the plot for each of the 3 years. The thin lines indicate ± 1 standard deviation of the measurements. Note that in 2013, region 2, SP2 samples averaged 0.65 ng g^{-1} rBC indicating that most of the 8 ng g^{-1} of the eBC estimate could be due to light absorption by dust.

surface samples to not be consistently related. During the 3 expeditions there were 92 paired measurements where a surface sample and a subsurface sample were collected. Of those samples, in 56 cases, the surface sample contained higher eBC values than the subsurface samples while for the remaining 36 samples the opposite was true. The average ratio (averaged in log-space) was 0.96 suggesting that on the average, the surface and subsurface samples were statistically similar. The average difference between the two samples was a factor of 2.17 with a maximum difference of a factor of 12. Based on the fact that it does snow from time to time in the dry season, and that samples are never collected near the zone of ablation, the differences are thought to be due to short term changes in the snow pack rather than annual variability. Figure 7 shows a probability distribution function of the ratio between the surface measurement to the subsurface measurement for the 92 pairs of measurements.

4.2 By region

Figure 8 shows a map of the Cordillera Blanca region and the eBC values in snow determined from filter samples using the LAHM analysis technique for filter samples collected in different areas during each of the three expeditions. For the three expeditions, the data are grouped into five different valleys or mountains that were sampled each of the 3 years. Regions 1 and 2 lie on the north end of the range and consist of the mountains in the Santa Cruz and Paron valleys (region 1, 70 km from Huaraz), and Llanganuco valley (region 2, 54 km from Huaraz). These regions are characterized by higher precipitation and lower nearby population densities. The mountains around Ishinca valley (region 3, 21 km from Huaraz), Vallunaraju mountain (region 4, 14 km from Huaraz), and the mountains around the Quilcayhuanca valley (region 5, 24 km from Huaraz) are all clustered near Huaraz, the largest population center in the region. The eBC values shown in Fig. 8 are determined by taking an average of all measurements of surface and subsurface snow in each of the regions. As the relationship between the surface and subsurface values was highly variable (as shown in Fig. 7), the averaged values are thought to better represent the typical conditions for the top 10 cm of snow. Standard deviations of the measurements in each region are shown in Fig. 8. Data from all 3 years show a distinct trend, with the southern regions having 2–3 times more light-absorbing particles compared to the northern regions. The SP2 measurements discussed in the calibration section were collected in 2013 in region 2 and 4. The region 2 SP2 measurements averaged 0.65 ng g^{-1} while the region 4 SP2 values averaged 31.0 ng g^{-1}. The LAHM technique suggests that the eBC values in region 1 and 2 would be on the

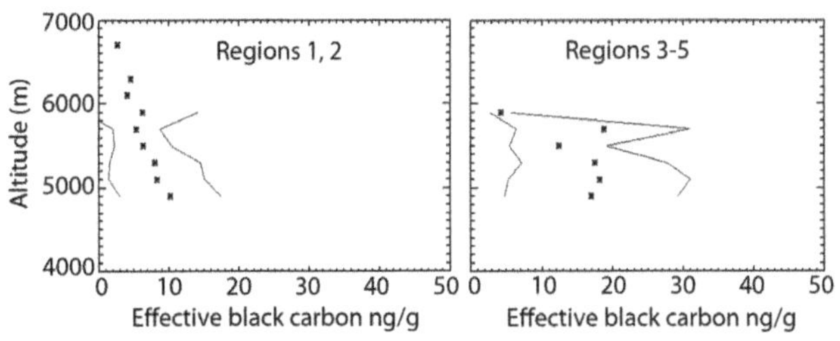

Figure 9. LAHM determined eBC values averaged by altitude bins plotted with +1 standard deviation (lines) by altitude for the northern (regions 1 and 2) mountains and the southern (regions 3–5) mountains in the Cordillera Blanca. Absorbing particles in snow in regions 1 and 2 are likely primarily dust while regions 3–5 are strongly influenced by pollution from the city of Huaraz.

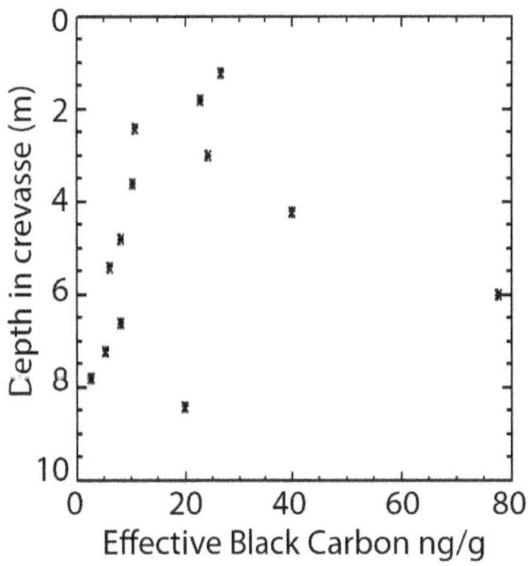

Figure 10. eBC from snow samples collected from the walls of a crevasse on Vallunaraju (region 4) the nearest high mountain to Huaraz. A decreasing trend with depth is noted. These lowest samples were visually the cleanest snow samples from the crevasse walls, and were likely to have accumulated during the wet season. The points at 4, 6, and 8.5 m were thin noticeably dark layers.

order of $10\,ng\,g^{-1}$ while the SP2 rBC was much lower. This leads us to the hypothesis that the bulk of the temperature increase on the filters from region 1 and 2 is likely due almost exclusively to dust rather than BC, while in regions 3–5, near the city of Huaraz, BC likely makes up a substantial portion of the absorptive particle matter present on the filters because the eBC values and the SP2 rBC values are similar in magnitude in region 4. Differences observed from year to year in each region are likely due to sampling being weighted to different altitudes or mountains rather than substantial changes in pollutant levels.

Figure 9 shows the mean value and standard deviation for data collected from all 3 years separated by altitude. Data are binned into two groups: region 1 and 2 in one group and 3–5 in the second group for all 3 years combined. Note the relatively linear decrease in eBC with altitude for both regions. Thin lines show the standard deviation calculated for the more numerous measurements. In general, eBC concentration levels decrease with altitude, though it should be noted that there are few measurements at altitudes above 6000 m and those are mostly in region 2.

4.3 By depth

During the 2012 expedition, samples were collected from the wall of a newly opened crevasse on Vallunaraju mountain in region 4. Vallunaraju is the nearest glaciated mountain to Huaraz and has some of the highest eBC measurements using the LAHM technique as well as the highest concentration observed by the SP2 (75 ng-rBC g-$H2O^{-1}$). Figure 10 shows the eBC values for the samples collected from the crevasse wall as a function of depth in the crevasse. There were several dark streaks at relatively uniform spacing down the crevasse wall which are suspected to be surface accumulations during the dry seasons (local glaciology experts agreed with this assessment). The eBC measurements higher than $20\,ng\,g^{-1}$ in Fig. 10 were from the darker areas while the lower values were from cleaner ice that dominated the crevasse wall. It is suspected that the cleaner samples were from wet season storms. A distinct decrease of eBC values with depth can be seen in the suspected wet season samples. While this might indicate that black carbon values have been increasing in recent wet seasons, the values are low enough that the trend is within the expected uncertainty range.

5 Conclusions

It is well documented that tropical glaciers are melting rapidly with concurrent effects on critical water supplies. Numerous factors could be contributing to low latitude glacier mass loss, including larger concentrations of light-absorbing particles on glacier surfaces. Because melting glacial water is an important natural resource in the region for agriculture, hydroelectric power, and drinking water, an increased understanding of the pressures on this resource could aid regional planners in adapting to future regional climate changes. This paper presents the results of 3 years of measurements of light-absorbing particles on the glaciers in the Cordillera Blanca in Peru. Samples were collected by volunteers participating in American Climber Science Program expeditions. Snow samples collected from glacier surfaces were melted, and the light-absorbing particles collected on filters.

A new technique (called the LAHM technique herein) to quantify the amount of visible radiation the light-absorbing particles can absorb has been developed and calibrated using filters with a known amount of black carbon. These "effective black carbon" estimates from the LAHM technique were found to be well correlated ($r^2 = 0.92$) with refractory black carbon mass measurements made with the Single Particle Soot Photometer instrument for a subset of the measurements. The LAHM technique provides an accurate, useful, and cost effective way to measuring light-absorbing particles in snow.

In the Cordillera Blanca, the concentration of light-absorbing particles was highest near Huaraz, the largest city in the region, while samples from more remote regions contained substantially lower amounts of absorptive material. The measured levels of light-absorbing particles in the snow near Huaraz absorb a similar amount of visible radiation as snow that contains between 20 to 80 $ng\,g^{-1}$ of black carbon. This amount of black carbon in pristine fresh snow can result in a decrease in spectral albedo of greater than 1 % (SNICAR-online: Snow, Ice, and Aerosol Radiation model, Flanner et al., 2007).

Acknowledgements. This work would not have been possible without the help of the American Climber Science Program volunteers for the 2011, 2012, and 2013 expeditions. The authors specifically wish to thank Frank Nederhand for initiating the program that would lead to the ACSP and Chris Benway at La Cima Logistics in Huaraz, Peru for helping to organize the expeditions. Thanks to Darrel Baumgardner for helping to facilitate the SP2 measurements as well as for helping to bring this work to the attention of the Pollution and its Impacts on the South American Cryosphere (PISAC) Initiative. Thanks to Karl Froyd and Jin Liao for their helpful comments and analysis with the PALMS instrument. Thanks also to Huascaran National Park, Jesus Gomez park director, for helping ensure the long term success of this research. Additional thanks go to the many generous donors who have made the ACSP expeditions possible. Also, the American Alpine Club whose early assistance was invaluable.

Edited by: A. Klein

References

Baraer, M., Mark, B. G., McKenzie, J. M., Condom, T., Bury, J., Huh, K.-I., Portocarrero, C., and Gomez, J.: Glacier recession and water resources in Peru's Cordillera Blanca, J. Glaciol., 58, 134–150, doi:10.3189/2012JoG11J186, 2012.

Barnett, T. P., Adams, J. C., and Lettenmaier, D. P.: Potential impacts of a warming climate on water availability in snow-dominated regions, Nature, 438, 303–309, doi:10.1038/nature04141, 2005.

Baumgardner, D., Popovicheva, O., Allan, J., Bernardoni, V., Cao, J., Cavalli, F., Cozic, J., Diapouli, E., Eleftheriadis, K., Genberg, P. J., Gonzalez, C., Gysel, M., John, A., Kirchstetter, T. W., Kuhlbusch, T. A. J., Laborde, M., Lack, D., Müller, T., Niessner, R., Petzold, A., Piazzalunga, A., Putaud, J. P., Schwarz, J., Sheridan, P., Subramanian, R., Swietlicki, E., Valli, G., Vecchi, R., and Viana, M.: Soot reference materials for instrument calibration and intercomparisons: a workshop summary with recommendations, Atmos. Meas. Tech., 5, 1869–1887, doi:10.5194/amt-5-1869-2012, 2012.

Bradley, R. S., Vuille, M., Diaz, H. F., and Vergara, W.: Threats to water supplies in the tropical Andes, Science, 312, 1755–1756, doi:10.1126/science.1128087, 2006.

Bradley, R. S., Keimig, F. T., Dias, H. F., and Hardy, R. D.: Recent changes in freezing level heights in the tropics with implications for the deglacierization of high mountain regions, Geophys. Res. Lett., 36, L17701, doi:10.1029/2009GL037712, 2009.

Doherty, S. J., Warren, S. G., Grenfell, T. C., Clarke, A. D., and Brandt, R. E.: Light-absorbing impurities in Arctic snow, Atmos. Chem. Phys., 10, 11647–11680, doi:10.5194/acp-10-11647-2010, 2010.

Doherty, S. J., Grenfell, T. C., Forsström, S., Hegg, D. L., Brandt, R. E., and Warren, S. G.: Observed vertical redistribution of black carbon and other insoluble light-absorbing particles in melting snow, J. Geophys. Res.-Atmos., 118, 5553–5569, doi:10.1002/jgrd.50235, 2013.

Flanner, M. G., Zender, C. S., Randerson, J. T., and Rasch, P. J.: Present day climate forcing and response from black carbon in snow, J. Geophys. Res., 112, D11202, doi:10.1029/2006JD008003, 2007.

Grenfell, T. C., Perovich, D. K., and Ogren, J. A.: Spectral albedos of an alpine snowpack, Cold Reg. Sci. Technol., 4, 121–127, 1981.

Grenfell, T. C., Doherty, S. J., Clarke, A. D., and Warren, S. G.: Light absorption from particulate impurities in snow and ice determined by spectrophotometric analysis of filters, Appl. Optics, 50, 1–12, 2011.

Huang, J, Fu, Q., Zhang, W., Wang, X., Zhang, R., Ye, H., and Warren, S. G.: Dust and black carbon in seasonal snow across northern China, B. Am. Meteorol. Soc., 92, 175–181, doi:10.1175/2010BAMS3064.1, 2011.

Kaiser, J. W., Heil, A., Andreae, M. O., Benedetti, A., Chubarova, N., Jones, L., Morcrette, J.-J., Razinger, M., Schultz, M. G., Suttie, M., and van der Werf, G. R.: Biomass burning emissions estimated with a global fire assimilation system based on observed fire radiative power, Biogeosciences, 9, 527–554, doi:10.5194/bg-9-527-2012, 2012.

Kalnay, E., Kanamitsu, M., Kistler, R., Collins, W., Deaven, D., Gandin, L., Iredell, M., Saha, S., White, G., Woollen, J., Zhu, Y., Chelliah, M., Ebisuzaki, W., Higgins, W., Janowiak, J., Mo, K. C., Ropelewski, C., Wang, J., Leetmaa, A., Reynolds, R., Jenne, R., and Joseph, D.: The NCEP/NCAR 40-year reanalysis project, B. Am. Meteorol. Soc., 77, 437–471, 1996.

Kaser, G.: A review of the modern fluctuations of tropical glaciers, Global Planet Change, 22, 93–103, 1999.

Lamarque, J.-F., Bond, T. C., Eyring, V., Granier, C., Heil, A., Klimont, Z., Lee, D., Liousse, C., Mieville, A., Owen, B., Schultz, M. G., Shindell, D., Smith, S. J., Stehfest, E., Van Aardenne, J., Cooper, O. R., Kainuma, M., Mahowald, N., McConnell, J. R., Naik, V., Riahi, K., and van Vuuren, D. P.: Historical (1850–2000) gridded anthropogenic and biomass burning emissions of reactive gases and aerosols: methodology and application, Atmos. Chem. Phys., 10, 7017–7039, doi:10.5194/acp-10-7017-2010, 2010.

Mark, B. G. and Seltzer, G. O.: Tropical glacier meltwater contribution to stream discharge: a case study in the Cordillera Blanca, Peru, J. Glaciol., 49, 271–281, 2003.

Painter, T. H., Deems, J. S., Belnap, J., Hamlet, A. F., Landry, C. C., and Udall, B.: Response of Colorado river runoff to dust radiative forcing in snow, P. Natl. Acad. Sci., 107, 17125–17230, doi/10.1073/pnas.0913139107, 2010.

Painter, T. H., Skiles, S. M., Deems, J. S., Bryant, A. C., and Landry, C. C.: Dust radiative forcing in snow of the upper Colorado river basin: 1. A 6 year record of energy balance, radiation, and dust concentrations, Water Resour. Res., 48, W07521, doi:10.1029/2012WR011985, 2012.

Painter, T. H., Flanner, M. G., Kaser, G., Marzeion, B., VanCuren, R. A., and Abdalati, W.: End of the Little Ice Age in the Alps forced by industrial black carbon, P. Natl. Acad. Sci., 110, 15216–15221, doi:10.1073/pnas.1302570110, 2013.

Rabatel, A., Francou, B., Soruco, A., Gomez, J., Cáceres, B., Ceballos, J. L., Basantes, R., Vuille, M., Sicart, J.-E., Huggel, C., Scheel, M., Lejeune, Y., Arnaud, Y., Collet, M., Condom, T., Consoli, G., Favier, V., Jomelli, V., Galarraga, R., Ginot, P., Maisincho, L., Mendoza, J., Ménégoz, M., Ramirez, E., Ribstein, P., Suarez, W., Villacis, M., and Wagnon, P.: Current state of glaciers in the tropical Andes: a multi-century perspective on glacier evolution and climate change, The Cryosphere, 7, 81–102, doi:10.5194/tc-7-81-2013, 2013.

Schauwecker, S., Rohrer, M., Acuna, D., Cochichin, A., Davila, L., Frey, H., Giraldez, C., Gomez, J., Huggel, C., Jacques-Coper, M., Loarte, E., Salzmann, N., and Vuille, M.: Climate trends and glacier retreat in the cordillera blanca, Peru, revisited, Global Planet. Change, 119, 85–97, 2014.

Schwarz, J. P., Doherty, S. J., Li, F., Ruggiero, S. T., Tanner, C. E., Perring, A. E., Gao, R. S., and Fahey, D. W.: Assessing Single Particle Soot Photometer and Integrating Sphere/Integrating Sandwich Spectrophotometer measurement techniques for quantifying black carbon concentration in snow, Atmos. Meas. Tech., 5, 2581–2592, doi:10.5194/amt-5-2581-2012, 2012.

Sergent, C., Pougatch, E., and Sudul, M.: Experimental investigation of optical snow properties, Ann. Glaciol., 17, 281–287, 1993.

Torres, A., Bond, T. C., Lehmann, C. M. B., Subramanian, R., and Hadley, O. L.: Measuring organic carbon and black carbon in rainwater: evaluation of methods, Aerosol Sci. Tech., 48, 239–250, doi:10.1080/02786826.2013.868596, 2014.

Vuille, M., Francou, B., Wagnon, P., Juen, I., Kaser, G., Mark, B. G., and Bradley, R. S.: Climate change and tropical Andean glaciers: past, present and future, Earth-Sci. Rev., 89, 79–96, doi:10.1016/j.earscirev.2008.04.002, 2008.

Warren, S. G. and Wiscombe W. J.: A model for the spectral albedo of snow II: Snow containing atmospheric Aerosols, J. Atmos. Sci., 37, 2734–2745, 1980.

Ye, H., Zhang, R., Shi, J., Huang, J., Warren, S. G., and Fu, Q.: Black carbon in seasonal snow across northern Xinjiang in northwestern China, Environ. Res. Lett., 7, 1–9, doi:10.1088/1748-9326/7/4/044002, 2012.

7

A wavelet melt detection algorithm applied to enhanced-resolution scatterometer data over Antarctica (2000–2009)

N. Steiner and M. Tedesco

The City College of New York at the City University of New York, New York, USA

Correspondence to: N. Steiner (nsteiner@ccny.cuny.edu)

Abstract. Melting is mapped over Antarctica at a high spatial resolution using a novel melt detection algorithm based on wavelets and multiscale analysis. The method is applied to Ku-band (13.4 GHz) normalized backscattering measured by SeaWinds onboard the satellite QuikSCAT and spatially enhanced on a 5 km grid over the operational life of the sensor (1999–2009). Wavelet-based estimates of melt spatial extent and duration are compared with those obtained by means of threshold-based detection methods, where melting is detected when the measured backscattering is 3 dB below the preceding winter mean value. Results from both methods are assessed by means of automatic weather station (AWS) air surface temperature records. The yearly melting index, the product of melted area and melting duration, found using a fixed threshold and wavelet-based melt algorithm are found to have a relative difference within 7 % for all years. Most of the difference between melting records determined from QuikSCAT is related to short-duration backscatter changes identified as melting using the threshold methodology but not the wavelet-based method. The ability to classify melting based on relative persistence is a critical aspect of the wavelet-based algorithm. Compared with AWS air-temperature records, both methods show a relative agreement to within 10 % based on estimated melt conditions, although the fixed threshold generally finds a greater agreement with AWS. Melting maps obtained with the wavelet-based approach are also compared with those obtained from space-borne brightness temperatures recorded by the Special Sensor Microwave/Image (SSM/I). With respect to passive microwave records, we find a higher degree of agreement (9 % relative difference) for the melting index using the wavelet-based approach than threshold-based methods (11 % relative difference).

1 Introduction

The future response of the Antarctic Ice Sheet to a changing climate is one of the largest uncertainties in the estimates and predictions of global sea-level rise over the coming decades (Hughes, 1981; Joughin and Alley, 2011; Overpeck et al., 2006; Shepherd and Wingham, 2007; Bromwich and Nicolas, 2010; Dowdeswell, 2006; Lemke et al., 2007). As temperature increases at high latitudes (e.g., Comiso, 2010; Hansen et al., 2010) the rate of surface melting is expected to increase (Ohmura, 2001). Analysis of long-term trends in weather station air temperatures indicates a strongly positive trend in the duration of melting conditions over the Antarctic Peninsula (Vaughn, 2006; Barrand et al., 2013). Efforts to quantify mass balance indicate a positive trend in mass loss for much of the West Antarctic Ice Sheet and the Antarctic Peninsula but a negative trend for much of the East Antarctic Ice Sheet (e.g., Chen et al., 2011; Rignot and Thomas, 2002; Shepherd and Wingham, 2007; Rignot et al., 2011). Recent results using a combined climate modeling and satellite observational approach suggest that in the period 1992 through 2011, the East Antarctic Ice Sheet gains mass at a rate of $14 \pm 43\,\mathrm{Gt\,yr^{-1}}$, while the West Antarctic Ice Sheet and the Antarctic Peninsula exhibit a mass loss of -65 ± 26, and $-20 \pm 14\,\mathrm{Gt\,yr^{-1}}$, respectively (Shepherd et al., 2012).

In the recent past, the Antarctic Peninsula has experienced episodes of dramatic ice-shelf break-up, for example the 2002 Larsen B Ice Shelf and the 2008 Wilkins Ice Shelf collapses. Both were the consequence of a suite of physical surface processes as well as ocean–ice-shelf interaction mechanisms, but with surface melting likely playing a key role. Indeed, the formation of meltwater ponds on the surface of the ice shelves caused by the accumulation of runoff might have contributed to the process of disintegration through hydrofracturing (e.g., MacAyeal et al., 2003; Scambos et al., 2009). Instability caused by ice-shelf loss has been shown to increase observed ice flow velocity in related glacial tributaries (Rignot et al., 2004; Scambos et al., 2004; Rott et al., 2011). Melting also drives firn densification and compaction (Holland et al., 2011).

Direct measurements of melting are not available from in situ data. Additionally, quantities necessary to fully solve the surface energy balance are often unavailable. Therefore, surface melting is generally estimated from near-surface air temperature measurements performed by automatic weather stations (AWS) when and where available. Such measurements are sparse over Antarctica and mostly performed around coastal areas and at low elevations. Moreover, data measured from AWS represent only local conditions and are difficult to extrapolate or estimate melting with at large spatial scales without adding biases or increasing uncertainty.

Active and passive microwave spaceborne instruments are used to monitor melting over snow-covered areas due to the insensitivity to atmospheric and illumination conditions and high sensitivity to liquid water (e.g., Abdalati and Steffen, 2001; Ashcraft and Long, 2006; Liu et al., 2005; Mote et al., 1993; Nghiem et al., 2001, 2007; Steffen et al., 2004; Tedesco et al., 2007; Torinesi et al., 2003; Wang et al., 2008). For vegetation-free snow-covered areas, the volume scatter component will be dominant for radar backscatter measurements at Ku-band frequencies (Ulaby et al., 1982). Backscatter loss due to the presence of liquid water in snow during melting is responsible for a rapid and considerable decrease in the Ku-band normalized microwave backscatter, σ^0, with respect to dry snow conditions. This is because of the increased imaginary component in the bulk complex dielectric constant of wet snow relative to dry snow (e.g., Ngheim et al., 2001; Stiles and Ulaby, 1980). This same emissivity change will create near-blackbody emission characteristics for wet snowpacks leading to a marked increase in brightness temperatures (Ulaby et al., 1982; Stiles and Ulaby, 1980).

Various melt detection algorithms have been developed and applied to active microwave time series to estimate seasonal melt. Often, a threshold value of absolute magnitude signal change either constant or regionally variable is used to detect melt-related changes. (e.g., Ashcraft and Long, 2006; Trusel et al., 2012; Wang et al., 2008; Zwally and Fiegles, 1994). Generally, this threshold value is chosen as an approximation of the expected microwave response during snowmelt with respect to a baseline referring to wintertime dry snow conditions (e.g., Ashcraft and Long, 2006). We refer to all methods that consider a constant value threshold in the following as fixed-threshold approaches. In contrast, approaches employing physically based temporally or spatially variable threshold values will be referred to as dynamic threshold approaches (Mote and Anderson, 1995; Tedesco, 2009). Algorithms that rely on the intrinsic properties of a measurement within time series have also been applied to snowmelt detection (Joshi et al., 2001; Liu et al., 2005; Wang and Yu, 2011). These approaches are dynamic in that they are based on the magnitude of relative change within each individual time series. A dynamic approach assumes that large changes in the microwave region are associated with melting events so that "edges" are created in the backscattering time series. These edges can be identified and used to estimate the timing of melting events. Edge-detection algorithms of this type have been developed using derivative-of-Gaussian, (e.g., Joshi et al., 2001) or multiscale wavelet edge detection (Liu et al., 2005).

Building on previous dynamic melt detection approaches (Joshi et al., 2001; Liu et al., 2005), we introduce a wavelet-based melt detection algorithm based on multiscale analysis of wavelet transforms to identify melting events using singularity detection (Mallat and Hwang, 1992). Such method identifies points of substantial transition in backscatter time series using no a priori information. In addition to this, a measure of the signal regularity at the point of transition can classify the transition "type", which allows the separation of persistent melting events (melting lasting continuously for a certain number of days) from transient or sporadic melting events. Besides the wavelet-based approach, we also consider a fixed-threshold algorithm-derived melting record from the same data set. This fixed-threshold record is to evaluate the wavelet-based method where alternative validation data, such as in situ weather-station measurements, are not available. The fixed-threshold method is performed as in Ashcraft and Long (2006) and Barrand et al. (2013). All results of the current dynamic algorithm are shown relative to this fixed-threshold algorithm approach.

We apply these two melt detection algorithms to an enhanced spatial resolution QuikSCAT (Ku band) scatterometer data set distributed by the Microwave Earth Remote Sensing (MERS) Laboratory with an effective resolution of 5 km gridded to 2.225 km (Early and Long, 2001; Spencer et al., 2000). The enhanced spatial resolution allows for the discrimination of melting patterns at sub-ice-shelf resolution and resolves melting patterns and trends that are not apparent in coarse-spatial-resolution products. High-resolution data sets are ideal for the Antarctic Peninsula, a region of high elevation contrast where the coarse resolution of some microwave observations may underestimate melt due to a large sub-pixel elevation gradient. We focus on melt onset date, melt-off date, melt extent and melt duration for the whole Antarctic continent. Results from both dynamic wavelet-based and fixed-threshold approaches applied to the enhanced-resolution QuikSCAT data record are also compared with those obtained from spaceborne microwave brightness temperatures from the Special Sensor Microwave/Imager (SSM/I) and melt estimates obtained from the analysis of surface air-temperature values recorded by automated weather stations.

2 Methodology

A general overview of wavelets and the specific methodology applied here is presented in Sect. 2.1. This is followed by a more detailed discussion of the mathematics and examples of wavelet melt detection in Sect. 2.2. An overview of the processing steps and operation of the melt detection algorithm is presented in Sect. 2.3.

2.1 Melt detection using wavelets

The duration of seasonal melting over Antarctica is estimated by means of a wavelet analysis. A wavelet transform unfolds a one-dimensional time series of σ^0 into a two-dimensional power spectrum of position and scale (i.e., inverse frequency). The wavelet transform can evaluate localized variability of a backscatter time series using a series of convolutions with a dilating and translating wave-like function (Daubechies, 1992). In this study, we use the wavelet transform as a differential operator, in that it is able to approximate the derivative of a smoothed data series at each time location. Melting and refreezing events will cause large variations in σ^0 and therefore appear as local maxima or minima in the wavelet transform.

Many studies in the natural sciences have used the wavelet transform to detect changes in one-dimensional time series, for example the detection of tropical convection anomalies (Weng and Lau, 1994) and geomagnetic jerks (Alexandrescu et al., 1995). Wavelet analysis methods have also been applied to snowmelt detection: specifically, Liu et al. (2005) apply a wavelet-based methodology to identify large changes in measured brightness temperature values associated with melting events over Antarctica.

We apply an approach similar to Liu et al. (2005), but with several key differences. First, we apply this approach to active microwave (Ku band, 13.4 GHz) measurements. Additionally, we use no a priori information, such as statistical or physically based thresholding of wavelet coefficients, for any of the pixel locations. Melt and refreezing events are both identified and classified in the framework of singularity detection as introduced Mallat and Hwang (1992). Continuous wavelet transforms are used to detect melting events that appear as discontinuous events in the backscatter time series and to eliminate those melting events that are determined to be sporadic in time using multiscale analysis (Mallat and Hwang, 1992; Mallat, 1999; Alexandrescu, 1995). This methodology is novel, and to our knowledge, it is the first time that such an approach has been applied to remote sensing of snow and ice.

2.2 Continuous wavelet transform and multiscale analysis

The continuous wavelet transform (CWT) of the seasonal (June of one year through May of the successive year) backscatter, $\sigma^0(t)$ is defined by the convolution product

$$W\sigma^0(u,s) = \int_{-\infty}^{+\infty} \sigma^0(t) \frac{1}{\sqrt{s}} \psi\left(\frac{t-u}{s}\right) \mathrm{d}t = \left(\psi_s^* \sigma^0\right)(u), \quad (1)$$

where ψ is the wavelet function, u is the translation parameter, s is the scaling factor and $*$ is the convolution operator (Mallat, 1999). The analyzing wavelet, ψ, is a real-valued, localized zero-mean function with a vanishing integral (e.g., the integral of ψ is zero) (Mallat, 1999; Holschneider, 1995). The analyzing wavelet function used in this study is defined using the first derivative of a Gaussian function:

$$\psi(t) = (-1)^n \frac{\mathrm{d}^n \theta(t)}{\mathrm{d}t^n} \quad (2)$$

and

$$\theta(t) = \frac{1}{\sqrt{2\pi}} \exp\left(-\frac{t^2}{2}\right), \quad (3)$$

with order $n = 1$ (Mallat, 1999).

A wavelet function with a Gaussian base is necessary to ensure that wavelet coefficient maxima will be continuous from large to small scales (Mallat, 1999). This allows for the tracing and association of wavelet maxima across scales, a process that is key for multiscale analysis. As an example, a microwave backscattering time series recorded over the Larsen Ice Shelf AWS during the 2006–2007 season is plotted in Fig. 1a, with the corresponding CWT plotted in Fig. 1b. The magnitude of local maxima or minima in $W\sigma^0$ (black and white areas, respectively) are correlated with the magnitude of backscatter change (Liu et al., 2005). This is expected as ψ is equivalent to the first derivative of a smoothing function as indicated in Eqs. (2) and (3). In Fig. 1b the conical field of negative $W\sigma^0$ intercepts the scale axis at its narrowest width at a position ($u = 185$) coincident with a $\sim$20 dB decrease in σ^0 over several days. Fields of elevated $W\sigma^0$ in a CWT, or "cones of influence" (Mallat, 1999), are located where the data in the time series (or its derivative) behave as a discontinuous function (Mallat and Hwang, 1992; Holschneider, 1995; Mallat, 1999). Melting or refreezing events therefore create cones of influences that will converge at fine scales to the position of signal discontinuity (Mallat and Hwang, 1992; Mallat, 1999; Liu et al., 2005). We refer to these positions as singularities (e.g., Mallat and Hwang, 1992).

The values contained in a cone of influence compose the scalar components (the signal change relative to the temporal scale) of that transition and are used in multiscale analysis. We define the set of all CWT coefficient maxima along the "ridge" of the maxima cone as the wavelet transform modulus maxima line (WTMML) for each singularity. These positions are found where $(\delta \backslash \delta u) W\sigma^0(u,s) = 0$ and are connected across adjacent scales from large duration to small (Hermann, 2001; Mallat, 1999; Mallat and Hwang, 1992; Alexandrescu, 1995). The CWT, although redundant in its

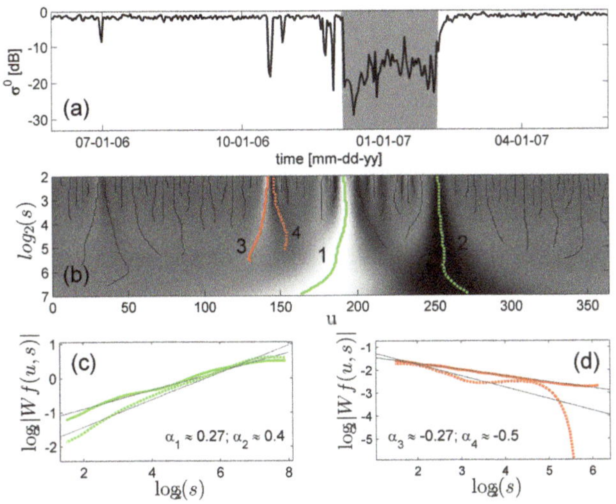

Fig. 1. **(a)** QuikSCAT σ^0 time series with the melt duration (MD) estimated using the CWT method shaded in grey. **(b)** The CWT of the time series in Fig. 1a, where dark-grey to black values indicate positive $W\sigma^0$ and light-grey to white values negative $W\sigma^0$. WTMMLs are indicated as black, green or red lines associated with melt onset (MO, WTMML 1) melt refreeze (MF, WTMML 2) and sporadic, early season melt (WTMML 3, WTMML 4). **(c)** The $W\sigma^0$ along WTMML plotted over all scales, (s), and the associated α for the MO and MF events. **(d)** Same as Fig. 1c but for an early season sporadic melting.

frequency or scale representation of any signal, is necessary in this application as these positions are not easily connected if determined only at discrete dyadic scales (e.g., Liu et al., 2005). Additionally, tracing CWT maxima to the finest temporal scale ensures the accurate localization of melting events as these positions will be shifted with increasing scale, as evident in Fig. 1b.

The WTMML is then used to characterize the nature of σ^0 change at each singularity using multiscale analysis (e.g., Mallat and Zhong, 1992; Le Gondidec et al., 2002). The WTMMLs for several singularities in σ^0 are highlighted in Fig. 1b and are labeled 1 through 4. The apparent regularity of a σ^0 time series in the immediate neighborhood of each singularity can be estimated from the wavelet coefficient values that compose each WTMML (e.g., Mallat, 1999). The Holder exponent, a, is a measure of estimated regularity at the terminating position of each WTMML (Mallat and Zhong, 1992). An estimate of a is found as (Mallat, 1999)

$$\left|W\sigma^0(u,s)\right| \leq Ae^a \tag{4}$$

so that

$$\ln\left|W\sigma^0(u,s)\right| \leq a \ln a + \ln A. \tag{5}$$

Here, $\left|W\sigma^0\right|$ correspond to the wavelet coefficients that compose the WTMML. Using Eq. (4) and Eq. (5), a as well as the coefficient A are determined using a linear least-squares regression.

Each WTMML can be then classified by a using several theoretical transition types whose a value is known. For example, a step-like or Heaviside function will result in a near $0\ a$ (Le Gonidec et al., 2002), while a ramp-like or smoothed transition has a of 1 or greater (Mallat, 1999; Le Gonidec et al., 2002). For wet snow detection, we assume melt onset transitions (e.g., similar to $u = 185$ in Fig. 1b) can be approximated by either a step-like or smoothed transition. Here the corresponding value of a is $a_1 = 0.27$. Due to the complex nature of a time series, a values determined by Eq. (5) are not expected to match the theoretical case since nearby singularities will have contributions to the WTMML (i.e., cones of influence will overlap if transitions are close in time) (Mallat, 1999). In Fig. 1a we observe a more gradual (smoothed) σ^0 transition, coincident with a refreezing event near $u = 250$, where the determined regularity ($a_2 = 0.4$, Fig. 1c).

Spike-like or cusp-like transitions will produce negative a in multiscale analysis (Le Gonidec et al., 2002; Mallat, 1999). In Fig. 1a, a σ^0 fluctuation, corresponding to position $u = 140$ in Fig. 1b illustrates a transition that is sporadic in time. This transition produces two WTMMLs, shown in Fig. 1d, labeled 3 and 4, having negative and positive $W\sigma^0$ components, respectively. From Eq. (2) it is determined that both can be approximated with negative a values ($a_3 = \sim -0.27$, $a_4 = \sim -0.4$). Negative a values associated with a WTMML indicate the signal is both discontinuous and non-differentiable at that position, here $u = 140$ (Mallat and Hwang, 1992). In terms of melt detection, by removing all negative a transitions, we eliminate sporadic σ^0 changes that return to "dry" conditions rapidly relative to the reference scale. This creates a melt detection process that eliminates sporadic melting events.

For locations that do not experience melting events, changes in backscatter associated with snow properties changes are of low magnitude compared to changes in liquid water content, but produce positive a in multiscale analysis. To reduce the influence of falsely classified melting events, we set a minimum threshold for $|W\sigma^0|$at each temporal scale along the WTMML corresponding to one order of magnitude ($10\times$) greater than that observed during winter (June, July and August). This is a conservative threshold and this choice does not appear to influence classification for areas experiencing melt.

2.3 Melt detection process

The WTMMLs from each backscattering time series are evaluated as a possible melting event according to the following criteria:

1. The WTMML must have a scale component that extends above a set minimum scale. Signal noise and noisy transition in backscatter produce WTMMLs

only at small scales (i.e., high frequency). We set the minimum scale at 2^5. We observed that using a scale of 2^3 similar to Liu et al. (2005) could not eliminate all noisy transitions during the melting season. This increased minimum scale relative to previous studies is expected since the enhanced-resolution remote sensing product used in this study is noisier than the coarser resolution one (Ashcraft and Long, 2006; Spencer et al., 2000).

2. All $|W\sigma^0|$ along the WTMML must have a value one order of magnitude ($10\times$) greater than those measured during the winter season. For areas experiencing melt, this condition does not have a large effect, since $|W\sigma^0|$ produced at all scales for a ~ 3 dB transition in backscatter is many orders of magnitude greater than observed wintertime conditions. However, areas that experience snow property changes but no seasonal melt will produce WTMMLs with large scalar components since these changes are not "noisy" transitions and are temporally sustained. Since changes in LWC produce greater differences in backscattering values compared to snow property changes over a relatively short period (Stiles and Ulaby, 1980), we set the threshold with the $|W\sigma^0|$ values of a WTMML, at all spatial scales, with a value an order of magnitude ($10\times$) greater than observed wintertime conditions.

3. The Holder exponent, α, of each WTMML must be equal to or greater than zero. For reasons discussed in the previous section, we eliminate spike-like transitions in backscatter using a test of point-wise signal regularity.

All signal transitions that meet the three criteria above are considered to be either melt onset or refreeze events. To define periods of melt we select the WTMML that extends to the largest scales that has the greatest mean $|W\sigma^0|$. This transition is matched with a WTMML greatest mean $|W\sigma^0|$ in the set of transitions of opposite magnitude. This defines one seasonal melting event. We assume that refreeze must follow melting. This process is repeated with the remaining melt-related WTMMLs to define additional periods of melting or periods of sustained refreeze within melting. The melt onset (MO) is defined as the first day of melting, and melt-off (MF) is defined as the last melting day plus one. The melt duration (MD) at any pixel location is the total number of days when melting occurs.

A fixed-threshold melting record is created for comparison, hereafter referred to as FT3. In the FT3 record, melting is classified as where the enhanced-resolution $\sigma^0(t)$ is at or below 3 dB minus winter (June–July–August) mean backscatter, equivalent to the expected backscatter loss from a 2.8 cm layer of 1 % volumetric water content as in Ashcraft and Long (2006). Melting events whose durations are shorter than three continuous days are removed from the melting record at each pixel following Tedesco et al. (2007).

For our study, we use a MATLAB® wavelet analysis software library (WaveLab 850) distributed by Stanford University and available at http://statweb.stanford.edu/~wavelab/. Because of the computationally intensive nature of the problem and of the high number of pixels necessary to cover the entire Antarctic continent at the spatial resolution considered here (2.225 km), we also made use of the MATLAB's parallel computing toolbox, running on a dedicated server using eight processing cores. In this configuration, one continent-scale melting season requires between 24 and 48 h of processing.

3 Data sets

3.1 SeaWinds on QuikSCAT

Enhanced spatial resolution melting maps (gridded at 2.225 km) are derived from the enhanced-resolution SeaWinds (commonly referred to by the satellite name QuikSCAT) scatterometer data set distributed by MERS at Brigham Young University (http://www.mers.byu.edu/). The SeaWinds instrument is a conically scanning pencil-beam-width scatterometer that records normalized radar cross-section measurements in the Ku band (13.4 GHz). This instrument records both horizontal and vertical polarizations at a fixed incidence angle of 46° for horizontal polarizations and 54.1° for vertical polarizations (Long and Hicks, 2000). The effective $\sim 7\,\text{km} \times 30\,\text{km}$ instantaneous field of view of the QuikSCAT range-Doppler-processed "slice" product (Spencer et al., 2000) is enhanced through a scatterometer image reconstruction algorithm (SIR) developed at BYU. By combining multiple observations that are weighTed within with the senor's antenna function, the SIR algorithm can recover information attenuated at the antenna side lobe, thereby effectively increasing the spatial resolution of the standard scatterometer product (Early and Long, 2001). This resolution enhancement is done at the expense of increased signal noise. In view of the strong change in measured backscatter following snowmelt, the increase in signal noise does not affect melt detection considerably, hence allowing for the generation of a surface-melting product at a gridded spatial resolution of 2.225 km (Trusel et al., 2012; Barrand et al., 2013). In addition, the CWT method is well suited for noisy data sets since high-frequency, noisy transitions in backscatter are not likely to create a WTMML that extends to large scales (e.g., Liu et al., 2005) and those that do will decrease in $|W\sigma^0|$ with increasing scale, leading to negative Holder exponents and subsequently being removed in the melt detection methodology.

3.2 Automated weather stations

Automated weather station data from the Antarctic Meteorological Research Center and Automatic Weather Station program, maintained by the Space Science and Engineering Center at the University of Wisconsin, Madison, are used to evaluate the results of the melt detection algorithms. For our comparison, we use the hand-corrected 3-hourly air temperature records. The AWS used in this study are Larsen Ice Shelf (lat. 67.01 S, long. 61.55 W, elev. 17 m), Uranus Glacier (lat. 71.43 S, long. 68.93 W, elev: 780 m), Fossil Bluff (lat. 71.33 S, long. 68.28 W, elev. 63 m), Butler Island (lat. 72.21 S, long. 60.17 W, elev. 91 m), Pegasus South (lat. 77.99 S, long. 166.57 E, elev. 5 m) and Limbert (lat. 75.91 S, long. 59.26 W, elev. 40 m). For each AWS, air-temperature measurements and backscatter are spatially and temporally coregistered using overpass times available from MERS-SCP. Melting is determined from AWS air-temperature measurements where there are at least 2–3 h daily above-zero measurements. For Antarctica, extreme fluctuations in daily temperature often prevent the daily mean temperature from exceeding 0 °C, though satellites observations indicate that melting is likely taking place. A threshold below 0 °C is often used to account for this fact (Tedesco and Monaghan, 2009; Van den Broeke et al., 2010), or in those cases where additional measurements are available, such as surface shortwave and long-wave radiative fluxes, melting may be determined using a simple thermodynamic model (Van den Broeke, 2005). From the lack of a defined sub-zero threshold for each AWS station, as well as of sufficient surface measurements for modeling approaches, we choose a temporal threshold of at least six hours of above-zero measurements in one day to establish melting conditions from AWS measurements and acknowledge that is a source of uncertainty in the AWS validation data set. This is equivalent to at least two daily above-zero measurements for the 3 h AWS air temperature data set. Once melt is estimated from AWS data, we study the number of days when the remote-sensing- and air-temperature-based estimates agree (true positive); the omission error, computed as the percentage of days when air temperature indicates melting but the remote sensing-based approach does not (true negative); and the commission, computed as the percentage of days when the satellite data indicate melting but this is not occurring from the analysis of air temperature (false positive).

4 Results

4.1 Comparison between QuikSCAT-derived melting and analysis of automated weather stations

Results of the FT3 and dynamic wavelet-based (CWT) approaches are compared with estimates of melting derived from surface air temperatures measured by AWS. Because of the lack of in situ liquid water content or snow temperature measurements, melt is estimated from AWS air temperatures where the temperature is above zero for at least six hours per day. The time series of coregistered backscattering, air temperature and positive temperatures for the stations considered in this study are plotted in Fig. 2. We evaluate the agreement, commission and omission in percentage at each station, and the results are presented in Table 1. Agreement is defined as the percentage of cases where surface temperature and spaceborne-based estimates both detect melting. Commission is defined as the percentage of cases in which remote sensing algorithms indicate melting but AWS does not. Lastly, omission indicates the percentage of cases in which AWS suggests melting but the remote sensing algorithms do not.

For stations that experience at least 10 days of melt per year (Butler Island, Fossil Bluff and Larsen Ice Shelf stations), we find the highest rates of agreement. For these stations the FT3 approach performs with more agreement (~ 10 %) than the CWT approach. The Larsen Ice Shelf and Fossil Bluff stations show the highest overall agreement, averaging 87 % (77 %) and 75 % (66 %) for the FT3 (CWT) approach for all years. Both of these stations have large magnitude and sustained change in backscatter over the melting season (Fig. 2a and b). For the Limbert, Butler Island and Uranus Glacier stations, shown in Fig. 2c, d and e, respectively, the difference in agreement for the FT3 and CWT methods is greater, up to 43 % (FT3 > CWT) between methods. At these stations, most of the melting occurs as short-duration events. This is not the case during the 2002–2003 and 2003–2004 seasons for the Limbert and Uranus Glacier stations, respectively, where the backscattering time series shows a substantial decrease in backscatter. In this case both remote sensing methods are in good agreement with AWS estimates. Figure 2a and b show that most of the difference between methods at these stations occurs during transitional periods, meaning near the seasonal MO or MF. For the Pegasus South station, the CWT method is in better agreement with AWS compared to FT3, with a 10 % greater mean agreement. At the Pegasus South station the apparent backscatter response to melting, as shown in Fig. 2f, is below detection by the FT3. Additionally, the Pegasus South station has the highest mean air temperature during melting conditions among these stations at +2.16 °C and a mean loss of +1.86 dB (from winter mean) during AWS defined melting. Only the Uranus Glacier station has a smaller backscatter change, +1.80 dB, with an average temperature of +0.79 °C during melting. In contrast, the Larsen Ice Shelf averages a loss of +15.07 dB with a mean temperature of +1.06 °C during AWS-defined melting.

Omission errors for all stations show an average of 12 % fewer false positives for the FT3 compared to the CWT. For the Limbert AWS shown in Fig. 2e the CWT method has a maximum mean omission error of 83 %, compared to the FT3 at 40%. The Limbert station has a short melting season,

Table 1. Stations used in the AWS evaluation. Shown are the agreement (true positive), omission (false positive) and commission (true negative) relative differences from the comparison of melting estimated from AWS air temperatures and melt duration (MD) time series estimated using the FT3 and CWT melt algorithms.

AWS Station Name	Agreement		Omission Error		Commission Error	
	FT3	CWT	FT3	CWT	FT3	CWT
Butler Island	50 %(92/184)	30 % (55/184)	50 % (92/184)	70 % (129/184)	2 % (59/2708)	21 % (38/2708)
Fossil Bluff	75 % (123/164)	66 % (109/164)	25 % (41/164)	34 % (55/164)	10 % (91/923)	10 % (89/923)
Larsen Ice Shelf	87 % (413/477)	77 % (366/477)	13 % (64/477)	23 % (111/477)	10 % (285/2771)	9 % (238/2771)
Limbert	60 % (21/35)	17 %(6/35)	40 %(14/35)	83 %(29/35)	0 %(7/3225)	0 %(5/3225)
Pegasus South	38 % (46/121)	48 % (58/121)	62 % (75/121)	52 % (63/121)	1 % (14/2421)	1 % (28/2421)
Uranus Glacier	36 % (57/160)	17 % (27/160)	64 % (103/160)	83 % (133/160)	1 % (11/1641)	0 % (3/1641)
Total	66 % (752/1141)	54 % (621/1141)	34 % (389/1141)	46 % (520/1141)	3 % (467/13689)	3 % (401/13689)

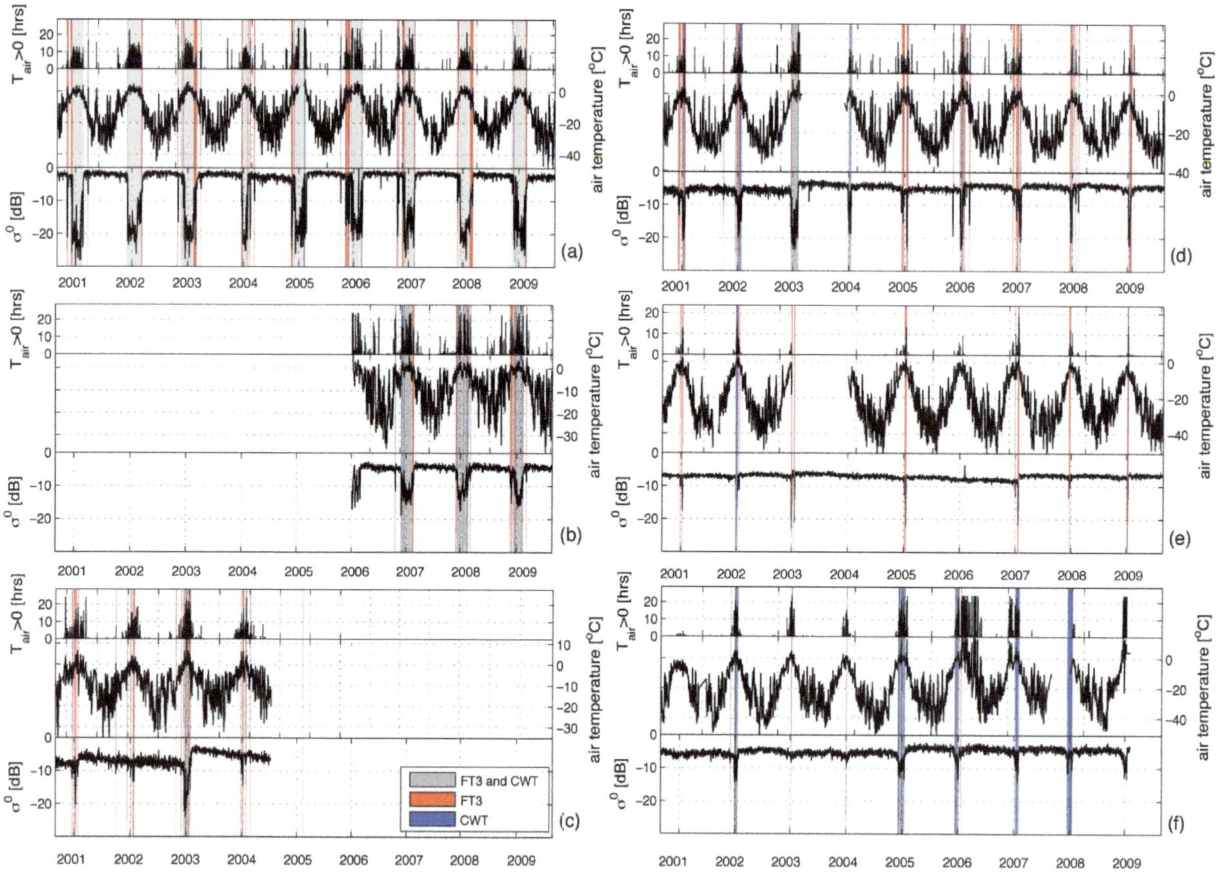

Fig. 2. AWS temperature records for **(a)** Larsen Ice Shelf, **(b)**, Fossil Bluff, **(c)**, Uranus Glacier, **(d)** Butler Island, **(e)** Limbert and **(f)** Pegasus South for 2000 to 2009 plotted with the corresponding QuikSCAT backscattering values. Times where the FT3 or CWT approaches estimate melting are shaded red and blue, respectively, while times where both methods agree are shaded grey.

averaging close to 4 days yr^{-1} in length from AWS-derived melting. The majority of these melting events are omitted using the CWT. Conversely, the FT3 approach detects these events with relatively high agreement (83 %). These findings are consistent with the basic principles of the CWT algorithm, which aims to detect only sustained melting events. For the Larsen Ice Shelf, we find the lowest rate of omission, averaging 34 % (25 %) for the CWT (FT3). Commission errors are found to be small for both methods, averaging 3 % for all stations for both the FT3 and CWT, respectively.

4.2 Comparison between QuikSCAT melting algorithms

The mean continent-scale seasonal (2000 through 2009) MD is shown in Fig. 3a and b for the FT3 and CWT approach, respectively. Figure 3c shows a map of the difference between

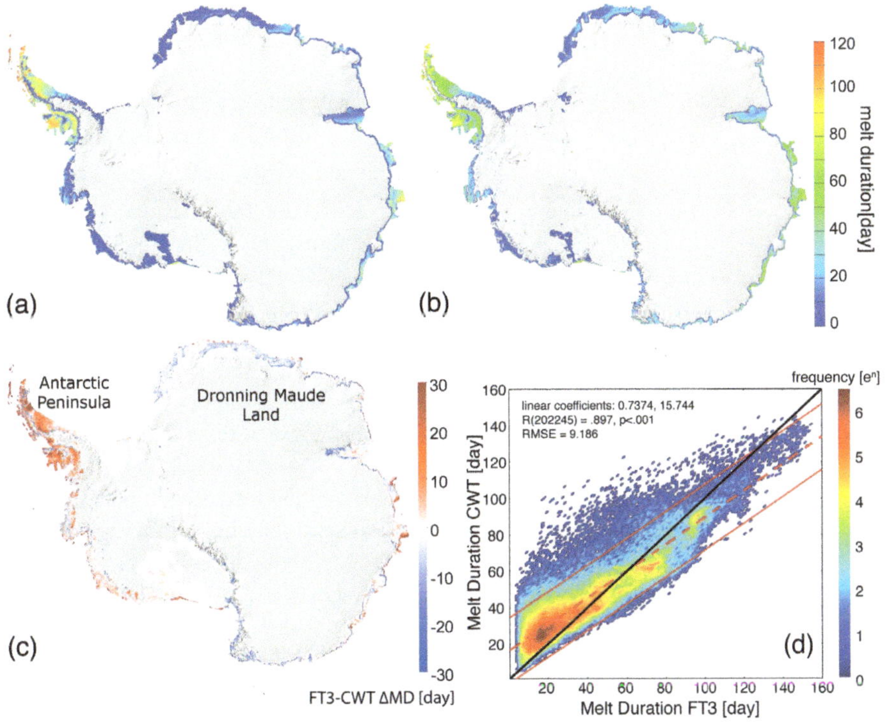

Fig. 3. **(a)** A map of mean (2000 through 2009) seasonal melt duration estimated from the SeaWinds sensor on QuikSCAT using a 3 dB below winter mean threshold (FT3) **(b)** and a continuous wavelet-based method (CWT). **(c)** The difference between mean MD from both methods is indicated as ΔMD (FT3-CWT). The location of the Antarctic Peninsula (AP) and Dronning Maud Land (D) are labeled. **(d)** Autocorrelation analysis using a log-transformed two-axis histogram, where mean MD is compared for all areas where melt is detected by both methods.

MD approaches where $\Delta MD = MD_{FT3} - MD_{CWT}$. Although the spatial patterns of MD from the two methods are similar, systematic differences exists regionally and with elevation change. Results are presented and discussed in the context of Antarctica as a whole, specific regions and trends in ΔMD with elevation.

4.2.1 Continent-scale results

For all areas and years, the CWT algorithm produces an average MD value of 41 days vs. 28 days obtained with the FT3 algorithm. The mean continent-scale melt index (MI), defined as the area subject to melting times the number of melting days, for the two methods is similar although larger for the CWT at 2.971×10^7 day km^2 than the FT3 at 2.813×10^7 day km^2. The MI detected by the CWT is larger than the FT3 because melting events detected by the CWT are continuous and do not contain short-duration intermittent non-melting classifications as in the FT3 melting records. This is also a factor in MD differences; however, the difference in MD can also be explained by the omission of smaller duration melting events by the CWT as shown in air-temperature comparisons at the Butler Island, Uranus Glacier and Limbert AWS. The inclusion of short-duration events in the FT3 record will greatly reduce the mean MD. The total melting extent, defined as the total surface area of Antarctica subject to at least one day of melting per season, is found to be 6.23 % when using the CWT finds and 8.14 % from the FT3.

The average MO, expressed as day of year, determined by the CWT approach is day 347 (e.g., 13 December for non-leap years) and in the case of the FT3 algorithm is day 352. The mean MF date for the whole of Antarctica between methods differs by 6 days, with the FT3 suggesting a later refreezing (day 28) than the CWT approach (day 22). The difference (5 days MO and 6 days MF) between the seasonal MO and MF obtained with the two approaches is small compared to the standard deviation (σ) of either method, being $\sigma_{FT3} = 26$ days and $\sigma_{CWT} = 16$ days for MO and $\sigma_{FT3} = 28$ days and $\sigma_{CWT} = 17$ in MF over all locations and years.

The 10 yr mean MD value per grid cell over Antarctica obtained with the FT3 method is plotted vs. the CWT method 10 yr mean MD in Fig. 3d. Results of a linear least-squares regression and correlation analysis show that the results of the two algorithms are highly correlated ($r = .897$, $p < 0.001$) with a root-mean-squared error (RMSE) of ~ 9 days. The CWT method finds a lesser MD than the FT3 approach over areas that experience a melting season shorter than the mean value of 34 days of both methods. The opposite is true

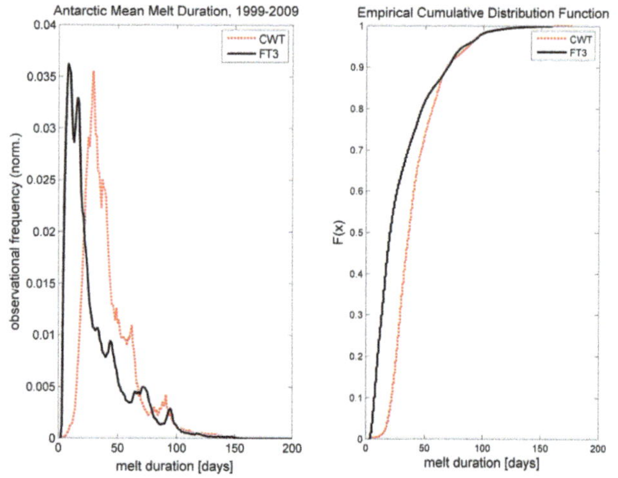

Fig. 4. (**a**) A histogram of all observed seasonal melt duration in days for the years 2000 through 2009 for the wavelet-based melt detection method (CWT) and a fixed-threshold method (FT3). It is found that the observations for the CWT are shifted towards longer durations due to the exclusion of short-duration melting events. (**b**) An empirical cumulative density function of observed melting duration.

for MD greater than 34 days, where the CWT method shows longer melt durations.

An analysis of the distribution of melt-duration occurrences, as shown in Fig. 4a, shows that the CWT algorithm finds more longer MDs relative to the FT3 and that the FT3 is much more likely to detect short-duration events. Close to 5 % of MD values found by the FT3 are 10 days or less, while for the CWT method this accounts for less than 1 % of all detected melting events (Fig. 4b). The inclusion of short-duration melting events using a fixed threshold leads to decreased mean MD value. The majority (> 50 %) of MDs for the FT3 method are under 27 days, compared to a 37-day MD for the CWT method. Both methods show better agreement for longer MD, where about 20 % of all MDs for both methods are 60 days or longer.

4.2.2 Regional results

Detailed results are presented in this section for the Antarctic Peninsula and the coastal area of Dronning Maud Land (as shown in Fig. 3c). These regions are chosen because they exhibit the largest differences in regionally integrated ΔMD or season length between algorithms.

Over the Antarctic Peninsula, seasonal mean MD values estimated by the CWT and FT3 are similar. The mean MD value obtained from the CWT approach is 55 days and 51 days from the FT3 approach. The spatial distribution of ΔMD values obtained with the two approaches (Fig. 3c) indicates that the CWT shows generally larger MD values over the ice shelves of the Antarctic Peninsula than those obtained with the FT3 approach. The similarity in mean MD in the light of a visible difference in MD over large ice shelves can be attributed to the greater melting extent found by the FT3, compared to the CWT, over areas experiencing short melting durations. The melting index over the Antarctic Peninsula is 8.3 % larger for the FT3 than the CWT (the relative difference is defined as $|MI_1 - MI_2| \Big/ \frac{1}{2}(MI_1 + MI_2)$).

When observing the seasonal melt initiation over the Antarctic Peninsula, we find that the mean MO date from the FT3 to be day 1 (e.g., 1 January) and the mean MO for the CWT is day 342. The MF dates for the CWT and FT3 over the peninsula are day 46 and day 33, respectively. The FT3 approach estimates a substantially longer (22 days) mean melting season for the Antarctic Peninsula compared to the CWT and most of this difference is due to the estimates of MO.

An analysis of selected backscattering time series during the melting season along with regionally integrated melting is presented to show difference between the CWT and FT3 algorithms. For the Antarctic Peninsula during the 2004–2005 melting season, FT3 values (Fig. 5a, solid grey line) exhibit an early season (November, 2004) peak in ME reaching an extent of up to ~ 80 % of the yearly maximum for a period of ~ 8 days. This transient melting event is not observed in the time series of melting extent from the CWT approach as melt (Fig. 5a, dotted black line). A time series of backscatter from within the region and over this same period is shown in Fig. 5c. Melting is indicated by a shaded region for the CWT method, and by the location of the 3 dB threshold for the FT3 approach in Fig. 5c, we observe that the CWT excludes several melting events at the beginning of the season (outside of the shaded area) that correspond to the large increase in region-wide melting extent as shown by the FT3 (Fig. 5a). This omission by the CWT here is representative of the differences in melting extent observed regionally. Much of difference between methods over the Peninsula can be attributed to non-sustained short-term melting/refreezing events, as shown in Fig. 5a, before and after the period when most of the melting occurs. Additionally, it is clear that the MO date for most of the Antarctic Peninsula determined using the FT3 approach (defined as the first melting event greater than three days) will correspond to the brief November melting event that, as shown in Fig. 5a, may be several weeks prior to the main melting events. The omissions of melting events during transitional periods by the CWT method is also indicated in the AWS validation of the Butler Island, Limbert and Uranus Glacier AWS records.

The maximum in absolute ΔMD between the FT3 and CWT method (Fig. 3c) is observed over areas of east Antarctica near the Donning Maud Land. Here ΔMD averages −25 days. This difference can be attributed to both differences in the in mean season length (defined as the continuous period between MO and MF) and the time that melting is sustained through a season. Differences in mean seasonal length account for only ~ 14 % of the observed ΔMD

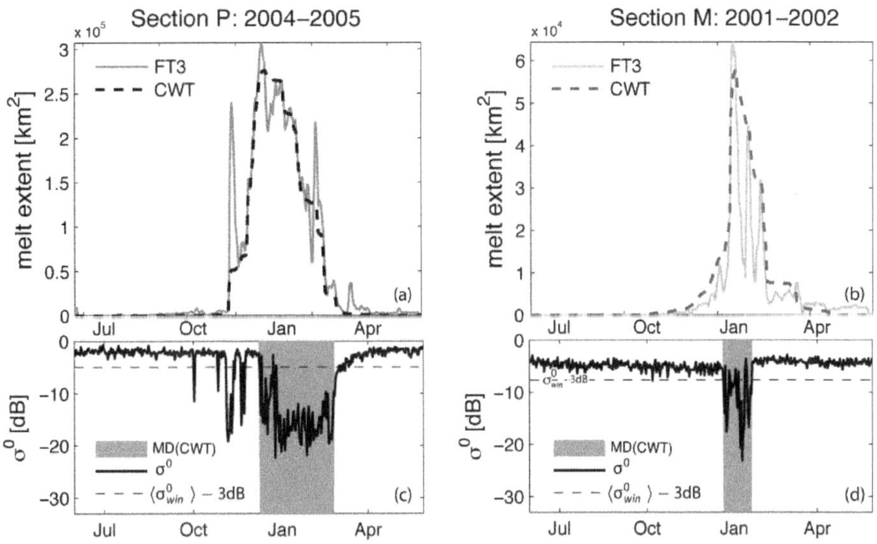

Fig. 5. (a) A time series of total integrated melt extent for 2004–2005 estimated using the FT3 and CWT methods over the Antarctic Peninsula (Section P) and **(b)** during the 2001–2002 melting season for Dronning Maud Land (Section M). Backscattering time series are plotted for **(c)** a representative pixel location for the Antarctic Peninsula during the 2004–2005 austral summer and **(d)** Dronning Maud Land during the 2001–2002 austral summer to illustrate the results of melt classification by of CWT (shaded) and FT3 methods (threshold indicated).

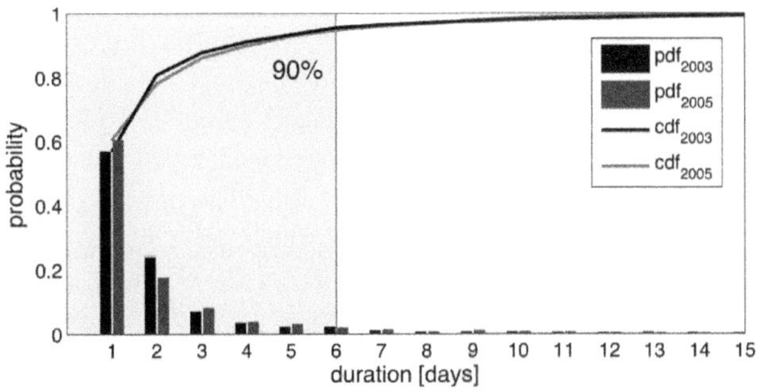

Fig. 6. The empirical probability density function (bar) and cumulative density function (line) of continuous melting periods per season observed over Antarctica (not the total per-pixel seasonal difference) for the 2003–2004 and 2005–2006 season that are found using a FT3 but rejected in by CWT.

found between methods. The percentage of melt days per melting season length (calculated as (MF-MO)/MD) is 65 % for the FT3 method and close to 95 % for the CWT approach. This important difference can be partly attributed to threshold values that underestimate the actual backscattering response to increasing liquid water content and as a result backscatter with melting is close in magnitude to the to seasonal in backscatter threshold (e.g., 3 dB). With signal noise, a fixed-threshold algorithm will alternate between classifications of melting and refreeze as backscatter fluctuates around the backscatter threshold. Alternatively, similar fluctuations in backscatter could be attributed to rapid melt–refreeze events in areas where the liquid water content is not sustained through diurnal refreeze cycles. These cases are difficult to validate without further information and are a weakness of fixed-threshold algorithms in varying snow property and temperature regimes. In the CWT methodology, sustained deviations from wintertime conditions will be detected and classified as melt until an additional refreeze transition occurs, regardless of intermittent non-sustained (negative a) fluctuations in backscatter.

A time series of ME and σ^0 for Dronning Maud Land are shown in Fig. 5b and d. Here, the CWT method finds a greater regionally integrated MI with respect to the FT3 method. A time series of σ^0 chosen from within Dronning Maud Land, as in Fig. 5c. This backscatter series indicates that the FT3 method will classify multiple melt–refreeze events as σ^0 changes rapidly around the threshold value, while the CWT will record a single melting event. This will lead to estimates of a shorter MD by the FT3. It is found that this case is representative of the majority of regions with $-\Delta$MD. Melting events as shown in Fig. 5d will not result in considerable differences in MO or MF dates.

An estimate of the average length of MD omitted by the CWT (i.e., the minimum number of days where the CWT will detect melting) is determined empirically from melting records over the Antarctic Peninsula. Figure 6 shows the cumulative sum (lines) and histogram (bars) of the length of melting days obtained from the two approaches over the Antarctic Peninsula for two selected years of 2003–2004 and 2005–2006; these seasons are selected as an example of a relatively low and high mean melting index, respectively. It is important to note that this is not an analysis of total MD per season, as shown in Fig. 4, but rather the length of a period during the melting season where FT3 detects any melting

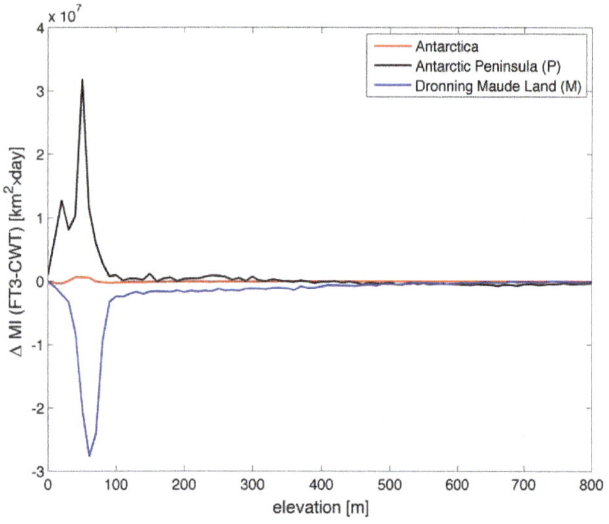

Fig. 7. An analysis of the difference in melting index ($\Delta MI=MI_{FT3}-MI_{CWT}$) estimates from the FT3 and CWT methods at a range of elevations for the entire Antarctic continent (red line) as well as the Antarctic Peninsula (black line) and Dronning Maud Land (blue line).

event and CWT does not. During the 2003–2004 season, ~60 % of MD values differ by only one day, ~20 % show differences of two days, with the remaining values differing by more than six days (~90 % of observations are six days or less). This six-day duration is similar to the temporal filtering methods used in previous melting studies to eliminate transient melting events (e.g., three days; Tedesco et al., 2007).

4.2.3 Elevation

Figure 7 shows results concerning the difference between the outputs of the two approaches as a function of elevation for the Antarctic Peninsula, Dronning Maud Land and the whole of Antarctica. For the Dronning Maud Land region, the melt index difference between the two methods ($\Delta MI = MI_{FT3}-MI_{CWT}$) is negative, independent of elevation. For the Antarctic Peninsula, however, the mean difference between methods is positive, and will vary with elevation. For areas below 400 m a.s.l. the CWT method underestimates MI with respect to the FT3 method. Conversely, above 400 m a.s.l., the MI difference becomes negative. Building on our previous analysis in this paper, we infer that for the Antarctic Peninsula at lower elevations we find that the melt difference is due to the omission of short-duration melting events, while at higher elevations we find a that melt detection differences are more closely related to either short melt–refreeze cycles at high elevations or a lesser backscatter response to snow wetness. This may indicate the need for terrain correction when using fixed-threshold methods.

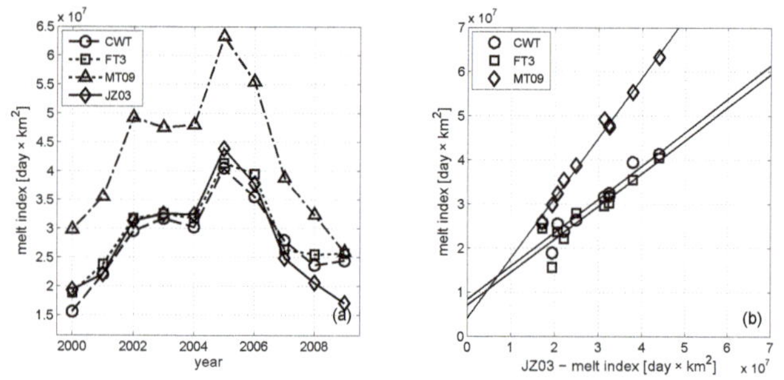

Fig. 8. **(a)** The time series of total melt index (MI), in day km^2, for the Antarctic continent plotted for the years 1999 through 2009 derived from the FT3 and CWT methods on an enhanced-resolution QuikSCAT active microwave data set along with estimates from the MT09 and M+K30 methods and an SSM/I data set. **(b)** The correlation between the FT3, QuikSCAT data set MI estimates and the CWT, M+K30 and MT09 methods is indicated using a linear regression.

4.3 Comparison with results from passive microwave measurements

We compare melting records from passive microwave SSM/I observations using approaches proposed in the literature with the outputs of both the CWT and FT3 methods. The SSM/I-derived melting is produced using two methods. The first is a fixed-threshold approach as in Zwally and Fiegles (1994), here denoted M+30K. The second data set is produced using a dynamic electromagnetic modeling-based detection approach as in Tedesco (2009), here denoted MT09. Outputs from both methods are projected onto the 2.225 km QuikSCAT grid using nearest-neighbor interpolation. Enhanced-resolution passive microwave brightness temperatures generated using the SIR algorithm are available for Antarctica (Long and Stroeve, 2011). Melting records based on these data would reduce scaling discrepancies between the active and passive microwave melt records reported here and provide a more consistent analysis. This will be explored in further studies.

The values of seasonally integrated MI for Antarctica for both active and passive microwave methods are plotted in Fig. 8a. The M+30K and CWT approaches show the most similar magnitude in seasonal MI, where the relative difference is 9 %. This is only 2 % greater than the difference between the CWT and FT3 methods (7 %). There is a slightly greater relative difference between the FT3 and M+30K of 11 %. The MT09 method finds on average a 40 % greater MI than the FT3 method, and 35 % greater than the CWT.

For individual seasons, all methods find a MI maximum during the 2004–2005 period. The active and passive methods, however, do not agree on the year of minimum MI. Indeed, both the CWT and FT3 approaches find a minimum MI during the 1999–2000 seasons, while the passive methods find a minimum during the 2008–2009 melting season, where most of this difference is from the Antarctic Peninsula. Seasonally integrated melting indices derived using the passive microwave methods are well correlated with the QuikSCAT-derived melting records. This relationship is shown in Fig. 8b. The best agreement between active and passive microwave records is found between the CWT and M+30K, as indicated in Table 2a. Further, both MT09 and M+30K have a greater degree of correlation and a lower RMSE for the CWT than compared to the FT3.

The spatial differences between the mean MD estimated using the M+30K approach and the CWT and FT3 is shown in Fig. 9a and b, respectively. These maps indicate spatial patterns in the magnitude of difference between methods ($\Delta MD_{FT3} = M + 30K\text{-}FT3$). Over large ice shelves (e.g., the Larsen and Amery ice shelves), ΔMD_{FT3} is close to +10 days (Fig. 9a). Over the same regions, the ΔMD between the M+30K and the CWT algorithms is close to +20 days (Fig. 9b). The tendency for the passive microwave data to overestimate active microwave estimates (close to +10 or +20ΔMD in most cases) is usually bordered, to some degree, by an area of underestimation by the passive measurements. The majority of areas that exhibit melting in Antarctica are generally at ice–ocean boundaries. For the Antarctic Peninsula (among other places) areas of melt occur adjacent to sharp contrasts in elevation. For many of these areas sub-pixel mixing will likely lead to a decrease in the observed brightness temperatures of some melting areas. For example, over the coastal regions of Dronning Maud Land we find that the systematic occurrence of positive ΔMD bordered negative ΔMD. These negative ΔMDs are found adjacent to ocean pixels and high elevations and are likely due to a sub-pixel mixing effect. In another case, the relatively narrow King George VI (lat. −71.965, long. −67.807) Ice Shelf, located roughly east of Wilkens Ice Shelf (lat. −67.525, long. −62.775), appears to increase the apparent MD of the surrounding high-elevation areas for the passive microwave case, resulting in a large positive ΔMD. Since mixed pixel effects dominate the spatial differences between methods, it is difficult to determine a relationship between MD methods over similar areas other than the positive ∼10 days (FT3) and ∼20 days (CWT) reported previously.

An analysis of correlation between colocated melting durations from active microwave data sets and the M+30K method indicates that the FT3 has a higher degree of correlation with the passive microwave M+30K method Table 2b. It is also found that AMW methods find an RMSE of ∼18 days in comparison with both the M+30 and MT09. The relationship between spatially colocated data records are illustrated the Fig. 9c and d. A first-order least-squares regression between CWT and M+30K shows an ∼13-day positive bias. There is an ∼3-day positive bias between the FT3 and M+30K from similar analysis. Artifacts that appear as vertical striping of data points indicate a high degree of variability for many pixels with similar MD as detected using the M+30K. These are likely a result of a high degree of sub-pixel variability due to the difference in spatial resolution as previously discussed. In Fig. 9c and d, we see that a larger component of the melt data set for the FT3 method falls along the 1 : 1 line. The CWT method appears to underestimate MD as compared to M+30K, or vice versa. It also appears that the CWT method finds greater MD for shorter durations, evident by a cluster of points 0 to 50 MD for CWT and 0 to 20 days for M+30K. This relationship is similar to observation between active microwave methods. The relatively strong agreement between active and passive melting indices, shown in Fig. 8a, indicates that the high sub-pixel variability is averaged out when integrated over the entire data set.

5 Summary and conclusion

The use of a combined continuous wavelet transform and multiscale analysis is able to detect changes in the backscattering signal upon an increase in liquid water content. This methodology does not require an estimation of the expected response to snowmelt and is therefore applicable to spatially variable snow characteristics and across instrumentation. It has also been found to be effective in the light of increased signal noise. Multiscale analysis provides a quantifiable measure of the nature of transitions in backscatter in terms of relative persistence and rate of transition. This can be used in classification. Here we use multiscale analysis to separate sporadic from persistent melting events.

Estimated mean MD derived from the novel wavelet approach and a more standard fixed-threshold method are very well correlated, $r = .897$ ($p < 0.001$) with an RMSE of ∼9 days. In mean MD, we found that there is as a 19 % relative difference between methods, and the CWT method averages 13 days longer MD than the FT3. This difference in MD is largely due to the omission of shorter duration melting events, most of which are six days or less.

Measures of MI from both measures have a 5.5 % relative difference, the CWT greater than the FT3. Areas where the CWT is greater in mean MD are found to have intermittent refreezing events or a backscatter response to liquid water content that is close to or below the 3 dB threshold value with signal noise. This is found over much of Dronning Maud Land and at high elevations over the Antarctic Peninsula.

At a continental scale the mean melting season length (MF-MO) estimated in both methodologies agrees to within two days. On the Antarctic Peninsula, shorter duration melting events during the transitional periods are omitted by the CWT leading to a melt onset 12 days earlier detected by the FT3 method. This is illustrated during the 2004–2005 season, where the melt extent reaches ∼80 % of its maximum seasonal value for close to 8 days and the CWT method does not detect melting over an equivalent extent until several weeks later. The MO found by the CWT is closer in time to the start of a sustained melting period. The FT3 defines

Table 2. The correlation (r) and the root-mean-squared error (RMSE) between surface melting data sets for **(a)** seasonally integrated melting index **(b)** and spatially coincident mean melting duration.

(a)	CWT	FT3	M+30K	MT09
CWT	r	0.972 ($p < 0.001$)	0.934 ($p < 0.001$)	0.931 ($p < 0.001$)
	RMSE [day × km^2]	1.89×10^6	2.74×10^6	1.56×10^7
FT3	r	–	0.922 ($p = 0.01$)	0.911 ($p = 0.002$)
	RMSE [day × km^2]	–	3.10×10^6	1.41×10^7
(b)	**CWT**	**FT3**	**M+30K**	**MT09**
CWT	r	0.897 ($p < 0.001$)	0.664 ($p < 0.001$)	0.663 ($p < 0.001$)
	RMSE [day]	9	19	18
FT3	r	–	0.782 ($p < 0.001$)	0.672 ($p < 0.001$)
	RMSE [day]	–	17	17

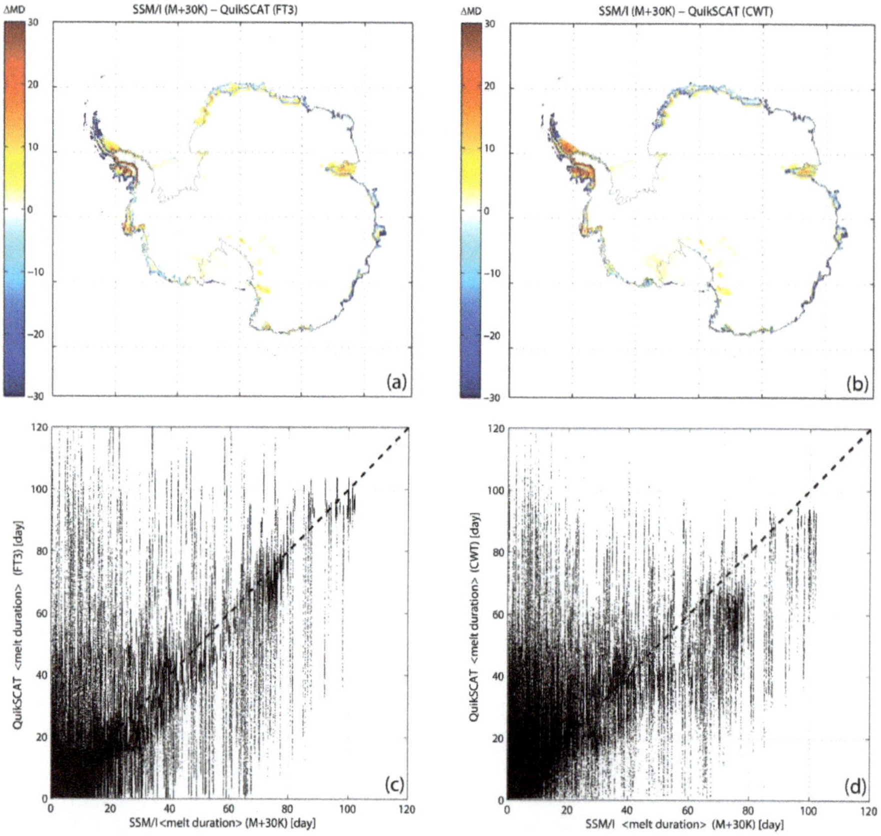

Fig. 9. The difference in average melt duration ($\Delta MD = MD_{SSM/I} - MD_{QuikSCAT}$) over the period of 1999–2009 between the up-sampled passive microwave SSM/I data set, where melt is estimated using the M+K30 approach, and the enhanced-resolution QuikSCAT scatterometer data set, where melt is estimated using the **(a)** FT3 and **(b)** CWT methods. Per-pixel scatterplots of average melt duration between SSM/I (M+30K) and the **(c)** FT3 and **(d)** CWT approaches.

MO coincident to the start of a transitional period of multiple freeze–thaw cycles.

From comparison with AWS we find that the FT3 has higher overall level of agreement with air-temperature measurements, with a 66 % total agreement compared to the 54 % from the CWT method. This is true for all AWS apart from the Pegasus South station, where the CWT has a 10 % greater agreement than the FT3 method. The FT3 method has a better agreement in general because of its ability to measure short-duration melting events, which are common for several stations (e.g., Limbert). For the Pegasus South AWS, it appears that as the melting season occurs with a relatively low loss in backscatter with melting conditions, it is therefore never consistently under the 3 dB threshold.

Many of these above-zero temperature events are not long lasting in nature, and to evaluate the CWT method solely using agreement assessments does not indicate the true utility of this approach to create a record of sustained melting events.

Compared to M+30K, a passive microwave-derived approach, both active microwave melting records find a similar yearly MI, where the FT3 is within 11 % and the CWT within 9 % relative difference. These methods find a greater level of disagreement with a dynamic thresholding approach, MT09. The CWT methods find a greater overall agreement in temporal trend for both PMW methods, for the M+30K method $r = .943$ ($p < 0.001$) with CWT and $r = .922$ ($p = 0.001$) for the FT3 method. Spatially, we find that comparison is dominated by a mixed pixel effect, making it difficult to determine the difference between melt duration on a per-pixel basis. The FT3 method has a higher degree of spatial correlation with passive microwave (M+30K) approaches than the CWT, with significant variability in AMW-derived MD for similar passive microwave measurements. Since PMW pixels are ~ 10 times the scale of the spatially enhanced QuikSCAT product, we attribute this variability to mixed pixel effects. Based on the strong correlation in yearly MI totals, the differences in spatial variability appear to be averaged out over the total area.

In conclusion, the CWT methodology is an alternative to a fixed-threshold method that works well on data with increased signal noise, an application where threshold-based methods find difficulty. This approach is also able to classify melting events based on relative duration. This allows for the focused measurements of persistent melting, rather than total integrated melting, a utility that may be useful studying long-term trends in melting

Acknowledgements. The authors acknowledge the support of NSF grant ANT-1141993. The authors acknowledge D. Long and the NASA Jet Propulsion Laboratory PO.DAAC for providing the enhanced-resolution scatterometer products. The authors appreciate the support of the Automatic Weather Station program and the Antarctic Meteorological Research Center for the data set, data display, information and further assistance (M. Lazzara, NSF grant numbers ARC-0713843, ANT-0944018, and ANT-1141908). This research was supported in part by National Science Foundation grants CNS-0958379 and CNS-0855217 and the City University of New York High Performance Computing Center at the College of Staten Island.

Edited by: J. Stroeve

References

Abdalati, W. and Steffen, K.: Greenland ice sheet melt extent: 1979–1999, J. Geophys. Res.-Atmos., 106, 33983–33988, doi:10.1029/2001jd900181, 2001.

Alexandrescu, M., Gibert, D., Hulot, G., Lemouel, J. L., and Saracco, G.: Detection of Geomagnetic Jerks Using Wavelet Analysis, J. Geophys. Res.-Sol. Ea., 100, 12557–12572, doi:10.1029/95JB00314, 1995.

Ashcraft, I. S. and Long, D. G.: Comparison of methods for melt detection over Greenland using active and passive microwave measurements, Int. J. Remote Sens., 27, 2469–2488, doi:10.1080/01431160500534465, 2006.

Barrand, N. E., Vaughan, D. G., Steiner, N., Tedesco, M., Munneke, P. K., Van den Broeke, M. R., and Hosking, J. S.: Trends in Antarctic Peninsula surface melting conditions from observations and regional climate modeling, J. Geophys. Res.-Earth, 118, 315–330, 2013.

Bromwich, D. H. and Nicolas, J. P.: Sea-Level Rise Ice-sheet uncertainty, Nat. Geosci., 3, 596–597, doi:10.1038/ngeo946, 2010.

Chen, J. L., Wilson, C. R., and Tapley, B. D.: Interannual variability of Greenland ice losses from satellite gravimetry, J. Geophys. Res.-Sol. Ea., 116, B07406, doi:10.1029/2010jb007789, 2011.

Comiso, J.: Polar oceans from space, Springer, New York, 2010.

Daubechies, I.: Ten lectures on wavelets, Society for Industrial and Applied Mathematics, Philadelphia, Pa., 1992.

Dowdeswell, J. A.: Atmospheric science – The Greenland Ice Sheet and global sea-level rise, Science, 311, 963–964, doi:10.1126/science.1124190, 2006.

Early, D. S. and Long, D. G.: Image reconstruction and enhanced resolution imaging from irregular samples, IEEE T. Geosci. Remote, 39, 291–302, doi:10.1109/36.905237, 2001.

Hansen, J., Ruedy, R., Sato, M., and Lo, K.: Global Surface Temperature Change, Rev. Geophys., 48, Rg4004, doi:10.1029/2010rg000345, 2010.

Herrmann, F. J.: Singularity characterization by monoscale analysis: Application to seismic imaging, Appl. Comput. Harmon. A., 11, 64–88, doi:10.1006/acha.2000.0349, 2001.

Holland, P. R., Corr, H. F. J., Pritchard, H. D., Vaughan, D. G., Arthern, R. J., Jenkins, A., and Tedesco, M.: The air content of Larsen Ice Shelf, Geophys. Res. Lett., 38, L10503, doi:10.1029/2011GL047245, 2011.

Holschneider, M.: Wavelets: an analysis tool, Clarendon Press, Oxford University Press, Oxford, New York, 1995.

Hughes, T. J.: The Weak Underbelly of the West Antarctic Ice-sheet, J. Glaciol., 27, 518–525, 1981.

Joshi, M., Merry, C. J., Jezek, K. C., and Bolzan, J. F.: An edge detection technique to estimate melt duration, season and melt extent on the Greenland ice sheet using passive microwave data, Geophys. Res. Lett., 28, 3497–3500, doi:10.1029/2000GL012503, 2001.

Joughin, I. and Alley, R. B.: Stability of the West Antarctic ice sheet in a warming world, Nat. Geosci., 4, 506–513, doi:10.1038/ngeo1194, 2011.

Le Gonidec, Y., Gibert, D., and Proust, J. N.: Multiscale analysis of waves reflected by complex interfaces: Basic principles and experiments, J. Geophys. Res.-Sol. Ea., 107, 2184, doi:10.1029/2001JB000558, 2002.

Lemke, P., Ren, J., Alley, R. B., Allison, I., Carrasco, J., Flato, G., Fujii, Y., Kaser, G., Mote, P., Thomas R. H., and Zhang, T.: Observations: Changes in Snow, Ice and Frozen Ground, in: Climate Change 2007: The Physical Science Basis. Contribution of Working Group I to the Fourth Assessment Report of the Intergovernmental Panel on Climate Change, edited by: Solomon, S.,

Qin, D., Manning, M., Chen, Z., Marquis, M., Averyt, K. B., Tignor, M., and Miller, H. L., Cambridge University Press, Cambridge, UK and New York, NY, USA, 2007.

Liu, H., Wang, L., and Jezek, K. C.: Wavelet-transform based edge detection approach to derivation of snowmelt onset, end and duration from satellite passive microwave measurements, Int. J. Remote Sens., 26, 4639–4660, 2005.

Long, D. G. and Hicks, B. R.: Standard BYU QuikScat/SeaWinds land/ice image products, Brigham Young Univ., Provo, UT, QuikScat Image Product documentation, 2000.

Long, D. G. and Stroeve, J.: Enhanced-Resolution SSM/I and AMSR-E Daily Polar Brightness Temperatures, Boulder, Colorado, USA: NASA DAAC at the National Snow and Ice Data Center, 2011.

MacAyeal, D. R., Scambos, T. A., Hulbe, C. L., and Fahnestock, M. A.: Catastrophic ice-shelf break-up by an ice-shelf-fragment-capsize mechanism, J. Glaciol., 49, 22–36, doi:10.3189/172756503781830863, 2003.

Mallat, S. G.: A wavelet tour of signal processing, 2nd Edn., Academic Press, San Diego, 1999.

Mallat, S. G. and Hwang, W. L.: Singularity Detection and Processing with Wavelets, Ieee T. Inf. Th., 38, 617–643, doi:10.1109/18.119727, 1992.

Mallat, S. G. and Zhong, S.: Characterization of Signals from Multiscale Edges, Ieee T. Pattern Anal., 14, 710–732, doi:10.1109/34.142909, 1992.

Mote, T. L. and Anderson, M. R.: Variations in Snowpack Melt on the Greenland Ice-sheet Based on Passive-microwave Measurements, J. Glaciol., 41, 51–60, 1995.

Mote, T. L., Anderson, M. R., Kuivinen, K. C., and Rowe, C. M.: Passive microwave-derived spatial and temporal variations of summer melt on the Greenland ice sheet, Ann. Glaciol., 17, p. 233, 1993.

Nghiem, S. V., Steffen, K., Kwok, R., and Tsai, W. Y.: Detection of snowmelt regions on the Greenland ice sheet using diurnal backscatter change, J. Glaciol., 47, 539–547, doi:10.3189/172756501781831738, 2001.

Nghiem, S. V, Steffen, K., Neumann, G., and Huff, R.: Snow accumulation and snowmelt monitoring in Greenland and Antarctica, in: Dynamic Planet, 31–38, Springer, 2007.

Ohmura, A.: Physical basis for the temperature-based melt-index method, J. Appl. Meteorol., 40, 753–761, doi:10.1175/1520-0450(2001)040<0753:pbfttb>2.0.co;2, 2001.

Overpeck, J. T., Otto-Bliesner, B. L., Miller, G. H., Muhs, D. R., Alley, R. B., and Kiehl, J. T.: Paleoclimatic evidence for future ice-sheet instability and rapid sea-level rise, Science, 311, 1747–1750, doi:10.1126/science.1115159, 2006.

Rignot, E. and Thomas, R. H.: Mass balance of polar ice sheets, Science, 297, 1502–1506, 2002.

Rignot, E., Casassa, G., Gogineni, P., Krabill, W., Rivera, A., and Thomas, R.: Accelerated ice discharge from the Antarctic Peninsula following the collapse of Larsen B ice shelf, Geophys. Res. Lett., 31, L18401, doi:10.1029/2004GL020697, 2004.

Rignot, E., Velicogna, I., Van den Broeke, M. R., Monaghan, A., and Lenaerts, J.: Acceleration of the contribution of the Greenland and Antarctic ice sheets to sea level rise, Geophys. Res. Lett., 38, L05503, doi:10.1029/2011gl046583, 2011.

Rott, H., Muller, F., Nagler, T., and Floricioiu, D.: The imbalance of glaciers after disintegration of Larsen-B ice shelf, Antarctic Peninsula, Cryosphere, 5, 125–134, doi:10.5194/tc-5-125-2011, 2011.

Scambos, T. A., Bohlander, J. A., Shuman, C. A., and Skvarca, P.: Glacier acceleration and thinning after ice shelf collapse in the Larsen B embayment, Antarctica, Geophys. Res. Lett., 31, L18402, doi:10.1029/2004gl020670, 2004.

Scambos, T. A., Fricker, H. A., Liu, C. C., Bohlander, J., Fastook, J., Sargent, A., Massom, R., and Wu, A. M.: Ice shelf disintegration by plate bending and hydro-fracture: Satellite observations and model results of the 2008 Wilkins ice shelf break-ups, Earth Planet. Sc. Lett., 280, 51–60, doi:10.1016/j.epsl.2008.12.027, 2009.

Shepherd, A. and Wingham, D.: Recent sea-level contributions of the Antarctic and Greenland ice sheets, Science, 315, 1529–1532, doi:10.1126/science.1136776, 2007.

Shepherd, A., Ivins, E. R., Geruo, A., Barletta, V. R., Bentley, M. J., Bettadpur, S., Briggs, K. H., Bromwich, D. H., Forsberg, R., and Galin, N.: A reconciled estimate of ice-sheet mass balance, Science, 338, 1183–1189, 2012.

Spencer, M. W., Wu, C. L., and Long, D. G.: Improved resolution backscatter measurements with the SeaWinds pencil-beam scatterometer, Ieee T. Geosci. Remote, 38, 89–104, doi:10.1109/36.823904, 2000.

Steffen, K., Nghiem, S. V, Huff, R., and Neumann, G.: The melt anomaly of 2002 on the Greenland Ice Sheet from active and passive microwave satellite observations, Geophys. Res. Lett., 31, L20402, doi:10.1029/2004GL020444, 2004.

Stiles, W. H. and Ulaby, F. T.: The Active and Passive Microwave Response to Snow Parameters .1. Wetness, J. Geophys. Res.-Oc. Atmos., 85, 1037–1044, doi:10.1029/JC085iC02p01037, 1980.

Tedesco, M.: Assessment and development of snowmelt retrieval algorithms over Antarctica from K-band spaceborne brightness temperature (1979–2008), Remote Sens. Environ., 113, 979–997, doi:10.1016/j.rse.2009.01.009, 2009.

Tedesco, M. and Monaghan, A. J.: An updated Antarctic melt record through 2009 and its linkages to high-latitude and tropical climate variability, Geophys. Res. Lett., 36, L18502, doi:10.1029/2009GL039186, 2009.

Tedesco, M., Abdalati, W., and Zwally, H. J.: Persistent surface snowmelt over Antarctica (1987–2006) from 19.35 GHz brightness temperatures, Geophys. Res. Lett., 34, L18504, doi:10.1029/2007gl031199, 2007.

Torinesi, O., Fily, M., and Genthon, C.: Variability and trends of the summer melt period of Antarctic ice margins since 1980 from microwave sensors, J. Climate, 16, 1047–1060, doi:10.1175/1520-0442(2003)016<1047:VATOTS>2.0.CO;2, 2003.

Trusel, L. D., Frey, K. E., and Das, S. B.: Antarctic surface melting dynamics: Enhanced perspectives from radar scatterometer data, J. Geophys. Res.-Earth, 117, 2156–2202, doi:10.1029/2011JF002126, 2012.

Ulaby, F. T. and Stiles, W. H.: The Active and Passive Microwave Response to Snow Parameters 2: Water Equivalent of Dry Snow, J. Geophys. Res.-Oc. Atm., 85, 1045–1049, doi:10.1029/JC085iC02p01045, 1980.

Ulaby, F. T., Moore, R. K., and Fung, A. K.: Microwave Remote Sensing: Active and Passive Volume II: Radar Remote Sensing and Surface Scattering and Emission Theory, Ann Arbor, Artech House., 1982.

Van den Broeke, M.: Strong surface melting preceded collapse of Antarctic Peninsula ice shelf, Geophys. Res. Lett., 32, L12815, doi:10.1029/2005gl023247, 2005.

Van den Broeke, M., Bus, C., Ettema, J., and Smeets, P.: Temperature thresholds for degree-day modelling of Greenland ice sheet melt rates, Geophys. Res. Lett., 37, L18501, doi:10.1029/2010gl044123, 2010.

Vaughan, D. G.: Recent trends in melting conditions on the Antarctic Peninsula and their implications for ice-sheet mass balance and sea level, Arct. Antarct. Alp. Res., 38, 147–152, doi:10.1657/1523-0430(2006)038[0147:rtimco]2.0.co;2, 2006.

Wang, L. and Yu, J.: Spatiotemporal Segmentation of Spaceborne Passive Microwave Data for Change Detection, IEEE Geosci. Remote S., 8, 909–913, doi:10.1109/LGRS.2011.2140312, 2011.

Wang, L., Derksen, C., and Brown, R.: Detection of pan-Arctic terrestrial snowmelt from QuikSCAT, 2000–2005, Remote Sens. Environ., 112, 3794–3805, doi:10.1016/j.rse.2008.05.017, 2008.

Weng, H. Y. and Lau, K. M.: Wavelets, Period-doubling, and Time-frequency Localization with Application to Organization of Convection over the Tropical Western Pacific, J. Atmos. Sci., 51, 2523–2541, doi:10.1175/1520-0469(1994)051<2523:WPDATL>2.0.CO;2, 1994.

Zwally, H. J. and Fiegles, S.: Extent and Duration of Antarctic Surface Melting, J. Glaciol., 40, 463–476, 1994.

8

A double continuum hydrological model for glacier applications

B. de Fleurian[1,2], O. Gagliardini[1,2,3], T. Zwinger[4], G. Durand[1,2], E. Le Meur[1,2], D. Mair[5], and P. Råback[4]

[1]CNRS, LGGE UMR5183, 38041 Grenoble, France
[2]Univ. Grenoble Alpes, LGGE, 38041 Grenoble, France
[3]Institut Universitaire de France, Paris, France
[4]CSC – IT Center for Science Ltd, Espoo, Finland
[5]Department of Geography and Environment, University of Aberdeen, Aberdeen, UK

Correspondence to: B. de Fleurian (basile.defleurian@uci.edu)

Abstract. The flow of glaciers and ice streams is strongly influenced by the presence of water at the interface between ice and bed. In this paper, a hydrological model evaluating the subglacial water pressure is developed with the final aim of estimating the sliding velocities of glaciers. The global model fully couples the subglacial hydrology and the ice dynamics through a water-dependent friction law. The hydrological part of the model follows a double continuum approach which relies on the use of porous layers to compute water heads in inefficient and efficient drainage systems. This method has the advantage of a relatively low computational cost that would allow its application to large ice bodies such as Greenland or Antarctica ice streams. The hydrological model has been implemented in the finite element code Elmer/Ice, which simultaneously computes the ice flow. Herein, we present an application to the Haut Glacier d'Arolla for which we have a large number of observations, making it well suited to the purpose of validating both the hydrology and ice flow model components. The selection of hydrological, under-determined parameters from a wide range of values is guided by comparison of the model results with available glacier observations. Once this selection has been performed, the coupling between subglacial hydrology and ice dynamics is undertaken throughout a melt season. Results indicate that this new modelling approach for subglacial hydrology is able to reproduce the broad temporal and spatial patterns of the observed subglacial hydrological system. Furthermore, the coupling with the ice dynamics shows good agreement with the observed spring speed-up.

1 Introduction

The flow of glaciers is a combination of viscous ice deformation and subglacial phenomena such as basal sliding and/or deformation of the sediment layer if it exists. The sliding component is particularly important for temperate glaciers, and can account for up to 90 % of the total surface speed (Cuffey and Paterson, 2010). As shown by earlier theoretical considerations (Weertman, 1957; Lliboutry, 1968; Schoof, 2005), water pressure is the key variable explaining most of the modulation of basal sliding. For high water pressure, the strength of the bed resistance is reduced, either by cavitation in the case of hard beds or by dilatation of the sediment layer for soft beds, inducing an increase in sliding speed. Observations of surface velocity and basal water pressure at the same location on glaciers confirm the importance of water pressure in controlling the flow of glaciers (e.g. Iken and Bindschadler, 1986; Mair et al., 2003; Harper et al., 2007; Fudge et al., 2009).

The aim of our work is to develop a subglacial hydrological model which can simulate basal water pressure and couple it to an ice flow model. The general framework is therefore in line with work by e.g. Pimentel et al. (2010) and Hewitt (2013) but with a different approach for the hydrological model. Nevertheless, as in these earlier works, supraglacial and englacial water systems are extremely simplified and our focus is on the subglacial system. The characterization of the subglacial system is rather ill-defined. Whatever the drainage systems, they can be classified into two main groups: (i) inefficient drainage systems, exhibiting high water pressure, such as water film (Weertman, 1972; Walder, 1982), linked cavities (Lliboutry, 1968; Walder, 1986) or diffusion in a sediment layer (Shoemaker, 1986) and (ii) efficient drainage systems, exhibiting lower water pressure, such as ice-walled channels (Röthlisberger, 1972), channels opened in the bedrock (Nye, 1976) or at the interface between ice and sediment (Walder and Fowler, 1994).

Drainage systems under glaciers are a combination of these inefficient and efficient systems (Shoemaker, 1986; Fountain and Walder, 1998; Boulton et al., 2007; Schoof, 2010a). These two types of system, as a consequence of their efficiency to drain water, have different impacts on water pressure and consequently on basal sliding. Inefficient drainage systems are highly pressurized, which results in relatively fast sliding speeds, whereas efficient drainage systems allow water to drain at lower pressures. These two types of system, where they coexist, are tightly coupled and the efficient drainage system will tend to drain water out of the inefficient one, which in turn induces a decrease in the basal velocities (Björnsson, 1974; Magnusson et al., 2010).

Recently published subglacial hydrological models take into account inefficient and efficient components for the drainage system (Pimentel et al., 2010; Schoof, 2012; Hewitt et al., 2012; Werder et al., 2013). Following the work initiated by Flowers and Clarke (2002a), a sediment layer is used to model the inefficient drainage system (IDS), and, rather than actually modelling a network of channels to represent the efficient drainage system, we use an equivalent porous layer (EPL). This approach, known as the *dual continuum porous equivalent approach* in hydrogeology has been developed for karstified aquifers (Teutsch and Sauter, 1991). Karstified and glaciological hydrological systems share some distinctive features that motivate this approach. Specifically, they both consist of systems with an inefficient drainage component and a more efficient one which is activated only under some water head conditions (Hubbard and Nienow, 1997; White, 1999; Gulley et al., 2012).

The model is tuned and validated by performing a set of three experiments of increasing complexity using data obtained on the Haut Glacier d'Arolla. Data sets containing both hydrological and ice flow observations are rare. The Haut Glacier d'Arolla data set is one of the most complete, although it suffers from some non-synchronous measurements. To our knowledge, this is the first time that results from a three-dimensional coupled hydrology–ice flow model are compared to a data set including both hydrology and ice flow data for a winter to summer transition.

The present paper describes the double continuum approach and the numerical methods that are used for its treatment in Sect. 2. The ice flow model equations and boundary conditions are introduced in Sect. 3. Section 4 presents the simulation settings and the results leading to the selection of hydrological parameters. Finally, the coupling between the hydrological and ice dynamic model is presented in Sect. 5.

2 Hydrological model

The basal drag of glaciers is strongly modulated by the effective pressure (e.g. Schoof, 2005), i.e. the difference between the overburden ice normal stress σ_{nn} and the water pressure p_{w}:

$$N = -\sigma_{\mathrm{nn}} - p_{\mathrm{w}}. \tag{1}$$

High water pressure induces low effective pressure, and where $N = 0$ the ice is locally floating. The effective pressure is the key variable for the coupling between glacier sliding and the subglacial hydrological system. In Eq. (1), the basal water pressure p_{w} is positive for compression, whereas the normal Cauchy stress σ_{nn} is negative for compression and defined as

$$\sigma_{\mathrm{nn}} = \boldsymbol{n} \cdot \boldsymbol{\sigma} \boldsymbol{n}, \tag{2}$$

with $\boldsymbol{n}$ the upward pointing vector normal to the bedrock and $\boldsymbol{\sigma}$ the Cauchy stress tensor.

Note that the definition of N in Eq. (1), using the normal Cauchy stress instead of the overburden hydrostatic ice pressure p, is more rigorous and fully accounts for the stress distribution at the base of the glacier. Moreover, solving the Stokes equations for the ice flow often results in a normal Cauchy stress at the base which differs slightly from the hydrostatic pressure, justifying the use of Eq. (1) to define the effective pressure.

Using the dual continuum porous equivalent approach, the inefficient and efficient drainage components are both modelled as sediment layers with the use of a specific activation scheme for the efficient drainage system. This approach defines in a continuous manner the location where the efficient drainage system is most likely to develop. This strategy has the advantage of requiring a lower spatial resolution than the discrete approaches describing each channel individually (Schoof, 2010b; Hewitt et al., 2012; Werder et al., 2013). Furthermore, the use of a diffusion equation to compute the water head distribution in both systems allows for the implementation of an implicit time-stepping scheme which yields to a rather stable system, which in turn reduces the computational cost.

2.1 Water distribution in a porous medium

Hereafter, the index j (subscript or superscript) may either refer to the IDS ($j = \mathrm{i}$) or to the EPL ($j = \mathrm{e}$), and the term *porous medium* is used to describe both systems. The two main assumptions of the model are that (i) the porous medium is always saturated with water and (ii) the aquifer is confined, assuming that the overlying glacier and underlying bedrock are impermeable.

Considering these assumptions, mass conservation for both the porous medium and the water has to be considered at each point of the porous medium (Bear, 1988). Mass

Table 1. Definition of the different variables, constants and parameters in the model. As stated in Sect. 2.1, the subscript j is used to refer to the porous media in general, and $j = \mathrm{i}$ for the sediment layer IDS and $j = \mathrm{e}$ for the EPL.

a	mass balance $[\mathrm{m\,s^{-1}}]$
A	fluidity parameter $[\mathrm{Pa^{-n}\,s^{-1}}]$
A_s	sliding parameter without cavity $[\mathrm{m\,Pa^{-n}\,s^{-1}}]$
C	friction law maximum value
e_j	thickness of the layer [m]
g	gravitational acceleration $[\mathrm{m\,s^{-2}}]$
h_j	water head of the porous media [m]
$\mathbf{k}_j$	intrinsic permeability of the porous media $[\mathrm{m^2}]$
$\mathbf{K}_j$	permeability of the porous media $[\mathrm{m\,s^{-1}}]$
n	flow law exponent
N	effective pressure [Pa]
p_j	water pressure in the media [Pa]
q_j	volumetric sink/source term $[\mathrm{s^{-1}}]$
Q_j	water flux by unit of surface $[\mathrm{m\,s^{-1}}]$
Q_t	water transfer between the two layers $[\mathrm{m\,s^{-1}}]$
S_s^j	specific storage coefficient $[\mathrm{m^{-1}}]$
S_j	storage coefficient of porous media
$\mathbf{T}_j$	transmissivity of porous media $[\mathrm{m^2\,s^{-1}}]$
$\boldsymbol{u}$	ice velocity vector $[\mathrm{m\,s^{-1}}]$
$u_{\mathrm{b}i}$	basal velocity $[\mathrm{m\,s^{-1}}]$
$\boldsymbol{U}_\mathrm{d}$	filtration velocity of water with respect to the porous media $[\mathrm{m\,s^{-1}}]$
$\boldsymbol{U}_j$	filtration velocity of porous media $[\mathrm{m\,s^{-1}}]$
$\boldsymbol{U}_\mathrm{w}$	filtration velocity of water $[\mathrm{m\,s^{-1}}]$
z	vertical coordinate [m]
α	compressibility of the solid $[\mathrm{Pa^{-1}}]$
β_s	compressibility of the sediment $[\mathrm{Pa^{-1}}]$
β_w	compressibility of the water $[\mathrm{Pa^{-1}}]$
$\dot{\boldsymbol{\varepsilon}}$	strain-rate tensor $[\mathrm{s^{-1}}]$
$\dot{\varepsilon}_\mathrm{e}$	strain-rate invariant $[\mathrm{s^{-1}}]$
η	effective viscosity of ice [Pa s]
μ_w	viscosity of water [Pa s]
ρ_w	density of water $[\mathrm{kg\,m^{-3}}]$
ρ_ice	density of ice $[\mathrm{kg\,m^{-3}}]$
ρ_j	density of the porous media $[\mathrm{kg\,m^{-3}}]$
$\boldsymbol{\sigma}$	Cauchy stress tensor [Pa]
$\boldsymbol{\tau}$	deviatoric stress tensor [Pa]
$\tau_{\mathrm{b}i}$	mean basal drag [Pa]
φ	leakage length scale [m]
ω_j	porosity of the media

conservation for the water is given as

$$\nabla\cdot(\rho_\mathrm{w}\boldsymbol{U}_\mathrm{w})+\frac{\partial}{\partial t}(\rho_\mathrm{w}\omega_j)-\rho_\mathrm{w}q_j=0. \tag{3}$$

For the porous medium, mass conservation reads

$$\nabla\cdot(\rho_j\boldsymbol{U}_j)+\frac{\partial}{\partial t}[\rho_j(1-\omega_j)]=0, \tag{4}$$

where $\boldsymbol{U}_\mathrm{w}$ refers to the filtration velocity of the fluid with respect to the fixed referential, while $\boldsymbol{U}_j$ takes into account the filtration velocity of the solid with respect to the same referential. The filtration velocity corresponds to the velocity that the material would have should it use all of the available section. This definition relies on the first experiments made by Darcy in 1856, where the velocities of the fluid and solid were computed as a flux divided by the surface of the sample. Densities of water and porous medium are ρ_w and ρ_j, respectively, and ω_j represents the porosity of the porous medium. Finally, q_j is an inflow (outflow if negative) by unit of volume which is due to the transfer of water from the inefficient to the efficient drainage system and/or the recharge of the inefficient system (see Sect. 2.2 and following). Darcy's law in its classical form (Darcy, 1856) reads

$$\boldsymbol{U}_\mathrm{d}=-\frac{\mathbf{k}_j}{\mu_\mathrm{w}}\left(\nabla p_j+\rho_\mathrm{w}g\nabla z_j\right), \tag{5}$$

where $\mathbf{k}_j$ is the tensor of intrinsic permeability of the porous media, μ_w is the viscosity of water, g is the norm of the acceleration of gravity, and p_j is water pressure in the porous medium j at the altitude z_j. In Eq. (5), $\boldsymbol{U}_\mathrm{d}$ is computed as the velocity of the water with respect to the porous medium in the fixed referential frame, which is obtained from the composition of $\boldsymbol{U}_\mathrm{w}$ and $\boldsymbol{U}_j$.

It is a common assumption in hydrology to consider that water density shows very limited spatial variations and that the velocity of the solid is negligible with respect to that of the liquid (Bear, 1988). Introducing these assumptions, the combination of conservation Eqs. (3) and (4) with Darcy's law Eq. (5) gives the diffusion equation for a confined aquifer as follows:

$$S_\mathrm{s}^j\frac{\partial h_j}{\partial t}-\nabla\cdot\left(\mathbf{K}_j\nabla h_j\right)=q_j. \tag{6}$$

In Eq. (6), the water pressure p_j is expressed in terms of water head h_j, the altitude of the water free surface for an equivalent unconfined aquifer. On a glacier, h_j would be the altitude of the water surface measured in a borehole connected to the subglacial drainage system. With z_j defined as the elevation of the observed point from a reference level (here the mean sea level), the relation between water head and water pressure is

$$p_j=\rho_\mathrm{w}g\,(h_j-z_j). \tag{7}$$

Equation (6) introduces the two main physical parameters for the porous media, namely the tensor of hydraulic conductivity $\mathbf{K}_j$ and the specific storage coefficient S_s^j. These two parameters are defined as

$$\mathbf{K}_j=\frac{\rho_\mathrm{w}\,g\,\mathbf{k}_j}{\mu_\mathrm{w}} \tag{8}$$

and

$$S_\mathrm{s}^j=\rho_\mathrm{w}\omega_j g\left[\beta_\mathrm{w}-\beta_\mathrm{s}+\frac{\alpha}{\omega_j}\right], \tag{9}$$

where β_s and α are the compressibility of the solid phase and the porous media, respectively, while β_w is the compressibility of water. The compressibility β_s defines the compressibility of the solid phase of the media (i.e., it can be assessed by a compression experiment on pure material), whereas the compressibility of the sediment, α, takes into account the compressibility due to the rearrangement of the grains. As it is usually done, the β_s term is dropped from the expression of the specific storage coefficient since it is negligible relative to the water compressibility ($\beta_s \approx 1/25\,\beta_w$; Freeze and Cherry, 1979).

Furthermore, Eq. (6) is vertically integrated so that the hydrological model is one dimension lower than the ice flow model and can be solved only over the bottom boundary of the ice flow model. Doing so, the problem simplifies to depend only upon the horizontal coordinates, which is consistent with the goal of simplicity of our approach. With the assumption of z-independent terms in Eq. (6), the integrated values reduce to

$$\mathbf{T}_j = \int_{z_{bj}}^{z_{tj}} \mathbf{K}_j \mathrm{d}z = e_j \mathbf{K}_j,$$
$$S_j = \int_{z_{bj}}^{z_{tj}} S_s^j \,\mathrm{d}z = e_j S_s^j, \qquad (10)$$
$$Q_j = \int_{z_{bj}}^{z_{tj}} q_j \mathrm{d}z = e_j q_j,$$

where z_{tj} and z_{bj} are, respectively, the altitudes of the top and base of the considered layer, and e_j is its thickness. Using these expressions, Eq. (6) is rewritten in its vertically integrated form

$$S_j \frac{\partial h_j}{\partial t} - \nabla \cdot (\mathbf{T}_j \nabla h_j) = Q_j. \qquad (11)$$

This last equation gives the water head at each point of the domain within a porous sediment layer under a given flux per unit of surface (Q_j), as a function of the layer transmissivity tensor $\mathbf{T}_j$ and storage coefficient S_j. The behaviour of both the inefficient and the efficient drainage systems are given by Eq. (11). In the following sections, the differences between the two systems are presented along with their coupling scheme.

2.2 Water routing through the inefficient drainage system

Darcy's law is commonly used in glaciology to express the drainage of water through a sediment layer (e.g. Boulton and Dobbie, 1993; Fischer et al., 2001). It describes inefficient drainage in connection with high water pressure (Walder and Fowler, 1994). Consistent with this last assumption, the subglacial flow in a sediment layer has an important impact on glacier sliding.

In our approach, the distinctive feature of the IDS is that the water head h_i is bounded by an upper limit h_{max} such that the effective pressure at the bedrock stays larger than or equal to zero ($N \geq 0$). Using the definition of the effective pressure, Eq. (1), and the definition of the water head, Eq. (7), the upper limit reads

$$h_{max} = \frac{-\sigma_{nn}}{\rho_w g} + z_j. \qquad (12)$$

To conserve water volume, the water flux generating water heads larger than h_{max} is transferred to the efficient layer as a flux Q_{xs}. The h_{max} limitation on the IDS water head is imposed as a Dirichlet boundary condition on the system for the nodes where $h_i \geq h_{max}$ and the corresponding residual is used to compute Q_{xs} (see Sect. 2.5 for numerical details). This leads to an iterative method similar to the one used by Zwinger et al. (2007) for the treatment of temperature fields in glaciers where the temperature is limited by the pressure melting point and the excess of energy used to melt ice into water.

2.3 Water drainage through the efficient drainage system

As stated before, the *dual continuum porous equivalent approach* adopted here relies on the use of equivalent porous medium to model the efficient drainage system. The modelling of an efficient drainage system by means of a system usually considered as inefficient in glaciology (e.g. Boulton and Hindmarsh, 1987; Hubbard and Nienow, 1997) is not straightforward, but we believe that this approach could lead to convincing results. In this approach, the efficient drainage system is used as an incidental system whose goal is to drain the excess of water from the sediment layer. Keeping that in mind, places where the EPL is activated are more representative of zones where efficient drainage is likely to occur than of actual channel positions.

The use of a diffusion equation, Eq. (11), to model the efficient drainage system requires the development of a special treatment to reproduce the characteristics of this system, namely, a low storing capacity and high conductivity. The physical parameters of the EPL (i.e. $\mathbf{T}_e$, S_e) are adjusted to account for the high hydraulic transmissivity and the low storage coefficient which characterize such efficient drainage systems (Hubbard and Nienow, 1997). Moreover, the solving of the equation itself is subject to certain conditions. Indeed, activation of the EPL is not needed if the sediment layer alone can drain all the water produced. Therefore, the EPL is activated only where the water head in the sediment layer first reaches the maximum water head h_{max} defined by Eq. (12), leading to two distinctive states for the EPL: (i) the EPL is not active where h_i has stayed below h_{max} for all the preceding times (see Fig. 1a) or (ii) the EPL is active where h_i has reach h_{max} at least one time in is history (see Fig. 1b and c).

The first of these two states could represent a winter configuration, when the amount of water driven to the base of the glacier is small enough to be solely drained by the sediment

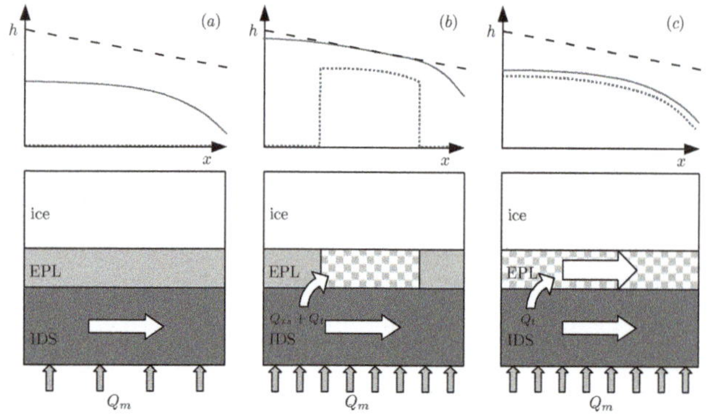

Fig. 1. Description of the coupling between the two layers of the hydrological model. The top panels represent the water load in the IDS (solid line), in the EPL (dotted line) and the flotation limit (dashed line). The lower panels show the routing of water, **(a)** when the EPL is not active, **(b)** when the EPL is active in a transitional state (grey and white chess) and **(c)** when the EPL is effective.

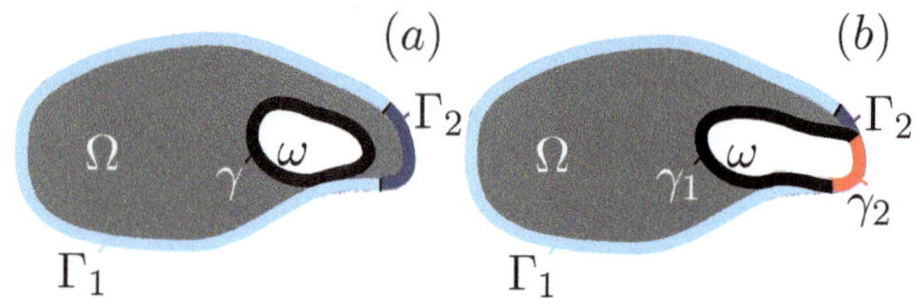

Fig. 2. Description of the evolution of the boundaries of the EPL. Panel **(a)** shows the transitional phase where Ω is the IDS domain with Γ_1 and Γ_2 the Neumann and Dirichlet boundary conditions, respectively; ω is the domain in which the EPL equation is solved with a zero flux boundary condition γ. In panel **(b)**, the EPL is effective; in this case the boundary is such that $\gamma = \gamma_1 + \gamma_2$ where $\gamma_2 = \gamma \bigcap \Gamma_2$ is a Dirichlet boundary condition and γ_1 is a zero flux boundary condition.

layer. Once h_i reaches h_max at some places, the efficient drainage system is activated and starts to fill up from its initial head, which is given by the water head at the snout of the glacier. However, the EPL passes through a transitional state before being able to drain water from the sediment layer leading to two sub-states for the activated EPL. The transitional state represents the time required for the efficient drainage system to extend enough to reach an infinite sink. In glaciology, an infinite sink can be a large subglacial lake, the snout of the glacier or the ocean. At these places, the water head is known, allowing a Dirichlet boundary condition to be imposed to the hydrological model. The transition between the two sub-states of the active EPL is illustrated in Fig. 2.

Considering a glacier of horizontal domain Ω (the ice–bedrock interface on which the hydrological system is defined), the hydrological boundary condition is of Neumann type on the sides (Γ_1) and of Dirichlet type for any infinite sink (Γ_2), such as the snout. In its active transitional state, the EPL diffusion equation is solved on a domain ω with a *zero flux* boundary condition on all its boundaries (γ). The EPL becomes efficient when the boundary of its active domain ω reaches the Dirichlet boundary condition Γ_2. Considering this, the transition between the two sub-states of the active EPL is defined on the domain as

i. the EPL is in a transitional state if $\gamma \bigcap \Gamma_2 = 0$ (Fig. 2a),

ii. the EPL is in an effective state if $\gamma \bigcap \Gamma_2 \neq 0$ (Fig. 2b).

The transitional state of the EPL represents a growing phase during which the water head in the EPL is maintained at a high level. These high water heads are due to the incoming water flux from the IDS which is not evacuated due to the zero flux boundary condition on the EPL.

The spreading of the EPL is controlled by the maximum water head h_max. Once the water head in the EPL h_e reaches h_max, the neighbouring closed EPL element with the lowest IDS water head is activated. The solver is then iterated on the new domain to control that the new EPL water head is below the upper limit, if this is not the case, a new element is activated further down the hydropotential gradient until the upper limit condition is verified on all nodes of the domain where the EPL is in a transitional state.

Once the EPL becomes effective, its functioning is the same as that of the inefficient system. At this point, the only differences between the two systems are the source flux and the value of the physical parameters $\mathbf{T}_j$ and S_j. The method used for the estimation of the source flux Q_t for the EPL is presented in the next section.

So far, the EPL parameters are fixed throughout the simulation, which does not allow taking into account the modification in the draining capacity of the efficient drainage system observed on the field. The evolution of the draining capacity of the EPL, ultimately leading to its closure would be required to perform pluri-annual simulation. This feature is not yet included in the current version of the model, and therefore the applications are restricted to the opening phase of the efficient drainage system. Nevertheless, future work will focus on making the capacity of the EPL non-constant.

2.4 Coupling of the inefficient and efficient drainage systems

Once the EPL is activated, a transfer flux is established between the two different systems. This flux Q_t is illustrated in Fig. 1b and c. The expression of the transfer term is a function of the water head in the two systems, of the characteristics of the inefficient drainage system (thickness e_i and transmissivity $\mathbf{T}_\mathrm{i}$) and of the leakage length scale φ, such that

$$Q_\mathrm{t} = \frac{\mathbf{T}_\mathrm{i} S_j}{\varphi e_\mathrm{i}} (h_\mathrm{i} - h_\mathrm{e}). \tag{13}$$

The leakage length scale φ is a characteristic distance that the water has to cross to pass from one to the other drainage system. The introduction of the storage coefficient S_j $(j = \mathrm{i}, \mathrm{e})$ is needed to convert the water head into volume due to the characteristic of the confined aquifer in which the storage is due to a compression of water and porous medium. The value is dependent on the source of the water; thus if the water comes from the inefficient drainage system, S_i is used, and if the water drains from the EPL to the IDS S_e is used instead. Once the EPL is in an effective state, the establishment of the transfer flux allows the water head in the IDS to be lowered due to the highest transmissivity of the EPL which yields lowest water head in this system.

2.5 Numerical methods

The diffusion equation (Eq. 11) is solved using the finite element method. The variational formulation is obtained by integrating over the domain Ω Eq. (11) and multiplying it by the test function ϕ, such that

$$\int_\Omega S\frac{\partial h_j}{\partial t}\phi \,\mathrm{d}\Omega - \int_\Omega \nabla\cdot\left(\mathbf{T}_j \nabla h_j\right)\phi \,\mathrm{d}\Omega = \int_\Omega Q_j \phi \,\mathrm{d}\Omega. \quad (14)$$

This equation is further transformed by applying Green's theorem to the second term, so that

$$\int_\Omega S\frac{\partial h_j}{\partial t}\phi \,\mathrm{d}\Omega + \int_\Omega \mathbf{T}_j \nabla h_j \cdot \nabla\phi \,\mathrm{d}\Omega \quad (15)$$
$$= \int_\Gamma \mathbf{T}_j \nabla h_j \cdot \boldsymbol{n}\phi \,\mathrm{d}\Gamma + \int_\Omega Q_j \phi \,\mathrm{d}\Omega,$$

where Γ is the boundary surface of the domain Ω. Discretization of this system finally leads to a formulation of the problem such that

$$\mathbf{M}\frac{\partial \boldsymbol{H}_j}{\partial t} + \mathbf{A}_j \boldsymbol{H}_j = \boldsymbol{B}_j, \quad (16)$$

where $\boldsymbol{H}_j$ is the solution vector, $\mathbf{M}$ a *mass* matrix, $\mathbf{A}_j$ is the system matrix defined by the second term in Eq. (15) and $\boldsymbol{B}_j$ is the force vector constituted by the two last terms. A backward difference formula is then applied to discretize the time derivative,

$$\mathbf{X}_j \boldsymbol{H}_j^{(p+1)} = \boldsymbol{F}_j, \quad (17)$$

with

$$\mathbf{X}_j = \mathbf{A}_j + \frac{1}{\Delta t}\mathbf{M} \quad (18)$$

being the new matrix of the system and

$$\boldsymbol{F}_j = \boldsymbol{B}_j^{(p+1)} + \frac{1}{\Delta t}\mathbf{M}\boldsymbol{H}_j^p \quad (19)$$

the new force vector. $\boldsymbol{H}_j^p$ and $\boldsymbol{B}_j^p$ are the solution and force vector at time step p and Δt is the time step. The treatment of this equation is the same for the two systems as long as h_j remains lower than h_max. The way the upper limit is imposed for each system requires two different treatments.

For the IDS, as stated in Sect. 2.2, a Dirichlet method is applied to the water head to limit its height to the value h_max. After the first iteration of the system, if any element $h_{\mathrm{i}p}$ of the solution vector $\boldsymbol{H}_\mathrm{i}$ is greater than h_max, then the system is manipulated such that

$$\mathbf{X}_\mathrm{i}' \boldsymbol{H}_\mathrm{i} = \boldsymbol{F}_\mathrm{i}', \quad (20)$$

where the $\mathbf{X}_\mathrm{i}'$ matrix is the same as the $\mathbf{X}_\mathrm{i}$ matrix except for the pth line, which is fixed to zero apart from the value on the diagonal, which is fixed to unity. Similarly, the force vector $\boldsymbol{F}_\mathrm{i}'$ is equal to the $\boldsymbol{F}_\mathrm{i}$ vector except for its pth value which is fixed to h_max. From this new system, a residual vector $\boldsymbol{R}_\mathrm{i}$ is computed such that

$$\boldsymbol{R}_\mathrm{i} = \mathbf{X}_\mathrm{i}\boldsymbol{H}_\mathrm{i} - \boldsymbol{F}_\mathrm{i}. \quad (21)$$

This is repeated until the relative change of $\boldsymbol{H}_\mathrm{i}$ falls below a given threshold. For the converged solution, the residual $\boldsymbol{R}_\mathrm{i}$ represents the necessary sink per node needed to keep the local water head below its maximum limit. Due to the assumption that all the water is drained into the effective layer, the residual $\boldsymbol{R}_\mathrm{i}$ is then treated as a source term in the EPL equation as follows:

$$\mathbf{X}_\mathrm{e}\boldsymbol{H}_\mathrm{e} = \boldsymbol{F}_\mathrm{e} + \boldsymbol{R}_\mathrm{i}. \quad (22)$$

For the EPL, the volume of water above the given maximum limit is used to increase the size of the efficient drainage system. After the first iteration of the system, if any value h_{ep} of the solution vector $\boldsymbol{H}_\mathrm{e}$ is greater than h_max, the EPL domain (see Fig. 2) is increased in the direction of the lowest hydrological potential and the system is iterated until each element of $\boldsymbol{H}_\mathrm{e}$ satisfies the fixed upper limit.

The coupling of the two hydrological systems requires iteration between the two drainage systems to achieve stability. This outer iteration loop is completed by two inner ones on each subsystem, which ensures that the water head in each system does not overflow the maximum boundary h_max. These inner loops are also needed to compute the water transfer between the two systems. A schematic view of this iterative process is presented in Fig. 3.

3 Ice flow model

The finite element code Elmer/Ice is used to solve both the hydrological and ice flow equations. The governing equations for the ice flow model are briefly summarized before describing in detail the basal boundary condition which links the hydrological and ice flow models. Further information on the numerics and capabilities of Elmer/Ice can be found in Gagliardini et al. (2013).

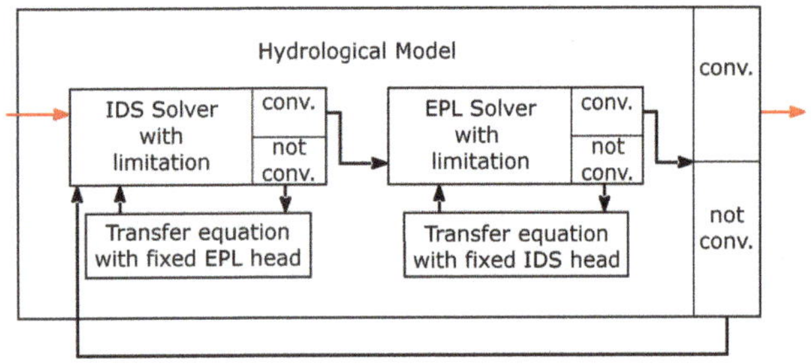

Fig. 3. Schematic description of the iteration scheme of the hydrological model. The outer box represents the entire model and each of the inside boxes is a component of the system which is solved with information from the other components. The convergence or not of each system is indicated with its iterative loop. The red arrows represent the start and end of a hydrological time step.

3.1 Governing equations

The problem to be solved is the one of a gravity-driven flow of incompressible and non-linear viscous ice sliding over a rigid bedrock.

The ice rheology is given by Glen's flow law, defined as

$$\boldsymbol{\tau} = 2\eta\dot{\boldsymbol{\varepsilon}}, \tag{23}$$

where $\boldsymbol{\tau}$ is the deviatoric stress tensor, $\dot{\varepsilon}_{ij} = (u_{i,j} + u_{j,i})/2$ are the components of the strain-rate tensor, and $\boldsymbol{u}$ is the ice velocity vector.

The effective viscosity η in Eq. (23) is expressed as a function of the fluidity parameter A as

$$\eta = \frac{1}{2} A^{-1/n} \dot{\varepsilon}_{\mathrm{e}}^{(1-n)/n}, \tag{24}$$

where $\dot{\varepsilon}_{\mathrm{e}}^2 = \mathrm{tr}(\dot{\boldsymbol{\varepsilon}}^2)/2$ is the square of the second invariant of the strain-rate tensor. Ice is assumed to be isothermal so that the fluidity parameter A is a uniform constant. Moreover, the commonly used value for the exponent, $n = 3$, is adopted.

Ice flow is governed by the Stokes equations that consist of the conservation of mass for incompressible fluids

$$\mathrm{tr}\dot{\boldsymbol{\varepsilon}} = \mathrm{div}\boldsymbol{u} = 0 \tag{25}$$

and the conservation of linear momentum

$$\mathrm{div}\boldsymbol{\sigma} + \rho_{\mathrm{ice}}\boldsymbol{g} = 0. \tag{26}$$

In Eq. (26), ρ_{ice} is the ice density, $\boldsymbol{g}$ the gravitational acceleration vector and the Cauchy stress tensor $\boldsymbol{\sigma}$ is linked to the deviatoric stress tensor such that $\boldsymbol{\sigma} = \boldsymbol{\tau} - p\mathbf{I}$, where $p = -\mathrm{tr}\boldsymbol{\sigma}/3$ is the isotropic pressure. More details regarding the numerics of the ice flow model can be found in Gagliardini and Zwinger (2008).

Solving for changes in the upper surface elevation $z = z_{\mathrm{s}}(x, y, t)$ involves a local transport equation which reads

$$\frac{\partial z_{\mathrm{s}}}{\partial t} + u_x(x, y, z_{\mathrm{s}})\frac{\partial z_{\mathrm{s}}}{\partial x} + u_y(x, y, z_{\mathrm{s}})\frac{\partial z_{\mathrm{s}}}{\partial y} - u_z(x, y, z_{\mathrm{s}}) = a, \tag{27}$$

where a is the accumulation/ablation function given as a vertical flux at the upper surface. Due to the duration of the simulation performed here (less than a year), we further assume $a = 0$.

3.2 Boundary conditions

Upper and lateral boundaries are treated as stress-free surfaces. The bedrock boundary is used to couple the ice dynamics with the subglacial hydrology. This coupling is achieved by computing the effective pressure N from the water pressure in the IDS. The pressure in the EPL is not taken into account as it represents a local pressure which is not likely to have a strong effect on glacier sliding (Hewitt and Fowler, 2008). In return, the mass redistribution derived by the ice flow model influences the hydrological model by modifying the Cauchy normal stress at the bedrock.

The relation between the mean basal drag $\tau_{\mathrm{b}i}$, basal velocity $u_{\mathrm{b}i}$ and effective pressure N was first introduced by Lliboutry (1968). Recent studies from Schoof (2005) and Gagliardini et al. (2007) provide a friction law based on three parameters which depend only on the bedrock geometry. The proposed formulation fulfils the upper limit on the basal drag for a finite sliding velocity known as Iken's bound (Iken, 1981).

In the simplified case where the post-peak decrease exponent is equal to one, this Coulomb-type friction law reads

$$\frac{\tau_{\mathrm{b}i}}{N} = C\left(\frac{\chi_i}{1+\chi_i}\right)^{1/n}, \tag{28}$$

with

$$\chi_i = \frac{u_{\mathrm{b}i}}{C^n N^n A_{\mathrm{s}}}. \tag{29}$$

In these relations, $\tau_{\mathrm{b}i} = \boldsymbol{t}_i \cdot (\boldsymbol{\sigma}\boldsymbol{n})|_{\mathrm{b}}$ is the basal shear stress in the tangential direction i ($i = 1, 2$), and $u_{\mathrm{b}i} = \boldsymbol{u}(z_b) \cdot \boldsymbol{t}_i$ ($i = 1, 2$) is the basal tangential velocity in direction i, with $\boldsymbol{n}$ being the upward-pointing normal to the bedrock surface. The parameters A_{s} and n are the sliding parameter without cavities and the friction law exponent ($n = 3$ taken as in Glen's flow law), respectively. With the assumption of a post-peak exponent equal to 1, $\tau_{\mathrm{b}i}/N$ monotonously increases to its upper bound C.

The coupling between the two systems through this friction law and the input of the normal stress in the hydrological model needs further iteration between the two to achieve stability. Regarding the different timescale of the involved processes, the hydrological and ice flow component of the model are solved with the time step required for their specific needs. A typical simulation would then have time steps of the order of an hour to solve the hydrological equations, and the ice dynamics will then be solved on a daily time step.

For each time step when the hydrological and ice flow components are solved, iterations are performed between the two to achieve stability of the coupled system.

4 Field site and methods

Haut Glacier d'Arolla is an alpine glacier located in Switzerland (Fig. 4). This glacier is relatively small with a surface of 6.33 km^2 (Sharp et al., 1993) at altitudes ranging from 2560 to 3500 m a.s.l. The glacier is believed to be warm based and resting on unconsolidated sediments (Copland et al., 1997; Hubbard et al., 1995). The bed and surface of the glacier were mapped in 1989 by Sharp et al. (1993), and several surface DEMs have been created since then. In our study, we will use the 1989 bedrock DEM along with the surface elevation from 1993.

The main interest in this glacier for our study are the hydrological investigations that have been undertaken on it (e.g. Arnold et al., 1998; Gordon et al., 2001; Kulessa et al., 2003). These studies give a sound knowledge of the hydrological configuration in the area beneath the main tongue of the glacier, about 1.5 km from its snout and labelled *borehole array* in Fig. 4. A study by Hubbard et al. (1995) in the same area gives a range of values for the sediment hydraulic conductivity in the vicinity. Some other studies, involving dye-tracing experiments (e.g. Mair et al., 2002a; Nienow et al., 1998) give a good insight to the evolution of the subglacial drainage system during the melt season.

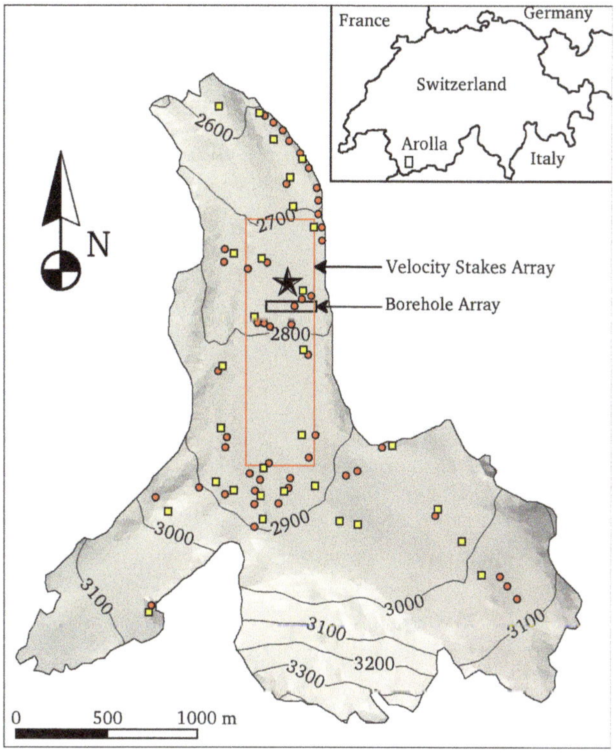

Fig. 4. Map of Haut Glacier d'Arolla with the position of the borehole and velocity stakes arrays. The star is the position of the reference point used in Sect. 5. The glacier surface elevation is contoured every 100 m. Red circles are the position of the moulins used for the dye-tracing experiments in 1989 and 1990; the yellow squares are the position of the moulins recorded in 1993 and used for the modelling.

4.1 Strategy to estimate the hydrological parameters

The hydrological model has been designed to rely on a limited number of parameters. As presented in Table 2, most of the hydrological parameters are well defined with the exception of the transmitivities of both layers $\mathbf{T}_j$ $(j = \mathrm{i}, \mathrm{e})$, their thicknesses e_j $(j = \mathrm{i}, \mathrm{e})$, and the leakage length scale φ. We further assume that the transmissivity of both systems is isotropic and is therefore a scalar value T_j. Estimating the parameter values for the hydrological model leads to two distinct problems. First, measurements of subglacial sediment transmissivity are rare and encompass a large range of values (Freeze and Cherry, 1979). Second, the use of an equivalent layer for the treatment of the efficient drainage system prevents us from directly using parameter values that would characterize a true discrete channel. The EPL transmissivity, leakage length scale and layer thickness then have to be estimated by comparing the model results directly to observations of the hydrological systems.

We therefore adopted the following strategy. As a first approximation, a broad range of values for the unknown parameters are estimated from the available measurements. These ranges are large enough to produce very different model results. Then, these ranges of values are decreased by comparing the model results to large-scale features of the hydrological system of Haut Glacier d'Arolla. The comparison to large-scale features allows the local variability of water head observed in neighbouring boreholes to be discarded. The strategy for evaluating the layers thickness is rather different. The presented simulations are performed for layer thickness of $e_\mathrm{i} = 20$ m for the IDS and $e_\mathrm{e} = 1$ m for the EPL; for each experiment, simulations are performed with values around these references to assess the sensitivity of the model.

The sediment transmissivity is estimated using the hydraulic diffusivity at the bed of Haut Glacier d'Arolla measured by Hubbard et al. (1995). The hydraulic diffusivity D represents the velocity of a pressure pulse through the media and is given as

$$D = \frac{T_j}{S_j}. \tag{30}$$

Hubbard et al. (1995) measured hydraulic diffusivities ranging from 4×10^1 $m^2 s^{-1}$ near an efficient drainage zone down to 7×10^{-2} $m^2 s^{-1}$ 70 m away from this zone. From these values, Eq. (30) and the physical parameters needed to evaluate S_i (given in Table 2), the IDS transmissivity T_i is estimated to range from 1.4×10^{-4} to 8×10^{-2} $m^2 s^{-1}$. The choice of

Table 2. Values of the parameters used in the hydrological model along with their sources. Well-known parameters are referenced as wkp and poorly known parameters are labelled pkp.

Parameter	Value	Source
e_i	20 m	pkp
e_e	1 m	pkp
g	$9.81\ \mathrm{m\,s^{-2}}$	wkp
β_w	$5.04 \times 10^{-10}\ \mathrm{Pa^{-1}}$	wkp
ρ_w	$1000\ \mathrm{kg\,m^{-3}}$	wkp
ρ_{ice}	$917\ \mathrm{kg\,m^{-3}}$	wkp[a]
α	$10^{-8}\ \mathrm{Pa^{-1}}$	wkp[b]
ω_j	0.4	pkp

[a] Cuffey and Paterson (2010).
[b] Flowers and Clarke (2002b); Björnsson (2003).

the EPL transmissivity is more complicated. It cannot be directly measured as it represents the mean behaviour of a number of discrete channels. However, based on the previous measurements, the lowest value of the EPL transmissivity is set to $4 \times 10^{-3}\ \mathrm{m^2\,s^{-1}}$, which corresponds to the value measured closest to the channel margin. The higher limit for the EPL transmissivity is then fixed at $8 \times 10^{-1}\ \mathrm{m^2\,s^{-1}}$, ten times larger than the maximum value of the IDS transmissivity. This scaling is consistent with the results of Nienow et al. (1998) which describe the differences between the mean flow velocity of the distributed and of the channelized drainage systems. Finally, the leakage length scale cannot be constrained by measurements, and so a large range of values, from 1 m up to 50 m, is adopted.

Assuming these broad ranges for the three unknown hydrological parameters (see values in Table 3), two configurations of the hydrological system are then constructed and compared to measurements. The first configuration is characteristic of an end-of-winter system, whereas the second reproduces the development of the drainage system during summer. The comparison is done using a large-scale feature of the hydrological system which will be comparable to the results given by the double continuum approach.

The metric is defined as the maximum length of the active EPL, which represents the development of the efficient drainage system. The EPL length can be compared to the maximum channel length estimated by dye-tracing experiments performed at different dates during the summer season (Nienow et al., 1998). The control observations are extracted from a number of dye-tracing experiments that were undertaken during summers 1990, 1991 and 1995 (Nienow et al., 1998; Mair et al., 2002b). The earliest dye-tracing measurements performed on Haut Glacier d'Arolla are dated around 10 June and show channel lengths of slightly less than 700 m. At this time of the year, the discharge at the snout of the glacier is already ten times higher than the base winter discharge used for the forcing of our simulations, and we would therefore expect to model an EPL length substantially lower than the recorded 700 m. The other constraint given by these observations is the timing of the up-glacier migration of the head of the channel system throughout the melt season. This dynamic aspect is compared to the evolution of the EPL length during the transient summer simulation.

The mean water head of the IDS in the borehole array shown in Fig. 4 further referenced as to *IDS water head* is also presented to help the comparison between simulation results.

4.2 End-of-winter configuration

The end-of-winter configuration is achieved by distributing the observed winter discharge at the snout of the glacier ($5 \times 10^{-2}\ \mathrm{m^3\,s^{-1}}$) over the whole glacier surface, giving an input of $\sim 8 \times 10^{-9}\ \mathrm{m\,s^{-1}}$ of water. This constant water flux is maintained until the water head of both the IDS and the EPL reach a steady state.

Figure 5 presents the IDS water pressure on the whole glacier and the extent of the active EPL at the end of winter for three different values of the IDS transmissivity. The length of the EPL, measured along the effective EPL from the snout of the glacier to the farthest source of the EPL, is depicted by the white thick line in Fig. 5. Comparison of steady-state configurations indicates that an increasing IDS transmissivity leads to a decreasing water pressure and a shorter EPL. As expected, the large range of values for the IDS transmissivity leads to a large spread in the model results. For $T_i = 1.6 \times 10^{-4}\ \mathrm{m^2\,s^{-1}}$ the drainage system is dominated by the EPL. Conversely, for $T_i = 1.6 \times 10^{-2}\ \mathrm{m^2\,s^{-1}}$ the drainage capacity of the IDS is such that it can drain all the input water and the development of the EPL is therefore not required. These two extreme cases indicate that the chosen range of IDS transmissivity covers all possible behaviour of the IDS. Nevertheless, excessively high values of the IDS transmissivity lead to unrealistic behaviour. Using the position of the head of the channelized component at the end of spring (Nienow et al., 1998; Mair et al., 2002b), the range of the IDS transmissivity can be constrained.

Thus, only transmissivity values that lead to an EPL length lower than 700 m will be considered as admissible. Figure 6 shows the evolution of the EPL length (a) and of the IDS water head (b) for IDS transmissivity ranging from 1.4×10^{-4} to $8 \times 10^{-2}\ \mathrm{m^2\,s^{-1}}$. As in Fig. 5, a decrease in IDS transmissivity leads to an increase in the EPL length. The adopted restriction on the EPL length is represented by the grey zone in Fig. 6a. This constraint leads to a new range for the admissible values of the IDS transmissivity from 3×10^{-4} to $8 \times 10^{-2}\ \mathrm{m^2\,s^{-1}}$. The thickness of the lines in Fig. 6 represents the scattering of the metrics in response to a modification of the EPL transmissivity. The relatively small line thickness indicates that, in the case of the steady-state configuration characterizing the end of winter, both drainage systems are quite insensitive to the EPL transmissivity.

Table 3. Values of the tunable hydrological parameters for the different steps of the selection procedure.

Parameter	Starting range	Range after end-of-winter selection	Final value
T_i [$\mathrm{m^2\,s^{-1}}$]	1.4×10^{-4}–8×10^{-2}	3×10^{-4}–8×10^{-2}	1.6×10^{-3}
T_e [$\mathrm{m^2\,s^{-1}}$]	4×10^{-3}–8×10^{-1}	4×10^{-3}–8×10^{-1}	7.9×10^{-2}
φ [m]	1–50	1–20	10

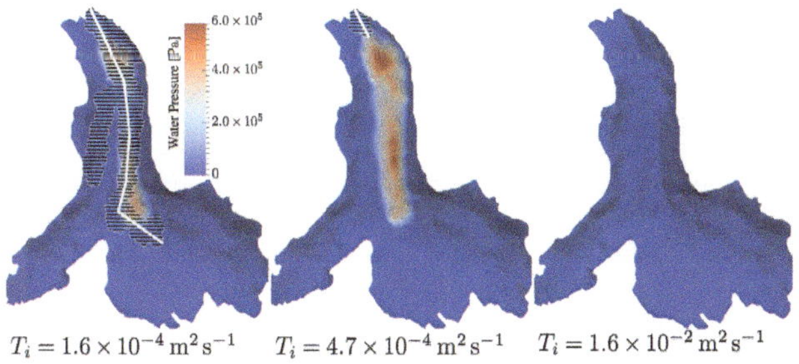

Fig. 5. Maps of Arolla Glacier showing the water pressure of the IDS and the development of the EPL (hatched zone) at the end of the winter season for three different IDS transmissivity values (T_i). For the highest IDS transmissivity, all the produced water is drained by the sediment layer, explaining the very low water pressure and the non-development of the EPL. The white thick line indicates how the length of the EPL is determined.

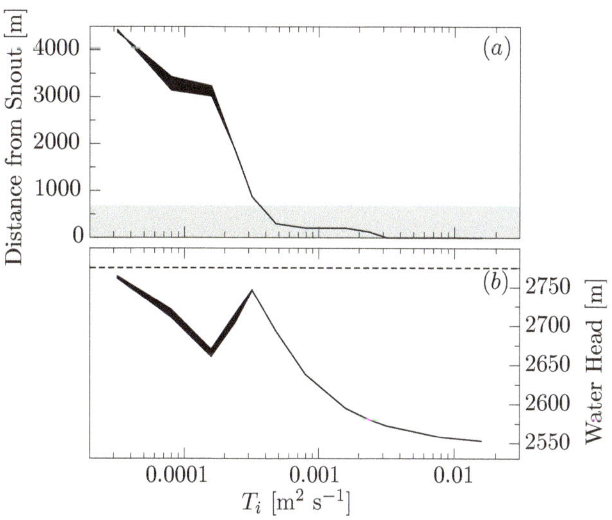

Fig. 6. Length of the EPL **(a)** and IDS water head in the borehole array **(b)** as a function of the IDS transmissivity T_i. The grey zone in **(a)** indicates the admissible values for the EPL maximum length. The dashed line in **(b)** represents the flotation limit. The spread of the curves represents the scattering due to EPL transmissivity ranging from 4×10^{-3} to $8\times10^{-1}\,\mathrm{m^2\,s^{-1}}$, with the higher transmissivity values leading to the lowest EPL length and IDS water head.

The IDS water head for its part increases with decreasing IDS transmissivity until the EPL extent is such that it can drain the borehole array (around 1200 m from the snout, as depicted in Fig. 4). For the configurations where the EPL reaches the borehole array, the water from the IDS can then be easily drained, explaining the decrease of the IDS water head. After reaching a minimum around $T_\mathrm{i} = 1.5\times10^{-4}\,\mathrm{m^2\,s^{-1}}$, the IDS water head increases again in response to the decrease in IDS transmissivity. The value of the local minimum is a function of the drainage efficiency of the EPL but is not sensitive to the EPL transmissivity. It should then be driven by the leakage length scale as discussed below.

Figure 7 shows the sensitivity experiments to the leakage length scale φ. As explained in Sect. 2.4, a large leakage length scale implies low efficiency of the water transfer between the two layers. This weak transfer triggers a larger extent of the EPL due to the higher water head in the IDS. The IDS water head is more sensitive to the leakage length scale than is the EPL length metric. As shown in Fig. 7, the drop of IDS water head is amplified for smaller leakage length scale. This amplitude variation is explained by the ability of the EPL to drain water from the IDS. The smaller the leakage length scales, the easier the transfer of water from the IDS system to the EPL system. For unrealistically large leakage length scales (i.e. $\varphi \geq 20$ m), even if the EPL is activated the water head in the IDS returns quickly to the value it has before the opening of the EPL. The range for the leakage length scale is therefore restricted from 1 to 20 m.

In summary, the end-of-winter configuration allows the range of both the IDS transmissivity and the leakage length scale to be decreased, but not the EPL transmissivity (values given in the second column of Table 3). Modifying the thickness of both layers in this experiment while keeping the transmitivities at the same values (an increase in thickness then leads to a decrease in conductivity) does not lead to any change in the observed results. However, if the same changes in thickness are done with a constant conductivity (an increase in thickness then leads to an increase of the transmissivity), then the response of the model is on the line of the one that is observed for varying transmitivities with a negligible impact of the thickness change. To continue, the second configuration corresponding to the development of the hydrological system during the summer is then used.

4.3 Transient summer configuration

The previous steady-state configuration corresponding to an end of winter is subsequently used as the initial state of the transient summer simulations. The transient response of

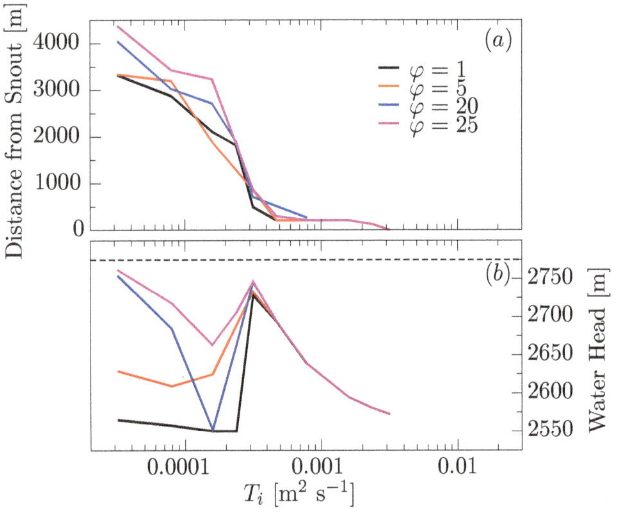

Fig. 7. Length of the EPL **(a)** and IDS water head in the borehole array **(b)** as a function of the IDS transmissivity T_i for four different values of the leakage length scale φ and an EPL transmissivity of $T_\mathrm{e} = 7.9 \times 10^{-2}\,\mathrm{m^2\,s^{-1}}$. The dashed line in **(b)** represents the flotation limit.

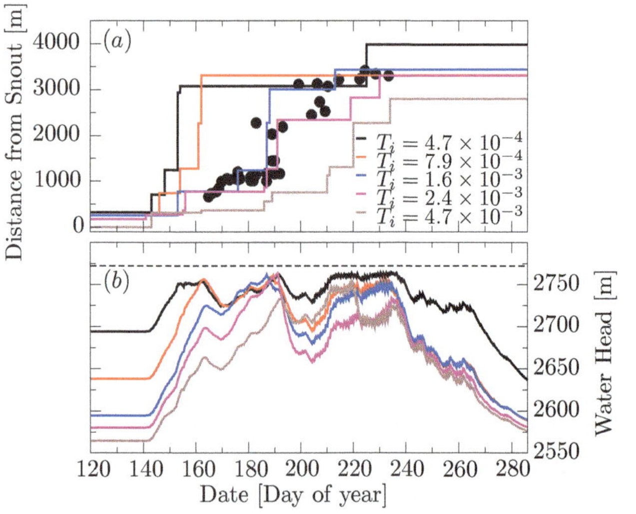

Fig. 8. Evolution with time for different values of the IDS transmissivity of **(a)** the maximum length of the modelled EPL (lines) and **(b)** the IDS water head. The position of the head of the channelized drainage system derived from observations (black dots) is presented in **(a)** for comparison. The dashed line in **(b)** represents the flotation limit. Simulations are performed with a constant EPL transmissivity $T_\mathrm{e} = 7.9 \times 10^{-2}\,\mathrm{m^2\,s^{-1}}$ and a constant leakage length scale $\varphi = 10\,\mathrm{m}$.

the model is obtained by imposing a time-dependent water flux at prescribed moulin positions. To this aim, we use the moulin positions recorded during the 1993 melt season and the associated influx modelled by Arnold et al. (1998) for the 1993 summer season. Each moulin is assumed to be perfectly vertical and is represented by a single node of the mesh.

Unfortunately, dye-tracing experiments were not performed during the 1993 season and the evolution of the channel drainage system is therefore compared against the 1990 measurements. Nevertheless, the comparison of the 1990, 1991 and 1995 melt seasons indicates that the changing extent of the hydrological system during the summer seasons develops at a similar rate and follows similar structures despite some variations in the spatial and temporal distribution of the water sources. Moreover, Fig. 4 shows that the moulin positions in 1993 are very similar to the 1990 positions. Comparing the model results to the 1990 melt season drainage system evolution is therefore a reasonable assumption, especially given the uncertainties in modelled moulin influxes.

Starting from the poorly developed EPL observed at the end of winter, a fully developed EPL draining the major part of the glacier bed is obtained by the end of the melt season. As for the end-of-winter configuration, the model results are strongly dependent on the IDS transmissivity value but rather insensitive to variations in EPL transmissivity. Because of this lack of sensitivity, only the simulation performed for $T_\mathrm{e} = 7.9 \times 10^{-2}\,\mathrm{m^2\,s^{-1}}$ will be presented in the following.

Figure 8 shows for various IDS transmitivities the evolution during the summer of the EPL length and IDS water head. In the range of applied transmissivity, an increase in IDS transmissivity leads to a delay in the spreading of the EPL during the melt season. This delay is induced by the postponing of the EPL opening induced by lower water head at the end of winter for large transmissivity, as can be shown in Fig. 8b. From this sensitivity study, an IDS transmissivity of $1.6 \times 10^3\,\mathrm{m^2\,s^{-1}}$ seems to best reproduce the observed development of the drainage system during summer.

The sensitivity of the model to a variation in the IDS layer thickness is similar to the response which is observed when varying the transmissivity. If the transmissivity is kept constant while the IDS thickness is increased, the increase of the IDS storing coefficient leads to a time lag in the opening of the EPL. The value of 20 m taken here is the one that fits best with the chosen IDS transmissivity value. The thickness of the EPL is limited by the fact that its storing coefficient should remain below that of the IDS. Modifying the thickness of the EPL while taking this limitation into account does not lead to significant differences in the model results. The EPL thickness is then kept at $e_\mathrm{e} = 1\,\mathrm{m}$.

The model's sensitivity to the leakage length scale is presented in Fig. 9; panel b indicates that a higher leakage length scale leads to a higher IDS water head because of the less efficient transfer from the IDS system to the EPL system. As shown in Fig. 9a, the EPL length metric shows little sensitivity to varying the leakage length scale except for very low leakage length scale values ($\varphi < 10\,\mathrm{m}$). As for the end-of-winter configuration, the IDS water head is more sensitive to the leakage length scale than is the EPL length. Gordon et al. (1998) have reported a decrease of the water head by

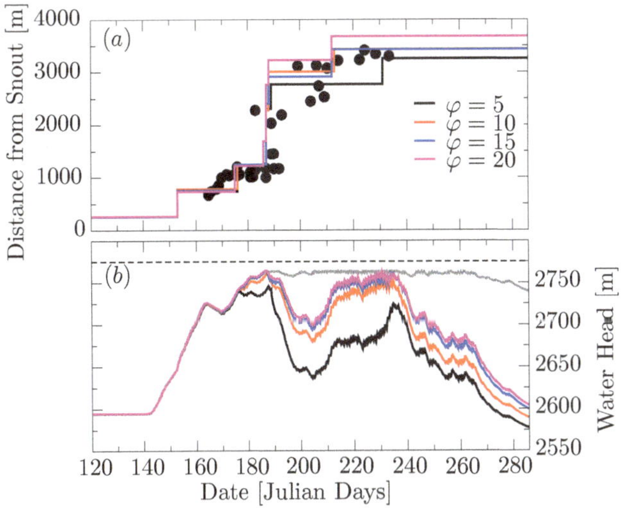

Fig. 9. Same as Fig. 8 but with varying leakage length scales. The grey line in **(b)** is the IDS water head for a simulation without the EPL. Simulations are performed with a constant IDS transmissivity $T_{\rm i} = 1.6 \times 10^{-3}\,{\rm m^2\,s^{-1}}$ and a constant EPL transmissivity $T_{\rm e} = 7.9 \times 10^{-2}\,{\rm m^2\,s^{-1}}$.

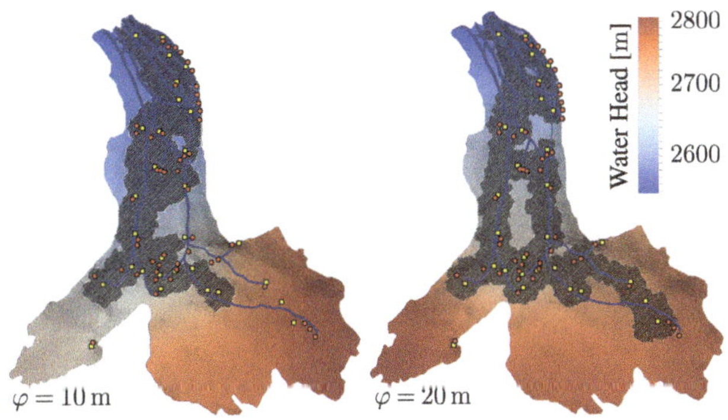

Fig. 10. Comparison between the modelled active EPL (black dashed zone) and the observed channel system (blue line) at the end of the melt season for two different values of the leakage length scale φ. The observed channel system and the corresponding moulin positions (red circles) for the summer of 1989 and 1990 are reproduced from Sharp et al. (1993). The moulins observed during the 1993 season which are used for the simulations are marked by yellow squares. The colour scale represents the water head of the IDS in metres. Parameters of the simulation are $T_{\rm i} = 1.6 \times 10^{-3}\,{\rm m^2\,s^{-1}}$ and $T_{\rm e} = 7.9 \times 10^{-2}\,{\rm m^2\,s^{-1}}$.

70 m at the opening of the efficient drainage system, in good agreement with the modelled drop for leakage length scale $\varphi \geq 10$ m.

Figure 10 compares the modelled EPL extent for $\varphi = 10$ and $\varphi = 20$ m to the reconstructed channel system at the end of the 1990 melt season (Sharp et al., 1993). This comparison shows a good agreement between the extent of the modelled active EPL and the channel system observed at the end of summer for both values of φ. Therefore, even if the two modelled EPL extents show some differences, they are too similar to identify an optimum leakage length scale value, bearing in mind that the *observed* channel system is itself a reconstruction from dye-tracing measurements. In the following, the value $\varphi = 10$ m is therefore adopted.

Figure 11 presents the evolution of the EPL length for the adopted values of the three hydrological parameters for different grid resolutions. Specific lengths of the element ranging from 25 to 100 m show similar results in the spreading velocity of the EPL. As in Werder et al. (2013), the results are also impacted by the position of the nodes, which could change the activation point of the EPL and then its general pattern. However, this sensitivity does not affect the global variables of the model. The estimated length of the channelized drainage system (black dots) is presented as a reference. Our results seem to indicate that the EPL extent is less smooth than the one proposed by Nienow et al. (1998) (black line) and it evolves by steps driven by the moulin positions and the bedrock topography, which is consistent with the interpretation of Mair et al. (2002a).

In summary, comparison between model results and observations for two distinct configurations of the hydrological system has allowed selection of the most reasonable values of the three hydrological parameters constrained by observations and independent interpretation. The adopted values are given in Table 3. This set of parameters is now used in the following section to model the coupling between the ice flow and the hydrological system throughout the melt season.

5 Modelling of spring speed-up events

Adopting the newly defined set of parameters, the ice flow and hydrological models are coupled with the aim of modelling the spring speed-up observed at Haut Glacier d'Arolla. Speed-up events were recorded during four melt seasons on this glacier, in 1994 (Mair et al., 2001), 1995 (Mair et al., 2002a), 1998 and 1999 (Mair et al., 2003). Again, unfortunately no velocity measurements are available from the 1993 melt season for which we have water inputs (Arnold et al., 1998). However, from these various speed-up observations, we can characterize them as being short-lived (three to four days) periods during which the surface velocities show a two- to four-fold increase with respect to their average values. We will further use these general characteristics for the comparison with the modelled velocities. The first spring speed-up of 1998, as documented by Mair et al. (2003), will be used as a representative speed-up event.

The hydrological and ice flow models are coupled through the friction law (Eq. 28) which depends on 4 parameters. As stated in Sect. 3.2, the post-peak decrease exponent and the friction law exponent n are fixed to 1 and 3, respectively. The latter is fixed at the usually accepted value of Glen's exponent, whereas the former is chosen to achieve a better

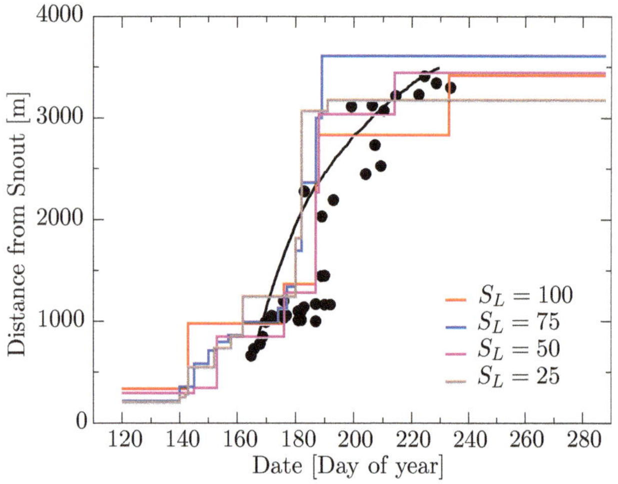

Fig. 11. Comparison of the evolution of the maximum length of the modelled EPL for different grid specific length (coloured lines). The position of the head of the channelized drainage system derived from observations (black dots) and the interpretation of the channel spreading by Nienow et al. (1998) (black line) are shown for comparison. The modelling is performed using the modelled water input of the 1993 melt season (Arnold et al., 1998) with $T_i = 1.6\times10^{-3}\,m^2\,s^{-1}$, $T_e = 7.9\times10^{-2}\,m^2\,s^{-1}$ and $\varphi = 10\,m$. The position of the head of the channelized system is computed following Nienow et al. (1998) using dye-tracing data from the 1990 melt season; the specific length (S_L) for each simulation is given in metres.

numerical stability. The sliding parameter in the absence of cavities is assumed to be uniform at the base of the glacier. The value $A_s = 1.6\times10^{-23}\,m\,Pa^{-3}\,s^{-1}$ is adopted to reproduce the observed winter velocity when the water pressure is low. In addition, the sliding for bed elevation above 3000 m is computed with an effective pressure of $N = 1.2$ MPa rather than the computed one. This limitation was necessary to stabilize the ice flow model on these parts of the glacier where the ice flow takes the form of steep ice falls (Hubbard et al., 2000). Contrary to what was done in previous studies (Hubbard et al., 1998), this steep area was kept in the modelled domain to avoid the use of fictitious boundary conditions, but, on the other hand, this requires constraining the friction law with a fixed effective pressure. Comparison between limited and not limited simulations shows that this upstream limitation does not impact the velocities on the lower part of the glacier on which we now focus.

The fourth parameter in the friction law represents the maximum value reached by τ_{bi}/N and should be smaller than the maximum value of the slope of the local obstacles m_{max}. In Pimentel et al. (2010), the value $C = 0.84m_{max}$ for a sinusoidal roughness distribution was adopted, with $m_{max} = 0.5$. Here we will test the model sensitivity to values of C from 0.5 to 0.9.

The results of the simulation performed with this set of parameters are presented in Fig. 12 for the reference point shown in Fig. 4. The effective pressure computed by the hydrological model controls the variability in the surface velocity throughout the melt season. A spring speed-up occurs between days 185 and 190 before the activation of the efficient system which is highlighted by the sudden increase of the EPL head. The speed-up continues until the efficient system is activated and the water pressure drops. In the 12 days prior to this major event, two minor speed-ups are modelled. After day 205, the effective pressure decreases again in response to a heavy water input which triggers a new increase in the glacier speed with a peak around day 235. In comparison to the first one, this second speed-up period is characterized by higher daily variability. By this stage of the season, the subglacial hydrological system has reached its maximum capacity and any melt occurring during the day induces a quasi-instantaneous increase in water pressure, explaining this higher daily variability. After this second speed-up period, the glacier enters a quieter regime due to the relatively low water input during fall.

Winter observations cannot be used to constrain the value of the parameter C, which has a very negligible impact on the velocity when the water pressure is low. However, during spring and summer, the model is quite sensitive to C due to higher water pressure. Figure 13 presents the evolution of the surface velocities obtained for different values of C during the spring speed-up event. The larger the value of C, the less marked the acceleration during the spring speed-up event. With $C = 0.9$, the speed-up is hardly distinguishable from the background velocity, whereas $C = 0.5$ gives the highest speed-up event with velocities that tend to stay higher in between the speed-up events.

The surface longitudinal velocity pattern for $C = 0.5$ during the first speed-up is presented in Fig. 14 for the lower part of the glacier. The evolution of the pattern during the spring speed-up matches the one that is observed on the glacier with a two-fold increase in the velocities during the speed-up. For comparison purposes, the velocities measured during the first 1998 spring speed-up event (Mair et al., 2003) are superimposed on the modelled velocities. The comparison with this specific event shows that the modelled speed-up is less pronounced than the observed one and that the maximum speed is shifted downstream by approximately 400 m. Considering the various assumptions in the model and the non-synchronous data sets used for velocity and water input, a complete agreement with the observations is not to be expected.

6 Discussion

The approach presented here gives a new alternative for the modelling of subglacial hydrology. The aim of this work is to propose a hydrological model which evaluates the water pressure at the base of large ice sheet outlet glaciers and allow the computation of sliding at their bases. The application

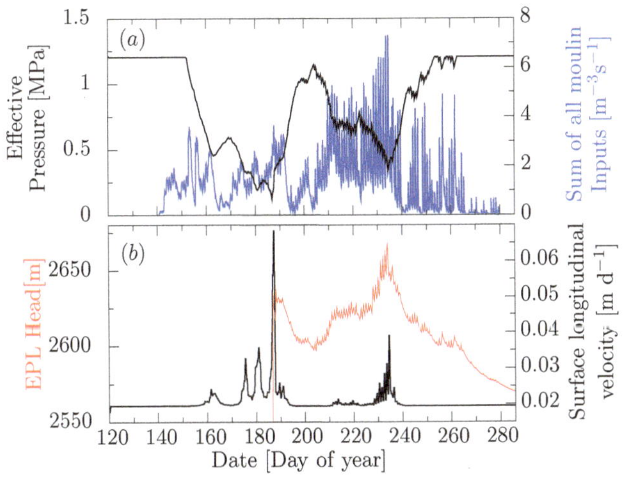

Fig. 12. Evolution throughout the melt season at the reference point depicted in Fig. 4 of **(a)** the effective pressure (left axis, black curve) and the water input (right axis, blue curve), **(b)** the surface longitudinal velocity. The simulation is performed with $A_S = 1.6 \times 10^{-23}\,\mathrm{m\,Pa^{-3}\,s^{-1}}$ and $C = 0.5$.

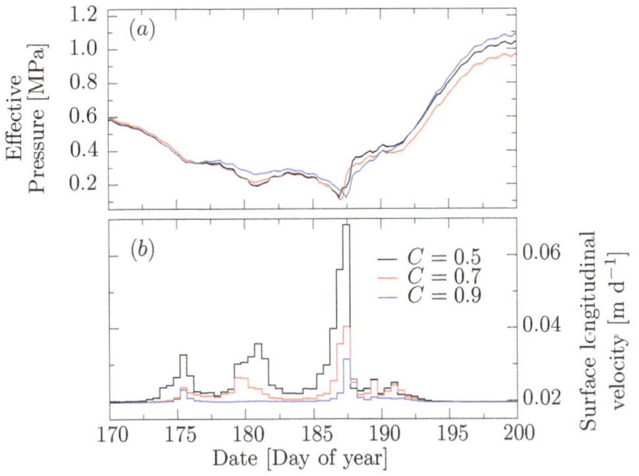

Fig. 13. Evolution during the spring speed-up event for different values of the parameter C of the effective pressure **(a)** and surface longitudinal velocity **(b)** at the reference point, depicted in Fig. 4.

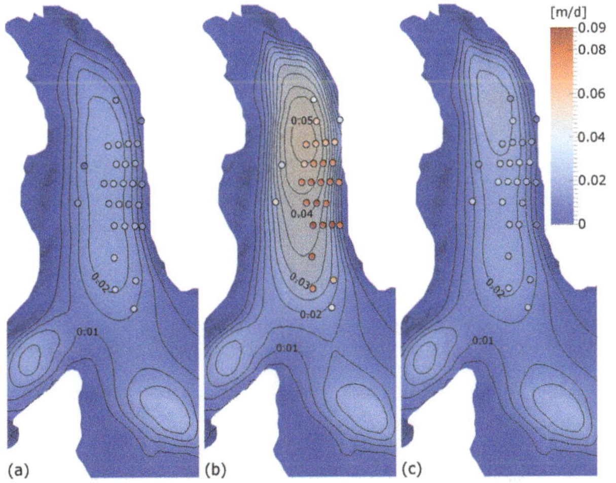

Fig. 14. Spatial pattern of modelled surface velocities on the tongue of Haut Glacier d'Arolla for the year 1993. Before the spring event **(a)** (days 182 to 185), during the speed-up **(b)** (days 186 to 188) and after the event **(c)** (days 188 to 190). The surface velocities are given by the colour scale in $\mathrm{m\,d^{-1}}$ and contoured every $0.005\,\mathrm{m\,d^{-1}}$. The dots on the three panels present the measured velocities of the first 1998 speed-up event as documented in Mair et al. (2003) with the same colour scale as that of the model results.

of this model to a small valley glacier such as the Haut Glacier d'Arolla might not be the best suited to present the advantages of our approach, but the data set available on this glacier allows the building of convincing validation experiments for this new approach. Compared to existing large-scale models (e.g. Le Brocq et al., 2009; Goeller et al., 2013), our model effectively computes the effective pressure. This feature makes it comparable to more physically based models (e.g. Schoof, 2010b; Hewitt et al., 2012; Werder et al., 2013) in which the efficient drainage system is modelled by channel, or the approach of Flowers and Clarke (2002a) in which a single porous layer with varying capacity accounts for both efficient and inefficient drainage systems.

In comparison to the model incorporating a discrete description of the channels (e.g. Schoof, 2010b; Hewitt et al., 2012; Werder et al., 2013), in the proposed approach the precise location of the efficient subglacial drainage system is not achieved. This prevents validation of the model against punctual water pressure measurements in borehole or precise modelling of dye tracer return. This first limitation is intrinsic to the model formulation but is not an issue as far as the model results can be compared to global variables as is done in this study.

A second limitation of the model, not intrinsic to its formulation, is the current lack of a closing mechanism for the EPL. This type of mechanism would require the implementation of an evolving transmissivity for the EPL. In the present study, the lack of this closing mechanism is of small concern as we focus on the opening of the EPL. However, the end of the summer simulation would probably take advantage of the implementation of an evolving draining capacity of the EPL, which would probably allow accommodation of the inputs that occur later in the season (days 220 to 240 of the coupled simulation). Introducing this mechanism will be part of future development of the model.

7 Conclusions

We have presented a new hydrological model especially designed to be coupled to an ice flow model. This hydrological model is based on a double continuum approach and solves the same set of equations using different parameters for both the inefficient and efficient drainage systems. The two systems are coupled so that the total amount of water is conserved, and an ad hoc scheme is proposed to activate the efficient drainage system where the water pressure exceeds the overburden ice pressure. In our approach, the channels are not represented individually but in a continuous manner, presenting the advantage of not requiring a very fine description of the basal topography. In this paper, the hydrological model and its coupling to an ice flow model are validated by performing a series of three applications of increasing complexity using the data set of Haut Glacier d'Arolla. A first application aiming for a steady-state configuration, corresponding to the winter state, is used to decrease the range of possible values for the three most poorly constrained parameters of the model, i.e. the transmissivity of both drainage systems and the leakage length scale. In the second experiment, the evolution of the drainage system during the spring and summer seasons is studied, starting from the previously obtained winter steady state. Again, the model sensitivity to the three most poorly constrained parameters is tested. The third and last application couples the hydrological and ice flow models, and results are compared to observed glacier speeds. Despite the use of non-synchronous data sets and the number of simplifying assumptions in the model, good agreement is obtained in terms both of the temporal and spatial drainage system evolution during the spring and summer seasons and of the magnitude and duration of the observed speed-ups. The largest uncertainties in the model's velocity response are in the values of the leakage length scale and the parameter C in the friction law. The other parameters in the model have proven to be relatively well constrained by observations or empirical interpretations (as for the IDS transmissivity or A_s), or the model has proven to be quite insensitive to them (as for the EPL transmissivity).

Acknowledgements. This work was supported by the DACOTA project (ANR-06-VULN-016-01) funded by the Agence Nationale de la Recherche (ANR). Parts of the work were carried out under the HPC-Europa2 project (project number 228398) with the support of the European Commission – Capacities Area – Research Infrastructures. The authors would like to thank Neil Arnold for providing the modelled water inputs on Haut Glacier d'Arolla. The three anonymous reviewers are thanked for their constructive comments that helped to improve the paper quality.

Edited by: I. M. Howat

References

Arnold, N., Richards, K., Willis, I., and Sharp, M.: Initial results from a distributed, physically based model of glacier hydrology, Hydrol. Process., 12, 191–219, doi:10.1002/(SICI)1099-1085(199802)12:2<191::AID-HYP571>3.0.CO;2-C, 1998.

Bear, J.: Dynamics of fluids in porous media, Dover Publications, 1988.

Björnsson, H.: Explanation of jokulhlaups from Grimsvotn, Vatnajokull, Iceland, Jokull, 24, 1–26, 1974.

Björnsson, H.: Subglacial lakes and jokulhlaups in Iceland, Global and Planetary Change, 35, 255–271, doi:10.1016/S0921-8181(02)00130-3, symposium on Subglacial Lakes, SAN FRANCISCO, CA, 2001, 2003.

Boulton, G. and Dobbie, K.: Consolidation of sediments by glaciers: relations between sediment geotechnics, soft-bed glacier dynamics and subglacial ground-water flow, J. Glaciol., 39, 26–44, 1993.

Boulton, G. S. and Hindmarsh, R. C. A.: Sediment deformation beneath glaciers – rheology and geological consequences, J. Geophys. Res., 92, 9059–9082, doi:10.1029/JB092iB09p09059, 1987.

Boulton, G. S., Lunn, R., Vidstrand, P., and Zatsepin, S.: Subglacial drainage by groundwater-channel coupling, and the origin of esker systems: part II-theory and simulation of a modern system, Quat. Sci. Rev., 26, 1091–1105, doi:10.1016/j.quascirev.2007.01.006, 2007.

Copland, L., Harbor, J., and Sharp, M.: Borehole video observation of englacial and basal ice conditions in a temperate valley glacier, Ann. Gla, 24, 277–82, international Symposium on Changing Glaciers, 24-27 June 1996, Fjaerland, Norway, 1997.

Cuffey, K. and Paterson, W. S. B.: The Physics of Glaciers, 4th Edition, Elsevier, Oxford, 2010.

Darcy, H.: Les fontaines publiques de la ville de Dijon, Dalmont, Paris, 1856.

Fischer, U., Porter, P., Schuler, T., Evans, A., and Gudmundsson, G.: Hydraulic and mechanical properties of glacial sediments beneath Unteraargletscher, Switzerland: implications for glacier basal motion, Hydrol. Process., 15, 3525–3540, doi:10.1002/hyp.349, joint Meeting of the Eastern Snow Conference/Canadian-Geophysical-Union-Hydrology-Section, OTTAWA, CANADA, MAY 17–19, 2001, 2001.

Flowers, G. and Clarke, G.: A multicomponent coupled model of glacier hydrology – 1. Theory and synthetic examples, J. Geophys. Res., 107, 1–17, doi:10.1029/2001JB001122, 2002a.

Flowers, G. and Clarke, G.: A multicomponent coupled model of glacier hydrology – 2. Application to Trapridge Glacier, Yukon, Canada, J. Geophys. Res., 107, 1–16, doi:10.1029/2001JB001124, 2002b.

Fountain, A. and Walder, J.: Water flow through temperate glaciers, Rev. Geophys., 36, 299–328, doi:10.1029/97RG03579, 1998.

Freeze, R. A. and Cherry, J. A.: Groundwater, Prentice Hall, 1979.

Fudge, T. J., Harper, J. T., Humphrey, N. F., and Pfeffer, W. T.: Rapid glacier sliding, reverse ice motion and subglacial water pressure during an autumn rainstorm, Ann. Glaciol, 50, 101–108, 2009.

Gagliardini, O. and Zwinger, T.: The ISMIP-HOM benchmark experiments performed using the Finite-Element code Elmer, The Cryosphere, 2, 67–76, doi:10.5194/tc-2-67-2008, 2008.

Gagliardini, O., Cohen, D., Raback, P., and Zwinger, T.: Finite-element modeling of subglacial cavities and related friction law, J. Geophys. Res.-Earth Surface, 112, 1–11, doi:10.1029/2006JF000576, 2007.

Gagliardini, O., Zwinger, T., Gillet-Chaulet, F., Durand, G., Favier, L., de Fleurian, B., Greve, R., Malinen, M., Martín, C., Råback, P., Ruokolainen, J., Sacchettini, M., Schäfer, M., Seddik, H., and Thies, J.: Capabilities and performance of Elmer/Ice, a new-generation ice sheet model, Geosci. Model Dev., 6, 1299–1318, doi:10.5194/gmd-6-1299-2013, 2013.

Goeller, S., Thoma, M., Grosfeld, K., and Miller, H.: A balanced water layer concept for subglacial hydrology in large-scale ice sheet models, The Cryosphere, 7, 1095–1106, doi:10.5194/tc-7-1095-2013, 2013.

Gordon, S., Sharp, M., Hubbard, B., Smart, C., Ketterling, B., and Willis, I.: Seasonal reorganization of subglacial drainage inferred from measurements in boreholes, Hydrol. Process., 12, 105–133, doi:10.1002/(SICI)1099-1085(199801)12:1<105::AID-HYP566>3.3.CO;2-R, 1998.

Gordon, S., Sharp, M., Hubbard, B., Willis, I., Smart, C., Copland, L., Harbor, J., and Ketterling, B.: Borehole drainage and its implications for the investigation of glacier hydrology: experiences from Haut Glacier d'Arolla, Switzerland, Hydrol. Process., 15, 797–813, doi:10.1002/hyp.184, 2001.

Gulley, J. D., Walthard, P., Martin, J., Banwell, A. F., Benn, D. I., and Catania, G.: Conduit roughness and dye-trace breakthrough curves: why slow velocity and high dispersivity may not reflect flow in distributed systems, J. Glaciol., 58, 915–925, doi:10.3189/2012JoG11J115, 2012.

Harper, J. T., Humphrey, N. F., Pfeffer, W. T., and Lazar, B.: Two modes of accelerated glacier sliding related to water, Geophys. Res. Lett., 34, 2–5, doi:10.1029/2007GL030233, 2007.

Hewitt, I. J.: Seasonal changes in ice sheet motion due to melt water lubrication, Earth Planet. Sci. Lett., 371–372, 16–25, doi:10.1016/j.epsl.2013.04.022, 2013.

Hewitt, I. J. and Fowler, A. C.: Seasonal waves on glaciers, Hydrol. Process., 22, 3919–3930, doi:10.1002/hyp.7029, workshop on Glaciers in Watershed and Global Hydrology, Obergurgl, AUSTRIA, AUG 27–31, 2007, 2008.

Hewitt, I. J., Schoof, C., and Werder, M. A.: Flotation and free surface flow in a model for subglacial drainage. Part 2. Channel flow, J. Fluid Mech., 702, 157–187, doi:10.1017/jfm.2012.166, 2012.

Hubbard, A., Blatter, H., Nienow, P., Mair, D., and Hubbard, B.: Comparison of a three-dimensional model for glacier flow with field data from Haut Glacier d'Arolla, Switzerland, J. Glaciol., 44, 368–378, 1998.

Hubbard, A., Willis, I., Sharp, M., Mair, D., Nienow, P., Hubbard, B., and Blatter, H.: Glacier mass-balance determination by remote sensing and high-resolution modelling, J. Glaciol., 46, 491–498, doi:10.3189/172756500781833016, 2000.

Hubbard, B. and Nienow, P.: Alpine subglacial hydrology, Quat. Sci. Rev., 16, 939–955, doi:10.1016/S0277-3791(97)00031-0, 1997.

Hubbard, B., Sharp, M., Willis, I., and Nielsen, M.K.and Smart, C.: Borehole water-level variations and the structure of the subglacial hydrological system of Haut Glacier d'Arolla, Valais, Switzerland, J. Glaciol., 41, 572–583, 1995.

Iken, A.: The effect of the subglacial water pressure on the sliding velocity of a glacier in an idealized numerical Model, J. Gl., 27, 407–421, 1981.

Iken, A. and Bindschadler, R. A.: Combined measurments of subglacial water-pressure and surface velocity of Findelengletsher, Switzerland, J. Glaciol., 32, 101–119, 1986.

Kulessa, B., Hubbard, B., Brown, G., and Becker, J.: Earth tide forcing of glacier drainage, Geophys. Res. Lett., 30, 1–4, doi:10.1029/2002GL015303, 2003.

Le Brocq, A., Payne, A., Siegert, M., and Alley, R.: A subglacial water-flow model for West Antarctica, J. Glaciol., 55, 879–888, 2009.

Lliboutry, L.: General theory of subglacial cavitation and sliding of temperate glaciers, J. Glaciol., 7, 21–58, 1968.

Magnússon, E., Björnsson, H., Rott, H., and Pálsson, F.: Reduced glacier sliding caused by persistent drainage from a subglacial lake, The Cryosphere, 4, 13–20, doi:10.5194/tc-4-13-2010, 2010.

Mair, D., Nienow, P., Willis, I., and Sharp, M.: Spatial patterns of glacier motion during a high-velocity event: Haut Glacier d'Arolla, Switzerland, J. Glaciol., 47, 9–20, doi:10.3189/172756501781832412, 2001.

Mair, D., Nienow, P., Sharp, M., Wohlleben, T., and Willis, I.: Influence of subglacial drainage system evolution on glacier surface motion: Haut Glacier d'Arolla, Switzerland, J. Geophys. Res., 107, 1–8, doi:10.1029/2001JB000514, 2002a.

Mair, D., Sharp, M., and Willis, I.: Evidence for basal cavity opening from analysis of surface uplift during a high-velocity event: Haut Glacier d'Arolla, Switzerland, J. Glaciol., 48, 208–216, doi:10.3189/172756502781831502, 2002b.

Mair, D., Willis, I., Fischer, U., Hubbard, B., Nienow, P., and Hubbard, A.: Hydrological controls on patterns of surface, internal and basal motion during three "spring events": Haut Glacier d'Arolla, Switzerland, J. Glaciol., 49, 555–567, doi:10.3189/172756503781830467, 2003.

Nienow, P., Sharp, M., and Willis, I.: Seasonal changes in the morphology of the subglacial drainage system, Haut Glacier d'Arolla, Switzerland, Earth Surf. Process. Landforms, 23, 825–843, doi:10.1002/(SICI)1096-9837(199809)23:9<825::AID-ESP893>3.0.CO;2-2, 1998.

Nye, J.: Water flow in glaciers: jokulhlaups, tunnels and veins, J. Glaciol., 17, 181–207, 1976.

Pimentel, S., Flowers, G. E., and Schoof, C. G.: A hydrologically coupled higher-order flow-band model of ice dynamics with a Coulomb friction sliding law, J. Geophys. Res., 115, 1–16, doi:10.1029/2009JF001621, 2010.

Röthlisberger, H.: Water pressure in intra-and subglacial channels, J. Glaciol., 11, 177–203, 1972.

Schoof, C.: The effect of cavitation on glacier sliding, Proc. R. Soc. A, 461, 609–627, doi:10.1098/rspa.2004.1350, 2005.

Schoof, C.: Coulomb Friction And Other Sliding Laws In A Higher-order Glacier Flow Model, Math. Models Methods Appl. Sci., 20, 157–189, doi:10.1142/S0218202510004180, 2010a.

Schoof, C.: Ice-sheet acceleration driven by melt supply variability, Nature, 468, 803–806, doi:10.1038/nature09618, 2010b.

Schoof, C.: Marine ice sheet stability, J. Fluid Mech., 698, 62–72, doi:10.1017/jfm.2012.43, 2012.

Sharp, M., Richards, K., Willis, I., Arnold, N., Nienow, P., Lawson, W., and Tison, J. L.: Geometry, bed topography and drainage structure of the Haut Glacier d'Arolla, Switzerland, Earth Surf. Process. Landforms, 18, 557–571, doi:10.1002/esp.3290180608, 1993.

Shoemaker, E. M.: Subglacial hydrology for an ice sheet resting on a deformable aquifer, J. Glaciol., 32, 20–30, 1986.

Teutsch, G. and Sauter, M.: Groundwater modeling in karst terranes: Scale effects, data acquisition and field validation, in: Third Conference on Hydrogeology, Ecology, Monitoring, and Management of Ground Water in Karst Terranes. National Ground Water Association, Dublin, Ohio, 17–35, 1991.

Walder, J. S.: Stablity of sheet flow of water beneath temperate glaciers and implications for glacier surging, J. Glaciol., 28, 273–293, 1982.

Walder, J. S.: Hydraulics of subglacial cavities, J. Glaciol., 32, 439–445, 1986.

Walder, J. S. and Fowler, A.: Channelized subglacial drainage over a deformable bed, J. Glaciol., 40, 3–15, 1994.

Weertman, J.: On the sliding of glaciers, J. Glaciol., 3, 33–38, 1957.

Weertman, J.: General theory of water flow at the base of glacier or ice sheet, Rev. Geophys. Space Phys., 10, 287–333, doi:10.1029/RG010i001p00287, 1972.

Werder, M. A., Hewitt, I. J., Schoof, C. G., and Flowers, G. E.: Modeling channelized and distributed subglacial drainage in two dimensions, J. Geophys. Res., 118, 1–19, doi:10.1002/jgrf.20146,, 2013.

White, W. B.: Conceptual models for karstic aquifers, Karst Modeling, 5, 11–16, 1999.

Zwinger, T., Greve, R., Gagliardini, O., Shiraiwa, T., and Lyly, M.: A full Stokes-flow thermo-mechanical model for firn and ice applied to the Gorshkov crater glacier, Kamchatka, Ann. Glaciol., 45, 29–37, 2007.

UAV photogrammetry and structure from motion to assess calving dynamics at Store Glacier, a large outlet draining the Greenland ice sheet

J. C. Ryan[1], A. L. Hubbard[2], J. E. Box[3], J. Todd[4], P. Christoffersen[4], J. R. Carr[1], T. O. Holt[1], and N. Snooke[5]

[1]Centre for Glaciology, Institute of Geography and Earth Sciences, Aberystwyth University, Aberystwyth, SY23 3DB, UK
[2]Department of Geology, University of Tromsø, 9037 Tromsø, Norway
[3]Geological Survey of Denmark and Greenland, Copenhagen, Denmark
[4]Scott Polar Research Institute, University of Cambridge, Cambridge, UK
[5]Department of Computer Science, Aberystwyth University, Aberystwyth, SY23 3DB, UK

Correspondence to: J. C. Ryan (jor44@aber.ac.uk)

Abstract. This study presents the application of a cost-effective, unmanned aerial vehicle (UAV) to investigate calving dynamics at a major marine-terminating outlet glacier draining the western sector of the Greenland ice sheet. The UAV was flown over Store Glacier on three sorties during summer 2013 and acquired over 2000 overlapping, geotagged images of the calving front at an $\sim$40 cm ground sampling distance. Stereo-photogrammetry applied to these images enabled the extraction of high-resolution digital elevation models (DEMs) with vertical accuracies of $\pm$1.9 m which were used to quantify glaciological processes from early July to late August 2013. The central zone of the calving front advanced by $\sim$500 m, whilst the lateral margins remained stable. The orientation of crevasses and the surface velocity field derived from feature tracking indicates that lateral drag is the primary resistive force and that ice flow varies across the calving front from 2.5 m d^{-1} at the margins to in excess of 16 m d^{-1} at the centreline. Ice flux through the calving front is 3.8×10^7 m^3 d^{-1}, equivalent to 13.9 Gt a^{-1} and comparable to flux-gate estimates of Store Glacier's annual discharge. Water-filled crevasses were present throughout the observation period but covered a limited area of between 0.025 and 0.24 % of the terminus and did not appear to exert any significant control over fracture or calving. We conclude that the use of repeat UAV surveys coupled with the processing techniques outlined in this paper have great potential for elucidating the complex frontal dynamics that characterise large calving outlet glaciers.

1 Introduction

Observational and modelling studies have demonstrated that Greenland's marine outlet glaciers have a complex and potentially non-linear response to both environmental forcing (e.g. Vieli et al., 2000; Benn et al., 2007; Holland et al., 2008; Howat et al., 2010; Hubbard, 2011; Joughin et al., 2012; Walter et al., 2012; Carr et al., 2013) and to changes in front position (Howat et al., 2007; Luckman et al., 2006; Joughin et al., 2008). To quantify these processes and feedbacks, regular and accurate high-resolution measurements are required to capture the key spatio-temporal linkages between rates of ice calving, flow, surface lowering and frontal advance/retreat. Despite significant advances in satellite remote sensing, limitations of spatial resolution (e.g. MODIS) and/or frequency of repeat imagery (e.g. Landsat or TerraSar-X) render detailed, day-to-day analysis of calving-front dynamics unfeasible. On the other hand, acquisition of digital imagery from unmanned aerial vehicles (UAVs) combined with the development of stereo-photogrammetry software has enabled the provision of high-resolution 3-D georeferenced data on demand for geoscience applications (e.g. d'Oleire-Oltmanns et al., 2012; Hugenholtz et al., 2012, 2013; Whitehead et al., 2013; Lucieer et al., 2014). This represents a cost-effective technique technique for acquiring aerial data in remote, hazardous and/or inaccessible regions and recent applications for emerging snow and ice investigation abound the web (e.g. see the highly informative site of Matt

Table 1. Attributes of the flight surveys and image acquisition of the UAV.

Flight no.	Date	Interval between pictures (s)	No. images	Glacier coverage (km^2)	Resolution of DEM (cm/pixel)
1	1 July	1.55	611	3.17	40
2	2 July	1.51	1051	4.95	38
3	23 August	2.36	567	5.02	39

Nolan; http://www.drmattnolan.org/photography/2013/). To date, published (peer-reviewed) application appears to be limited to the investigation of inter-annual changes of a land-terminating glacier on Bylot Island, Canadian Arctic (Whitehead et al., 2013).

Between July and August 2013, an off-the-shelf, fixed wing UAV equipped with a compact digital camera flew three sortics ovcr thc calving front of Store Glacier, West Greenland. The aerial photographs obtained during these flights were used to produce high-resolution (~ 40 cm; Table 1) digital elevation models (DEMs) and orthophotos of the glacier terminus. These data allowed for the investigation of the spatially complex and time-varying glaciological processes operating at the glacier's calving front. The aim of this paper is to

1. detail the UAV, in terms of its payload and camera settings, and its specific deployment to Store Glacier;
2. describe the techniques used for processing the aerial images and quantifying glaciological processes;
3. discuss the significance of the data we obtained which includes calving events, the character, orientation and morphology of crevasses, surface velocities, ice discharge and changes in thickness and position of the calving front.

2 Data and methods

2.1 Study site

Store Glacier is a large marine-terminating (tidewater) outlet glacier located in the Uummannaq district of West Greenland (Fig. 1). The calving front has a width of 5.3 km and an aerial calving front (freeboard) of up to 110 m a.s.l. (Ahn and Box, 2010). Aerial photography from 1948 onwards reveals that Store Glacier's frontal position has remained stable over the last 65 years (Weidick, 1995). Seasonally, the calving front exhibits advance and retreat of up to 400 m (Howat et al., 2010). The study here focuses specifically on glacier dynamics during the melt season under open-water, tidal modulation of ice flow.

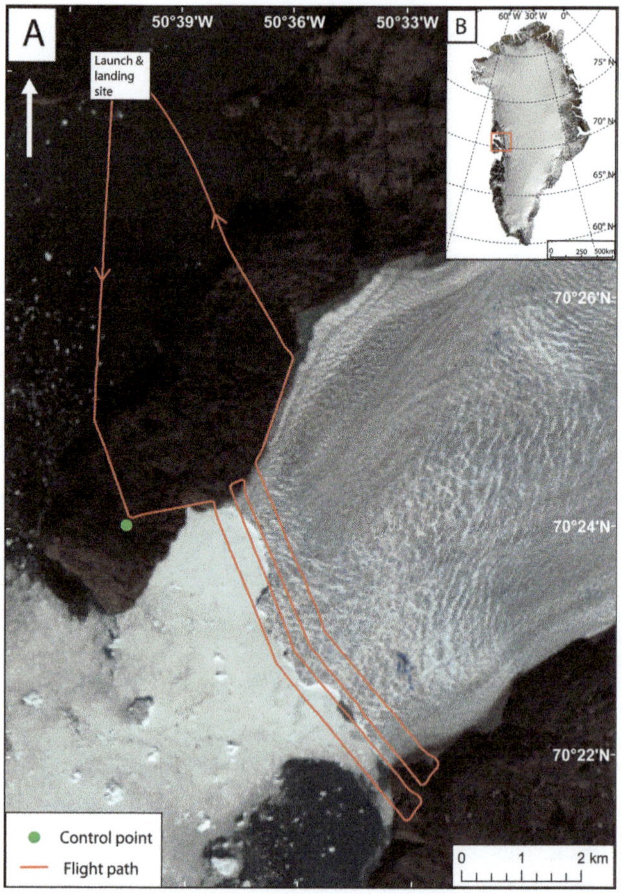

Figure 1. **(a)** A typical UAV sortie over Store Glacier. The background map is a Landsat 8 true colour image from 12 June 2013. The red line shows the UAV flight path on the 2 July 2013. **(b)** Location of Store Glacier in the Uummannaq region, West Greenland on a MODIS mosaic image of Greenland (Kargel et al., 2012).

2.2 UAV platform

The UAV airframe is an off-the-self Skywalker X8 (www.hobbyking.com) which has a wingspan of 2.12 m and is made from expanded polypropylene (EPP) foam (Fig. 2). For this deployment, the X8 was powered by two 5Ah four-cell (14.8 V) lithium polymer batteries driving a 910 W brushless electric motor turning an 11 × 7 foldable propeller. In this configuration, the X8 has a flying mass of ~ 3 kg (including 1 kg payload), which allows for a cruising speed of around 55–70 km per hour with a maximum range of ~ 60 km in benign conditions at constant altitude. A small propeller/high-revolution motor combination was chosen to provide maximum instantaneous thrust to ensure a clean launch (for novice operators) and to handle the potentially strong katabatic winds encountered during its 40 km sortie.

The autopilot is an open-source project called Ardupilot (http://ardupilot.com/) based on an Atmel 2560 8 bit microcontroller and standard radio control parts including 2.4 GHz

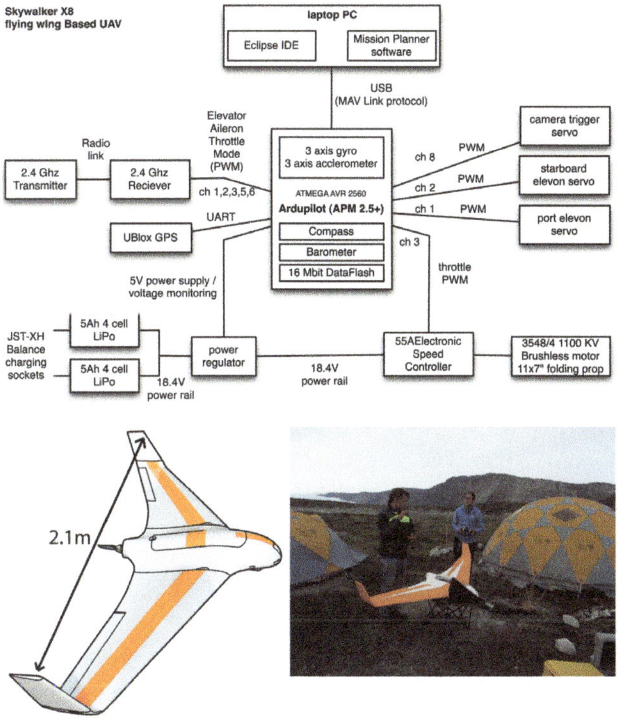

Figure 2. Flowchart of the control set-up and picture of the UAV at base camp with the relative novices.

radio control and pulse-width modulation (PWM) controlled servos for aileron and elevon control (Fig. 2). Ardupilot implements a dual-level proportional-integral-derivative (PID) controller architecture. The lower level controls flight stabilisation and the higher level controls based navigation. Tuning of the PID parameters is necessary to suit the mass and dynamics of the airframe to ensure accurate stabilisation without pitch/roll oscillation (lower-level controller) or flight-path weaving (higher-level controller). The autopilot allows the UAV to fly autonomously according to a pre-programmed flight path defined by a series of 3-D waypoints chosen by the user. The autopilot utilises a GPS for navigation, a triple axis accelerometer and gyro for stabilisation and a barometric pressure sensor for altitude control. These parameters are logged to memory at 10 Hz throughout the flight (Fig. 2).

The advantage of this package is that it can be assembled within a day from off-the-shelf parts and is cost effective at less than USD 2000. The X8 is also relatively straightforward to fly, robust, easily repairable and floats, all added bonuses when being deployed in remote areas by potential novices. Furthermore, the Ardupilot firmware is open source and hence can be programmed for specific requirements, for example camera triggering (see below).

Two lightweight digital cameras were tested at the field site: a Panasonic Lumix DMC-LX5 10.1 megapixel (MP) camera with a 24 mm wide-angle zoom lens and a 16.1 MP Sony NEX-5N with a 16 mm fixed focal length lens though results presented here are limited to the former. A SPOT GPS tracking device was also included in the payload to facilitate recovery should a mission fail (which it did). The focal length of the Lumix lens was adjusted to 5.1 mm (35 mm equivalent) to allow the widest possible coverage which gave the camera a 73.7° horizontal and 53.1° vertical field of view. A short exposure time of 1 / 1600 and a focal ratio of 8 were chosen to prevent overexposure and blurring of the ice surface. The Ardupilot open-source code was amended to trigger the camera automatically at user defined time or distance intervals at or between certain waypoints. The cameras were mounted pointing downwards within the airframe using neoprene and velcro straps to dampen vibration in a custom recessed aperture cut in the bottom with a UV filter to protect the lens and seal it.

2.3 Flight planning

The open-source software APM Mission Planner (http://plane.ardupilot.com/) was used for flight waypoint manipulation and planning in conjunction with the 30 m Greenland Mapping Project (GIMP) DEM (Howat et al., 2014). To optimise spatial coverage against required resolution, flight endurance and stability, the UAV was programmed to fly at a constant altitude of 500 m a.s.l. (Fig. 1). Based on the camera's focal length and field of view (53.1° by 73.7°), the ground (sea-level) footprint at 500 m a.s.l. for each photo was $\sim 450 \times 750$ m. To ensure coverage of the entire glacier terminus and overlap for successful photogrammetric processing, the four transects broadly parallel to the calving front were flown with ~ 250 m separation yielding a side overlap between photos of 70 % (Fig. 1). The mean ground speed of the UAV was ~ 70 km h^{-1} and the camera trigger interval was adjusted between surveys. On flights 1 and 2, the interval between camera triggers was 1.5 s corresponding to a forward overlap of 94 % and over 1000 geotagged images acquired. Flight 3 had a 2.4 s interval yielding a 90 % forward overlap and 581 images (Table 1).

UAV operations were based out of a field camp with the advantage of a 50 m area of flat alluvial terrace with relatively boulder and bedrock free ground for manual remote-control take-off and landing. This location did, however, require a ~ 10 km transit to the calving front over a 450 m high peninsula which significantly reduced the useful endurance over the target. Of the six sorties flown over outlet glaciers in the region during July and August 2013, the three over Store Glacier were most successful. Each sortie was 40 km long and ~ 35 min in duration after the UAV had attained its operating altitude at the start of the mission and was passed from manual remote-control mode into autopilot mode (Fig. 1). Visual and remote-control contact is lost within a few kilometres of the UAV being placed in autopilot mode; hence, validation of the mission plan is essential.

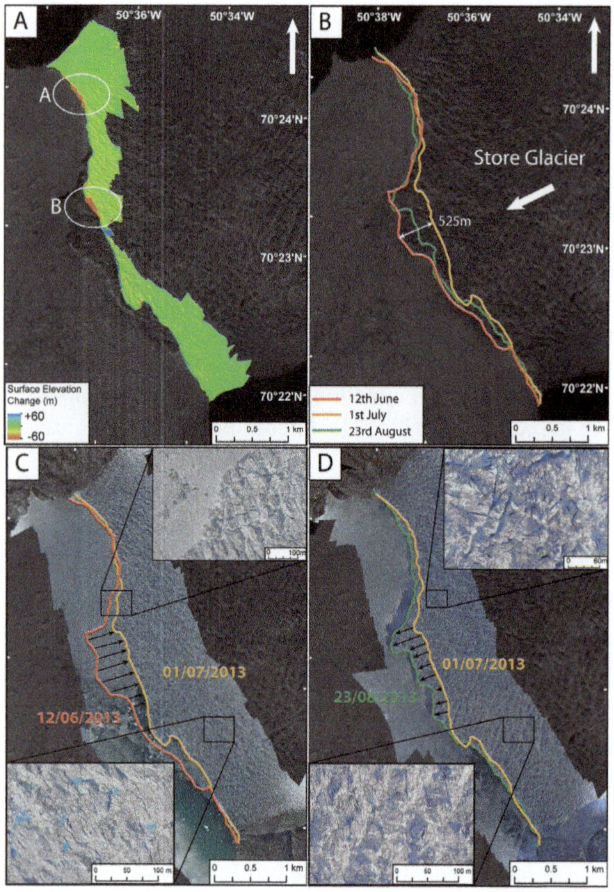

Figure 3. **(a)** Surface elevation difference between two DEMs collected on 1 July and 2 July. Red areas show elevation loss whilst blue areas show elevation gain. White circles highlight the calving events that occurred between the two UAV surveys. **(b)** The position of the calving front of Store Glacier during the summer of 2013. **(c)** Calving-front retreat observed between 12 June and 1 July. Inset is an orthorectified image of the water-filled crevasses observed on 1 July with a pixel resolution of 30 cm. **(d)** Calving-front advance observed between 1 July and 23 August. Inset is an orthorectified image showing water-filled crevasses observed on 23 August. The coverage and size of water-filled crevasses is smaller.

2.4 Three-dimensional model generation

Three-dimensional data were extracted from the aerial photos using Agisoft PhotoScan Pro software (Agisoft LLC, 2013). This software's strength lies in its ability to fully automate workflow and enables non-specialists to process aerial images and produce 3-D models which can be exported as georeferenced orthophotos and DEMs (e.g. Figs. 3 and 6). The first stage of processing is image alignment using the structure-from-motion (SfM) technique. SfM allows for the reconstruction of 3-D geometry and camera position from a sequence of two-dimensional images captured from multiple viewpoints (Ullman, 1979). PhotoScan implements SfM algorithms to monitor the movement of features through a sequence of multiple images and is used to estimate the location of high-contrast features (e.g. edges), obtain the relative location of the acquisition positions and produce a sparse 3-D point cloud of those features. The Ardupilot flight logs of the onboard navigation sensors allow the camera positions and the 3-D point cloud to be georeferenced within instrument precision. SfM also enables the camera calibration parameters (e.g. focal length and distortion coefficients) to be automatically refined; hence, there is no need to pre-calibrate the cameras and lens optics (Verhoeven, 2011).

Once the photos have been aligned, a multi-view reconstruction algorithm is applied to produce a 3-D polygon mesh which operates on pixel values rather than features and enables the fine details of the 3-D geometry to be constructed (Verhoeven, 2011). The user determines the precision of the final 3-D model based on image resolution and pixel footprint. A medium quality setting was chosen yielding DEMs with between 38 and 40 cm/pixel ground sampling resolution (GSD), which were resampled to a Cartesian 50 cm grid to enable intercomparison (Table 1). Higher resolutions (< 30 cm GSD) are attainable but the increase in computational time and the accuracy of georeferencing limits the benefits of such apparent precision.

Two problems of accuracy were encountered in DEM production: (1) PhotoScan failed to reconstruct a flat sea level of constant elevation, and (2) relative positional errors between the DEMs constructed from different sorties were up to 17.12 m horizontally and 11.38 m vertically. Positional errors were due to the specified limits of the onboard L1 GPS of ± 5.0 m horizontally and, when combined with the barometric sensor, to a similar accuracy vertically. These were compounded by the time lag between the camera triggering and actual photograph acquisition. Hence, a secondary stage of processing was carried out which involved 3-D co-registration of the DEMs. To do this, the horizontal and vertical coordinates of common control points (CPs) based on distinct features such as cliff bases, large boulders and promontories were extracted from the georeferenced orthoimages. The CPs that were at sea level were nominally given elevation values of zero, re-imported into PhotoScan and subsequently reprocessed along with a geodetic GPS ground CP located at 70.401° N, −50.665° E and 335.85 m altitude on the bedrock headland overlooking the glacier's northern flank. During this secondary stage of processing, PhotoScan's optimisation procedure was run to correct for possible distortions. After processing with the CPs, a flat sea level across the glacier front was produced and the relative errors between the three DEMs were reduced to ± 1.41 m horizontally and ± 1.90 m vertically. The georeferenced 3-D DEMs and orthophotos were then exported at 50 cm pixel size for further analysis in ArcGIS and ENVI software packages.

2.5 Analysis

Changes in calving-front positions were obtained from these data combined with a Landsat 8 panchromatic image obtained on 12 June (Fig. 3b). Each calving-front position was digitised according to the procedure outlined by Moon and Joughin (2008), whereby a polygon of the calving-front retreat or advance is digitised and divided by the width of the glacier. This method has been used in previous studies (e.g. Howat et al., 2010; Schild and Hamilton, 2013) and enables intercomparison of results. Surface elevation change was calculated from the residual difference of the DEMs (Fig. 3a).

Ice flow across the terminus region was calculated by feature tracking performed on successive DEMs using the ENVI Cosi-CORR software module (Fig. 4b). These velocities were then used to estimate ice flux through the calving front for the same period under the assumption of plug flow (uniform velocity profile with depth) and using a calving-front cross section obtained from Xu et al. (2013) and modified by single and multi-beam echo sounder bathymetry obtained by S/V *Gambo* in 2010 and 2012 (Chauché, unpublished). The frontal cross section was divided into 10 m vertical strips and under the plug-flow assumption, each was assigned its corresponding horizontal velocity (Fig. 4a). The floatation depth and buoyancy ratio across the calving front was calculated using the ice surface (freeboard) elevation and total ice thickness with a value for the density of ice of 917 kg m^{-3} and for sea water of 1028 kg m^{-3} (Fig. 5a).

To investigate the distribution and patterns of crevassing, each DEM was Gaussian filtered at 200 pixels (100 m) in ArcGIS and subtracted from the original DEM to yield the pattern of negative surface anomalies. These anomalies were converted into polygons to map and hence quantify crevasse distribution and character (Fig. 6a). The resulting polygons were enclosed by a minimum bounding rectangle, which allowed the orientation, width, length and depth of crevasses to be extracted (Fig. 6a, Table 2). Water-filled crevasses were automatically located in the ENVI package using the supervised maximum likelihood classification (MLC) method. Representative training samples for water-filled areas were chosen from the colour composite orthophoto (Fig. 6b). The trained tool then classifies pixels that are interpreted as water into the desired class. The resulting raster image was converted into a shapefile and used to mask and define the area of the water-filled crevasses across the terminus. These procedures allow thousands of crevasses in multiple orthoimages and DEMs to be quantified easily without the difficulties and dangers associated with direct field measurements.

2.6 Uncertainties and limitations

The relative horizontal uncertainties between the DEMs were investigated by feature tracking the stationary bedrock at the sides of the glacier. The root mean square (rms) horizontal displacement was ± 1.41 m which provides us with an

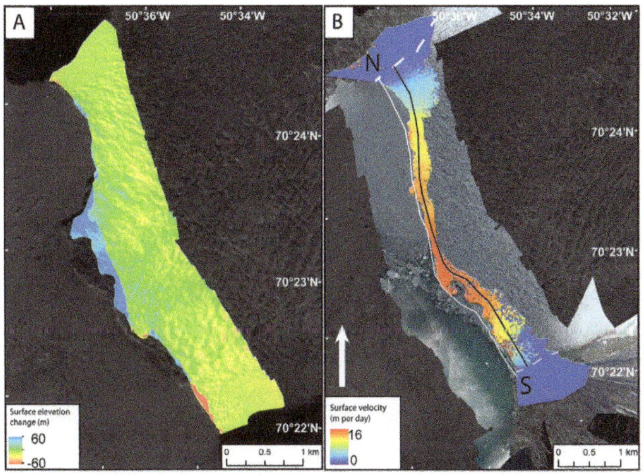

Figure 4. (a) Surface elevation changes between 2 July and 23 August. An average thinning of 0.12 m d^{-1} was estimated for the surveyed area. **(b)** The ice-flow speed structure of the terminus of Store Glacier between 1 and 2 July 2013. The centre of the glacier flows at approximately 16 m d^{-1} whilst the margins flow less than 5 m d^{-1}. Dotted white lines show the lateral margins of the glacier. The black line represents the locations of the horizontal velocity and surface elevation values that were used to estimate ice flux. The white line represents the location of the depth values used to estimate ice flux. The cross section of the calving front derived from these profiles is displayed in **(a)**.

approximate error estimate. The relative vertical uncertainties between the DEMs were estimated by calculating elevation differences between bedrock areas, which revealed an error estimate of ± 1.9 m. The two-stage procedure outlined in Sect. 2.4 therefore enabled us to improve the relative positional uncertainties from nearly 20 m to less than 2 m. For future studies, it is thought that several CPs on the bedrock either side of the glacier front would further reduce these uncertainties. A telemetric differential GPS deployed on or near the calving front, which is sufficiently large/bright to identify within the aerial imagery would allow further ground control in the centre of DEMs, away from bedrock CPs.

Due to the lack of reflected light from deep crevasse recesses, the DEM generation process cannot quantify the narrowest sections of all fractures and resultant crevasse depths are therefore a minimum estimate. The technique is also clearly limited to line of sight precluding narrow fractures which extend for tens of centimetres horizontally and potentially up to a few metres vertically (Hambrey and Lawson, 2000; Mottram and Benn, 2009).

Finally, there are a number of practical difficulties when operating an autonomous aircraft in remote and inaccessible environments. Mission planning is critical; knowledge of the local weather conditions, as well as up-to-date satellite imagery and DEMs are a prerequisite.

Table 2. Attributes of mean crevasse width, length and orientation in each zone labelled in Fig. 5. Orientations are measured along the long axis of each crevasse with respect to the direction of flow which is 0°.

Zone	Mean width (m)	Mean length (m)	Mean orientation (°)
Zone 1	3.6	9.4	9.2
Zone 2	4.8	14.0	36.7
Zone 3	10.5	32.6	85.1
Zone 4	6.5	17.8	60.4
Zone 5	3.5	8.5	10.8

3 Results

Three successful UAV sorties were flown over Store Glacier calving front providing imagery, orthophotos and DEMs on 1 and 2 July and the 23 August, herein referred to as flights and associated products 1 to 3, respectively (Table 1). The interval between flights 1 and 2 was 19 hours and comparison between these outputs enables identification of processes operating over a daily (short) timescale, be it a very specific snapshot. The third sortie was flown 52 days later and comparison between these outputs enables investigation of late-seasonal change. The footprint of the four cross-glacier transects flown extends just over 1 km upstream from the calving front and herein is referred to as the terminus.

3.1 Short timescale calving and surface elevation change

Residual elevation change between 1 and 2 July (Fig. 3a) reveals that the front retreated in two sections by up to 50 and 80 m, respectively. The more northerly calving event (A) resulted in a 450 m wide section of the terminus retreating by between 20 and 50 m, whilst event B produced between 20 and 80 m of retreat across a 400 m section (Fig. 3a). In addition to these two calving events (which are discussed in section 3.6), the central 4.5 km frontal section advanced between 12 to 16 m (Fig. 3a). At its lateral margins, the calving front shows no discernible systematic change though there are isolated, small calving events, for example, within 50 m of the southern flank (Fig. 3a). Upstream of the calving front, there is no net change in mean surface elevation away from the front and the dappled pattern of residual elevation change is a result of the advection of crevasses and seracs. Successive long profiles of the terminus between the 1 and 2 July reveal specific down-glacier crevasse advection with flow (Fig. 6) at a rate of 5 and 16 m d^{-1} on profile 1 and 2, respectively. These results provide corroboration for the surface velocities derived by feature tracking in Sect. 3.4.

3.2 Seasonal timescale calving-front position and surface elevation change

Over the entire melt season, larger fluctuations in calving-front position are observed (Fig. 3b). Over the 19-day period from 12 June to 1 July, mean frontal retreat was 160 m (Fig. 3c) and between 2 July and 23 August, the calving front advanced by an average of ~110 m to a position similar to that in 12 June (Fig. 3d). These mean values, however, do not convey the full extent and detail of the changes observed in the calving front. For example, the central section of the calving front retreated by up to 525 m between the 12 June and 1 July and advanced by up to 450 m between 2 July and 23 August (Fig. 3b). Furthermore, the lateral margins of Store Glacier (the southern 850 m and the northern 1.5 km) are relatively stable with < 50 m change in position. Over the 52-day period between 2 July (Flight 2) and 23 August (Flight 3) widespread surface lowering of 6.1 m (or 0.12 m d^{-1}) was observed across Store Glacier terminus (Fig. 4a), which is significantly larger than the estimated vertical uncertainties of the DEMs (± 1.9 m). Despite the same dappled patterns caused by local advection of crevasses and seracs, we infer this to be associated with dynamic thinning 1 km upstream of the calving front, which is discussed in Sect. 4.2.

3.3 Bathymetry

The deepest sector of the calving front is located 1 km south of the centreline and exceeds 540 m below sea level (Fig. 5a). This 200 m wide sector also corresponds to the greatest thickness of ~600 m. To the south of this deepest point, the bottom rises rapidly to a 200 m deep shelf located 500 m from the flank. To the north of the deepest point, the bottom shallows more gently to within 400 m where it becomes steeper towards the fjord wall.

3.4 Surface velocities

Maximum surface-flow velocities of 16 m d^{-1} between 1 and 2 July are consistent with results obtained in previous studies using other techniques, such as feature tracking images from a land-based time-lapse camera (between 11 and 15 m d^{-1}) (Ahn and Box, 2010; Walter et al., 2012). The spatial pattern of surface flow from feature tracking of images between the 1 and 2 July varies considerably across the terminus of Store Glacier (Fig. 4b) attaining velocities of 16 m d^{-1} (5.8 km a^{-1}) near the centre of the glacier down to 2.5 m d^{-1} at the lateral flanks. Surface velocities are related to slope, depth, thickness and distance from the lateral margins (Fig. 5c, d). As would be expected, maximum velocities correlate with maximum depth and towards the northern flank are linearly correlated ($R^2 = 0.90$) with frontal depth (Fig. 5c). Towards the southern flank the relationship is less apparent especially between 200 and 350 m depths. There

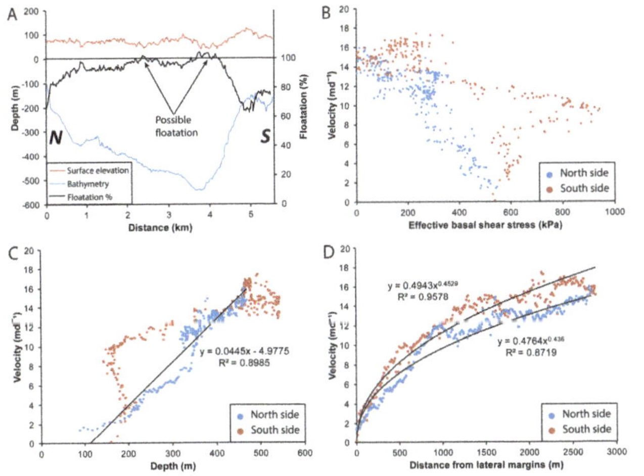

Figure 5. **(a)** Profiles showing the sea floor bathymetry and ice surface elevation at the calving front. These data were combined with surface velocities to estimate the ice flux of Store Glacier. Where the floatation percentage is over 100 %, it is assumed that the ice is not thick enough to be fully grounded in hydrostatic equilibrium. **(b)** The relationship between effective basal shear stress and velocity. **(c)** The relationship between depth and velocity. At depths deeper than 400 m, velocities are fairly constant. The two differing relationships between 150 and 350 m represent velocities from different sides of the glacier. **(d)** Relationship between velocity and distance from the lateral margins. The positive correlation demonstrates the importance of the resistance provided by the fjord walls.

is a strong correlation between velocities and distance from the lateral margins which can be approximated by a power function ($R^2 = 0.90$) (Fig. 5d). Although application of the floatation criteria reveals parts of calving front to be buoyant (Fig. 5a), side-scan sonar observations reveal that the glacier toe was resting on the fjord bed (Chauché, unpublished). When the surface flow pattern is combined with frontal bathymetric data we estimated that the mass flux through the calving front of Store Glacier was 3.8×10^7 m^3 d^{-1}, equivalent to $\sim$ 13.9 Gt a^{-1}.

Seasonal flow patterns were not obtainable between 2 July and 23 August as the majority of any matching features within the study area required for tracking had already calved into the ocean. Furthermore, it is likely that the morphology of many crevasses and seracs will have changed significantly through melt and deformation and would not be recognised by the cross-correlation procedure.

3.5 Crevassing

The morphology and orientation of crevasses varies markedly across the terminus (Fig. 6). The largest crevasses occur in a sector south of the glacier centre line in zone 4 (Fig. 6, Table 2). Here, crevasses have mean minimum depths of 18 m, lengths of 68 m and widths of 31 m. The largest crevasses are up to 30 m deep, over 500 m long and nearly 200 m wide but no crevasses that penetrated below sea level were identified. Most crevasses in this region are arcuate with limbs pointing towards the calving front and are orientated obliquely to the direction of ice flow (Fig. 6). This arcuate morphology of crevasses continues across the central 3 km of the terminus in zone 3 (Fig. 6). Here, crevasses have mean a depth of 10.5 m, length of 50 m and widths of 18 m (Table 2). In zone 2, 300 to 500 m from the northern flank, crevasses are aligned obliquely to the direction of ice flow (30–45°). Up to the fjord walls in zones 1 and 5, crevasses are generally orientated parallel to the ice flow ($> 15°$) (Fig. 6, Table 2) and are much smaller with a mean lengths of 22 m and width of 8 m (Table 2). No discernible difference in average crevasse depths, lengths or widths was observed between early July and late August and the pattern and character of crevassing was also similar.

Water-filled crevasses were clustered in zone 4, coinciding with the sector of larger crevasses (Fig. 6b). Water-filled crevasses covered 12 000 m^2 or 0.24 % of the survey area (to $\sim$ 1 km from the calving front) on 2 July (Table 1). Some 42 individual water-filled crevasses were identified with the largest having an area of 1200 m^2. By 23 August, the number, size and total area of water-filled crevasses were lower: only 10 water-filled crevasses could be identified, the largest of which was 400 m^2 and with a total area of 1230 m^2 (0.025 % of the survey area). We were not able to ascertain the depth of water in the crevasses as no common crevasses could be identified which drained or filled between flights but this would be a specific aim of future studies which, with regular sorties, could potentially determine the depth of a crevasse before filling or after drainage or otherwise exploit the light reflectance relationship with water depth (e.g. Fitzpatrick et al., 2014).

Successive profiles of the terminus from 1 and 2 July demonstrate how the UAV surveys are capable of capturing the displacement of crevasses, which advect downstream at a rate of 5 and 16 m d^{-1} in profiles 1 and 2, respectively (Fig. 6). The techniques used in this study are therefore capable of identifying changes in crevasses geometry, particularly width and depth through time.

3.6 Calving events

The two calving events identified between 1 and 2 July appear to take place under contrasting conditions. Event A consisted of the calving of multiple, relatively small ice blocks with the glacier failing along two main crevasses located 30 and 50 m behind the calving front. These crevasses were between 8 and 10 m deep, respectively, and in this instance, the crevasses located closest to the front were the ones that failed. Event B appears to be a single large event caused by the fracturing of a series of parallel crevasses which were up to 14 m deep and 60 m behind the calving front. Unlike, calving event A, the crevasses that failed in event B were not the closest to the calving front. Indeed, there were other crevasses that were deeper and located nearer to the front, yet did not calve. Water was not observed in any of the crevasses along which calving took place.

4 Discussion

4.1 Changes occurring over a daily timescale

The orientation of crevasses suggests that lateral drag is an important resistive stress on Store Glacier. The lateral margins of Store Glacier are characterised by crevasses that are orientated parallel to the direction of flow which suggests that they have formed in response to simple shear stresses associated with the drag of the fjord walls (Fig. 6) (Benn and Evans, 2010). The importance of lateral drag is further demonstrated by the morphology of crevasses found near the glacier flow line (Fig. 6). Their arcuate nature indicates that the principal tensile stresses operating on the ice have been rotated by lateral gradients in ice velocity. These gradients are caused by the simple shear stress between the fjord walls and the margins of the glacier which cause the ice to flow slower (Fig. 4b) (Benn and Evans, 2010).

The simple shearing caused by velocity gradients is further demonstrated by the differing relationship between velocity and depths between the north and south side of the glacier (Fig. 5c, d). On the north side, the velocity increases gradually from the fjord wall to the centre of the glacier, reflecting the gradual deepening of bathymetry and the resulting decrease of basal and lateral drag. On the south side, the velocities are higher than the north side for given depths and distances from the lateral margins (Fig. 5c, d). We hypothesise that, because the deepest part of the glacier is situated 1 km south of the centreline, the ice on south side is more influenced by faster flowing ice which exerts a simple shear stress on the shallower, adjacent ice (250–400 m thick). This causes the shallow ice to flow faster than ice with similar thicknesses and distance from the lateral margins on the north side (Fig. 5c).

The mass flux through the calving front was calculated at $3.8 \times 10^7\,m^3\,d^{-1}$ which needs to be balanced by three main frontal processes: calving, submarine melting and advective advance. Both calving and advance were observed in this study but it is likely that submarine melting also has a large role in ice output at a daily timescale. For example, Xu et al. (2013) used oceanographic data to calculate a melt water flux of between 0.5 and $1.1 \times 10^7\,m^3\,d^{-1}$ from Store Glacier in August 2010 equivalent to 13–29 % of the mass flux calculated by our study. For comparison, Rink glacier has an ice flux of $3.0 \times 10^7\,m^3\,d^{-1}$ of which 27 % is estimated to be lost through submarine melting each day (Enderlin and Howat, 2013).

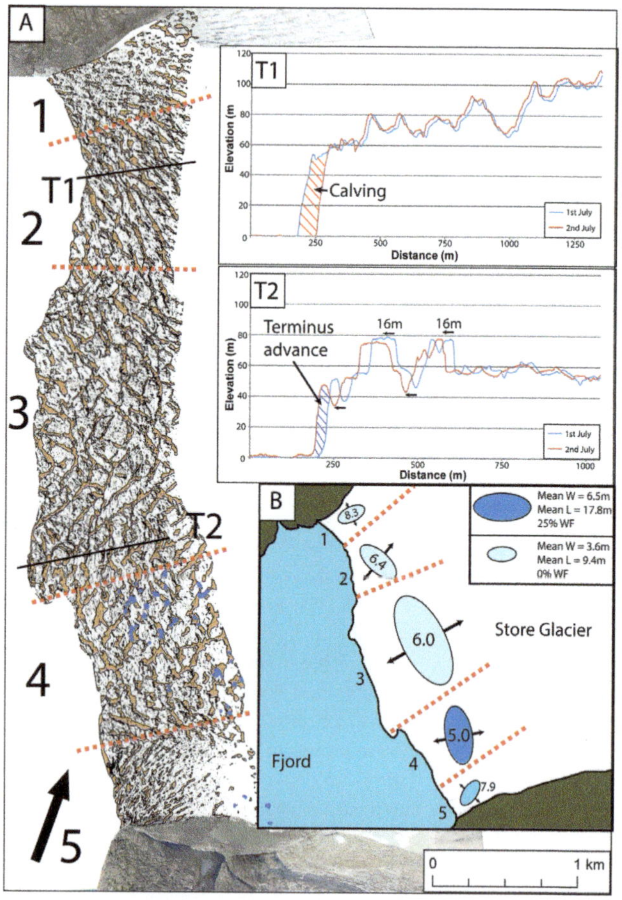

Figure 6. (a) Distribution and patterns of crevasses on Store Glacier. Dry crevasses, which are large structural features, are shown in orange. Narrower crevasses that are observed in the orthorectified images but whose 3-D geometry is not constructed are shown in black. The areas of water-filled crevasses are shown in blue and occur almost exclusively in zone 4. The regions of the terminus discussed are designated by the dotted red lines and are referred to as the black numbers. Transects 1 and 2 shown in the inset demonstrate how crevasses adverted downstream between 1 and 2 July. In T1, a series of calving events occurred which are discussed as calving event A. In T2, the calving front advanced 16 m. **(b)** Illustration of the terminus of Store Glacier with ellipsoids proportional to the average length, width and orientation of crevasses shown in **(a)** for the respective zones. The colour of the ellipsoids represents the proportion of crevasses that are water filled in each zone where WF refers to water filled in the legend. The italicised numbers denote the density of crevasses per $10\,m^2$ in each zone. Arrows illustrate inferred direction of principal strain.

4.2 Changes occurring over a seasonal timescale

The lack of variation in the position of the lateral margins of the glacier shows that a balance is maintained between the ice flux input and submarine melting and calving output in this zone throughout the melt season. The balance could be explained by the mechanism of calving events. At the lateral margins calving is characterised by small, regular events such as calving event A (Fig. 3a). The regularity of these small events means that any small advance or retreat is regulated almost instantly by changes in calving rate which returns the lateral margins of the glacier to the same position. Calving rate could also be moderated by changes in the bathymetry. When the lateral margins advance, calving rates increase due to the abrupt deepening of the bathymetry seaward of the lateral margins of the glacier which cause basal drag to be reduced. Ice-flow acceleration can lead to increased longitudinal stretching and deeper crevassing, thereby increasing calving rate and leading to retreat to its original, bathymetrically pinned position.

The centre of the calving front is much more active with calving and submarine melt rates that vary on a seasonal timescale. We propose that the main cause of variability is due to calving rates which are highly irregular throughout the melt season (Jung et al., 2010). Our observations also support the suggestion that calving rates are dominated by major calving events which have a time interval of around 28 days (e.g. Jung et al., 2010). If the calving front advances for 28 days at 16 $m\,d^{-1}$, it will advance $\sim$448 m. A large, single calving event can therefore yield a retreat of $\sim$448 m and would explain the variation in the position of the calving front during the melt season (Fig. 3b). On 25 August 2013, a tabular iceberg with a length of $\sim$500 m was observed to calve from the central zone of Store Glacier.

Towards the end of the melt season (23 August), a widespread surface deflation of 0.12 $m\,d^{-1}$ was observed (Fig. 4a). Application of a simple degree-day model reveals that part of this lowering can be attributed to ablation. Average daily air temperatures were recorded at an automated weather station (AWS) located near the UAV launch site (Fig. 1) and, using a melt factor of 6–10 mm per degree per day (Hock et al., 2005), surface lowering due to ablation is estimated between 0.038 and 0.064 $m\,d^{-1}$. It follows that ablation alone cannot account for the entire lowering rate observed and, hence, we infer an additional component of dynamic thinning due to relative strain extension across this zone, related to reduced upstream delivery of flux and/or frontal kinematics associated with enhanced late-season submarine melting and/or calving rates. GPS measurements by Ahlstrom et al. (2013) tentatively support the former interpretation and reveal that surface velocities 8 km upstream of Store's calving front tend to decrease between July and August. However, this raises questions regarding the timescales over which dynamic thinning and surface melt occur and whether or not the flow regime across the terminus is, to some extent, isolated or operating independent from processes upstream supplying mass. Either way, these questions are beyond the scope of the data sets presented here and require a study of greater areal extent and temporal coverage.

Another important observation is the order of magnitude reduction of the area of water-filled crevasses between early July and late August (Fig. 6). Surface air temperatures directly influence the extent of water-filled crevasses. AWS data reveal that mean daily air temperature was $\sim$6 °C during the 4 days prior to the UAV sortie on the 2 July. In contrast, mean temperature was $\sim$3.5 °C on the 4 days prior to the UAV sortie on 23 August. Water-filled crevasses have been hypothesised to penetrate deeper than crevasses without water (Weertman, 1973; van der Veen, 1998) and hence act as mechanism for calving (Benn et al., 2007). The calving events observed in this study did not specifically fail at water-filled crevasses and hence our limited results show no support for this mechanism. However, studies of greater scope with daily coverage will be required to determine definitively if water-filled crevasses have any appreciable impact on calving dynamics at Store Glacier or elsewhere.

5 Conclusions and future directions

A UAV equipped with a commercial digital camera enabled us to obtain high-resolution DEMs and orthophotos of the calving front of a major tidewater glacier at an affordable price. Airborne lidar currently presents the only alternative method for acquiring DEMs with comparable accuracy and precision. However, to fly consecutive sorties in a remote environment is likely to be prohibitively expensive and with sufficient ground control points the digital photogrammetry approach may also exceed the accuracy of this technique.

The three sorties flown enabled key glaciological parameters to be quantified at sufficient detail to reveal that the terminus of Store Glacier is a complex system with large variations in crevasse patterns surface velocities, calving processes, surface elevations and front positions at a daily and seasonal timescale. Surface velocities vary across the terminus and are influenced by both basal and lateral drag (Figs. 4b and 5c, d). The oblique orientation and arcuate nature of crevasses suggests that the principal extending strain rate is orientated obliquely to the direction of flow and we therefore propose that resistive stresses at the terminus of Store Glacier are dominated by lateral drag (Fig. 6). With this in mind, the retreat of Store Glacier into a wider trough could significantly increase the ice discharge. We estimated that the ice flux through the calving front of Store Glacier was 13.9 $Gt\,a^{-1}$ and we observed a small terminus advance between 1 and 2 July (Figs. 3a and 5a). This advance reveals that, during this period, the sum of calving and submarine melt rates are less than the ice flux. Calving is an irregular process and that the position of the calving front returned to its 12 June position by 23 August suggests that over this timescale calving and submarine melting balance ice flux (Fig. 3b). Water-filled crevasses covered 0.24 % of the survey area on 2 July but this fell to 0.025 % on 23 August (Fig. 6). It remains to be seen whether water-filled crevasses are more likely to initiate calving events but our tentative results here indicate no support for this mechanism.

Future studies, with more frequent sorties could be used to compare and investigate further glaciological changes over a more continuous timespan. There is also the possibility of more sophisticated payloads with radiation, albedo and other multi-band sensors as well as radar and laser altimetry. There are many potential cryospheric applications for investigation, such as sea ice, marine and terrestrial-terminating glaciers and, with increased range, ice sheets, that can be achieved with the use of repeat UAV surveys. We have demonstrated that for calving outlet glaciers, a UAV carrying a high-resolution digital camera would be sufficient to investigate the following projects:

- analysis of the thickness and back stress exerted by the ice mélange during the winter and the effect of its break out on glacier flow, calving rate and character;
- seasonal changes in the depth, density, orientation and nature of crevassing and their impact on calving rate and character;
- the influence of daily to seasonal melt and supraglacial lake drainage on downstream dynamics and calving;
- analysis of daily to seasonal fluctuations in calving flux, terminus position and impact on upstream dynamics and thinning.

Acknowledgements. We thank Matt Nolan, Doug Benn and Mauri Pelto for their thorough and insightful reviews, and Anders Damsgaard for his short comments: all of which greatly improved the manuscript. Funding for the fieldwork was made possible by the UK Natural Environmental Research Council (NERC) grant NE/K005871/1 (Subglacial Access and Fast Ice Research Experiment (SAFIRE): Resolving the Basal Control on Ice Flow and Calving in Greenland). NERC also funded *The Cryosphere* page charges. J. C. Ryan is funded by an Aberystwyth University Doctoral Career Development Scholarship (DCDS). We are also indebted to the crew of S/V *Gambo* who worked tirelessly to provide logistical support, and to the Uummannaq Polar Institute and Children's Home who provided accommodation in Uummannaq.

Edited by: A. Kääb

References

AgiSoft LLC: AgiSoft PhotoScan, available at: http://www.agisoft.ru/products/photoscan/ (last access: 14 February 2014), 2013.

Ahlstrøm, A. P., Andersen, S. B., Andersen, M. L., Machguth, H., Nick, F. M., Joughin, I., Reijmer, C. H., van de Wal, R. S. W., Merryman Boncori, J. P., Box, J. E., Citterio, M., van As, D., Fausto, R. S., and Hubbard, A.: Seasonal velocities of eight major marine-terminating outlet glaciers of the Greenland ice sheet from continuous in situ GPS instruments, Earth Syst. Sci. Data, 5, 277-287, doi:10.5194/essd-5-277-2013, 2013.

Ahn, Y. and Box, J. E.: Glacier velocities from time-lapse photos: technique development and first results from the Extreme Ice Survey (EIS) in Greenland, J. Glaciol., 56, 723–734, doi:10.3189/002214310793146313, 2010.

Benn, D. I., Warren, C. R., and Mottram, R. H.: Calving processes and the dynamics of calving glaciers, Earth-Sci. Rev., 82, 143–179, doi:10.1016/j.earscirev.2007.02.002, 2007.

Benn, D. I. and Evans, D. J. A.: Glaciers and glaciation, London: Hodder Education, 2010.

Carr, J. R., Vieli, A., and Stokes, C. R.: Climatic, oceanic and topographic controls on marine-terminating outlet glacier behavior in north-west Greenland at seasonal to interannual timescales, J. Geophys. Res., 118, 1210–1226, 2013.

d'Oleire-Oltmanns, S., Marzolff, I., Peter, K. D., and Ries, J. B.: Unmanned Aerial Vehicle (UAV) for monitoring soil erosion in Morocco, Remote Sens., 4, 3390–3416, 2012.

Enderlin, E. M. and Howat, I. M.: Submarine melt rate estimates for floating termini of Greenland outlet glaciers (2000–2010), J. Glaciol., 59, 67–75, doi:10.3189/2013jog12j049, 2013.

Fitzpatrick, A. A. W., Hubbard, A. L., Box, J. E., Quincey, D. J., van As, D., Mikkelsen, A. P. B., Doyle, S. H., Dow, C. F., Hasholt, B., and Jones, G. A.: A decade (2002–2012) of supraglacial lake volume estimates across Russell Glacier, West Greenland, The Cryosphere, 8, 107–121, doi:10.5194/tc-8-107-2014, 2014.

Hambrey, M. J. and Lawson, W.: Structural styles and deformation fields in glaciers: A review, Deform. Glacial Mater., 176, 59–83, doi:10.1144/Gsl.Sp.2000.176.01.06, 2000.

Hock, R.: Glacier melt: a review on processes and their modelling, Progr. Phys. Geogr., 29, 362–391, 2005.

Holland, D. M., Thomas, R. H., De Young, B., Ribergaard, M. H., and Lyberth, B.: Acceleration of Jakobshavn Isbrae triggered by warm subsurface ocean waters, Nat. Geosci., 1, 659–664, doi:10.1038/Ngeo316, 2008.

Howat, I. M., Joughin, I., and Scambos, T. A.: Rapid changes in ice discharge from Greenland outlet glaciers, Science, 315, 1559–1561, doi:10.1126/science.1138478, 2007.

Howat, I. M., Box, J. E., Ahn, Y., Herrington, A., and McFadden, E. M.: Seasonal variability in the dynamics of marine-terminating outlet glaciers in Greenland, J. Glaciol., 56, 601–613, 2010.

Howat, I. M., Negrete, A., and Smith, B. E.: The Greenland Ice Mapping Project (GIMP) land classification and surface elevation data sets, The Cryosphere, 8, 1509–1518, doi:10.5194/tc-8-1509-2014, 2014.

Hubbard, A.: The Times Atlas and actual Greenland ice loss, Geol. Today, 27, 214–217, 2011.

Hugenholtz, C. H., Levin, N., Barchyn, T. E., and Baddock, M. C.: Remote sensing and spatial analysis of aeolian sand dunes: A review and outlook, Earth-Sci. Rev., 111, 319–334, doi:10.1016/j.earscirev.2011.11.006, 2012.

Hugenholtz, C. H., Whitehead, K., Brown, O. W., Barchyn, T. E., Moorman, B. J., LeClair, A., Riddell, K., and Hamilton, T.: Geomorphological mapping with a small unmanned aircraft system (sUAS): Feature detection and accuracy assessment of a photogrammetrically-derived digital terrain model, Geomorphology, 194, 16–24, doi:10.1016/j.geomorph.2013.03.023, 2013.

Joughin, I., Das, S. B., King, M. A., Smith, B. E., Howat, I. M., and Moon, T.: Seasonal speedup along the western flank of the Greenland ice sheet, Science, 320, 781–783, doi:10.1126/science.1153288, 2008.

Joughin, I., Smith, B. E., Howat, I. M., Floricioiu, D., Alley, R. B., Truffer, M., and Fahnestock, M.: Seasonal to decadal scale variations in the surface velocity of Jakobshavn Isbrae, Greenland:

Observation and model-based analysis, J. Geophys. Res.-Earth Surf., 117, F02030, doi:10.1029/2011jf002110, 2012.

Jung, J., Box, J. E., Balog, J. D., Ahn, Y., Decker, D. T., and Hawbecker, P.: Greenland glacier calving rates from Extreme Ice Survey (EIS) time lapse photogrammetry, C23B-0628, American Geophysical Union, San Francisco, 2010.

Kargel, J. S., Ahlstrøm, A. P., Alley, R. B., Bamber, J. L., Benham, T. J., Box, J. E., Chen, C., Christoffersen, P., Citterio, M., Cogley, J. G., Jiskoot, H., Leonard, G. J., Morin, P., Scambos, T., Sheldon, T., and Willis, I.: Brief communication Greenland's shrinking ice cover: "fast times" but not that fast, The Cryosphere, 6, 533–537, doi:10.5194/tc-6-533-2012, 2012.

Lucieer, A., Turner, D., King, D. H., and Robinson, S. A.: Using an Unmanned Aerial Vehicle (UAV) to capture micro-topography of Antarctic moss beds, Int. J. Appl. Earth Observ.Geoinform., 27, 53–62, 2014.

Luckman, A., Murray, T., de Lange, R., and Hanna, E.: Rapid and synchronous ice-dynamic changes in East Greenland, Geophys. Res. Lett., 33, L03503, doi:10.1029/2005GL025428, 2006.

Moon, T. and Joughin, I.: Changes in ice front position on Greenland's outlet glaciers from 1992 to 2007, J. Geophys. Res.-Earth Surf., 113, F02022, doi:10.1029/2007jf000927, 2008.

Mottram, R. H. and Benn, D. I.: Testing crevasse-depth models: A field study at Breioamerkurjokull, Iceland, J. Glaciol., 55, 746–752, doi:10.3189/002214309789470905, 2009.

Schild, K. M. and Hamilton, G. S.: Seasonal variations of outlet glacier terminus position in Greenland, J. Glaciol., 59, 759–770, doi:10.3189/2013jog12j238, 2013.

Ullman, S.: The interpretation of structure from motion, Proc. Roy. Soc. London, B203, 405–426, 1979.

van der Veen, C.: Fracture mechanics approach to penetration of bottom crevasses on glaciers, Cold Reg. Sci. Technol., 27, 213–223, 1998.

Verhoeven, G.: Taking computer vision aloft – archaeological three-dimensional reconstructions from aerial photographs with Photoscan, Archaeol. Prospection, 18, 67–73, 2011.

Vieli, A., Funk, M., and Blatter, H.: Tidewater glaciers: Frontal flow acceleration and basal sliding, Ann. Glaciol., 31, 217–221, doi:10.3189/172756400781820417, 2000.

Xu, Y., Rignot, E., Fenty, I., Menemenlis, D., and Flexas, M. M.: Subaqueous melting of Store Glacier, West Greenland from three-dimensional, high-resolution numerical modeling and ocean observations, Geophys. Res. Lett., 40, 4648–4653, doi:10.1002/Grl.50825, 2013.

Walter, J. I., Box, J. E., Tulaczyk, S., Brodsky, E. E., Howat, I. M., Ahn, Y., and Brown, A.: Oceanic mechanical forcing of a marine-terminating Greenland glacier, Ann. Glaciol., 53, 181–192, doi:10.3189/2012aog60a083, 2012.

Weertman, J.: Can a water-filled crevasse reach the bottom surface of a glacier?, IAHS Publ., 95, 139–145, 1973.

Whitehead, K., Moorman, B. J., and Hugenholtz, C. H.: Brief Communication: Low-cost, on-demand aerial photogrammetry for glaciological measurement, The Cryosphere, 7, 1879–1884, doi:10.5194/tc-7-1879-2013, 2013.

Weidick, A.: Greenland, with a section on Landsat images of Greenland, in, Satellite image atlas of glaciers of the world, edited by: Williams, R. S. and Ferrigno, J. G., US Geological Survey, Washington, DC, C1–C105 (USGS Professional Paper 1386-C), 1995.

10

Mapping the bathymetry of supraglacial lakes and streams on the Greenland ice sheet using field measurements and high-resolution satellite images

C. J. Legleiter[1], M. Tedesco[2], L. C. Smith[3], A. E. Behar[4], and B. T. Overstreet[1]

[1]Department of Geography, University of Wyoming, Dept. 3371, 1000 E. University Ave., Laramie, WY 82071, USA
[2]Department of Earth and Atmospheric Sciences, The City College of New York, Marshak MR-106, 160 Convent Avenue, New York, NY 10031, USA
[3]Department of Geography, University of California Los Angeles, Box 951524, Los Angeles, CA 90095-1524, USA
[4]Jet Propulsion Laboratory, 4800 Oak Grove Drive, Pasadena, CA 91109-8099, USA

Correspondence to: C. J. Legleiter (carl.legleiter@uwyo.edu)

Abstract. Recent melt events on the Greenland ice sheet (GrIS) accentuate the need to constrain estimates of sea level rise through improved characterization of meltwater pathways. This effort will require more precise estimates of the volume of water stored on the surface of the GrIS. We assessed the potential to obtain such information by mapping the bathymetry of supraglacial lakes and streams from WorldView2 (WV2) satellite images. Simultaneous in situ observations of depth and reflectance from two streams and a lake with measured depths up to 10.45 m were used to test a spectrally based depth retrieval algorithm. We performed optimal band ratio analysis (OBRA) of continuous field spectra and spectra convolved to the bands of the WV2, Landsat 7 (ETM+), MODIS, and ASTER sensors. The field spectra yielded a strong relationship with depth ($R^2 = 0.94$), and OBRA R^2 values were nearly as high (0.87–0.92) for convolved spectra, suggesting that these sensors' broader bands would be sufficient for depth retrieval. Our field measurements thus indicated that remote sensing of supraglacial bathymetry is not only feasible but potentially highly accurate. OBRA of spectra from 2 m-pixel WV2 images acquired within 3–72 h of our field observations produced an optimal R^2 value of 0.92 and unbiased, precise depth estimates, with mean and root mean square errors < 1 % and 10–25 % of the mean depth. Bathymetric maps produced by applying OBRA relations revealed subtle features of lake and channel morphology. In addition to providing refined storage volume estimates for lakes of various sizes, this approach can help provide estimates of the transient flux of meltwater through streams.

1 Introduction

Recent accelerated melting of the Greenland ice sheet (GrIS) (Tedesco et al., 2013), including a record melt season in 2012 (Nghiem et al., 2012; Bennartz et al., 2013), underscores the need to improve our understanding of the ice sheet's hydrology and thus constrain estimates of the current and future contributions of the GrIS to sea level rise. These estimates are subject to a number of significant uncertainties, many related to the role of meltwater on the surface of the ice sheet. More specifically, two fundamental questions remain unresolved: (1) how much water is stored in supraglacial lakes and streams? and (2) how is this water transferred from the surface to the interior and/or base of the ice sheet, and at what rate? The limited information available for supraglacial water bodies hinders efforts to determine the volume of meltwater present on the surface of the GrIS and hence to develop estimates of meltwater storage and flux. Incorporating this type of information into a hydrologic budgeting framework would help to quantify the proportion of melt that is retained within and/or beneath the ice sheet and that which propagates to proglacial rivers and ultimately the ocean (Rennermalm et al., 2013). Motivated by these challenges, this study used simultaneous field measurements and high spatial resolution commercial satellite image data to evaluate the feasibility of measuring the bathymetry of Greenland's supraglacial lakes and streams via spectrally based remote sensing.

Remote sensing is a valuable tool for monitoring the GrIS, with a variety of instruments used to detect and characterize melt (e.g., Nghiem et al., 2012; Tedesco et al., 2013, and references therein). Analysis of satellite images, for example, has provided long-term, synoptic information on patterns of meltwater storage and drainage, establishing the importance of supraglacial lakes as temporary reservoirs (e.g., Liang et al., 2012). Storage volumes have been inferred from remotely sensed data by identifying lakes, measuring their areas, and estimating their depths on the basis of empirical relationships between depth and reflectance (e.g., Fitzpatrick et al., 2013). Although several previous studies have mapped the locations of supraglacial streams (e.g., Colgan et al., 2011; McGrath et al., 2011; Yang and Smith, 2013) or the bathymetry of supraglacial lakes (e.g., Box and Ski, 2007; Sneed and Hamilton, 2007; Tedesco and Steiner, 2011) from optical image data, none have attempted to retrieve water depth in supraglacial streams.

Spectrally based methods of bathymetric mapping are well established in coastal environments, where decades of research have yielded insight on the radiative transfer processes governing the interaction of light and water (e.g., Lyzenga, 1978; Philpot, 1989; Maritorena et al., 1994). In addition to depth, the reflectance from a water body depends on the optical properties of the water column and the albedo of the bottom. Tedesco and Steiner (2011) quantified these two parameters by collecting field measurements of depth and reflectance from a supraglacial lake on the GrIS; these data were also used to relate bathymetry to spectral reflectance for individual bands of the Landsat 7 (ETM+) and Moderate Resolution Imaging Spectroradiometer (MODIS) satellites. A related study used in situ observations of depth and optical properties from a melt pond to verify the underlying assumptions of the Sneed and Hamilton (2007) depth retrieval algorithm: negligible amounts of sediment or organic matter; a water surface undisturbed by waves; a lack of inelastic scattering; a flat, homogeneous substrate; and optically deep water (e.g., from the ocean) for estimating the volume reflectance of the water column. Field data supported the first three assumptions, while the role of substrate heterogeneity was ambiguous and deep-water observations were shown to be unnecessary; the analysis confirmed the validity of estimating melt pond depth via remote sensing (Sneed and Hamilton, 2011). Similarly, Fitzpatrick et al. (2013) used echo sounding, along with topographic surveys of a drained supraglacial lake, to calibrate reflectance values extracted from MODIS image pixels to depth. The resulting equation was applied to water-classified pixels to yield depth estimates, which were then integrated over the supraglacial lake area to calculate water storage volumes.

Although these early studies were encouraging, previous research also identified a number of issues and limitations that must be addressed. For example, Tedesco and Steiner (2011) showed that bathymetry inferred from satellite images could be sensitive to variations in bottom reflectance and/or water column optical properties. Sneed and Hamilton (2011) also identified substrate heterogeneity (e.g., cryconite patches) as a potential complicating factor in passive optical depth retrieval. Moreover, the coarser spatial resolutions of other satellites such as Landsat 7 (30 m), MODIS (250 m), and the Advanced Spaceborne Thermal Emission and Reflection Radiometer (ASTER, 15 m) are too coarse for depth estimates along lake margins and can produce biased results due to the presence of ice within mixed pixels. Larger pixel sizes, particularly for the widely used MODIS instrument, also restrict bathymetric mapping to relatively large supraglacial lakes and precludes analysis of the streams that deliver meltwater into the ice sheet via moulins. Also, aside from the observations of Sneed and Hamilton (2011) and Tedesco and Steiner (2011) on lakes up to 3.0 m and 4.5 m deep, respectively, coincident in situ measurements of depth and reflectance for validating spaceborne bathymetry are lacking. Modeling studies provide an alternative means of estimating water storage volumes (e.g., Luthje et al., 2006; Leeson et al., 2012), but this approach is subject to various uncertainties as well (e.g., Tedesco et al., 2012).

This study addressed these shortcomings by applying spectrally based depth retrieval methods developed in shallow marine settings (e.g., Dierssen et al., 2003) and adapted to alluvial rivers (e.g., Legleiter and Overstreet, 2012) to the GrIS. More specifically, we evaluated the performance of a band ratio-based algorithm expected to be robust to variations in bottom albedo and water column optical properties (Stumpf et al., 2003; Legleiter et al., 2009). To assess the feasibility of this approach, we obtained concurrent, co-located field measurements of depth and reflectance in two meltwater streams and a large lake, with measured depths ranging 0.31–10.45 m. Two WorldView2 (WV2) satellite images acquired within 3–72 h of our ground-based observations were used to assess the accuracy of image-derived depth estimates. Because WV2 provides higher spatial resolution (2 m pixels for multispectral data) relative to Landsat 7, MODIS, or ASTER, this sensor could help refine estimates of supraglacial storage by providing depth information from small lakes. In addition, WV2 data offer the possibility of mapping supraglacial streams, which represent not just a static storage volume but rather a transient flux conveying meltwater to moulins.

Ultimately, we seek to support development of a flexible, spectrally based approach to mapping the bathymetry of supraglacial lakes and streams that can serve as a central component of integrated workflows for characterizing the hydrology of the GrIS. Our investigation was motivated by the following objectives, which serve to organize the paper:

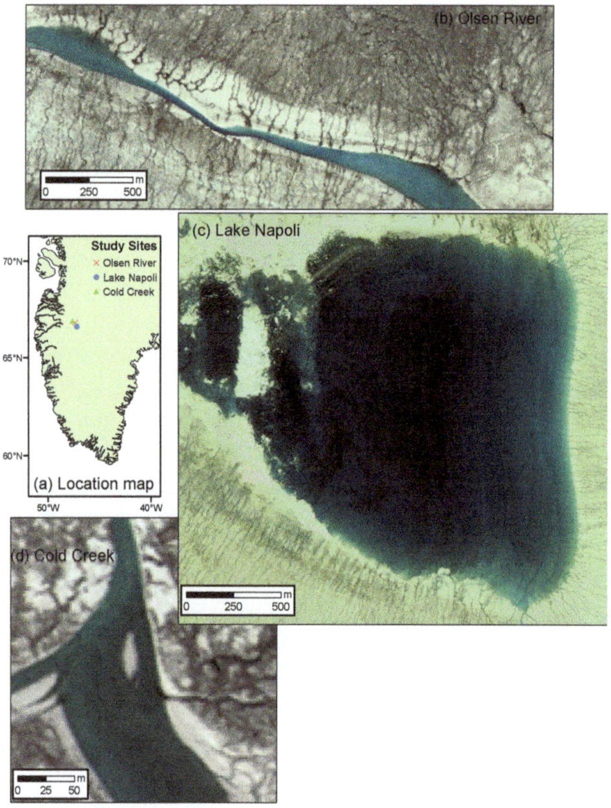

Fig. 1. Map **(a)** indicating the location of study area in southwestern Greenland. WorldView2 images of three primary field sites: **(b)** Olsen River, **(c)** Lake Napoli, and **(d)** Cold Creek. See Table 1 for more information on each site.

Fig. 2. Photographs of field sites on the surface of the Greenland ice sheet. **(a)** Remote-controlled drone boat used to collect simultaneous measurements of depth and reflectance, shown here along the Olsen River. **(b)** Olsen River, looking downstream toward the ice bar where the channel curves to the right. **(c)** Overview of Lake Napoli, taken from a helicopter, showing wave-cut platforms associated with former shorelines on the up-glacier (left) side of the lake. **(d)** Cold Creek, looking downstream.

1. Use field spectra and depth measurements from the surface of the GrIS to assess the feasibility of spectrally based bathymetric mapping of supraglacial lakes and streams.

2. Develop robust algorithms for retrieving depth information, and hence water storage volumes and fluxes, from high-resolution satellite image data.

3. Assess the accuracy of image-derived depth estimates and identify the potential limitations associated with this approach.

2 Methods

2.1 Study area

Our study area was located in a portion of the ablation zone in southwestern Greenland (Fig. 1a) that generally features numerous supraglacial lakes and streams during the summer melt season. Field data and satellite images were acquired from three primary field sites described in Table 1. The first stream we sampled, referred to as the Olsen River, consisted of a broad, shallow channel where meltwater exited a lake transitioning to a narrower, deeper morphology confined by high banks of ice (Fig. 1b, 2a). The channel also featured a curve around an "ice bar", shown in Fig. 2b, analogous to the point bars found along meandering terrestrial rivers. Our second site, referred to as Lake Napoli, was roughly circular in shape, with an area of $3.05 \times 10^6\,m^2$ and a maximum measured depth of 10.45 m. A series of parallel, north–south trending ridges evident on the lakebed in a WV2 scene (Fig. 1c) were interpreted as wave-cut platforms associated with former shorelines on the east (up-glacier) side of the lake (Fig. 2c). The western shore of the lake was more irregular, with a complex bottom topography. Our final sampling location, referred to as Cold Creek (Fig. 1d), was a shallow, wide, slow-flowing outlet channel from a small lake; several mid-channel ice bars were present in this stream.

2.2 Field data collection and analysis

Field observations from these three supraglacial water bodies were obtained using a small, unmanned boat similar to that employed by Tedesco and Steiner (2011) but with a number of significant upgrades. The new motorized drone boat, a specially customized Oceanscience Z-Boat 1800 (Fig. 2a), was operated from shore via a remote control, which ensured safety of personnel in case of a sudden drainage event. This platform was small enough to fit inside a helicopter for transport onto the ice sheet but large enough to accommodate the instrument payload. Designed as a hydrographic survey vessel, the drone boat could be operated at up to 1 km distance

Table 1. Study sites on the surface of the Greenland ice sheet (GrIS) from which field-based measurements of depth and reflectance and WorldView2 satellite images were acquired.

Site	Latitude Longitude Elevation (m a.s.l.)	Meltwater surface area (m^2)	Mean channel width (m)	Number of field data	Mean depth (m)	Max depth (m)	Field data collection	Image acquisition
Olsen River	67°7′5.7″ −48°19′19.3″ 1504	30 785	32.3	3264	1.08	3.16	7/20/2012	7/18/2012
Lake Napoli	66°54′9.0″ −48°6′2.7″ 1589	3 048 258	n/a	4383	4.53	10.45	7/21/2012	7/18/2012
Cold Creek	67°10′49.5″ −48°45′53.5″ 1335	14 155	67.5	1164	0.93	1.66	7/23/2012	7/23/2012

from the remote control, and the rechargeable battery system provided sufficient power for deployments up to several hours in duration. Maneuverability of the boat was defined by maximum speeds of 4–5 m s^{-1} and minimum turning radius of 4 m. Owing to the potential interference of the boat's hull with spectral measurements, the plastic hull was given a special flat black finish to minimize reflectance. Measurement locations were determined using a Trimble R8 GPS mounted on the boat. The GPS data collector also communicated with an Ohmex SonarMite 235 KHz echo sounder that measured water depths with an accuracy of ±0.025 m at a sampling frequency of 1 Hz. Because the drone boat was driven at low velocities (~ 1 m s^{-1}), this system provided very dense bathymetric coverage.

Measurements of upwelling spectral radiance $L_u(\lambda)$ at each wavelength λ were made using an Analytical Spectral Devices (ASD) FieldSpec3 spectroradiometer with a spectral sampling interval of 1 nm for visible and near-infrared (NIR) wavelengths from 350 to 1025 nm; only data from 400 to 900 nm were considered due to low signal-to-noise outside this range (Tedesco and Steiner, 2011). The instrument was secured inside the boat's watertight hull and connected via fiber optic cable to a fore optic with a 5° field of view mounted on a boom extending ahead of the boat to minimize self-shading effects. We attempted to navigate so as to avoid configurations for which the boat might cast a shadow within the field of view of the spectroradiometer, but in a small number of cases shadows could not be avoided. The fore optic was placed immediately above the water surface and pointed downward to ensure a consistent nadir-viewing geometry. Our sampling strategy thus differed from that of Tedesco and Steiner (2011), who made reflectance measurements underwater to avoid the confounding effects of reflectance from and transmittance through the air–water interface. In this study, we reasoned that because remotely sensed data also are subject to these effects, field spectra recorded from above the water surface would provide a more direct means of assessing the potential for mapping bathymetry from satellite images. The boat-based spectroradiometer was re-calibrated periodically during each deployment by pointing the fore optic at a Spectralon reference panel, which served to set the integration time, account for the instrument's dark current, and establish the conversion from raw digital counts to spectral radiance units.

A simultaneous time series of downwelling spectral irradiance $E_d(\lambda)$ was recorded on shore by an ASD HandHeld2 spectroradiometer. This instrument was located within 400 m of all boat-based measurements from Olsen River and Cold Creek and within 800 m of the boat on Lake Napoli. Over these sub-kilometer distances, differences in solar zenith angle between the shore- and boat-based sensors were negligible. To capture incoming radiation from throughout the upper hemisphere, the on-shore spectroradiometer was equipped with a cosine receptor oriented skyward. Data from the boat- and shore-based ASD's were synchronized and remote sensing reflectance $R_{rs}(\lambda)$ calculated as the ratio of $L_u(\lambda)$ measured on the boat to $E_d(\lambda)$ recorded on shore. Reflectance $R(\lambda)$ was obtained by multiplying R_{rs} by π, assuming a Lambertian bidirectional reflectance distribution function. Time stamps were used to pair these reflectance data with the depth measurements and GPS positions.

2.3 Image data acquisition and processing

To assess the potential for spaceborne mapping of supraglacial bathymetry, we examined two WV2 satellite images. This sensor provides higher spatial and spectral resolution than more widely used platforms such as Landsat 7 (ETM+), MODIS, or ASTER with 2 m pixels for each of eight spectral bands. In addition to standard blue, green, red, and NIR bands, WV2 includes coastal (400–450 nm), yellow (585–625 nm), and red edge (705–745 nm) bands that were expected to be useful for depth retrieval from shallow meltwater streams. WV2's longest wavelength band

(860–1040 nm) might be helpful for discriminating water from ice along lake and channel margins but was excluded from the depth retrieval analysis because strong absorption by pure water in this portion of the spectrum implies small $L_u(\lambda)$ values and poor signal-to-noise for water pixels, particularly in deeper areas.

WV2 images from our study sites were acquired on 18 and 23 July 2012, within 3–72 h of field data collection (Table 1). Images were orthorectified and projected to a polar stereographic coordinate system using a code developed by the Polar Geospatial Center at the University of Minnesota. Radiometric coefficients for each band were included in the WV2 metadata and used to convert raw digital counts to radiance. Atmospheric correction was performed using the Fast Line-of-sight Atmospheric Analysis of Spectral Hypercubes (FLAASH) algorithm based on the MODTRAN4 radiation transfer code (Adler-Golden et al., 1999). Inputs to FLAASH included the radiance image, scene date and location, ground elevation and sensor altitude, spectral band configuration, visibility (40 km), and standard models for the atmosphere (sub-arctic summer) and aerosols (rural). By performing a series of iterative calculations to estimate column water vapor, FLAASH generated atmospherically corrected images in units of apparent surface reflectance $R(\lambda)$.

Linking specific pixels from these images to individual field measurements required precise spatial referencing of the remotely sensed data. Orthorectification of the WV2 images for our Lake Napoli and Cold Creek sites was highly accurate, with all of our boat-based field measurements plotting within the water bodies depicted in the scene. For the Olsen River, however, the initial geo-referencing was in error by several meters, with many of our field data located outside the stream on the image. To refine the spatial referencing, we manually digitized a channel polygon on the image and compared this feature to a polygon enveloping the field data. An iterative transformation procedure described by Legleiter (2012) was then used to shift the image into alignment with our field observations.

To isolate the meltwater present in each image, we created masks based on NIR reflectance thresholds. The longest wavelength NIR band was displayed as a grayscale image and the contrast stretch adjusted interactively to identify an appropriate cutoff value for distinguishing water from ice. The initial threshold-based mask was refined through morphological opening, segmentation, and closing (Legleiter et al., 2011). The resulting masks were applied to the reflectance images and a 3 × 3 pixel Wiener filter used to smooth the in-stream portion of each scene (Legleiter, 2013).

2.4 Spectrally based depth retrieval

Quantitative relationships between water depth d and spectral reflectance $R(\lambda)$ provide a basis for mapping supraglacial bathymetry from passive optical image data. Although the upwelling radiance recorded by a remote detector also is influenced by several other factors, the effect of depth can be isolated using measurements from multiple spectral bands. More specifically, under appropriate conditions, outlined below, and for certain combinations of wavelengths, the image-derived quantity

$$X = \ln\left[\frac{R(\lambda_1)}{R(\lambda_2)}\right] \tag{1}$$

is linearly related to depth. The physical principles underlying depth retrieval via band ratios were described, for example, by Dierssen et al. (2003) for coastal settings and Legleiter et al. (2004, 2009) for gravel-bed rivers; the premise of this study is that similar radiative transfer processes operate in the supraglacial environment as well.

To summarize, the total radiance $L_T(\lambda)$ measured above an optically shallow water body is the sum of contributions from the bottom, water column, water surface, and atmosphere. Of these components, only the radiance reflected from the bottom $L_B(\lambda)$ is directly related to depth. $L_B(\lambda)$ also depends on the reflectance $R_b(\lambda)$ of the bottom itself, but for an appropriate pair of bands the ratio $R_b(\lambda_1)/R_b(\lambda_2)$ tends to be nearly constant across bottom types. Moreover, spectral differences in $R_b(\lambda)$ for a given substrate are small, on the order of a few percent, relative to the order-of-magnitude increase in attenuation by the water column as wavelength increases from the blue into NIR. The transfer of radiant energy through the water column is controlled by various absorption and scattering processes that impart a color, or volume reflectance ,$R_c(\lambda)$, to the water itself; these processes are summarized in terms of an effective attenuation coefficient, $K(\lambda)$. In clear water free from high concentrations of suspended sediment or organic material, $K(\lambda)$ is driven primarily by absorption by pure water (Legleiter et al., 2009). Analysis of water samples collected from an East Greenland melt pond by Sneed and Hamilton (2011) confirmed that suspended or dissolved, organic or inorganic, particulate matter was minimal and had no significant influence on radiative transfer within the water column.

Legleiter et al. (2009) presented a scaling argument that in shallow, clear water bodies with highly reflective substrates, $L_T(\lambda)$ is dominated by $L_b(\lambda)$ and the other radiance components can be considered negligible. Beginning from an expression for $L_T(\lambda)$ presented by Philpot (1989), Legleiter et al. (2009) showed that under these circumstances the logarithm of the ratio of the radiances (or, equivalently, reflectances since $R = \pi L_T/E_D$) measured in two spectral bands, denoted by numeric subscripts, can be simplified to yield a linear equation relating the radiometric quantity X to the water depth d:

$$X = \ln\left[\frac{L_{T1}}{L_{T2}}\right] \approx (K_2 - K_1)d + \ln\left[\frac{(R_{b1} - R_{c1})}{(R_{b2} - R_{c2})}\right] + A, \tag{2}$$

where the dependence on wavelength of all quantities except for depth has been suppressed for clarity. The slope term in this equation depends on the difference in attenuation between the two bands; X increases with depth for $K_2 > K_1$ because the radiance in the more strongly absorbing band λ_2 decreases more rapidly as depth increases than does L_{T1}. The intercept term accounts for the "bottom contrast" $R_b - R_c$ between the substrate and water column as well as a constant A that depends on the downwelling spectral irradiance $E_D(\lambda)$ and transmission across the air–water interface and through the atmosphere. If the optical properties of the water column remain uniform throughout the lake or stream reach, the only quantity in Eq. (2) expected to vary spatially on a pixel-by-pixel basis is the one of interest, d; the remotely sensed variable X is thus well suited for bathymetric mapping (Legleiter et al., 2009).

Depth retrieval via this method involves identifying an appropriate pair of wavelengths and calibrating a relation between d and X. A technique called optimal band ratio analysis, or OBRA (Legleiter et al., 2009), accomplishes both of these objectives by performing regressions of d on X for all possible combinations of numerator λ_1 and denominator λ_2 bands. The optimal band ratio is the one yielding the highest R^2, with the corresponding regression equation serving to calibrate X values to depth. Because regressions are performed for all possible pairs of wavelengths, OBRA also allows spectral variations in the strength of the d vs. X relationship to be visualized as a matrix of $R^2(\lambda_1, \lambda_2)$ values.

We conducted OBRA for both field spectra recorded from the boat and image spectra from the WV2 satellite. For the field spectra, we used depths measured by the echo sounder at the same time as the reflectance data. To assess the feasibility of retrieving depth from multispectral image data with relatively broad bands, we also convolved the original field spectra, sampled every 1 nm, with the spectral response functions of WV2, Landsat 7, MODIS, and ASTER and repeated the OBRA for the convolved spectra. Although this study focused on WV2 images, conducting OBRA for the other, more widely used sensors provided a means of assessing whether strong depth–reflectance relations could be obtained from coarser spectral resolution data. For OBRA of the WV2 image spectra, all depth measurements located within a given pixel were averaged and a mean depth assigned to the pixel center. Reflectance spectra for the field data locations were then extracted from the masked, filtered WV2 image and used as input to OBRA. Because Lake Napoli featured depths over 10 m, we also performed a quadratic version of OBRA that included both X and X^2 terms in each regression; Dierssen et al. (2003) found that such a quadratic formulation provided more accurate depth estimates in deeper areas.

Linear and quadratic OBRA were performed for the original and convolved field spectra and for the WV2 image spectra for each site individually and for a merged data set aggregated across all three sites. For the merged field data set and each of the images, calibration relationships were derived from a randomly selected subset (50 %) of the reflectance spectra and collocated depth measurements. The other half of each data set was retained and used to assess the accuracy of image-derived depth estimates in terms of mean error, root mean square errors (RMSE), and regression of observed (field-based) vs. predicted (image-derived) depths (Pineiro et al., 2008). In addition, we applied OBRA relations to the WV2 images to produce bathymetric maps, which were inspected for realistic spatial patterns and general agreement with our field observations.

3 Results

3.1 Depth–reflectance relations from field spectra

To assess the feasibility of inferring the depth of supraglacial water bodies from their spectral characteristics, we used field measurements of depth and reflectance to perform OBRA for each field site individually and for a merged data set (Table 2). Moderately strong X vs. d relations were observed for both streams, but correlations between band ratio values and water depth were much greater for Lake Napoli and the merged data set. The weaker OBRA relations for the streams could in part be a consequence of the limited range of depths measured in these shallow channels (0.31–4.45 m), particularly Cold Creek (0.31–1.69 m), whereas the Lake Napoli and merged data sets encompassed depth observations up to 10.45 m. In addition, reflectance spectra from the streams might have been influenced to a greater degree by factors other than depth, including the localized effects of cryconite substrates with darker bottom albedo, cracks in the ice bed, variable water surface roughness, shadows, and floating pieces of slush.

For all three sites, the numerator and denominator wavelengths identified via linear OBRA were very close to one another, implying that the greatest sensitivity to depth occurred in narrow portions of the spectrum, ranging from 580 nm for Lake Napoli to 660 nm for the Olsen River. Inspection of the OBRA matrices shown in Fig. 3, however, indicated that a broader range of wavelengths throughout the visible portion of the spectrum would provide X vs. d relations nearly as strong as the optimal band ratios. For the Olsen River and Cold Creek, blue and green wavelengths less than 550 nm were of limited value as denominator bands due to efficient penetration of light through water in these bands. In these shallow channels, changes in depth translated into only small changes in reflectance in this portion of the spectrum, whereas stronger absorption of red and NIR light resulted in greater decreases in reflectance for a given increase in depth. Conversely, greater depths in Lake Napoli dictated that shorter wavelengths capable of penetrating to the bottom were more useful than red and NIR bands in which the radiance signal saturated due to nearly complete absorption in deeper water.

For the merged data set comprised of observations from shallow streams as well as deeper areas of the lake, the

Table 2. Summary of optimal band ratio analysis (OBRA) of field spectra and depth measurements from three field sites on the surface of the GrIS and for the merged data set aggregated over the three sites.

Site	Calibration data (count)	Spectrum type[a]	Sensor	OBRA type[b]	OBRA R^2	λ_1^c (nm)	λ_2^d (nm)	SE[e] (m)	SE (%)
Olsen River	2371	Cont.	ASD[f]	Linear	0.80	655	659	0.30	28
Cold Creek	781	Cont.	ASD	Linear	0.61	593	594	0.16	18
Lake Napoli	2424	Cont.	ASD	Linear	0.90	581	582	0.38	8
Merged	2788[g]	Cont.	ASD	Linear	0.92	583	604	0.47	31
		Cont.	ASD	Quad.	0.94	584	599	0.43	29
		Conv.	WV2	Linear	0.89	545	605	0.55	37
		Conv.	WV2	Quad.	0.92	545	605	0.49	33
		Conv.	Landsat[h]	Linear	0.85	499	669	0.65	43
		Conv.	Landsat	Quad.	0.88	499	669	0.60	40
		Conv.	MODIS	Linear	0.88	550	658	0.60	40
		Conv.	MODIS	Quad.	0.89	550	658	0.56	37
		Conv.	ASTER	Linear	0.86	531	638	0.65	43
		Conv.	ASTER	Quad.	0.87	531	638	0.61	41

[a] Cont.: continuous field spectra, or Conv.: field spectra convolved to sensor bands.
[b] Linear or quadratic formulation of OBRA.
[c] λ_1: numerator wavelength for optimal band ratio.
[d] λ_2: denominator wavelength for optimal band ratio.
[e] Regression standard error for optimal band ratio.
[f] Analytical Spectral Devices spectroradiometer used to measure field spectra.
[g] For merged data set, half of the data were used for calibration and half reserved for validation.
[h] More specifically, "Landsat" refers to the Landsat 7 (ETM+) sensor.

addition of a quadratic term to the OBRA formulation produced a marginal improvement in the strength of the X vs. d relation ($R^2 = 0.94$). In this case, a pair of wavelengths in the yellow portion of the spectrum, 584 and 599 nm, were identified as optimal, but the OBRA matrix in Fig. 3d indicated that many other band combinations had R^2 values nearly as high; $\lambda_1 < 600$ nm yielded strong correlations between X and d when paired with $470 < \lambda_2 < 720$ nm.

These results suggested that broader wavelength bands, as opposed to essentially continuous field spectra, would be sufficient for depth retrieval. To test this possibility, we convolved the field spectra from the merged data set to match the spectral response functions of the WV2, Landsat 7, MODIS, and ASTER sensors and performed OBRA of the convolved spectra. For all four sensors and for both the linear and quadratic formulations of OBRA, the reduced spectral resolution resulted in only a slight decrease in the strength of the relationship between X and d (Table 2 and Fig. 4). The WV2 sensor's green and yellow bands were optimal, consistent with the wavelengths selected from the original field spectra. Even for the Landsat 7, MODIS, and ASTER sensors having only three or four visible/NIR bands, the OBRA R^2 values were nearly as high as for the 7-band WV2 system. Figure 4 indicates that for the latter three instruments, defining X using green and red bands yielded the strongest linear relations with depth. These results imply that multispectral imaging systems, including Landsat 7, MODIS, and ASTER as well as WV2, have adequate spectral resolution for bathymetric mapping. Because the OBRA R^2 values for all three sensors were similar, the primary advantage of WV2 would be enhanced spatial resolution, but our results indicate that robust depth–reflectance relations can be derived for Landsat 7, MODIS, and ASTER data as well. Although these sensors have coarser pixel sizes that would restrict analysis to larger water bodies, the broader spatial and more frequent temporal coverage of Landsat 7 and MODIS make them useful for scaling up and examining more extensive areas of the GrIS over longer time periods. Irrespective of the remote sensing system employed, our field measurements of depth and reflectance provided direct, on-the-ground evidence confirming the feasibility of spectrally based depth retrieval from supraglacial water bodies.

3.2 Mapping supraglacial bathymetry from satellite image data

To more directly evaluate whether reliable depth information could be inferred from passive optical image data, we coupled field measurements of depth with spectra extracted from two WV2 images that were acquired within 3–72 h of field data collection. Table 3 and Fig. 6 summarize the results of linear and quadratic OBRA performed for each site individually and for a merged data set aggregated across the three sites. Strong to very strong relationships between the image-derived quantity X and water depth d were observed in all cases, with OBRA R^2 values ranging from 0.79 to

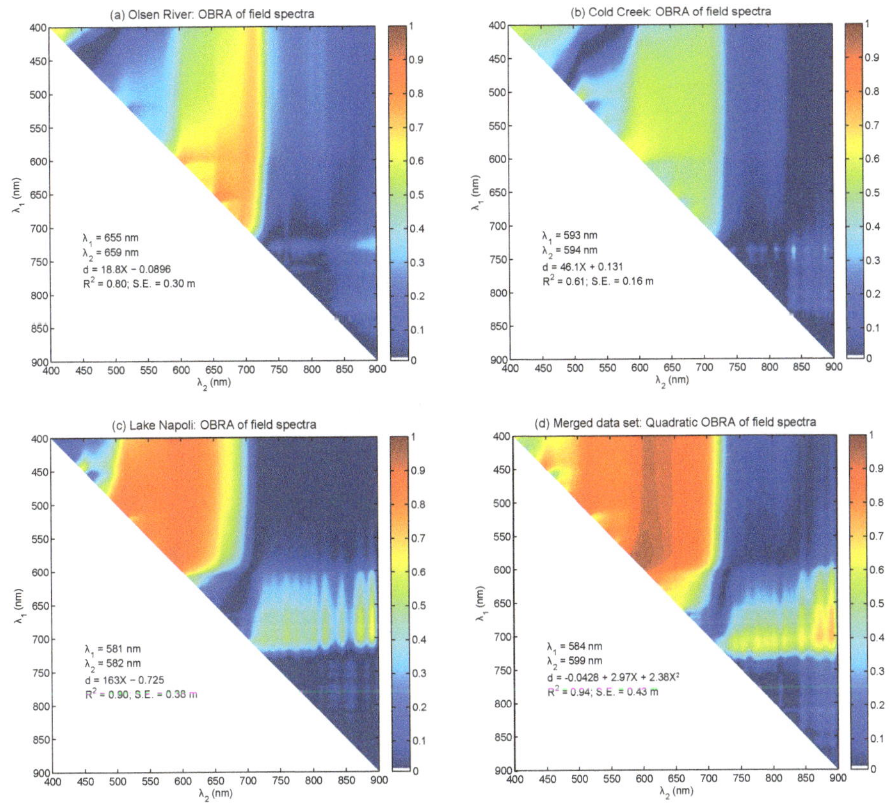

Fig. 3. Optimal band ratio analysis (OBRA) of field spectra for each study site and for a merged data set aggregated across the three sites.

0.93. Accuracy assessment involved applying OBRA regression equations to a validation sub-sample of each data set. The difference between field-measured and image-derived depths was less than 1 % of the mean depth in all cases, indicating that depth estimates were unbiased on average. Small root mean square errors, ranging from 10 to 26 % of the mean depth for each site, implied that depth estimates were precise as well. Similarly, regressions of observed vs. predicted depths showed that remote bathymetric mapping of supraglacial lakes and streams was accurate, with R^2 values from 0.78 to 0.93. Moreover, OP regression intercept and slope coefficients near 0 and 1, respectively, indicated that image-derived depths were not subject to any kind of systematic bias. The results of our depth retrieval accuracy assessment compare favorably to those reported by Fitzpatrick et al. (2013). In that study, a depth–reflectance relation based on a single MODIS band yielded an OP regression R^2 value of 0.79 and a typical error of 1.47 m for four lakes with depths up to 12 m.

For the individual sites, linear OBRA yielded R^2 values nearly as high as the quadratic version of the algorithm, implying that the addition of an X^2 term was not necessary on a site-by-site basis. For the merged data set, however, the broader range of field-measured depths, including some in excess of 10 m, introduced mild curvature to the relation between X and d (Fig. 5a); a quadratic function thus provided a better overall fit. Introducing an X^2 term prevented negative depth estimates in shallow water, a common problem in terrestrial rivers (Legleiter et al., 2009), and severe under-prediction of depth in deeper areas. Validation of the quadratic OBRA relation for the merged data set via the OP regression shown in Fig. 5b showed good overall agreement between observed and predicted depths ($R^2 = 0.93$) but considerable scatter about the 1-to-1 line, with a tendency to underestimate the greatest depths. The distinct clustering of points from the Olsen River vs. Cold Creek on both the calibration and validation plots in Fig. 5 was due at least in part to the location of the two streams on different WV2 images, which were processed separately to units of apparent surface reflectance. Because these sites were located on separate images and the majority of the merged data set was drawn from the deeper waters of Lake Napoli, the quadratic OBRA relation based on data from all three sites yielded low OP R^2 values when applied to the data from each individual stream: 0.25 for the Olsen River and 0.14 for Cold Creek. Applying the merged relation to the data from Lake Napoli alone resulted in a much higher OP R^2 of 0.90.

A potentially important advantage of OBRA relative to other depth retrieval approaches that are based on individual bands is an explicit focus on spectral variations in the nature and strength of the relationship between the image-derived quantity X and water depth. The information present in the

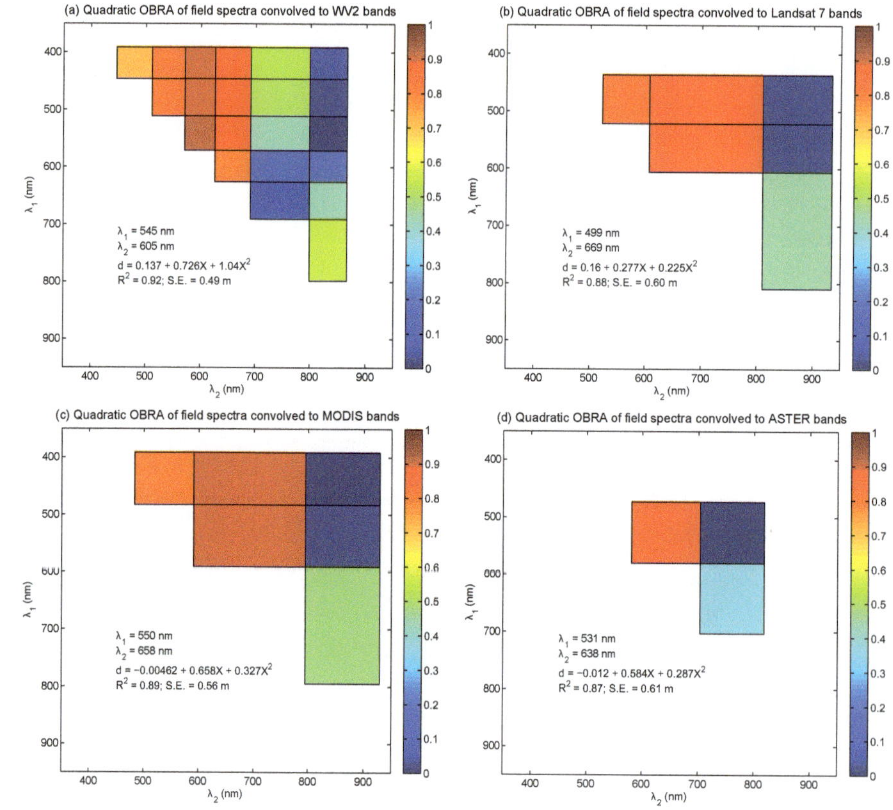

Fig. 4. Quadratic optimal band ratio analysis (OBRA) of field spectra aggregated across the three sites and convolved to the sensor bands for **(a)** WV2, **(b)** Landsat 7, **(c)** MODIS, and **(d)** ASTER.

OBRA matrices shown in Fig. 6, for example, can help to guide selection of an appropriate combination of bands. For our data from the GrIS, a different set of bands was identified as optimal for each site and for the merged data set (Table 3). Moreover, the OBRA matrices showed that the number of bands useful for depth retrieval and the utility of specific bands varied appreciably among data sets.

For the Olsen River the red / NIR ratio was optimal but was associated with an X vs. d relation with a negative slope, whereas a positive relation was expected. The selection of a NIR band also was not anticipated due to strong absorption of these wavelengths by pure water. In this case, both the strong correlation with NIR reflectance and the negative slope of the X vs. d relation can be explained in terms of adjacency effects. The presence of bright ice next to dark water, scattering by the atmosphere, and mixed, 2 m pixels representing a large fraction of the mean channel width all contributed to higher NIR reflectance near the banks, where depths were shallow. In the absence of such confounding effects, NIR reflectance would be much lower and would not vary significantly with changes in depth. In fact, OBRA of field spectra, which were less influenced by these pixel-scale, atmospheric issues, indicated that NIR wavelengths were not useful for depth retrieval. For image spectra subject to adjacency effects, however, NIR reflectance acted as a proxy for distance from the bank and was thus correlated with depth.

For Cold Creek, the red edge and NIR bands did not yield high X vs. d regression R^2 values because most of our observations from this broad, shallow channel were collected farther from the banks. Because the Olsen River was much narrower, a greater proportion of pixels from that site were subject to adjacency effects, which were most pronounced in the NIR because reflectance from the water was lowest at these longer wavelengths. Figure 6a also indicates, however, that the blue/yellow band ratio would yield an R^2 value nearly as high and would be less influenced by adjacency effects.

Even for the larger Lake Napoli, the optimal band ratio included a NIR denominator and yielded a negative slope for the X vs. d relation, suggesting that for some images, such as that which included the Olsen River and Lake Napoli, adjacency effects associated with atmospheric scattering could be pervasive. For the merged data set pooled across the three field sites and two images, some NIR band combinations yielded moderate R^2 values, but the ratio of the WV2 sensor's coastal and green bands stood out as optimal, with a positive relation between X and d (Fig. 5a). Because these wavelengths penetrated more efficiently through the water,

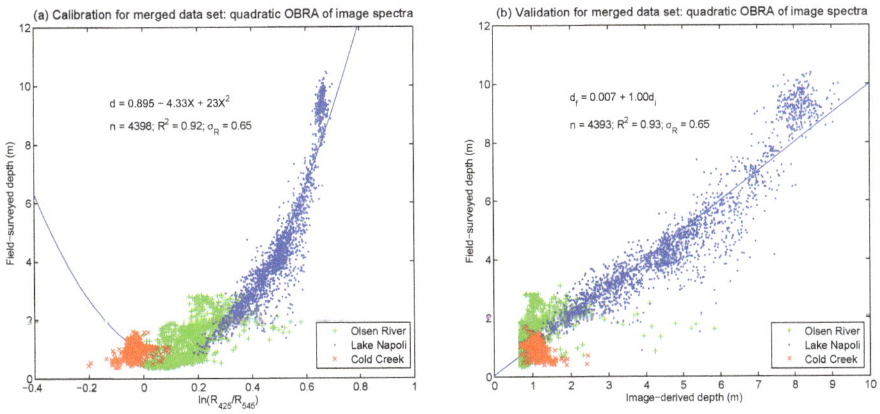

Fig. 5. **(a)** Calibration of the image-derived quantity X to water depth for the merged data set drawn from all three field sites, distinguished by the symbols shown in the legend. **(b)** Validation of image-derived depth estimates via regression of observed vs. predicted depths for a subset of the merged data set.

Table 3. Accuracy assessment for depth retrieval from WV2 satellite images of supraglacial water bodies on the GrIS.

Site	Calibration data (count)	OBRA[a] type[b]	OBRA R^2	λ_1^c (nm)	λ_2^d (nm)	ME (m)	ME (%)	RMSE (m)	RMSE (%)	OP[e] R^2	OP int.[f] (m)	OP slope
Olsen	1632	Linear	0.79	660	832	0.007	0.62	0.28	26	0.79	0.01	1.00
River		Quad.	0.82	480	605	0.006	0.56	0.27	25	0.80	0.01	0.99
Cold	574	Linear	0.79	425	660	−0.001	−0.12	0.10	11	0.78	0.00	1.00
Creek		Quad.	0.80	425	660	−0.002	−0.16	0.10	10	0.79	−0.01	1.01
Lake	2192	Linear	0.91	545	725	−0.031	−0.67	0.73	16	0.90	−0.01	1.00
Napoli		Quad.	0.93	545	832	−0.012	−0.25	0.61	13	0.93	−0.02	1.00
Merged	4398	Linear	0.75	425	545	0.023	0.80	1.17	42	0.77	0.01	1.01
		Quad.	0.92	425	545	−0.002	−0.05	0.65	23	0.93	0.007	1.00

[a] OBRA: optimal band ratio analysis.
[b] Linear or quadratic formulation of OBRA.
[c] λ_1: numerator wavelength for optimal band ratio.
[d] λ_2: denominator wavelength for optimal band ratio.
[e] OP: observed vs. predicted regression.
[f] int.: intercept of regression equation.

reflectances within the stream or lake were relatively high and more similar to the banks, making these bands less sensitive to adjacency effects.

By allowing for curvature in the X vs. d relation, quadratic OBRA could provide robust relationships less susceptible to these issues and more useful for depth retrieval. In addition to the quantitative accuracy assessment summarized above, we also inspected bathymetric maps produced by applying OBRA regression equations to the WV2 images. An example from the Olsen River is shown in Fig. 7, along with the field measurements used for calibration. For this site, the band ratio selected via linear OBRA, which included a NIR denominator, resulted in a noisy, speckled depth map due to low reflectance and thus poor signal-to-noise in the NIR band. Quadratic OBRA, in contrast, identified the blue/yellow band combination as optimal and resulted in a much smoother, spatially coherent bathymetric map. The red tones in Fig. 7 indicate shallow depths in the broad upper reach and over the ice bar where the channel curves to the right (Fig. 2b). The close-up in Fig. 7 also highlighted a deeper pool along the opposite bank just downstream of this bar, reminiscent of the bar-pool morphology common in terrestrial rivers. Similarly, a transect (Fig. 8a) across a straight segment of the Olsen River near the lower end of our study reach captured the gross, U-shaped morphology of the channel. Agreement between image-derived depths and field surveys was generally good for this cross section but the remotely sensed bathymetry was biased shallow and failed to detect a couple of small streambed irregularities evident in the field data. Nevertheless, our spectrally based approach, applied to high-resolution images, appeared capable of providing detailed information on the depth and morphology of supraglacial streams. A second transect (Fig. 8b) drawn from Lake Napoli exhibited very close correspondence between field surveys and image-derived depths along a traverse from the shore toward the center of the lake that spanned a range of depths up

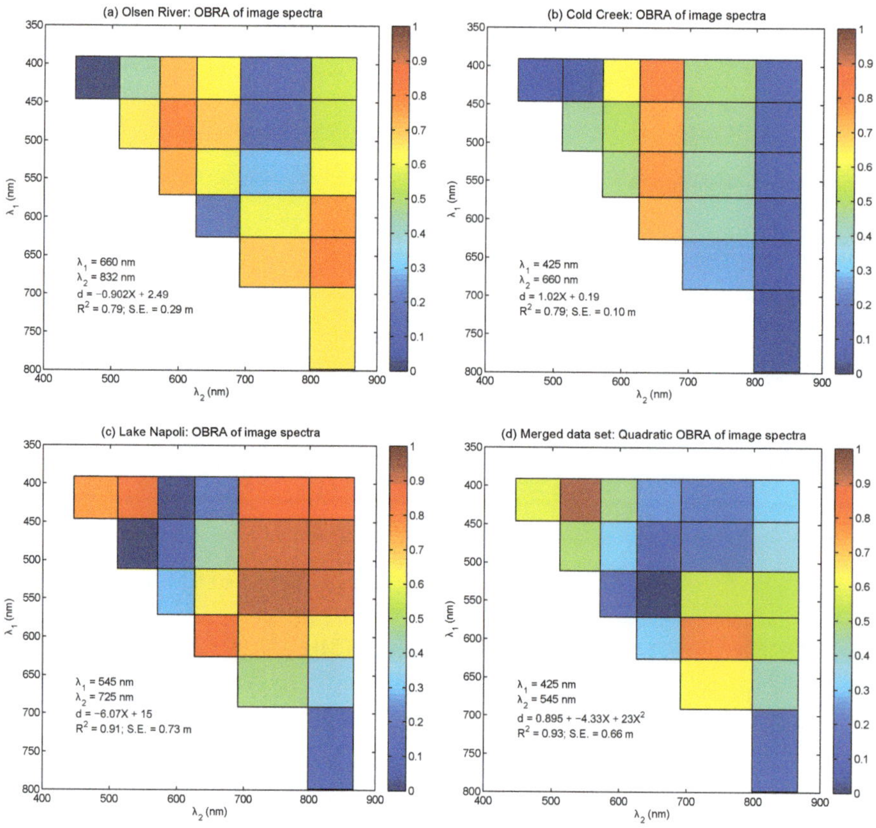

Fig. 6. OBRA of WV2 image spectra for each study site and for a merged data set aggregated across the three sites.

to 10 m; these results indicated that accurate depth retrieval was possible in deeper melt ponds than reported in previous field studies (e.g., Tedesco and Steiner, 2011; Sneed and Hamilton, 2011; Fitzpatrick et al., 2013). The combination of improved spatial resolution and more reliable bathymetric information across a broader range of depths can thus enable refined estimates of meltwater storage volumes on the surface of the Greenland ice sheet.

4 Discussion

The volume, spatial distribution, and flux of meltwater on the surface of the ice sheet are important aspects of Greenland's hydrologic system. Improved methods of characterizing these components could thus help to constrain estimates of the contribution of the GrIS to sea level rise. In this study, we evaluated the potential to retrieve information on water depth, and hence storage volumes, in supraglacial lakes and streams from multispectral satellite images. To make this assessment, we made coupled field observations of depth and reflectance in coordination with the near-simultaneous acquisition of WV2 images. A spectrally based depth retrieval algorithm developed for terrestrial rivers, called optimal band ratio analysis (OBRA), provided strong relationships between radiometric quantities and water depth for (1) essentially continuous spectra measured directly in the field; (2) field spectra convolved to match the response functions of the WV2, Landsat 7, MODIS, and ASTER sensors; and (3) spectra extracted from specific WV2 image pixels. Validation of image-derived depth estimates indicated that remotely sensed data can be used to measure supraglacial bathymetry with a high degree of accuracy (ME $< 1\,\%$ of mean depth) and precision (RMSE 10 – 23 % of mean depth, Table 3) across a broad range of depths up to 10.45 m. Similar results were reported by Fitzpatrick et al. (2013), who estimated depths in large lakes from MODIS band 1 reflectance values, but the use of higher resolution satellite images in this study allowed more detailed depth information to be retrieved from smaller water bodies. For example, bathymetric maps of meltwater streams produced from WV2 images were spatially coherent and hydraulically reasonable, revealing subtle features of the channel morphology. In addition to enabling refined estimates of storage volumes in lakes, this approach also opens up new possibilities for characterizing the transient flux of meltwater through stream channels to moulins and hence the interior and/or base of the ice sheet.

Though encouraging, our results also pointed to several key limitations of remote sensing techniques. Ideally, a general, robust relationship between depth and reflectance could

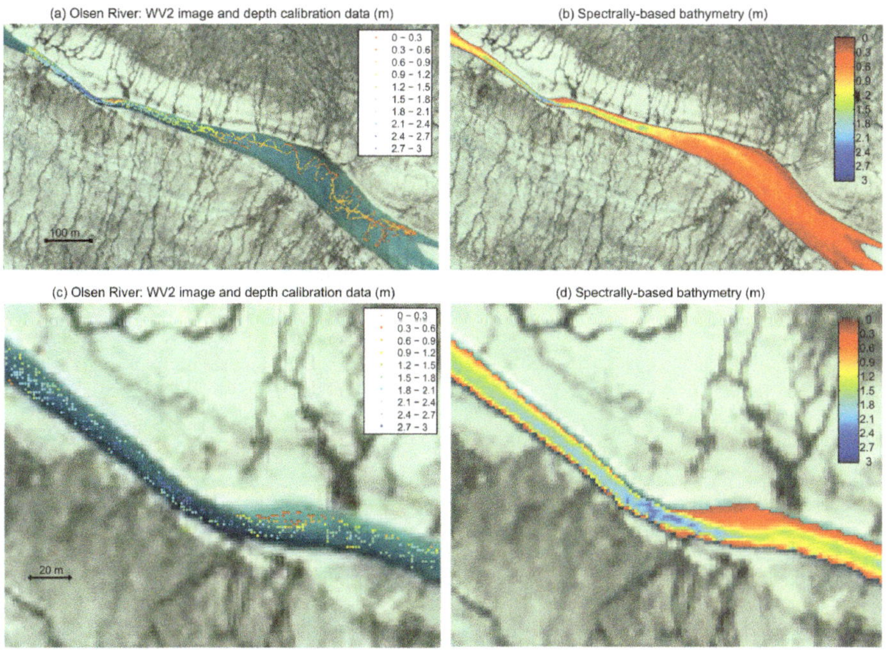

Fig. 7. Field measurements of depth from the Olsen River used for calibration **(a, c)** and image-derived bathymetry produced by applying the calibration equation from quadratic OBRA to the WV2 image **(b, d)**. Flow is from right to left.

be identified, calibrated, and applied to the entire ice sheet. This goal remains elusive, however, due to a number of complicating factors. For example, the disparity between OBRA of field spectra and OBRA of image spectra implies that these depth–reflectance relations are scale dependent. Contrary to our expectations, the optimal band ratio was more strongly related to depth for spectra extracted from 2 m image pixels than for spectra measured directly in the field. This finding suggests that some of the factors considered by Sneed and Hamilton (2011), such as heterogenous substrates, water surface roughness, shadows, and slush, were localized effects that introduced variability to the field spectra but were essentially averaged out over the scale of an image pixel. In general, agreement between field and image spectra was poor, even after atmospheric correction of the image data, with the image spectra tending to have higher reflectance values. In addition, the unexpected selection of NIR bands and negative X vs. d relations resulting from OBRA of image spectra suggest that adjacency effects were salient in the remotely sensed data. These effects can be mitigated by choosing shorter wavelength band combinations in which reflectance from the water is greater, but establishing a calibration relation applicable across images could prove difficult. For example, we found that applying a quadratic OBRA relation based on data drawn from all three of our sites yielded low OP R^2 values when applied to either of two streams located on separate images. These results imply that a general, image-based depth–reflectance relation might not provide reliable bathymetric information for individual sites, particularly shallow streams, and that site-specific calibration would be preferable, if not necessary. In theory, field spectra collected across a range of depths could be used to define a general depth–reflectance relationship in the absence of mixed pixels or atmospheric effects. We were unable to achieve this goal in this initial study, however. Instead, our results point to the need for careful calibration of field measurements and image pixel values to consistent radiometric quantities and for accurate atmospheric correction of remotely sensed data.

If these issues can be resolved, our results indicate that spectrally based remote sensing of supraglacial bathymetry could provide information on meltwater storage and flux with an unprecedented level of detail. This study demonstrated the ability to resolve subtle morphologic features, such as the former shorelines of Lake Napoli evident in the WV2 image shown in Fig. 1c or the ice bar on the Olsen River prominent in Fig. 7d, from high spatial resolution image data. Capturing the associated variations in depth, rather than averaging over coarser pixel sizes, would enable more precise estimates of meltwater storage and flux and thus support focused investigations that emphasize the spatial distribution of lakes, streams, and moulins for a given time period. Similarly, the finding, based on OBRA of convolved field spectra, that Landsat 7, MODIS, and ASTER have sufficient spectral resolution for bathymetric mapping bodes well for larger scale, longer term studies. Although these sensors have coarser spatial resolution that would limit such analyses to larger water bodies, our results indicate that accurate depth retrieval

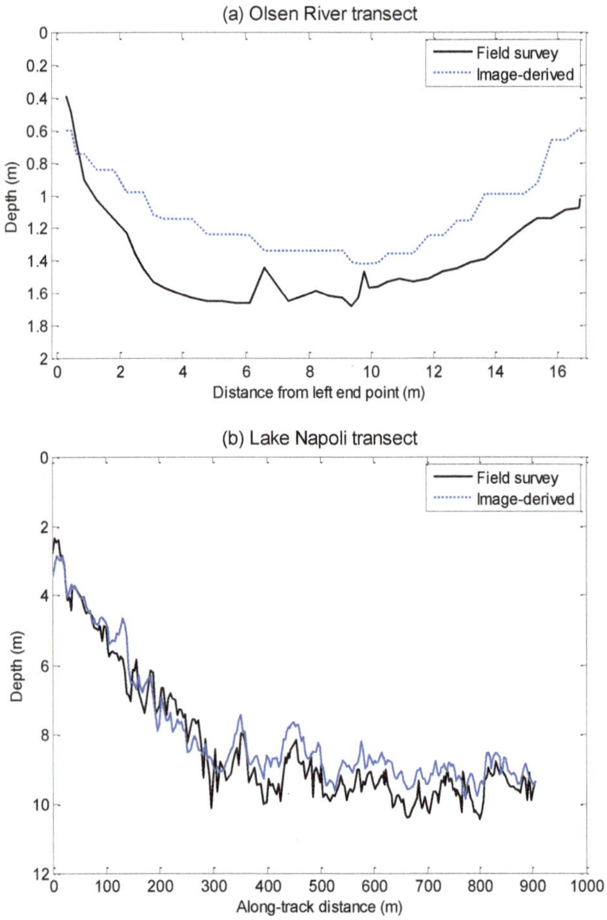

Fig. 8. Transects comparing field surveys and image-derived depth estimates obtained via quadratic OBRA for **(a)** a cross section of the Olsen River and **(b)** a boat track from the shore toward the center of Lake Napoli.

should be possible and could facilitate monitoring meltwater on the surface of the GrIS.

5 Conclusions

This study demonstrated, for the first time, the feasibility of spectrally based depth retrieval from high spatial resolution commercial satellite images of supraglacial lakes and streams on the GrIS. Given instrumentation with sufficient spatial resolution, optical remote sensing can thus be used to estimate accurately the volume of water stored in not only large lakes but also smaller melt ponds that might go undetected by sensors with coarser pixel sizes. In addition, such data can be used to measure the depth of supraglacial streams and hence characterize the transient flux of meltwater through channels. Future work will focus on addressing the calibration issues identified above, establishing a general relation between depth and reflectance, developing an integrated workflow for scaling up this type of analysis to larger areas of the GrIS, and incorporating the resulting information on meltwater storage and flux into a hydrologic budgeting framework.

Acknowledgements. This investigation was supported by the NASA Cryosphere Program (grant NNX11AQ38G managed by T. Wagner). C. Legleiter and B. Overstreet received additional support through a grant from the Office of Naval Research Littoral Geosciences and Optics Program (N000141010873 managed by T. Drake). M. Tedesco was supported by the NSF grant ARC0909388. Logistical support was provided by CH2M Hill Polar Field Services, the Kangerlussuaq International Science Station, and Air Greenland pilot T. Olsen. Asa Rennermalm loaned the ASD HandHeld2 spectroradiometer. The Polar Geospatial Center of the University of Minnesota assisted with image acquisition and processing. Field assistance was provided by V. Chu, R. Forster, C. Gleason, S. Moustafa, T. Olsen, L. Pitcher, and A. Rennermalm. A. Pope, G. Hamilton, and an anonymous reviewer provided useful comments that helped to improve our original manuscript.

Edited by: I. M. Howat

References

Adler-Golden, S. M., Matthew, M. W., Bernstein, L. S., Levine, R. Y., Berk, A., Richtmeier, S. C., Acharya, P. K., Anderson, G. P., Felde, J. W., Gardner, J. A., Hoke, M. L., Jeong, L. S., Pukall, B., Ratkowski, A. J., and Burke, H. K.: Atmospheric correction for short-wave imagery based on MODTRAN 4, SPIE Proc. Ser., 3753, 61–69, 1999.

Bennartz, R., Shupe, M. D., Turner, D. D., Walden, V. P., Steffen, K., Cox, C. J., Kulie, M. S., Miller, N. B., and Pettersen, C.: July 2012 Greenland melt extent enhanced by low-level liquid clouds, Nature, 496, 83–86, 2013.

Box, J. E. and Ski, K.: Remote sounding of Greenland supraglacial melt lakes: implications for subglacial hydraulics, J. Glaciol., 53, 257–265, 2007.

Colgan, W., Steffen, K., McLamb, W. S., Abdalati, W., Rajaram, H., Motyka, R., Phillips, T., and Anderson, R.: An increase in crevasse extent, West Greenland: Hydrologic implications, Geophys. Res. Lett., 38, L18502, doi:10.1029/2011gl048491, 2011.

Dierssen, H. M., Zimmerman, R. C., Leathers, R. A., Downes, T. V., and Davis, C. O.: Ocean color remote sensing of seagrass and bathymetry in the Bahamas Banks by high-resolution airborne imagery, Limnol. Oceanogr., 48, 444–455, 2003.

Fitzpatrick, A. A. W., Hubbard, A. L., Box, J. E., Quincey, D. J., van As, D., Mikkelsen, A. P. B., Doyle, S. H., Dow, C. F., Hasholt, B., and Jones, G. A.: A decade of supraglacial lake volume estimates across a land-terminating margin of the Greenland Ice Sheet, The Cryosphere Discuss., 7, 1383–1414, doi:10.5194/tcd-7-1383-2013, 2013.

Leeson, A. A., Shepherd, A., Palmer, S., Sundal, A., and Fettweis, X.: Simulating the growth of supraglacial lakes at the western margin of the Greenland ice sheet, The Cryosphere, 6, 1077–1086, doi:10.5194/tc-6-1077-2012, 2012.

Legleiter, C. J.: Remote measurement of river morphology via fusion of LiDAR topography and spectrally based bathymetry, Earth Surf. Proc. Land., 37, 499–518, 2012.

Legleiter, C. J.: Mapping river depth from publicly available aerial images, River Res. Appl., 29, 760–780, 2013.
Legleiter, C. J. and Overstreet, B. T.: Mapping gravel bed river bathymetry from space, J. Geophys. Res.-Earth, 117, doi:10.1029/2012jf002539, 2012.
Legleiter, C. J., Roberts, D. A., Marcus, W. A., and Fonstad, M. A.: Passive optical remote sensing of river channel morphology and in-stream habitat: Physical basis and feasibility, Remote Sens. Environ., 93, 493–510, 2004.
Legleiter, C. J., Roberts, D. A., and Lawrence, R. L.: Spectrally based remote sensing of river bathymetry, Earth Surf. Proc. Land., 34, 1039–1059, 2009.
Legleiter, C. J., Kinzel, P. J., and Overstreet, B. T.: Evaluating the potential for remote bathymetric mapping of a turbid, sand-bed river: 2. Application to hyperspectral image data from the Platte River, Water Resour. Res., 47, W09532, doi:10.1029/2011wr010592, 2011.
Liang, Y.-L., Colgan, W., Lv, Q., Steffen, K., Abdalati, W., Stroeve, J., Gallaher, D., and Bayou, N.: A decadal investigation of supraglacial lakes in West Greenland using a fully automatic detection and tracking algorithm, Remote Sens. Environ., 123, 127–138, 2012.
Luthje, M., Pedersen, L. T., Reeh, N., and Greuell, W.: Modelling the evolution of supraglacial lakes on the West Greenland ice-sheet margin, J. Glaciol., 52, 608–618, 2006.
Lyzenga, D. R.: Passive Remote-Sensing Techniques for Mapping Water Depth and Bottom Features, Appl. Optics, 17, 379–383, 1978.
Maritorena, S., Morel, A., and Gentili, B.: Diffuse-Reflectance of Oceanic Shallow Waters – Influence of Water Depth and Bottom Albedo, Limnol. Oceanogr., 39, 1689–1703, 1994.
McGrath, D., Colgan, W., Steffen, K., Lauffenburger, P., and Balog, J.: Assessing the summer water budget of a moulin basin in the Sermeq Avannarleq ablation region, Greenland ice sheet, J. Glaciol., 57, 954–964, 2011.
Nghiem, S. V., Hall, D. K., Mote, T. L., Tedesco, M., Albert, M. R., Keegan, K., Shuman, C. A., DiGirolamo, N. E., and Neumann, G.: The extreme melt across the Greenland ice sheet in 2012, Geophys. Res. Lett., 39, L20502, doi:10.1029/2012gl053611, 2012.
Philpot, W. D.: Bathymetric Mapping with Passive Multispectral Imagery, Appl. Optics, 28, 1569–1578, 1989.
Pineiro, G., Perelman, S., Guerschman, J. P., and Paruelo, J. M.: How to evaluate models: Observed vs. predicted or predicted vs. observed?, Ecol. Model., 216, 316–322, 2008.
Rennermalm, A. K., Moustafa, S. E., Mioduszewski, J., Chu, V. W., Forster, R. R., Hagedorn, B., Harper, J. T., Mote, T. L., Robinson, D. A., Shuman, C. A., Smith, L. C., and Tedesco, M.: Understanding Greenland ice sheet hydrology using an integrated multi-scale approach, Environ. Res. Lett., 8, 015017, doi:10.1088/1748-9326/8/1/015017, 2013.
Sneed, W. A. and Hamilton, G. S.: Evolution of melt pond volume on the surface of the Greenland Ice Sheet, Geophys. Res. Lett., 34, L03501, doi:10.1029/2006gl028697, 2007.
Sneed, W. A. and Hamilton, G. S.: Validation of a method for determining the depth of glacial melt ponds using satellite imagery, Ann. Glaciol., 52, 15–22, 2011.
Stumpf, R. P., Holderied, K., and Sinclair, M.: Determination of water depth with high-resolution satellite imagery over variable bottom types, Limnol. Oceanogr., 48, 547–556, 2003.
Tedesco, M. and Steiner, N.: In-situ multispectral and bathymetric measurements over a supraglacial lake in western Greenland using a remotely controlled watercraft, The Cryosphere, 5, 445–452, doi:10.5194/tc-5-445-2011, 2011.
Tedesco, M., Luthje, M., Steffen, K., Steiner, N., Fettweis, X., Willis, I., Bayou, N., and Banwell, A.: Measurement and modeling of ablation of the bottom of supraglacial lakes in western Greenland, Geophys. Res. Lett., 39, L02502, doi:10.1029/2011GL049882, 2012.
Tedesco, M., Fettweis, X., Mote, T., Wahr, J., Alexander, P., Box, J. E., and Wouters, B.: Evidence and analysis of 2012 Greenland records from spaceborne observations, a regional climate model and reanalysis data, The Cryosphere, 7, 615–630, doi:10.5194/tc-7-615-2013, 2013.
Yang, K. and Smith, L. C.: Supraglacial Streams on the Greenland Ice Sheet Delineated From Combined Spectral-Shape Information in High-Resolution Satellite Imagery, IEEE Geosci. Remote Sens., 10, 801–805, 2013.

11

Glacier change in the Cariboo Mountains, British Columbia, Canada (1952–2005)

M. J. Beedle, B. Menounos, and R. Wheate

Geography Program, University of Northern British Columbia, Prince George, BC, Canada

Correspondence to: M. J. Beedle (beedlem@unbc.ca)

Abstract. We applied photogrammetric methods with aerial photography from 11 different years between 1946 and 2005 to assess changes in area and volume of 33 glaciers in the Cariboo Mountains of British Columbia for the latter half of the 20th century. These are used to identify changes in extent and elevation primarily for the periods 1952–1985, 1985–2005, and 1952–2005. All glaciers receded during the period 1952–2005; area retreat averaged $-0.19 \pm 0.05\,\%\,a^{-1}$. From 1952 to 1985, nine glaciers advanced; following 1985, retreat rates accelerated to $-0.41 \pm 0.12\,\%\,a^{-1}$. Thinning rates of a subset of seven glaciers likewise accelerated, from -0.14 ± 0.04 m w.e. a^{-1} (1952–1985) to -0.50 ± 0.07 m w.e. a^{-1} for the period 1985–2005. Temperatures increased from the earlier to the latter period for the ablation (+0.38 °C) and accumulation (+0.87 °C) seasons, and average precipitation decreased, particularly in the accumulation season (−32 mm, −3.2 %). Our comparison of surface area change with glacier morphometry corroborates previous studies that show primary relations between extent change and surface area. We also find that the strength and sign of these relations varied for different epochs. Our results also indicate that the 1985 glacier extent for the study area reported previously by other studies may be slightly overestimated due to errant mapping of late-lying snow cover.

1 Introduction

Glaciers are important components of the hydrologic system. They contribute meltwater along with organic and inorganic materials to freshwater systems (Bogdal et al., 2009; Hood et al., 2009; Moore et al., 2009; Marshall et al., 2011). These inputs, along with downstream impacts, are altered when glaciers undergo changes in extent and volume. As glaciers and ice caps (GIC; all glaciers outside of the Antarctic and Greenland ice sheets) lose volume, sea level rises (Radić and Hock, 2011). The largest GIC contributions to sea level rise include the Canadian Arctic Archipelago and Gulf of Alaska regions (Gardner et al., 2011; Berthier et al., 2010), but recent work demonstrates that, collectively, small glaciers ($< 1\,km^2$) contribute a significant amount to total GIC volume (Bahr and Radić, 2012).

Recent studies of glacier change use a variety of remotely sensed products that include aerial photography (Koblet et al., 2010; Tennant et al., 2012), satellite imagery (Paul et al., 2004; Berthier et al., 2010), and laser altimetry (Sapiano et al., 1998; Arendt et al., 2002). Typically these studies employ a combination of geomatic data to assess changes in glaciers for a particular region (Schiefer et al., 2007; Abermann et al., 2009; Bolch et al., 2010). Comprehensive inventories of glacier extent and change (e.g., Bolch et al., 2010), as well as measurement of glacier volume change of large regions (e.g., Schiefer et al., 2007) primarily rely on satellite-borne instruments. Such analysis is therefore limited to the last 30 years.

Aerial photography can be used to extend glacier change documentation by up to three decades prior to the beginning of the satellite era. In British Columbia, repeat aerial surveys have been approximately decadal, and thus an opportunity exists to assess multi-decadal changes in the area and volume of alpine glaciers. DeBeer and Sharp (2007), for example, combined aerial photography and satellite imagery to assess changes in glacier cover in the southern Canadian Cordillera over the second half of the 20th century. In particular, they noted negligible change in small glaciers in their study. Our study builds on DeBeer and Sharp (2007) by expanding the spatial domain over which glacier change is evaluated and by

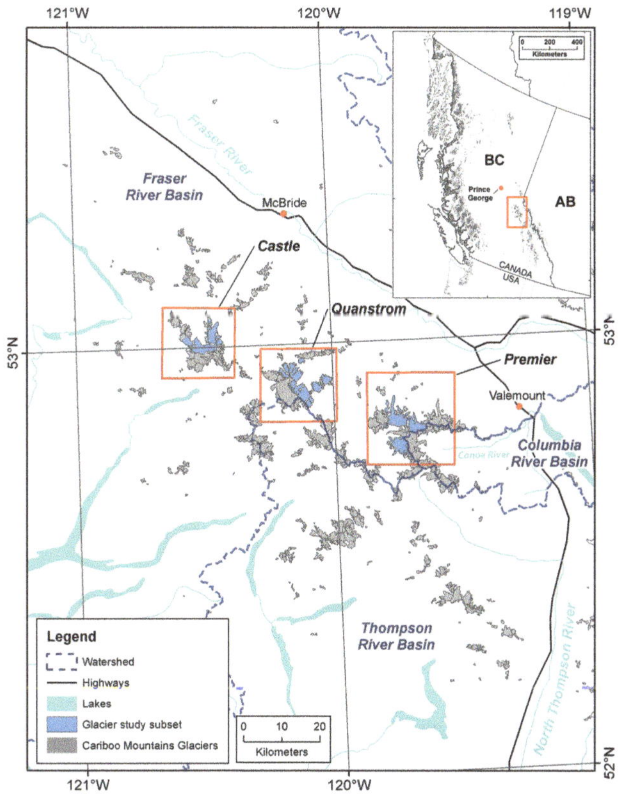

Figure 1. Study area: the three red rectangles indicate subregions in this study (Castle, Quanstrom, and Premier) and the location of the three maps in Fig. 2.

increasing the number of epochs over which these changes can be compared to the instrumental record.

In this study we use aerial photogrammetry to investigate the extent and volume change of a subset of glaciers in western Canada for the periods 1952–1970, 1970–1985, and 1985–2005. The use of aerial photography thus provides a method to temporally extend the glacier inventory studies of Schiefer et al. (2007) and Bolch et al. (2010) and, in addition, an independent check against glacier mapping using lower-resolution satellite imagery.

The objectives of this study are to (1) document the extent and volume change of a subset of glaciers in the Cariboo Mountains of British Columbia (BC) and (2) assess the climatological conditions that could explain observed glacier change in the study area.

2 Study area and previous work

The Cariboo Mountains are the northernmost range of the Columbia Mountains of BC, Canada (Fig. 1). The climate of the Cariboo Mountains is transitional, wetter than the Rocky Mountains to the east, and drier than the Coast Mountains. Total annual precipitation (1971–2000 climate normals) averages 1014 and 679 mm on the windward (Barkerville, BC) and lee (McBride, BC) sides of the Cariboo Mountains, and annual temperature averages 1.9 and 4.4 °C, respectively (Environment Canada, 2012).

Annual length change in Castle Creek Glacier in the Cariboo Mountains is more closely related to changes of glaciers in the Coast Mountains than those of the Rocky Mountains (Beedle et al., 2009). Meltwater from the majority of glaciers in the Cariboo Mountains contributes to the Fraser River; however, some glaciers in the Premier Range (a Cariboo Mountains subrange) contribute to the Columbia River (Fig. 1). Maurer et al. (2012) found that glaciers of the Cariboo Mountains nearly reached their Holocene maximum extents around 2.73–2.49 ka and that major retreat of the glaciers did not begin until the early 20th century. Luckman et al. (1987) showed that some glaciers of the Premier Range advanced in the 1960s and 1970s in response to decreased temperatures and increased winter precipitation. Bolch et al. (2010) completed the first satellite inventory of glaciers in western Canada, identifying 536 glaciers in the Cariboo Mountains (2005) with a total surface area of 731 km^2. Cariboo Mountains glaciers lost 7.06 km^3 of ice at a rate of -0.58 m water equivalent (w.e.) a^{-1} during the period 1985–1999 (Schiefer et al., 2007).

3 Methods

3.1 Imagery and supplemental data

We measured glacier extent and surface-elevation change from aerial photographs obtained from the British Columbia Government and Canada National Air Photo Library (Table 1). Imagery prior to 2005 was scanned at a resolution of 12–14 μm from diapositives or negatives using a photogrammetric scanner, whereas imagery from 2005 was available as digital aerial triangulation (AT) scans (digital photos with available exterior orientation; BC Ministry of Environment, 1998). Ground sampling distance ranges from 0.2 to 1.1 m depending on scanning resolution and image scale. Most photos were taken in late summer, but dates range from mid-July to late September. We used all images to estimate glacier length and area change, whereas only those from 1946, 1952, 1984, 1985, and 2005 were used to determine surface-elevation change. As photos for each subregion were taken in different years, we thus report a mean area-weighted year: 1952 for photos taken in 1946, 1952, or 1955; 1970 for those of 1967, 1970, or 1971; and 1985 for those of 1984 or 1985. All annual rates of extent or volume change are calculated using the actual duration between images. We report extent change of glaciers within three intervals (1952–1970, 1970–1985, and 1985–2005) and surface-elevation change for two epochs (1952–1985 and 1985–2005).

We use ClimateWNA (Wang et al., 2012) to assess the variability in temperature and precipitation during the periods for which we measure glacier ex-

Table 1. Aerial photography used to derive glacier extent and thickness change data.

Date	Region	Roll ID	# of Images	Scale	Resolution (m)	Contrast	Snowcover
10 Sep 1946	C	BC320	15	1 : 31 680	0.38	Poor	Good
25 Sep 1952	C	A13538	5	1 : 60 000	0.84	Fair	Good
9 Sep 1952	Q	A13538	5	1 : 60 000	0.84	Fair	Good
30 Aug 1955	P	A14930	7	1 : 70 000	0.98	Poor	Fair
15 Aug 1967	C	BC7019	15	1 : 15 840	0.19	Good	Fair
19 Aug 1970	P	BC5394	5	1 : 80 000	0.96	Fair	Good
15 Jul 1971	Q	A21587	4	1 : 80 000	1.12	Poor	Fair
24 Jul 1977	C	A24743	8	1 : 70 000	0.98	Poor	Poor
31 Jul 1984	C	BC84073	4	1 : 60 000	0.72	Fair	Fair
16 Aug 1985	Q	BC85071	5	1 : 60 000	0.72	Fair	Fair
16 Aug 1985	Q, P	BC85075	7	1 : 60 000	0.72	Fair	Fair
11 Sep 1996	C	BCB96106	6	1 : 40 000	0.48	Poor	Fair
25 Aug 2005	C	BCC05111	10	1 : 20 000	0.24	Good	Good
25 Aug 2005	P	BCC05108	18	1 : 20 000	0.24	Good	Good
25 Aug 2005	Q, P	BCC05109	16	1 : 20 000	0.24	Good	Good
25 Aug 2005	Q	BCC05110	6	1 : 20 000	0.24	Good	Good

Regions: C: Castle Creek, Q: Quanstrom, P: Premier Range. Roll ID: BC: Provincial aerial photos, A: Federal aerial photos. Contrast: Subjective determination of image contrast on a scale from poor to excellent. Snowcover: Subjective determination of snowcover extent. Poor refers to more snowcover.

tent and surface-elevation change. The ClimateWNA v4.72 program (http://www.genetics.forestry.ubc.ca/cfcg/ClimateWNA/ClimateWNA.html) interpolates and extracts climate data for specific locations for western North America, relying on downscaled PRISM (Daly et al., 2002) and historical data (Mitchell and Jones, 2005) to generate monthly time series of temperature and precipitation. For details, refer to Wang et al. (2012). We use averaged ClimateWNA output for points along the centerline every 100 m in elevation for all 33 glaciers ($n = 212$) to estimate Cariboo Mountains glacier climatology, and for glaciers of the Castle Creek region ($n = 47$), Quanstrom region ($n = 62$), and Premier region ($n = 103$) to estimate subregional climatology. We compare monthly temperature precipitation data from ClimateWNA with measurements at Castle Creek Glacier (Déry et al., 2010) for 2009–2011 to estimate the ability of ClimateWNA to represent conditions at high elevations in the Cariboo Mountains.

To investigate dominant synoptic patterns over the Cariboo Mountains, we use the National Centers for Environmental Protection/National Center for Atmospheric Research (NCEP/NCAR) reanalysis data (Kalnay et al., 1996) of 700 hPa geopotential height anomalies. The 700 hPa geopotential surface provides a good indicator of general atmospheric conditions in the middle troposphere (~ 3000 m), an elevation just above the uppermost extent of most glaciers in the Cariboo Mountains. We assess climatic conditions for balance years (October–September), and for ablation and accumulation seasons, defined as June–September and October–May, respectively.

3.2 Photogrammetry

We used the Vr Mapping photogrammetry software suite (Cardinal Systems LLC) to create stereo-models from aerial photos. Exterior orientation for the models was derived from tie points and common ground control points (GCP) horizontally and vertically distributed throughout the photos (Schiefer and Gilbert, 2007; Barrand et al., 2009). Within each subregion we used common GCPs to reduce systematic positional errors (Kääb and Vollmer, 2000; Schiefer and Gilbert, 2007; Schiefer et al., 2007). GCPs consisted of stable bedrock features or boulders obtained from aerial triangulated stereo-models (BC Ministry of Environment, 1998).

Eleven check patches, located on stable surfaces around each glacier region (Fig. 2), were used to quantify the relative accuracy of the stereo-models (Fig. 3). Each check patch is comprised of 25 individual checkpoints in a 5 m grid; we estimated systematic bias of stereo-models from the mean residuals of the check patches. Trend surfaces were created from these 11 mean residuals and used to apply a correction for glacier surface-elevation measurements.

3.3 Glacier subset selection

With the exception of 2005, glaciers of the Cariboo Mountains were not comprehensively photographed in any given year, and we thus concentrate our study on three subregions of the Cariboo Mountains with suitable photographic coverage: (1) the Castle Creek Glacier area, (2) the Mt. Quanstrom area, and (3) the Premier Range (Figs. 1 and 2). We selected our subset of glaciers based on (1) availability of imagery, both in terms of temporal resolution and spatial scale; (2) snow cover and contrast; and (3) representativeness

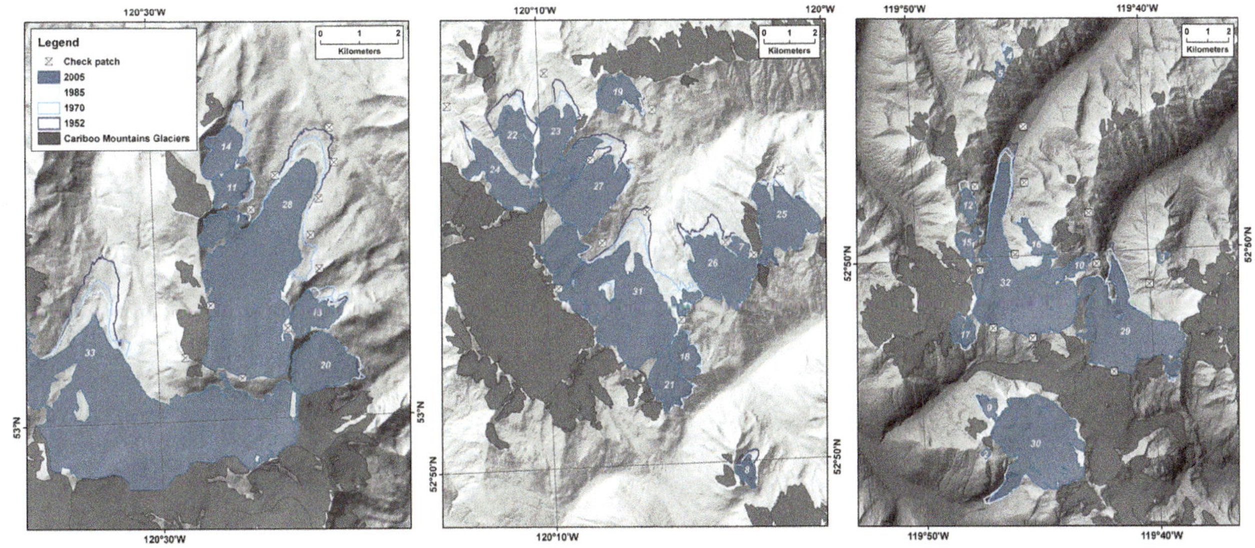

Figure 2. Cariboo Mountains subregions showing subset of 33 glaciers, including extents for 1952, 1970, 1985, and 2005. Panels show from left to right the Castle, Quanstrom, and Premier subregions. Numerical glacier identification, numbered by 2005 surface area from smallest to largest, corresponds to that of Tables 2 and 3.

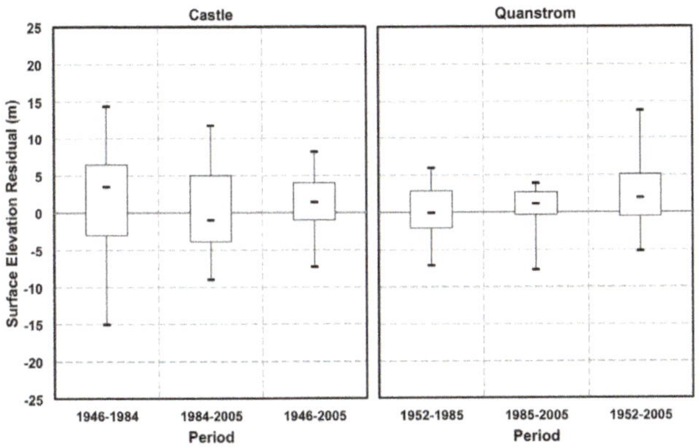

Figure 3. Box-and-whisker plots showing the maximum, interquartile range, median, and minimum of surface-elevation residuals of 275 checkpoints. Checkpoints are used to determine relative accuracy of stereo-models and bias correction for measurement of surface-elevation change of glaciers in the Castle and Quanstrom regions in three periods.

based on average glacier morphometry of Cariboo Mountains glaciers presented in Bolch et al. (2010). Snow cover and poor contrast of older aerial photography reduce potential glaciers for study (Table 1). Sun angle with respect to glacier aspect and slope leads to local areas of high reflectivity and an absence of contrast, particularly for some south-facing glaciers. We omitted glaciers with these issues in our selection of a subset.

Within the three Cariboo Mountains subregions, we selected 33 glaciers representing five size classes (0.1–0.5, 0.5–1.0, 1.0–5.0, 5.0–10.0, and $> 10.0\,\text{km}^2$). We used surface area as our primary criterion to select glaciers as many studies find it to be the key morphometric determinant of glacier extent change (e.g., Serandbrei-Barbero et al., 1999; Hoelzle et al., 2003; Paul et al., 2004; Andreassen et al., 2008; Bolch et al., 2010; Paul and Andreassen, 2009). Our analysis oversampled glaciers in the largest size class as these glaciers likely play a dominant role in regional volume change and meltwater contributions to their respective watersheds (e.g., Arendt et al., 2006; DeBeer and Sharp, 2007; Paul and Haeberli, 2008). All ice masses with the exception of Kiwa Glacier (glacier 32 in Fig. 2 and Table 3) are land-terminating glaciers.

3.4 Data collection and analysis

We collected glacier extents and point measurements of surface elevation directly from the stereo-models following the methods described elsewhere (Tennant and Menounos, 2013; Beedle et al., 2014). This approach allowed us to collect elevation and points while viewing the 3-D models with polarized glasses and a shutter screen. The primary advantage of this approach is that it minimizes common errors that can occur with automated digital elevation model extraction methods (e.g., pixel mismatch and interpolation errors). Complete glacier outlines were mapped for 2005, and glacier extents were updated for previous years only below the transient snow line (TSL), thus eliminating errantly mapping seasonal snow cover as glacierized area (e.g., Koblet et al., 2010). Delineation of planimetric glacier area above the TSL is problematic as seasonal snow cover masks the glacier margin, and can erroneously inflate estimates of glacier surface area. Mapping of ice divides relied on surface elevation and surface features such as crevasses and runnels; these divides were held constant in order to compare dimensional changes for a given ice body through time. Extent-change analysis was made for different glacier subsets by epoch based on aerial photo coverage and snow cover. Collectively, our analysis allows us to compare area change for 33 glaciers for the epochs 1952–1985, 1985–2005, and 1952–2005 (33-glacier subset), as well as for 26 glaciers in the additional epochs 1952–1970 and 1970–1985 (26-glacier subset).

Poor contrast and snow cover limit our analysis of glacier surface-elevation change to seven glaciers for the periods 1952–1985, 1985–2005, and 1952–2005. For these glaciers, we measured surface elevation on a 100 m grid for large glaciers ($> 1\,km^2$) and on a 50 m grid for small glaciers ($< 1\,km^2$). Poor contrast in the earliest photographs inhibited data collection over portions of glacier accumulation zones for four of the seven glaciers. In these areas, we extrapolate from the three highest elevation bins where we have measurements, assigning the average surface-elevation change of these observations to the bins with missing values. This assumption was necessary only for periods that rely on the earliest year of photography (1952); missing data total a maximum of 27 % of total glacier surface area.

To calculate glacier-wide volume change from point measurements, we multiplied average elevation change for each 50 m elevation bin and summed over the entire glacier surface. Elevation bins and glacier surface were derived from epoch-specific glacier hypsometries created from the same point measurements. Where poor contrast led to an absence of measurements in the accumulation area, we used surface-elevation measurements for that elevation bin from the stereo-model of a prior or subsequent year. This approach assumes that surface elevation and extent underwent negligible change, an assumption that would be inappropriate in the rapidly changing ablation zones, but one that is more likely in the accumulation zones where this assumption was applied. We calculated glacier-wide average thickness change based on the average of the two extents that define a given period (e.g., Arendt et al., 2002; Barrand et al., 2010). All values presented in water equivalent (w.e.) are calculated by assuming a density of $900\,kg\,m^{-3}$ in the ablation area and $750\,kg\,m^{-3}$ in the accumulation area (e.g., Huss, 2013; Beedle et al., 2014); ablation and accumulation areas are defined by each glacier's median elevation (Table 2).

To estimate regional (Cariboo Mountains) glacier extent change, we extrapolated from measurements of our subset of studied glaciers using the average rates of relative extent change by size class for different periods. We used two methods to estimate Cariboo Mountains surface-elevation change: first, the average surface-elevation change of our subset of seven glaciers; second, the average gradient of surface-elevation change with elevation, defined as the average of all measurements within each 50 m bin, and integrated it with the Cariboo Mountains glacier hypsometry of 1985. We compare extrapolated values with results from recent work that included volume change (Schiefer et al., 2007) and extent change (Bolch et al., 2010) of Cariboo Mountains glaciers.

3.5 Error analysis

We estimated error in glacier extent and extent change based on a buffer approach (Granshaw and Fountain, 2006). Our buffer (± 5 pixels) yields widths of ~ 5.5 to 1.3 m, depending on image resolution. Error in extent change is calculated as the root-mean-squared error (RMSE) of the error estimates for the years that define a given period. Average errors for the 33-glacier subset range from 0.8 to 2.4 % for extents, and from 2.3 to 3.6 % for extent change. These errors vary depending on image resolution (see Table 1) and glacier dimensions.

We used the standard deviation of 275 check points in 11 check patches to estimate error in our measurements of surface-elevation change ($E_{\Delta Z}$). To account for greater uncertainty over surfaces with reduced contrast, we added an error of ± 5 m for measurements above the TSL. In cases where measurements are absent due to poor contrast, we increased this added error term to ± 10 m (e.g., Tennant and Menounos, 2013). Error in measurements of surface-elevation change in the ablation area is 1σ of check points, whereas for the accumulation zone it is the quadrature sum of 1σ of check points (E_1) and the added error of ± 5 or 10 m for reduced contrast (E_1):

$$E_{\Delta Z} = \sqrt{E_1^2 + E_2^2}. \tag{1}$$

Converting to water equivalent units through density assumptions imparts an additional error term. To estimate error in the density assumption of $750\,kg\,m^{-3}$ for the accumulation zone, we used a range of possible values for accumulation-zone density ($600–900\,kg\,m^{-3}$). We estimate error in our measurements of surface-elevation change in water-equivalent units ($E_{\Delta Z\ \mathrm{w.e.}}$) as

$$E_{\Delta Z\ \mathrm{w.e.}} = \sqrt{E_{\Delta Z}^2 \rho^2 + E_\rho^2 \Delta Z^2}, \tag{2}$$

where ρ is density expressed as a water-equivalent conversion factor (0.9 in the ablation zone, 0.75 in the accumulation zone) and E_ρ is the error in our density assumptions, assumed to be 0 in the ablation area and 0.15 in the accumulation zone.

We estimate volume change error (E_{VOL}) after Barrand et al. (2010) as

$$E_{\mathrm{VOL}} = \sqrt{\sum_1^{\mathrm{bin}} (E_{\mathrm{bin}} A_{\mathrm{bin}})}, \tag{3}$$

where E_{bin} is the error ($E_{\Delta Z\ \mathrm{w.e.}}$) and A_{bin} is the surface area of each 50 m bin.

Previous work indicated a spatial correlation of 1000 m in photogrammetric DEMs (Rolstad et al., 2009), and others

Table 2. Morphometric properties of the 33-glacier subset.

Id (Name)	Region	Length (m)	Area (km^2)	Slope	Aspect	Mean Z (m)	Median Z (m)	Min. Z (m)	Max. Z (m)	Range Z (m)
1	C	513	0.109	22.7	E	2472	2471	2360	2590	230
2	P	452	0.135	19.5	W	2587	2593	2469	2682	213
3	P	641	0.155	27.5	NE	2607	2619	2413	2785	372
4	P	779	0.210	18.0	N	2485	2480	2379	2651	272
5	P	922	0.264	21.1	N	2523	2528	2324	2718	394
6	Q	731	0.276	22.2	SE	2532	2545	2256	2692	436
7	Q	952	0.422	21.3	NW	2386	2383	2228	2621	393
8	Q	1366	0.563	18.3	N	2283	2279	2093	2494	401
9	P	1386	0.648	25.7	NW	2420	2415	2117	2968	851
10	P	1552	0.650	26.6	NE	2406	2400	1990	2751	761
11	C	1173	0.653	21.2	SE	2452	2445	2246	2733	487
12	P	1671	0.830	21.6	N	2399	2407	2169	2646	477
13	C	1017	0.860	20.7	NE	2363	2343	2135	2720	585
14	C	1419	0.913	16.2	NE	2322	2314	2175	2580	405
15	P	1327	0.983	23.8	NE	2665	2662	2248	3157	909
16	P	2027	0.995	24.5	NW	2699	2716	2171	3093	922
17	P	1584	1.145	26.7	S	2728	2724	2313	3188	875
18	Q	1209	1.349	16.7	E	2362	2364	2172	2515	343
19	Q	2199	1.609	10.8	SE	2469	2478	2187	2633	446
20	C	1870	1.816	18.5	E	2387	2398	2044	2749	705
21	Q	1796	2.188	15.7	S	2421	2449	2052	2669	617
22	Q	2965	2.464	12.3	N	2438	2436	2198	2803	605
23	Q	3002	2.655	18.5	N	2482	2490	2054	3019	965
24	Q	3562	3.049	17.3	NW	2352	2359	1991	2789	798
25	Q	2300	4.280	14.9	N	2463	2488	2061	2752	691
26	Q	3294	4.625	11.9	NW	2417	2425	2137	2708	571
27	Q	3062	7.037	12.8	NE	2416	2408	2056	2917	861
28 (Castle Creek)	C	5875	9.645	11.1	NE	2387	2397	1881	2827	946
29 (Tête)	P	5971	12.674	16.9	N	2520	2565	1649	3357	1708
30 (David)	P	5228	13.533	19.3	SW	2609	2616	1352	3339	1987
31	Q	4010	14.782	13.2	NE	2319	2334	1868	2745	877
32 (Kiwa)	P	9597	17.154	17.2	N	2617	2705	1491	3509	2018
33	C	7302	19.094	12.8	N	2273	2288	1803	2910	1107

Slope and elevation data from TRIM DEM and 2005 extent. Length and area data from 2005 extent.

have calculated a number of independent measurements (n) assuming a correlation scale of 1000 m (Barrand et al., 2010). We calculated degrees of freedom for each glacier from point measurements of surface-elevation change in epochs with complete spatial coverage. Using the Incremental Spatial Autocorrelation tool in ArcGIS 10.1, we calculated correlation distances that range from 350 to 2000 m, yielding effective degrees of freedom that range from 1 to 5.

We used cross-validation to test the accuracy of our extrapolation of Cariboo Mountains glacier extent and surface-elevation change. In our test of glacier extent extrapolation, we partitioned the 26-glacier subset into four scenarios where we withheld 75, 50, and 25 % of the individual glaciers as a validation set and retained the remainder as a training set. Set selection was randomized within each size class, ensuring each set retained glaciers of varying size. Each of the three scenarios was applied to observations in five epochs: 1952–1970, 1970–1985, 1985–2005, 1952–1985, and 1952–2005. We used the standard deviation of these 15 test cases (±4.7 %) as an estimate of error in our extrapolation of the extent of all Cariboo Mountains glaciers. To estimate error in our extrapolation of surface-elevation change, we partitioned our seven glacier subset into two scenarios where we withheld three and four individual glaciers as a validation set and retain the remainder as a training set. Each scenario was applied to observations in three epochs: 1952–1985, 1985–2005, and 1952–2005. We used the variability of these six test cases to estimate error in our extrapolation of surface-elevation change by means of two methods: (1) mean annual rate of surface-elevation change in the glacier subset (± 0.077 m w.e. a^{-1}) and (2) mean surface-elevation change within each collective 50 m elevation bin for the glacier subset (± 0.089 m w.e. a^{-1}).

4 Results

4.1 Regional representativeness

As expected, our glacier subset undersamples ice masses of the Cariboo Mountains in all size classes except the largest size class (Fig. 4). Most glaciers in the Cariboo Mountains (40 %) fall in the 1.0–5.0 km^2 size class compared to 20 % in our sample. In contrast, our subset contains 61 % glaciers exceeding 5.0 km^2, compared to 21 % for the Cariboo Moun-

Table 3. Glacier extent and area change data for glaciers and summed by subregion and size class.

		Extent				Successive periods						Cumulative periods			
Id	Region	Area 1952 (km²)	Area 1970 (km²)	Area 1985 (km²)	Area 2005 (km²)	Area change 52–70 (km²)	Area change 52–70 (%)	Area change 70–85 (km²)	Area change 70–85 (%)	Area change 85–05 (km²)	Area change 85–05 (%)	Area change 52–85 (km²)	Area change 52–85 (%)	Area Change 52–05 (km²)	Area change 52–05 (%)
1	C	0.117	No data	0.131	0.109	No data	No data	No data	No data	*−0.022*	*−16.8*	**0.014**	**12.0**	−0.008	−6.8
2	P	0.137	No data	0.136	0.135	No data	No data	No data	No data	−0.001	−0.7	−0.001	−0.7	−0.002	−1.5
3	P	0.157	0.158	0.161	0.155	0.001	0.6	0.003	1.9	−0.006	−3.7	0.004	2.5	−0.002	−1.3
4	P	0.301	0.298	0.269	0.210	−0.003	−1.0	*−0.029*	*−9.7*	*−0.059*	*−21.9*	*−0.032*	*−10.6*	*−0.091*	*−30.2*
5	P	0.283	0.310	0.303	0.264	**0.027**	**9.5**	−0.007	−2.3	*−0.039*	*−12.9*	**0.020**	**7.1**	*−0.019*	*−6.7*
6	Q	0.284	No data	0.291	0.276	No data	No data	No data	No data	*−0.015*	*−5.2*	0.007	2.5	−0.008	−2.8
7	Q	0.536	No data	0.478	0.422	No data	No data	No data	No data	*−0.056*	*−11.7*	*−0.058*	*−10.8*	*−0.114*	*−21.3*
8	Q	0.775	No data	0.626	0.563	No data	No data	No data	No data	*−0.063*	*−10.1*	*−0.149*	*−19.2*	*−0.212*	*−27.4*
9	P	0.662	0.728	0.709	0.648	**0.066**	**10.0**	−0.019	−2.6	*−0.061*	*−8.6*	**0.047**	**7.1**	−0.014	−2.1
10	P	0.665	0.700	0.707	0.650	0.035	5.3	0.007	1.0	*−0.057*	*−8.1*	**0.042**	**6.3**	−0.015	−2.3
11	C	0.682	0.746	0.785	0.653	0.064	9.4	**0.039**	**5.2**	*−0.132*	*−16.8*	**0.103**	**15.1**	*−0.029*	*−4.3*
12	P	0.853	0.858	0.991	0.830	0.005	0.6	**0.133**	**15.5**	*−0.161*	*−16.2*	**0.138**	**16.2**	−0.023	−2.7
13	C	1.070	1.109	0.918	0.860	0.039	3.6	*−0.191*	*−17.2*	−0.058	*−6.3*	*−0.152*	*−14.2*	*−0.210*	*−19.6*
14	C	1.272	1.235	No data	0.913	*−0.037*	*−2.9*	No data	No data	No data	No data	No data	No data	*−0.359*	*−28.2*
15	P	1.003	1.030	1.012	0.983	0.027	2.7	−0.018	−1.7	*−0.029*	*−2.9*	0.009	0.9	−0.020	−2.0
16	P	1.056	1.093	1.074	0.995	0.037	3.5	−0.019	−1.7	*−0.079*	*−7.4*	0.018	1.7	*−0.061*	*−5.8*
17	P	1.166	1.244	1.235	1.145	**0.078**	**6.7**	−0.009	−0.7	*−0.090*	*−7.3*	**0.069**	**5.9**	−0.021	−1.8
18	Q	1.432	1.436	1.444	1.349	0.004	0.3	0.008	0.6	*−0.095*	*−6.6*	0.012	0.8	*−0.083*	*−5.8*
19	Q	1.767	1.759	1.757	1.609	−0.008	−0.5	−0.002	−0.1	*−0.148*	*−8.4*	−0.010	−0.6	*−0.158*	*−8.9*
20	C	1.833	1.874	1.854	1.816	0.041	2.2	−0.020	−1.1	*−0.038*	*−2.0*	0.021	1.1	−0.017	−0.9
21	Q	2.199	2.248	2.287	2.188	0.049	2.2	0.039	1.7	*−0.099*	*−4.3*	**0.088**	**4.0**	−0.011	−0.5
22	Q	3.609	2.771	2.702	2.464	*−0.838*	*−23.2*	*−0.069*	*−2.5*	*−0.238*	*−8.8*	*−0.907*	*−25.1*	*−1.145*	*−31.7*
23	Q	3.464	3.154	3.068	2.655	*−0.310*	*−8.9*	*−0.086*	*−2.7*	*−0.413*	*−13.5*	*−0.396*	*−11.4*	*−0.809*	*−23.4*
24	Q	3.393	3.317	3.323	3.049	*−0.076*	*−2.2*	0.006	0.2	*−0.274*	*−8.2*	−0.070	−2.1	*−0.344*	*−10.1*
25	Q	4.440	4.582	4.555	4.280	**0.142**	**3.2**	−0.027	−0.6	*−0.275*	*−6.0*	**0.115**	**2.6**	*−0.160*	*−3.6*
26	Q	5.584	No data	5.156	4.625	No data	No data	No data	No data	*−0.531*	*−10.3*	*−0.428*	*−7.7*	*−0.959*	*−17.2*
27	Q	8.119	7.428	7.424	7.037	*−0.691*	*−8.5*	−0.004	−0.1	*−0.387*	*−5.2*	*−0.695*	*−8.6*	*−1.082*	*−13.3*
28	C	11.257	10.838	10.206	9.645	*−0.419*	*−3.7*	*−0.632*	*−5.8*	*−0.561*	*−5.5*	*−1.051*	*−9.3*	*−1.612*	*−14.3*
29	P	13.296	13.303	13.253	12.674	0.007	0.1	−0.050	−0.4	*−0.579*	*−4.4*	−0.043	−0.3	*−0.622*	*−4.7*
30	P	13.711	13.821	13.894	13.533	0.110	0.8	0.073	0.5	*−0.361*	*−2.6*	0.183	1.3	*−0.178*	*−1.3*
31	Q	18.592	17.688	16.796	14.782	*−4.9*	*−0.892*	*−5.0*	*−2.014*	*−12.0*	*−1.796*	*−9.7*	*−3.810*	*−20.5*	
32	P	18.234	18.042	17.661	17.154	−0.192	−1.1	*−0.381*	*−2.1*	*−0.507*	*−2.9*	*−0.573*	*−3.1*	*−1.080*	*−5.9*
33	C	21.283	20.446	20.038	19.094	*−0.837*	*−3.9*	*−0.408*	*−2.0*	*−0.944*	*−4.7*	*−1.245*	*−5.8*	*−2.189*	*−10.3*
	All	141.960	No data	135.244	126.852	No data	No data	No data	No data	*−8.392*	*−6.2*	*−6.716*	*−4.7*	*−15.108*	*−10.6*
	C	36.125	35.013	33.801	32.068	*−1.112*	*−3.1*	*−1.212*	*−3.5*	*−1.733*	*−5.1*	*−2.324*	*−6.4*	*−4.057*	*−11.2*
	Q	54.194	No data	49.907	45.299	No data	No data	No data	No data	*−4.608*	*−9.2*	*−4.287*	*−7.9*	*−8.895*	*−16.4*
	P	51.387	51.585	51.269	49.241	0.198	0.4	−0.316	−0.6	*−2.028*	*−4.0*	−0.118	−0.2	*−2.146*	*−4.2*
	0.1–0.5 km²	1.815	No data	1.769	1.571	No data	No data	No data	No data	*−0.198*	*−11.2*	−0.046	−2.5	*−0.244*	*−13.4*
	0.5–1.0 km²	5.991	6.264	6.196	5.619	**0.273**	**4.6**	−0.068	−1.1	*−0.577*	*−9.3*	0.205	3.4	*−0.372*	*−6.2*
	1.0–5.0 km²	23.303	22.385	22.225	20.555	*−0.918*	*−3.9*	−0.160	−0.7	*−1.670*	*−7.5*	*−1.078*	*−4.6*	*−2.748*	*−11.8*
	5.0–10 km²	19.376	18.266	17.630	16.682	*−1.110*	*−5.7*	*−0.636*	*−3.5*	*−0.948*	*−5.4*	*−1.746*	*−9.0*	*−2.694*	*−13.9*
	10–50 km²	85.116	83.300	81.642	77.237	*−1.816*	*−2.1*	*−1.658*	*−2.0*	*−4.405*	*−5.4*	*−3.474*	*−4.1*	*−7.879*	*−9.3*

Values in bold indicate glacier advance, values in italics indicate recession (beyond margin of error). Extents summed by size class are based on 2005 extent. Summed extents and area change may omit some individual glaciers: All: 14, C: 1 and 14, Q: none, P: 2, 0.1–0.5: none, 0.5–1.0: 8 and 14, 1.0–5.0: none, 5.0–10: none, 10–50: none.

tains. Eliminating 4 of the 5 largest glaciers from our 33-glacier subset (glaciers 30–33; Table 3) would result in a size-class distribution that is nearly identical to glaciers in the Cariboo Mountains.

Our glacier subset also oversamples higher elevations (Fig. 5). Area–altitude distribution of glaciers in the Premier Range, for example, is higher than the average elevation of glaciers in the Cariboo Mountains, whereas the hypsometry of our glacier subset from the Castle and Quanstrom areas are more similar to the total glacierized area of the Cariboo Mountains.

Glaciers in our study also more commonly face north, northeast, and southwest (Fig. 6). Cariboo Mountains glaciers have an average slope of 21.1°, whereas our subset has an average slope of 18.7°; these differences, however, are not statistically significant.

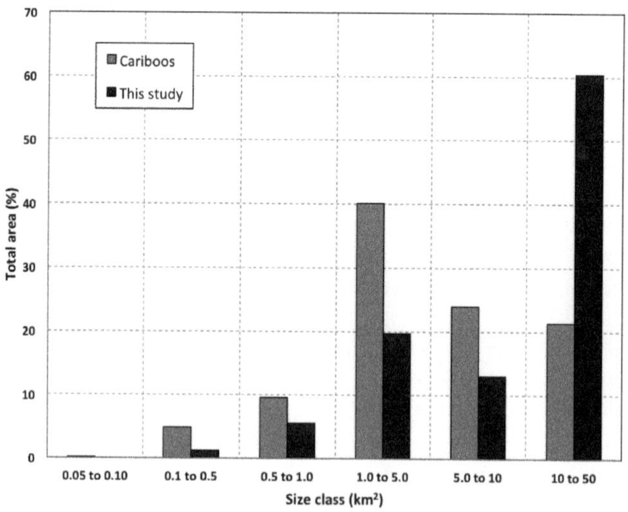

Figure 4. Fraction of total glacierized extent by size class for all Cariboo Mountains glaciers (gray) and this study (black).

4.2 Extent change

Between 1952 and 2005, glaciers lost $15.11 \pm 1.89\,km^2$ ($-10.6 \pm 2.9\,\%$) of their surface area (Table 3). Relative area change varied by subregion, with losses of $-11.2 \pm 2.7\,\%$, $-16.4 \pm 2.0\,\%$, and $-4.2 \pm 3.9\,\%$ for the Castle, Quanstrom, and Premier regions, respectively. Absolute area change by

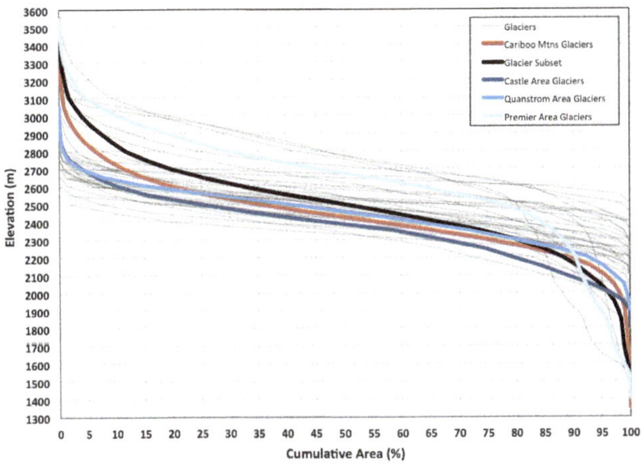

Figure 5. Hypsometries for our subset of glaciers (gray); average subset hypsometry (black); all Cariboo Mountains glaciers (red); and the Castle, Quanstrom, and Premier subregions (darker to lighter blue, respectively).

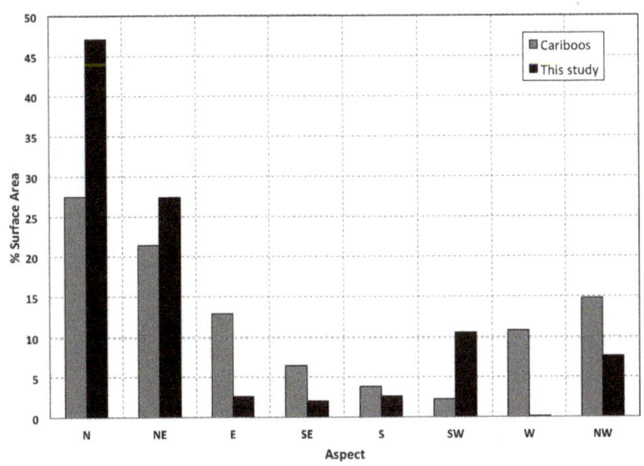

Figure 6. Fraction of total surface by aspect for all Cariboo Mountains glaciers (black) and our subset (gray).

size class over the same period varied from $-6.2\pm3.3\,\%$ (0.5–$1.0\,\mathrm{km}^2$) to $-13.9\pm1.1\,\%$ (5.0–$10\,\mathrm{km}^2$). Individual glacier area loss varied from -0.5 ± 1.6 to $-31.7\pm1.4\,\%$. Recession for 22 of the 33 glaciers exceeded the uncertainties in the data. Absolute extent change is dominated by three of the largest glaciers (glaciers 28, 31, and 33), which comprise 50 % of the total subset extent change for the period.

While glaciers on average shrank during the period 1952–2005, their change during intervening epochs was more complex (Table 3 and Fig. 7). Over the period 1952–1970, for example, 5 of 26 glaciers advanced, 7 receded, and 14 did not significantly change. Between the years 1970 and 1985, 2 of 26 glaciers advanced, 8 shrank, and the area of 16 glaciers did not change. For the cumulative period 1952–1985 nine glaciers advanced, 12 receded, and 11 had no discernable extent change.

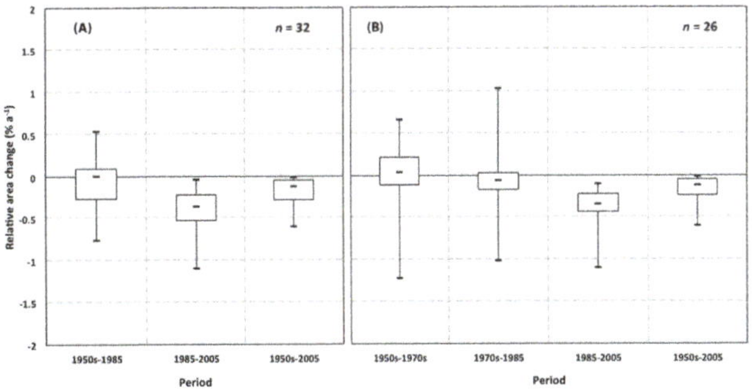

Figure 7. Box-and-whisker plots displaying maximum, interquartile range, median, and minimum of average annual rates of relative area change for **(a)** 32 glaciers for three periods and **(b)** for 26 individual glaciers for four periods.

Glaciers that advanced were generally shorter, smaller, steeper, and had higher median and minimum elevations than glaciers that receded (Table 4). Glaciers that negligibly changed tended to be steep, but were longer than those glaciers that advanced. Glaciers that changed little were also higher and flowed over a greater elevation range than advancing ones. Receding glaciers all had northerly aspects, whereas those that advanced or did not change had no dominant aspect.

After 1985, all glaciers throughout the region shrank (Fig. 7). The average rate of area change increased from $-0.05\pm0.10\,\%\,\mathrm{a}^{-1}$ between 1952 and 1985 to $-0.41\pm0.12\,\%\,\mathrm{a}^{-1}$ between 1985 and 2005. Rates of area change averaged $+0.03\pm0.18\,\%\,\mathrm{a}^{\ 1}$ for 1952–1970, $-0.08\pm0.13\,\%\,\mathrm{a}^{-1}$ for 1970–1985, and $-0.40\pm0.10\,\%\,\mathrm{a}^{-1}$ from 1985 to 2005. During the period 1985–2005, 30 glaciers receded, and the area of two glaciers did not change. Total surface area change in the period 1985–2005 was $-8.39\pm1.51\,\mathrm{km}^2$, or $-6.2\pm2.3\,\%$ (Table 3).

Extrapolating from the Bolch et al. (2010) inventory of 2005 ($731\,\mathrm{km}^2$) using relative area change of our 26-glacier subset and based on area loss per size class, we estimated that in 1952, the Cariboo Mountains contained a glacierized area of $824\pm39\,\mathrm{km}^2$. By 1970, glacier cover declined to $805\pm38\,\mathrm{km}^2$, and continued to shrink to $785\pm37\,\mathrm{km}^2$ by 1985.

We directly compare glacier extent and extent change (1985–2005) for 28 glaciers common to the Bolch et al. (2010) inventory and this study (Fig. 8). The Bolch et al. inventory relies on glacier extents mapped by the British Columbia Government Terrain Resource Inventory Management (TRIM) program from 1980s aerial photography and Landsat imagery from 2003 to 2007. The 2005 extents of this inventory, created through a semi-automated process, average 2 % larger than our extents manually digitized from aerial photographs with much higher resolution. However, the Bolch et al. (2010) inventory uses an area-weighted date (2005); the actual date of imagery covering the Cari-

Table 4. Median glacier morphometry for glaciers that underwent some advance, receded continuously, or exhibited no discernable change during the period 1952–1985.

	Count	Length (m)	Area (km^2)	Slope	Median Z (m)	Min. Z (m)	Max. Z (m)	Range Z (m)
Advance	9	1552	0.653	21.6	2449	2169	2733	617
Retreat	12	3032	3.640	15.2	2403	2075	2774	733
No Change	11	1949	1.406	20.8	2605	2210	2941	794

Table 5. Correlation table (Pearson-product moment) of relations of glacier area change with respect to glacier morphometry for three periods.

		Length	Area	Slope	Median Z	Min. Z	Max. Z	Range
1952–1985	Absolute Area Change	**−0.556**	**−0.650**	**0.598**	**0.358**	0.322	−0.065	−0.215
1952–1985	Relative Area Change	−0.235	−0.210	**0.518**	0.323	0.135	0.053	−0.046
1985–2005	Absolute Area Change	**−0.596**	**−0.765**	**0.523**	0.260	**0.513**	−0.193	**−0.393**
1985–2005	Relative Area Change	0.328	0.316	-0.060	0.282	**−0.364**	**0.412**	**0.431**
1952–2005	Absolute Area Change	**−0.597**	**−0.731**	**0.582**	0.312	**0.427**	−0.131	−0.309
1952–2005	Relative Area Change	−0.008	0.011	**0.440**	**0.487**	−0.087	0.310	0.221

Correlation coefficients in bold are significant ($p < 0.05$). $n = 32$ for the periods 1950s–1985 and 1985–2005. $n = 33$ for the period 1950s–2005.

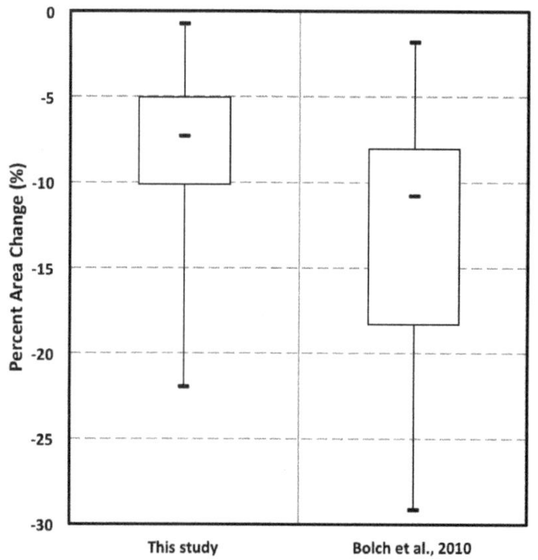

Figure 8. Box-and-whisker plots showing the maximum, interquartile range, median, and minimum percent area change for 28 glaciers from 1985 to 2005 for this study and that of Bolch et al. (2010).

boo Mountains is from 2006. Relative area change for the period 1985–2005 averaged $\sim 0.3\,\%\,a^{-1}$ for this 28-glacier subset, and thus the Bolch et al. (2010) inventory, based on 2006 imagery, will not significantly differ from our measurements from 2005 aerial photos. This provides evidence of the quality of the 2005 inventory and the precision of the method used by Bolch et al. (2010). The 1985 TRIM extents, however, digitized manually by the province of British Columbia from the same aerial photographs we use here, average 5 % larger than our extents. Late-lying snow is prevalent in the 1984 and 1985 photographs, but reference to additional photographs in years with a higher snow line helped us distinguish snowfields from glacierized area. The TRIM mapping included some late-lying snow as glacier, resulting in overestimation of 1985 extents in the TRIM data set, by approximately 5 %, and thus led to more surface-area loss reported in the Bolch et al. (2010) inventory than we find for the 28 glaciers of this comparison. However, the average change that we observe during the epoch 1985–2005 for our study and that of Bolch et al. (2010) is not statistically different.

Based on the extrapolation procedure described above, we estimate a net change of $-53\,km^2$ ($-6.8\,\%$) for the period 1985–2005 compared to $-114\,km^2$ ($-13.5\,\%$) of Bolch et al. (2010). We estimate Cariboo Mountains glacier extent to have been $785 \pm 37\,km^2$ in 1985, versus $845\,km^2$ reported by Bolch et al. (2010), a difference of 7.7 %. The Cariboo Mountains glacierized area for 1952 was $824 \pm 39\,km^2$, still 2.5 % less than the $845\,km^2$ of the 1985 TRIM extents used in previous work (Schiefer et al., 2007; Bolch et al., 2010). Our results thus demonstrate that glacier loss during the period 1985–2005 was higher than during previous epochs, but that the absolute surface area lost in this period may have been overestimated in previous studies.

The relation between area change and glacier morphometry temporally varies (Table 5). For three periods – 1952-1985, 1985-2005, and 1952-2005 – absolute area change correlates ($p < 0.05$) negatively with glacier length and area, indicating that the larger glaciers in absolute terms lost more surface area in all periods. The relation between relative area change and both length and area varies with time. For the

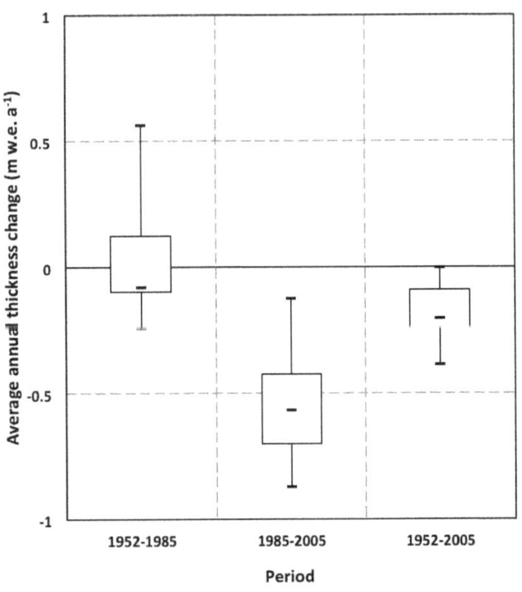

Figure 9. Box-and-whisker plots displaying maximum, interquartile range, median, and minimum of average annual thickness change in meters of water equivalent of seven glaciers for three periods.

period 1952–1985, we find r values of −0.235 and −0.210 for length and area, respectively, and these correlations for length and area change to 0.328 ($p < 0.10$) and 0.316 ($p < 0.10$) for the period 1985–2005. Large glaciers thus lost more relative area in the earlier period, whereas small glaciers lost more relative area in the latter period. Average glacier surface slope is correlated with absolute and relative area change in all periods, except for relative area change from 1985 to 2005, indicating steeper glaciers generally lost less absolute and relative surface area, except during the latter period when there was no correlation. Median glacier elevation is correlated with absolute area change for the period 1952–1985 ($r = 0.358$, $p < 0.05$) and relative area change for the period 1952–2005 ($r = 0.487$, $p < 0.05$), indicating that glaciers with more surface area at higher elevations experienced less area change in these two periods.

4.3 Volume change

Over the period 1952–2005, the seven glaciers of our study lost $-0.480 \pm 0.051\,\mathrm{km}^3$ w.e. Volume loss during the first period (1952–1985) totaled $-0.195 \pm 0.059\,\mathrm{km}^3$ w.e., and this loss accelerated to $-0.345 \pm 0.048\,\mathrm{km}^3$ w.e. during the later period (1985–2005). These volume losses correspond to respective mean thinning rates of $-0.14 \pm 0.04\,\mathrm{m\,w.e.\,a}^{-1}$ and $-0.50 \pm 0.07\,\mathrm{m\,w.e.\,a}^{-1}$ (Table 6 and Fig. 9). Net thinning rates over the period 1952 to 2005 averaged $-0.23 \pm 0.02\,\mathrm{m\,w.e.\,a}^{-1}$. Two glaciers, both of which advanced over the same period, markedly thickened between 1952 and 1985 (Table 6). Both of these glaciers thinned after 1985, but experienced little net change from 1952 to 2005. Five glaciers continuously thinned and receded.

Table 6. Average annual thickness change in meters of water equivalent.

Glacier Id	Thickness change 1952-1985 ($\mathrm{m\,w.e.\,a}^{-1}$)	Thickness change 1985-2005 ($\mathrm{m\,w.e.\,a}^{-1}$)	Thickness change 1952-2005 ($\mathrm{m\,w.e.\,a}^{-1}$)
1	0.56 ± 0.15	-0.87 ± 0.15	-0.04 ± 0.06
11	0.27 ± 0.07	-0.57 ± 0.08	-0.00 ± 0.04
19	-0.09 ± 0.16	-0.82 ± 0.22	-0.39 ± 0.14
24	-0.10 ± 0.22	-0.12 ± 0.22	-0.20 ± 0.18
26	-0.02 ± 0.25	-0.41 ± 0.30	-0.14 ± 0.21
28	-0.08 ± 0.06	-0.59 ± 0.08	-0.22 ± 0.04
31	-0.24 ± 0.20	-0.44 ± 0.23	-0.27 ± 0.17
Mean	-0.14 ± 0.04	-0.50 ± 0.07	-0.23 ± 0.02

Extrapolation of these thinning rates to unmeasured glaciers in the Cariboo Mountains yields volume losses of -4.027 ± 2.044 and $-7.580 \pm 1.167\,\mathrm{km}^3$ w.e. for the periods 1952–1985 and 1985–2005, respectively. We obtain lower estimates of volume change (-2.312 ± 2.170 and $-6.242 \pm 1.315\,\mathrm{km}^3$ w.e) using elevation-averaged surface-elevation change measurements from these seven glaciers.

4.4 Relation to climate

Surface temperature increased during the period of study, particularly after 1985 (Fig. 10). From the earlier (1952–1985) to the latter epoch (1986–2005), average temperatures increased by +0.38 and +0.87 °C for the ablation and accumulation season, respectively.

Accumulation season and annual precipitation decreased during the period of study, but little change occurred during the ablation season (Fig. 11). From the earlier (1952–1985) to the latter period (1986–2005), accumulation-season and annual precipitation decreased by −32 mm (−3.2 %) and −44 mm (−3.0 %), respectively, minor changes that are overwhelmed by large interannual variability.

Geopotential height (700 hPa) anomalies, based on 1952–2005 mean fields, reveal an area of persistent low pressure in western Canada during the accumulation and ablation seasons for the periods 1952–1970 and 1971–1985 (Fig. 12). These low-pressure anomalies were centered over the Yukon Territory and northern BC from 1952 to 1970. These areas of low pressure weakened during the period 1971–1985 with a prominent trough along the Coast Mountains (ablation season) and a diffuse area of anomalously low pressure over much of central North America (accumulation season). A marked change in circulation occurred for the later period with anomalously high pressure present during the period 1986–2005, with ridges over south central Alaska (ablation season) and off the coast of northern BC and southeastern Alaska (accumulation season).

These synoptic conditions broadly accord with the Pacific Decadal Oscillation (PDO; Mantua et al., 1997), a pattern of

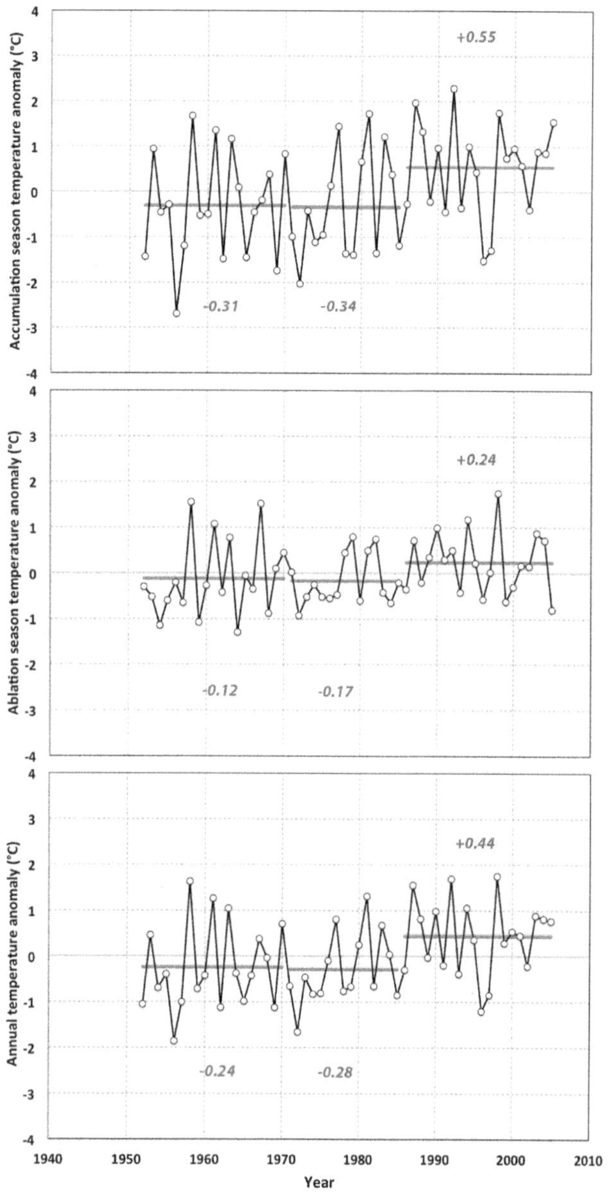

Figure 10. ClimateWNA temperature records of deviation from the long-term (1952–2005) mean of **(a)** average accumulation-season temperature, **(b)** average ablation season temperature, and **(c)** average annual temperature. Gray bars and values indicate averages for the periods 1952–1970, 1970–1985, and 1985–2005.

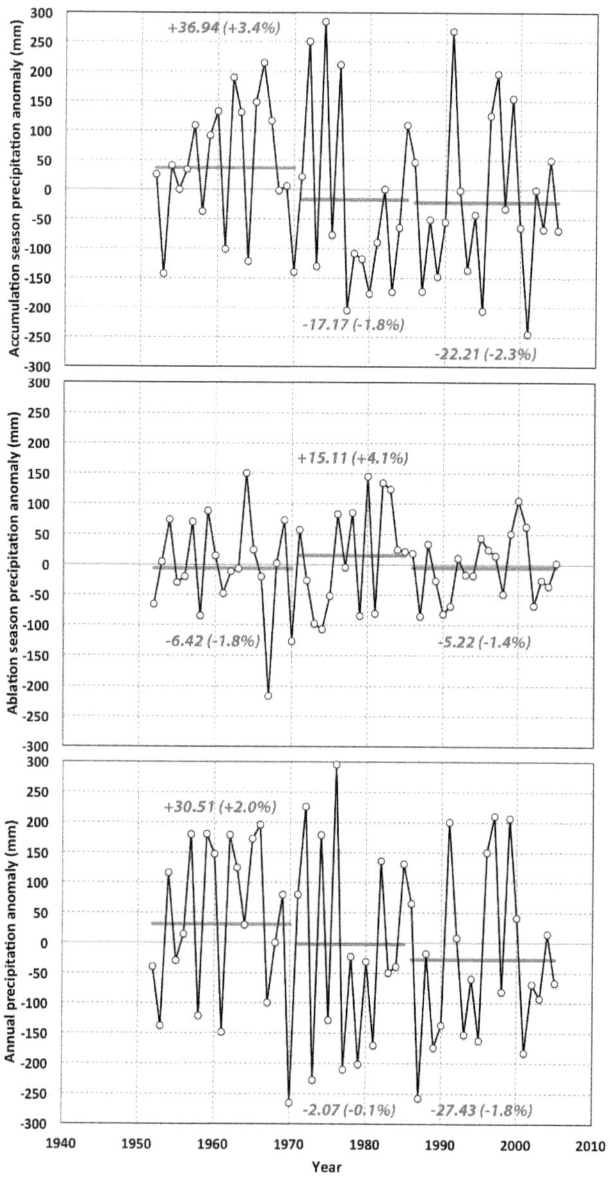

Figure 11. ClimateWNA precipitation records of deviation from the long-term (1952–2005) mean of **(a)** total accumulation-season precipitation, **(b)** total ablation season precipitation, and **(c)** total annual precipitation. Gray bars and values indicate averages for the periods 1952–1970, 1970–1985, and 1985–2005.

Pacific climate variability that persists for multiple decades and plays a prominent role in determining glacier mass balance in northwest North America (e.g., Bitz and Battisti, 1999). The earlier two periods (1950–1970 and 1971–1985) were largely or partially coincident with a cool phase of the PDO from 1947 to 1976. In contrast, the latest period (1985–2005) coincided with a warm phase of the PDO from 1977 to at least the mid-1990s.

5 Discussion

5.1 Regional representativeness

Our glacier subset represents a compromise between regional representativeness and the availability of suitable aerial photography. We oversampled large glaciers as they will contribute most to volume change for a given region. While glaciers of the 1–5 km^2 size class dominated volume change in the Cariboo Mountains for the period 1985–2005, losses from large glaciers ($> 5\,km^2$) were even more important for

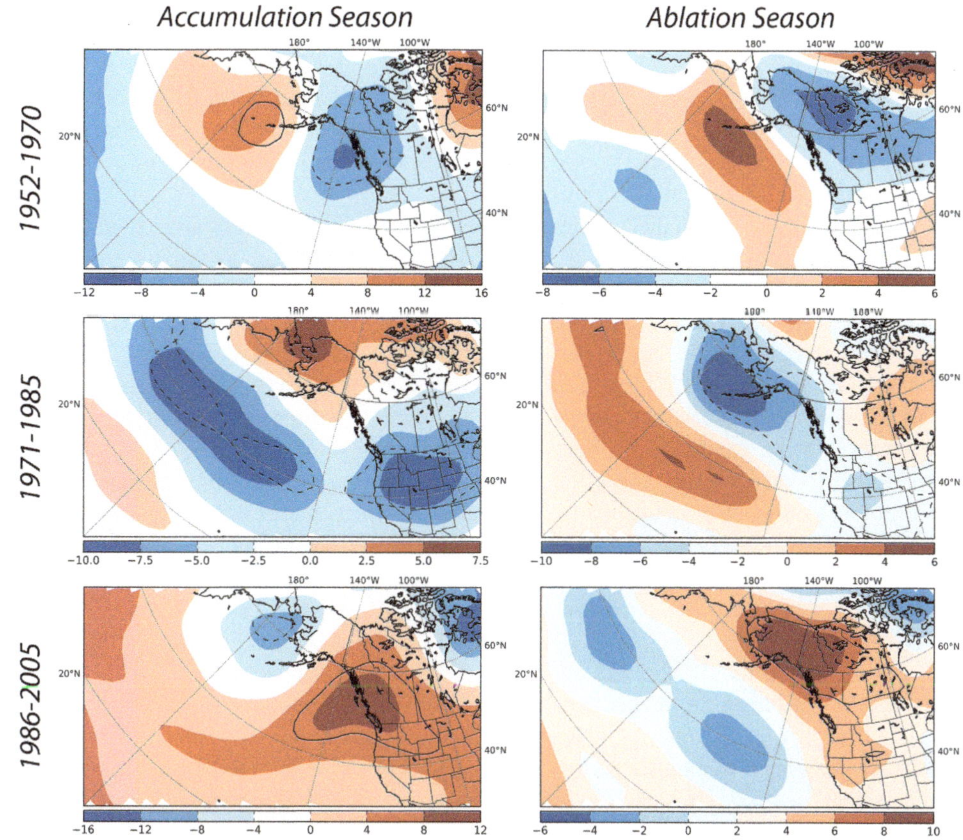

Figure 12. Geopotential height (700 hPa) anomalies (m) for the accumulation and ablation seasons. Color scale shows magnitude of anomaly. Data for the period 1952–2005 define the mean fields. Dashed and solid contours respectively denote negative and positive anomalies that are significantly different ($p < 0.05$) from the mean at a given grid point. Reanalysis data were downloaded from the National Oceanographic and Atmospheric Administration's Earth System Research Laboratory (http://www.esrl.noaa.gov/psd/data/gridded/tables/monthly.html).

previous epochs. The two largest glaciers for which we calculated volume change, for example, accounted for 94 % of the total volume change for seven glaciers from 1952 to 1985, and 75 % from 1985 to 2005. Selection of a representative subset of glaciers for a region is difficult, especially when glacier change is assessed over multiple epochs.

Most studies of glacier change note increased scatter of percent area change for smaller glaciers (Serandbrei-Barbero et al., 1999; Kääb et al., 2001; Paul, 2002; Paul et al., 2004; DeBeer and Sharp, 2007; Andreassen et al., 2008; Bolch et al., 2010; Paul and Andreassen, 2009). This scatter may arise from the influence of local topographic factors (DeBeer and Sharp, 2007; Paul and Andreassen, 2009). Kääb et al. (2001) concluded that this scatter could also be the result of using low-resolution satellite imagery, or from the fast response times of small glaciers. Late-lying seasonal snow may also impact the inferred relative area change in small glaciers, and could play a role in the observation of increased scatter for smaller glaciers. To assess this factor we compared the standard deviation in relative area change in the smallest glaciers as measured for 28 glaciers in both Bolch et al. (2010) and in our study, which, based on our analysis, is less affected by late-lying seasonal snow cover. The standard deviation of relative area change from 1985 to 2005 for the Bolch et al. (2010) inventory is 8.1 % for the smallest glaciers ($< 1.0\,\mathrm{km}^2$), whereas in our study it is 5.8 %. This difference is significant (F test, $p < 0.05$); it supports our hypothesis that late-lying seasonal snow plays a role in observations of increased scatter for smaller glaciers, and that the TRIM mapping of 1985 glacier extents errantly mapped seasonal snow cover as glacierized area.

5.2 Area change

Our results indicate net recession of Cariboo Mountains glaciers from 1952 to 2005, consistent with other studies of glacier change in BC (DeBeer and Sharp, 2007; Brewis, 2012; Tennant et al., 2012; Tennant and Menounos, 2013). While net glacier change in our 33-glacier subset over this period is negative ($-15.11 \pm 1.89\,\mathrm{km}^2$ or $-10.6 \pm 2.9\,\%$), recession for 11 of these glaciers did not exceed measurement error. Nine of these 11 glaciers advanced during the period 1952–1985.

Our finding of glacier advance in the period 1952–1985 accords with the advance in the Premier Range described by Luckman et al. (1987) and with glacier expansion elsewhere in BC during this period (Koch et al., 2009; Menounos et al., 2009; Brewis, 2012; Tennant and Menounos, 2013). Studies that only determine net recession over a multi-decadal period (e.g., DeBeer and Sharp, 2007, 2009; Jiskoot et al., 2009) may not capture minor advances like those of the present study.

During the period 1985–2005, glaciers shrank (mean relative area change per annum) 8 times faster than the period 1952–1985, and 5 times faster than 1970–1985 (Fig. 7). Annual recession of Castle Creek Glacier (glacier 28) increased from about $-10\,\mathrm{m\,a^{-1}}$ in the early to mid-1980s to about $-40\,\mathrm{m\,a^{-1}}$ in the early 1990s (Beedle et al., 2009). Tennant and Menounos (2013) also found increased recession rates for glaciers of the Columbia Icefield of BC and Alberta, which increased after 1979 and again after 2000.

5.3 Thickness change

Observed thinning rates during the period 1952–2005 averaged $-0.23 \pm 0.12\,\mathrm{m\,w.e.\,a^{-1}}$, resulting in a net loss of $-0.48 \pm 0.05\,\mathrm{km^3}$ w.e. (Table 6). This rate of thinning is considerably less then long-term thinning rates for glaciers of the Columbia Icefield ($-0.6 \pm 0.3\,\mathrm{m\,w.e.\,a^{-1}}$; Tennant and Menounos, 2013) and Alaskan glaciers ($-0.48 \pm 0.10\,\mathrm{m\,w.e.\,a^{-1}}$; Berthier et al., 2010). However, our average thinning rate is similar to those reported for nine North America glaciers for the period from the mid-1950s to the mid-1990s, which range from +0.01 to −0.61 and average $-0.21\,\mathrm{m\,w.e.\,a^{-1}}$ (Sapiano et al., 1998).

From 1952 to 1985, thinning rates averaged $-0.14 \pm 0.16\,\mathrm{m\,w.e.\,a^{-1}}$ and glaciers lost $-0.20 \pm 0.06\,\mathrm{km^3}$ w.e. The two smallest glaciers of our study thickened and advanced over the same period. Brewis (2012) also found that four of six glaciers in the Canoe River basin (Fig. 1) thickened and advanced during the period 1955–1970.

The post-1985 accelerated thinning of Cariboo Mountains glaciers agrees with other studies of glacier change in northwest North America (e.g., Rasmussen and Conway, 2004; Brewis, 2012; Tennant and Menounos, 2013). Schiefer et al. (2007) found that over the period 1985–1999, and using different density assumptions, glaciers of the Cariboo Mountains thinned at a rate of $-0.58\,\mathrm{m\,w.e.\,a^{-1}}$ and lost $-7.06\,\mathrm{km^3}$ w.e. Using our two different methods of extrapolation described in Sect. 3.3, our estimates yield comparable estimates of ice loss ($-6.24 \pm 1.32\,\mathrm{km^3}$ and $-7.58 \pm 1.17\,\mathrm{km^3}$) for the Cariboo Mountains.

Unlike the period 1985–2005, our methods of extrapolation to Cariboo Mountains glaciers yield significantly different volume change estimates for the period 1952–1985 ($-4.03 \pm 2.04\,\mathrm{km^3}$ versus $-2.31 \pm 2.17\,\mathrm{km^3}$). This discrepancy could arise from the time-varying magnitude of ice dynamics leading to extrapolation from a small subset performing poorly during epochs with higher surface mass balance and where some glaciers advanced. We recommend further study of extrapolation of regional volume change from a subset of glaciers, but with a more robust sample size to better constrain regional variability of surface-elevation change.

5.4 Relations of extent change to glacier morphometry

Glacier surface area is the single morphometric parameter that commonly correlates with extent change (Serandbrei-Barbero et al., 1999; Hoelzle et al., 2003; Paul et al., 2004; Andreassen et al., 2008; Bolch et al., 2010; Paul and Andreassen, 2009). From 1985 to 2005, we find average relative area change of −11.2, −9.3, −7.5, −5.4, and −5.4 % for our five size classes, arranged from smallest to largest (Table 3). Surface area and relative area change during the period 1985–2005 are weakly correlated ($p < 0.1$), indicating generally more relative area change for smaller glaciers (Table 5). However, from 1952 to 1985, a period with reduced rates of recession or advance, this does not apply; average relative area change was −2.5, +3.4, −4.6, −9.0, and −4.1 % for the same size classes, and the r value (not significant) is negative. It appears that the magnitude of glacier change is not simply related to dimensional attributes of a given glacier; they apparently vary with differing climatic regimes.

Studies have typically found no significant correlation among other morphometric parameters and area change (Paul, 2002; Granshaw and Fountain, 2006; Andreassen et al., 2008; Bolch et al., 2010; Paul and Andreassen, 2009). Other parameters include slope, aspect, and median, minimum, and maximum elevation. We, however, find correlation between slope and absolute area in all periods. Slope correlates with relative area change from 1952 to 1985, and on a net basis from 1952 to 2005, but not for the latter period 1985–2005 (Table 5). This relation, indicating that the steeper glaciers lost less absolute and relative surface area, may be spurious, however, as slope is highly correlated with glacier surface area and length (e.g., Hoelzle et al., 2003).

We also find that median, minimum, and maximum glacier elevations (along with elevation range) correlate with area change, but the strength of these associations is not constant through time (Table 5). A study with a larger sample size is required to assess whether the temporal changes we observed in this study are present for other mountain ranges, and how time-variable glacier dynamics might obscure relations between extent change and glacier geometry.

5.5 Relations to climate

Mean annual and seasonal temperatures increased from 1952 to 2005, particularly after 1985, whereas wintertime precipitation slightly decreased (Figs. 10 and 11). Observed mass and area loss coincided with intervals of increased (decreased) temperatures (precipitation), a finding in agreement with previous studies for glaciers in the Cariboo Mountains (Luckman et al., 1987; Brewis, 2012). Previous studies implicate increases in surface air temperatures as the recent cause for accelerated mass loss of North American glaciers in maritime environments (Rasmussen and Conway, 2004; Arendt et al., 2009). Tennant and Menounos (2013) also suggest that decreases in precipitation contribute to mass loss of the Columbia Icefield of the Canadian Rocky Mountains.

There was negligible change in surface temperatures from 1952–1970 to 1971–1986, but following 1985, climate warmed in the Cariboo Mountains (Fig. 10). After 1985, temperatures increased during the ablation season and doubled during the accumulation season.

Precipitation consistently decreased over the three periods of analysis for the accumulation season and annually (Fig. 11). Decreased annual precipitation is primarily an accumulation-season phenomenon, whereas average ablation season precipitation negligibly changed. Cool accumulation-season temperatures from 1971 to 1985 may have counteracted decreased precipitation, leading to snow accumulation comparable to that of the period 1952–1970. Some, but not all, glaciers thickened and advanced between 1952 and 1985, a result of both individual glacier hypsometry leading to more retained accumulation, and length and slope angle that led to a more rapid terminus response.

ClimateWNA records indicate that wet, cool periods coincide with anomalous low-pressure, whereas dry and warm intervals were associated with high-pressure anomalies (Fig. 12). Others note high-pressure anomalies over northwest North America after the late 1970s with coincident glacier recession and thinning (e.g., Bitz and Battisti, 1999; Hodge et al., 1998; Rasmussen and Conway, 2004). Glacier change in the Cariboo Mountains and its relation to broad scale changes in atmospheric circulation broadly accord with these previous studies, and, importantly, our work expands the spatial domain of these climate anomalies and how they influenced glacier change during the mid-20th to early 21st centuries.

Previous studies suggest that glaciologists need to further examine intra- and inter-regional mass-balance variability (Braithwaite, 2002), and conclude that there are no means of making an a priori selection of a regionally representative benchmark glacier (Fountain et al., 2009). The paucity of traditional mass-balance and glacier length change measurements limits assessments of regional representativeness. Widespread use of remotely sensed imagery to derive glacier extents and surface elevation, however, provides one method to document changes of glacier extent and volume change on decadal timescales. We demonstrate that comprehensive glacier inventories and assessments of decadal area and thickness change can be used to make an informed a priori selection of a representative subset of glaciers. Regional representativeness will be variable with respect to glacier morphometry, climate and glacier response, however. Selection of benchmark glaciers and representative glacier subsets, as well as testing of the regional representativeness of existing benchmark glaciers (e.g., Fountain et al., 2009), should be based on regional glacier morphometry, recent decadal glacier extent and thickness change, and with respect to varying glacier response in different climatic regimes.

6 Conclusions

During the period 1952–2005, glaciers of the Cariboo Mountains receded at a rate of $0.19 \pm 0.05\,\%\,a^{-1}$. Slow glacier recession rates characterized the epoch 1952–1985 ($-0.05 \pm 0.10\,\%\,a^{-1}$), and some glaciers advanced. After 1985, glacier recession rates increased 8-fold ($-0.41 \pm 0.12\,\%\,a^{-1}$). Thinning rates also accelerated from -0.143 ± 0.043 m w.e. a^{-1} (1952–1985) to -0.500 ± 0.070 m w.e. a^{-1} from 1985 to 2005. Temperatures increased from 1952 to 2005, primarily after 1985, whereas accumulation-season precipitation decreased over the period of study, indicating that both increases in temperature and decreases in precipitation played important roles in forcing the observed pattern of glacier change.

The inventory of Bolch et al. (2010) marginally overestimated the recession of Cariboo Mountains glaciers from 1985 to 2005. For 28 common glaciers, our extents average 2 % smaller than the 2005 extents generated with the semi-automated method of Bolch et al. (2010) and 5 % smaller than the 1985 extents of the TRIM data set, likely due to mapping of late-lying seasonal snow as glacierized area in the TRIM data set. When uncertainties are taken into account, however, the average difference between our subset and that of Bolch et al. (2010) is not significant.

Our results indicate that relations between glacier change and dimensional attributes of the glaciers of this study are not stable through time; this non-stationarity may be related to ice dynamics and the complex way glaciers respond to changes in surface mass balance.

Acknowledgements. We acknowledge research funding provided through the Western Canadian Cryospheric Network (from the Canadian Foundation for Climate and Atmospheric Sciences), the Natural Sciences and Engineering Research Council of Canada (NSERC), the Pacific Institute for Climate Solutions (Graduate Fellowship to M. Beedle), and the University of Northern British Columbia (graduate scholarship to M. Beedle). We also thank the province of British Columbia and Natural Resources Canada for access to aerial photography, Caillin Smith and Tyler Sylvestre for photogrammetric scanning, and Cardinal Systems for providing Vr Mapping software licenses and support. Stephen Déry, Peter Jackson, and Andrew Fountain provided comments that significantly improved our analysis and manuscript. We are appreciative of detailed reviews by Heidi Escher-Vetter and Chris DeBeer, and the efforts of editor Andreas Kääb.

Edited by: A. Kääb

References

Abermann, J., Lambrecht, A., Fischer, A., and Kuhn, M.: Quantifying changes and trends in glacier area and volume in the Austrian Ötztal Alps (1969-1997-2006), The Cryosphere, 3, 205–215, doi:10.5194/tc-3-205-2009, 2009.

Andreassen, L. M., Paul, F., Kääb, A., and Hausberg, J. E.: Landsat-derived glacier inventory for Jotunheimen, Norway, and deduced glacier changes since the 1930s, The Cryosphere, 2, 131–145, doi:10.5194/tc-2-131-2008, 2008.

Arendt, A. A., Echelmeyer, K. A., Harrison, W. D., Lingle, C. S., and Valentine, V. B.: Rapid wastage of Alaska glaciers and their contribution to rising sea level, Science, 297, 382–385, 2002.

Arendt, A., Echelmeyer, K., Harrison, W., Lingle, C., Zirnheld, S., Valentine, V., Ritchie, B., and Druckenmiller, M.: Updated estimates of glacier volume changes in the western Chugach Mountains, Alaska, and a comparison of regional extrapolation methods, J. Geophys. Res., 111, F03019, doi:10.1029/2005JF000436, 2006.

Arendt, A., Walsh, J., and Harrison, W.: Changes of glaciers and climate in northwestern North America during the late 20th century, J. Climate, 22, 4117–4134, 2009.

Bahr, D. B. and Radic, V.: Significant contribution to total mass from very small glaciers, The Cryosphere, 6, 763–770, doi:10.5194/tc-6-763-2012, 2012.

Barrand, N. E., Murray, T., James, T. D., Barr, S. L., and Mills, J. P.: Optimizing photogrammetric DEMs for glacier volume change assessment using laser-scanning derived ground-control points, J. Glaciol., 55, 106–116, 2009.

Barrand, N. E., James, T. D., and Murray, T.: Spatio-temporal variability in elevation changes of two high-Arctic valley glaciers, J. Glaciol., 56, 771–780, 2010.

Beedle, M. J., Menounos, B., Luckman, B. H., and Wheate, R.: Annual push moraines as climate proxy, Geophys. Res. Lett., 36, L20501, doi:10.1029/2009GL039533, 2009.

Beedle, M. J., Menounos, B., and Wheate, R.: An evaluation of mass balance techniques applied to Castle Creek Glacier, British Columbia, Canada, J. Glaciol., 60, 262–276, 2014.

Berthier, E., Schiefer, E., Clarke, G. K. C., Menounos, B., and Rémy, F.: Contribution of Alaskan glaciers to sea-level rise derived from satellite imagery, Nat. Geosci., 3, 92–95, doi:10.1038/NGEO737, 2010.

Bitz, C. M. and Battisti, D. S.: Interannual to decadal variability in climate and the glacier mass balance in Washington, western Canada, and Alaska, J. Climate, 12, 3181–3196, 1999.

Bogdal, C., Schmid, P., Zennegg, M., Anselmetti, F. S., Scheringer, M., and Hunberbühler, K.: Blast from the past: melting glaciers as a relevant source for persistent organic pollutants, Environ. Sci. Technol., 43, 8173–8177, 2009

Bolch, T., Menounos, B., and Wheate, R.: Landsat-based inventory of glaciers in western Canada, 1985–2005, Remote Sens. Environ., 114, 127–137, 2010.

Braithwaite, R. J.: Glacier mass balance: the first 50 years of international monitoring, Prog. Phys. Geog., 26, 76–95, 2002.

Brewis, T. B.: Glacier change and contribution to streamflow in the Canoe Basin, British Columbia, 1948–2005 (Master's Thesis), University of Northern British Columbia, 2012.

British Columbia (BC) Ministry of Environment: Specifications for aerial triangulation. Surveys and Resource Mapping, Province of British Columbia, Victoria, BC, 1998.

Daly, C., Gibson, W. P., Taylor, G. H., Johnson, G. L., and Pasteris, P.: A knowledge-based approach to the statistical mapping of climate, Clim. Res., 22, 99–113, 2002.

DeBeer, C. M. and Sharp, M. J.: Recent changes in glacier area and volume within the southern Canadian Cordillera, Ann. Glaciol., 46, 215–221, 2007.

DeBeer, C. M. and Sharp, M. J.: Topographic influences on recent changes of very small glaciers in the Monashee Mountains, British Columbia, Canada, J. Glaciol., 55, 1–10, 2009.

Déry, S. J., Clifton, A., MacLeod, S., and Beedle, M. J.: Blowing snow fluxes in the Cariboo Mountains of British Columbia, Canada. Arct. Antarct. Alp. Res., 42, 188–197, doi:10.1657/1938-4246-42.2.188, 2010.

Environment Canada: 1981–2010 Climate Normals & Averages. Meteorological Service of Canada, Environment Canada, available at: http://climate.weather.gc.ca/climate_normals/), 2012.

Fountain, A. G., Hoffman, M. J., Granshaw, F., and Riedel, J.: The "benchmark glacier" concept – does it work? Lessons from the North Cascade Range, USA, Ann. Glaciol., 50, 163–168, 2009.

Gardner, A. S., Moholdt, G., Wouters, B., Wolken, G. J., Burgess, D. O., Sharp, M. J., Cogley, J. G., Braun, C., and Labine, C.: Sharply increased mass loss from glaciers and ice caps in the Canadian Arctic Archipelago, Nature, 473, 357–360, doi:10.1038/nature10089, 2011.

Granshaw, F. D. and Fountain, A. G.: Glacier change (1958–1998) in the North Cascades National Park Complex, Washington, USA, J. Glaciol., 52, 251–256, 2006.

Hodge, S. M., Trabant, D. C., Krimmel, R. M., Heinrichs, T. A., March, R. S., and Josberger, E. G.: Climate variations and changes in mass of three glaciers in western North America, J. Climate, 11, 2161–2179, 1998.

Hoelzle, M., Haeberli, W., Dischl, M., and Peschke, W.: Secular glacier mass balances derived from cumulative glacier length changes, Global Planet. Change, 36, 295–306, 2003.

Hood, E., Fellman, J., Spencer, R. G. M., Hernes, P. J., Edwards, R., D'Amore, D., and Scott, D.: Glaciers as a source of ancient and labile organic matter to the marine environment, Nature, 462, 1044–1047, doi:10.1038/nature08580, 2009.

Huss, M.: Density assumptions for converting geodetic glacier volume change to mass change, The Cryosphere, 7, 877–887, doi:10.5194/tc-7-877-2013, 2013.

Jiskoot, H., Curran, C. J., Tessler, D. L., and Shenton, L. R.: Changes in Clemenceau Icefield and Chaba Group glaciers, Canada, related to hypsometry, tributary detachment, length-slope and area-aspect relations, Ann. Glaciol., 50, 133–143, 2009.

Kääb, A., Paul, F., Maisch, M., Hoelzle, M., and Haeberli, W.: The new remote sensing derived Swiss glacier inventory: II. First results, Ann. Glaciol., 34, 362–366, 2001.

Kääb, A. and Vollmer, M.: Surface geometry, thickness changes and flow fields on creeping mountain permafrost: automatic extraction by digital image analysis, Permafrost Periglac., 11, 315–326, 2000.

Kalnay, E. M., Kanamitsu, R., Kistler, W., Collins, W., Deaven, D., Gandin, L., Iredell, M., Saha, S., White, G., Woollen, J., Zhu, Y., Chelliah, M., Ebisuzaki, W., Higgins, W., Janowiak, J., Mo, K. C., Ropelewski, C., Wang, J., Leetmaa, A., Jenne, R., and Joseph, D.: The NCEP/NCAR 40-year reanalysis project, B. Am. Meteorol. Soc., 77, 3, 437–471, 1996.

Koblet, T., Gärtner-Roer, I., Zemp, M., Jansson, P., Thee, P., Haeberli, W., and Holmlund, P.: Reanalysis of multi-temporal aerial images of Storglaciären, Sweden (1959–99) – Part 1: Determination of length, area and volume changes, The Cryosphere, 4, 333–343, doi:10.5194/tc-4-333-2010, 2010.

Koch, J., Menounos, B., and Clague, J. J.: Glacier change in Garibaldi Provincial Park, southern Coast Mountains, British Columbia, since the Little Ice Age, Global Planet. Change, 66, 161–178, 2009.

Luckman, B. H., Harding, K. A., and Hamilton, J. P.: Recent glacier advances in the Premier Range, British Columbia, Can. J. Earth Sci., 24, 1149–1161, 1987.

Mantua, N. J., Hare, S. R., Zhang, Y., Wallace, J. M., and Francis, R. C.: A Pacific interdecadal climate oscillation with impacts on salmon production. B. Am. Meteorol. Soc., 78, 1069–1079, 1997.

Marshall, S. J., White, E. C., Demuth, M. N., Bolch, T., Wheate, R., Menounos, B., Beedle, M. J., and Shea, J. M.: Glacier water resources on the eastern slopes of the Canadian Rocky Mountains, Can. Water Resour. J., 36, 109–134, 2011.

Maurer, M. K., Menounos, B., Luckman, B. H., Osborn, G., Clague, J. J., Beedle, M. J., Smith, R., and Atkinson, N.: Late Holocene glacier expansion in the Cariboo and northern Rocky Mountains, British Columbia, Canada, Quaternary Sci. Rev., 51, 71–80, 2012.

Menounos, B., Osborn, G., Clague, J. J., and Luckman, B.: Latest Pleistocene and Holocene glacier fluctuations in western Canada, Quaternary Sci. Rev., 28, 2049–2074, 2009.

Mitchell, T. D. and Jones, P. D.: An improved method of constructing a database of monthly climate observations and associated high-resolution grids, Int. J. Climatol., 25, 693–712, 2005.

Moore, R. D., Fleming, S. W., Menounos, B., Wheate, R., Fountain, A., Stahl, K., Holm, K., and Jakob, M.: Glacier change in western North America: influences on hydrology, geomorphic hazards and water quality, Hydrol. Process., 24, 42–61, 2009.

Paul, F.: Changes in glacier area in Tyrol, Austria, between 1969 and 1992 derived from Landsat 5 Thematic Mapper and Austrian Glacier Inventory data, Int. J. Remote Sens., 23, 787–799, 2002.

Paul, F. and Andreassen, L. M.: A new glacier inventory for the Svartisen region, Norway, from Landsat ETM+ data: challenges and change assessment, J. Glaciol., 55, 607–618, 2009.

Paul, F. and Haeberli, W.: Spatial variability of glacier elevation changes in the Swiss Alps obtained from two digital elevation models, Geophys. Res. Lett., 35, L21502, doi:10.1029/2008GL034718, 2008.

Paul, F., Kääb, A., Maisch, M., Kellenberger, T., and Haeberli, W.: Rapid disintegration of Alpine glaciers observed with satellite data, Geophys. Res. Lett., 31, L21402, doi:10.1029/2004GL020816, 2004.

Radić, V. and Hock, R.: Regionally differentiated contribution of mountain glaciers and ice caps to future sea-level rise, Nat. Geosci., 4, 91–94, doi:10.1038/NGEO1052, 2011.

Rasmussen, L. A. and Conway, H.: Climate and glacier variability in western North America, J. Climate, 17, 1804–1815, 2004.

Rolstad, C., Haug, T., and Denby, B.: Spatially integrated geodetic glacier mass balance and its uncertainty based on geostatistical analysis: application to the western Svartisen ice cap, J. Glaciol., 55, 666–680, 2009.

Sapiano, J. J., Harrison, W. D., and Echelmeyer, K. A.: Elevation, volume and terminus changes in nine glaciers in North America, J. Glaciol., 44, 119–135, 1998.

Schiefer, E. and Gilbert, R.: Reconstructing morphometric change in a proglacial landscape using historical aerial photography and automated DEM generation, Geomorphology, 88, 167–178, 2007.

Schiefer, E., Menounos, B., and Wheate, R.: Recent volume loss of British Columbia glaciers, Canada, Geophys. Res. Lett., 34, L16503, doi:10.1029/2007GL030780, 2007.

Serandbrei-Barbero, R., Rabagliati, R., Binaghi, E., and Rampini, A.: Glacial retreat in the 1980s in the Breonie, Aurine and Pusteresi groups (eastern Alps, Italy) in Landsat TM images, Hydrol. Sci., 44, 279–296, 1999.

Tennant, C., Menounos, B., Wheate, R., and Clague, J. J.: Area change of glaciers in the Canadian Rocky mountains, 1919 to 2006, The Cryosphere, 6, 1541–1552, doi:10.5194/tc-6-1541-2012, 2012.

Tennant, C. and Menounos, B.: Glacier change of the Columbia Icefield, Canadian Rocky Mountains, 1919–2009, J. Glaciol. 59, 671–686, doi:10.3189/2013JoG12J135, 2013.

Wang, T., Hamann, A., Spittlehouse, D., and Murdock, T. N.: ClimateWNA – High-resolution spatial climate data for western North America, J. Appl. Meteorol. Clim., 61, 16–29, 2012.

12

Probabilistic parameterisation of the surface mass balance–elevation feedback in regional climate model simulations of the Greenland ice sheet

T. L. Edwards[1], X. Fettweis[2], O. Gagliardini[3,4], F. Gillet-Chaulet[3], H. Goelzer[5], J. M. Gregory[6,7], M. Hoffman[8], P. Huybrechts[5], A. J. Payne[1], M. Perego[9], S. Price[8], A. Quiquet[3], and C. Ritz[3]

[1]Department of Geographical Sciences, University of Bristol, Bristol BS8 1SS, UK
[2]Department of Geography, University of Liege, Laboratory of Climatology (Bat. B11), Allée du 6 Août, 2, 4000 Liège, Belgium
[3]Laboratoire de Glaciologie et Géophysique de l'Environnement, UJF – Grenoble 1/CNRS, 54, rue Molière BP 96, 38402 Saint-Martin-d'Hères Cedex, France
[4]Institut Universitaire de France, Paris, France
[5]Earth System Sciences & Departement Geografie, Vrije Universiteit Brussel, Pleinlaan 2, 1050 Brussels, Belgium
[6]NCAS-Climate, Department of Meteorology, University of Reading, Reading, UK
[7]Met Office Hadley Centre, Exeter, UK
[8]Fluid Dynamics and Solid Mechanics Group, Los Alamos National Laboratory, T3 MS B216, Los Alamos, NM 87545, USA
[9]Department of Scientific Computing, Florida State University, 400 Dirac Science Library, Tallahassee, FL 32306, USA

Correspondence to: T. L. Edwards (tamsin.edwards@bristol.ac.uk)

Abstract. We present a new parameterisation that relates surface mass balance (SMB: the sum of surface accumulation and surface ablation) to changes in surface elevation of the Greenland ice sheet (GrIS) for the MAR (Modèle Atmosphérique Régional: Fettweis, 2007) regional climate model. The motivation is to dynamically adjust SMB as the GrIS evolves, allowing us to force ice sheet models with SMB simulated by MAR while incorporating the SMB–elevation feedback, without the substantial technical challenges of coupling ice sheet and climate models. This also allows us to assess the effect of elevation feedback uncertainty on the GrIS contribution to sea level, using multiple global climate and ice sheet models, without the need for additional, expensive MAR simulations.

We estimate this relationship separately below and above the equilibrium line altitude (ELA, separating negative and positive SMB) and for regions north and south of 77° N, from a set of MAR simulations in which we alter the ice sheet surface elevation. These give four "SMB lapse rates", gradients that relate SMB changes to elevation changes. We assess uncertainties within a Bayesian framework, estimating probability distributions for each gradient from which we present best estimates and credibility intervals (CI) that bound 95 % of the probability. Below the ELA our gradient estimates are mostly positive, because SMB usually increases with elevation: 0.56 (95 % CI: −0.22 to 1.33) kg m^{-3} a^{-1} for the north, and 1.91 (1.03 to 2.61) kg m^{-3} a^{-1} for the south. Above the ELA, the gradients are much smaller in magnitude: 0.09 (−0.03 to 0.23) kg m^{-3} a^{-1} in the north, and 0.07 (−0.07 to 0.59) kg m^{-3} a^{-1} in the south, because SMB can either increase or decrease in response to increased elevation.

Our statistically founded approach allows us to make probabilistic assessments for the effect of elevation feedback uncertainty on sea level projections (Edwards et al., 2014).

1 Introduction

Over the past two decades the Greenland ice sheet (GrIS) has been losing mass at an increasing rate, on average $142 \pm 49\,\mathrm{Gt\,a^{-1}}$ with a total contribution to global sea level of about 8 mm (Shepherd et al., 2012). The GrIS has the potential to raise global sea level by several centimetres this century, and more in the next, with larger regional changes. The sensitivity of the GrIS to climate change is not well-known (IPCC, 2007), so it is important to improve estimates of its response and make projections of the resulting contribution to sea level over the next one to two centuries to inform policy and planning. Underestimating sea level rise would leave coastal cities around the globe at risk, while overestimating it could result in unwarranted expenditure on coastal defence. Projections should therefore include probabilistic assessments of uncertainty if they are to provide the most robust and complete information for making decisions.

Predictions of the GrIS response to projections of future climate change are made with physically based ice sheet models (ISMs) forced with climate model simulations. ISMs simulate both parts of ice sheet response: the flow of ice subject to its boundary conditions (dynamic); and surface mass balance (SMB), which is the sum of surface accumulation and surface ablation (broadly speaking, the balance of snowfall versus meltwater runoff). However, SMB models included in ISMs are usually rather simple. Most often they use an empirically derived positive degree-day (PDD) scheme, in which melting is parameterised as a function of the sum of daily air temperatures above melting point, and runoff is usually modelled from temperature and precipitation with a simple snow pack model (e.g. Janssens and Huybrechts, 2000). Daily climate means are often approximated from seasonal means to reduce the input data set size.

At the other end of the spectrum of model complexity are regional climate models (RCMs). These simulate the atmosphere and surface over a limited spatial domain, with higher spatial and temporal resolution than global climate models (GCMs), and are forced at their boundaries with GCM simulations or reanalysis data such as ERA-40. Some RCMs, such as MAR (Modèle Atmosphérique Régional: Fettweis, 2007) and RACMO2/GR (e.g. Ettema et al., 2009), include complex snow/ice schemes that represent many of the physical processes that govern SMB. Such RCMs have been shown to be quite successful in reproducing the current SMB of the GrIS (e.g. Ettema et al., 2009; Fettweis et al., 2011; Vernon et al., 2013). RCMs are computationally expensive so only short and/or a small number of simulations can be performed.

Some have suggested that PDD descriptions of ice sheet response are too sensitive to climate change (van de Wal, 1996; van de Berg, 2011). In contrast, comparisons made between RACMO2/GR and the Janssens and Huybrechts (2000) PDD model by Vernon et al. (2013) and Hanna et al. (2011) find the RCM is more sensitive. In an attempt to make the most robust comparison (e.g. using the same ice sheet extent and forcing from the same RCM), Goelzer et al. (2013) find that a PDD model underestimates sea level rise by 14–31 % compared to MAR. These large variations in response relative to RCMs may reflect the simplicity of the PDD scheme.

Ideally, then, we would prefer future projections of GrIS SMB to be made with the more complete representations in RCMs rather than simple parameterisations such as the PDD model (for example Rae et al., 2012; Fettweis et al., 2013). But the ice flow component of an ISM is still needed to simulate the dynamical response of the GrIS. ISMs are run at higher resolution than RCMs (kilometres rather than tens of kilometres), to better represent glacier flow at the ice sheet margin.

As the ice sheet evolves in response to climate change, it also affects the local climate through feedback processes. Some, like the ice albedo feedback, may be simulated within the RCM. Others relating to the dynamical response, including the evolving geometry of the ice sheet, can only be simulated by coupling the RCM and ISM, or else parameterising the feedback to adjust the input climate forcing throughout the simulation.

One important ice–climate feedback is the set of interactions between the atmosphere and the ice sheet surface elevation; here we focus on the feedback between the atmosphere and ice surface/snow pack. The two main parts of this SMB–elevation feedback are (i) temperature, where an initial increase in air temperature that leads to ice melting lowers the surface elevation and exposes the ice to warmer temperatures through the atmospheric lapse rate; and (ii) precipitation, where surface elevation changes affect air temperature and atmospheric circulation and therefore the location and amount of precipitation. Surface topography in RCMs is usually held constant, so they do not incorporate the elevation feedback at all. PDD schemes include a parameterisation of the temperature aspect of the feedback, using an atmospheric lapse rate to adjust the input temperature forcing as the ice sheet surface evolves. They do not represent the precipitation aspect of the feedback except, in some cases, through a scaling factor for temperature. Most PDD schemes assume constant feedbacks (temperature–elevation, i.e. atmospheric lapse rate correction; precipitation–elevation, i.e. scaling correction; and ice albedo) that do not vary across the ice sheet or with climate change (discussed by Robinson et al., 2010; Helsen et al., 2011; Stone et al., 2010), though there are exceptions (Tarasov and Peltier, 2002).

If we are to simulate SMB with an RCM and how that SMB is affected by ice topography changes (unlike Rae et al., 2012; Fettweis et al., 2013, who use RCMs with constant ice sheet topography), we must either couple an ISM to an RCM, or else force an ISM with RCM output using a parameterisation of the relationship in terms of an "SMB lapse rate". Coupling an ISM to an RCM or GCM is rarely done because it is technically challenging (one example is Ridley et al., 2005), and because the climate models, particularly RCMs, are too computationally expensive to simulate the timescales of long-term ice sheet response. The computational expense also drastically limits opportunities to perform multiple simulations to sample uncertainties in modelling choices.

The pragmatic solution is therefore to parameterise the SMB–elevation feedback. This allows us to explicitly simulate the SMB and dynamical responses without the technical challenges and substantial computational expense of coupling ISMs to RCMs. Provided the parameterisation adequately represents the feedback in MAR, this allows us to perform many simulations that we otherwise could not, because we can force ISMs with MAR that have not yet been coupled to it, and sample uncertainties in the feedback and ice sheet modelling with additional simulations that we would not otherwise have computational resources to perform.

Helsen et al. (2011) provide the first such parameterisation, for the relationship between SMB and height in RACMO2/GR, and use this to adjust the SMB forcing applied to an ISM. Franco et al. (2012) derive relationships between the individual components of SMB (snowfall, rainfall, meltwater runoff, and loss by sublimation and evaporation) and elevation changes in MAR, to correct low resolution SMB simulations onto a higher resolution ice sheet topography. Hakuba et al. (2012) study the SMB response to surface elevation changes in a version of the ECHAM5 GCM (Roeckner et al., 2003) by lowering the ice sheet topography to 75 %, 50 % and 25 % of the present day, though do not parameterise the relationship. We develop on these studies in method (presented here) and application (Edwards et al., 2014).

We derive a new parameterisation for the elevation feedback in MAR using a suite of simulations in which the MAR GrIS surface height is altered. The parameterisation is a set of four gradients that relate SMB changes to height changes. These can be used to adjust the input SMB forcing as the ice sheet geometry evolves (Edwards et al., 2014). The four gradients are used according to whether the adjusted mean SMB of the previous decade is positive or negative, and whether the grid cell is in the north or south of the ice sheet. Elevation feedback uncertainty can be sampled with different SMB lapse rates; with careful experimental design this can give a probabilistic assessment of the effect of elevation feedback uncertainty on sea level. ISM and GCM uncertainty can also be sampled by using different models. We present these results in a companion paper (Edwards et al., 2014).

2 Method

We derive the parameterisation from a set of MAR simulations in which the surface elevation is altered (Sect. 2.1). We try various choices for the parameterisation structure, judging them by their success in reproducing the SMB response in MAR and their flexibility and ease of implementation (Sect. 2.2). After deciding on the structure, we estimate probability distributions for the four SMB–elevation gradients (Sect. 2.3).

2.1 Climate simulations

The regional climate model MAR (Fettweis, 2007) has been adapted for simulating the climate over Greenland, with full coupling to a complex snow/ice model and relatively high horizontal resolution (25 km). MAR is one of the few RCMs (another is RACMO2/GR) that includes the positive feedback between ice surface albedo and melting (Fettweis, 2007), though this is only partially included because the ice sheet extent and elevation are constant (there is no change in the ice–tundra boundary). MAR has been shown to simulate GrIS SMB quite successfully (e.g. Fettweis et al., 2011).

We use a set of eight simulations, each 20 yr long, in which MAR is forced at the boundaries by the ECHAM5 GCM (Roeckner et al., 2003) under the SRES A1B emissions scenario (Nakićenović et al., 2000). Two are control simulations, using the default ice surface topography based on Bamber et al. (2001): they are the first two decades (2000–2019, t_1) and last two decades (2080–2099, t_2) of the MAR ECHAM5-A1B simulation described by Rae et al. (2012) and Fettweis et al. (2013). The other six are perturbation experiments, three for each time period, in which we alter the GrIS surface height. We use three types of height change: uniform lowering by 50 m ("−50 m"), uniform lowering by 100 m ("−100 m"), and NonUniform changes ("NonUn") derived from a GrIS simulation by Ridley et al. (2005). Ridley et al. (2005) couple GISM (Greenland Ice Sheet Model: Huybrechts and de Wolde, 1999) to the HadCM3 GCM (Gordon et al., 2000) so that the elevation feedback is included, and quadruple the atmospheric CO_2 concentrations from preindustrial values. We use the resulting GrIS surface height change after 140 yr, at which point the ice sheet has lost 10 % of its original volume. We interpolate these height changes from the GISM 20 km polar stereographic grid to the MAR grid, and add them to the default topography over ice sheet grid cells. The ice sheet area is not changed: no cells are changed from ice to tundra or vice versa. Any negative height values that result after applying the changes are set to zero, to avoid the ice surface being specified below the bedrock. Our analysis uses the mean of each two-decade simulation, over which the SMB time series is approximately stationary (Rae et al., 2012).

Figure 1 shows the default (control) topography and the height difference between the NonUn and control experiments. Figure 2 shows the SMB changes for the NonUn experiments and Fig. 3 the uniform height change experiments. These figures show that large decreases in elevation generally decrease SMB, due to increased melting and decreased snowfall (Franco et al., 2012). There are two main exceptions to this that arise from the complex effects of topography on local air circulation and precipitation. In the NonUn

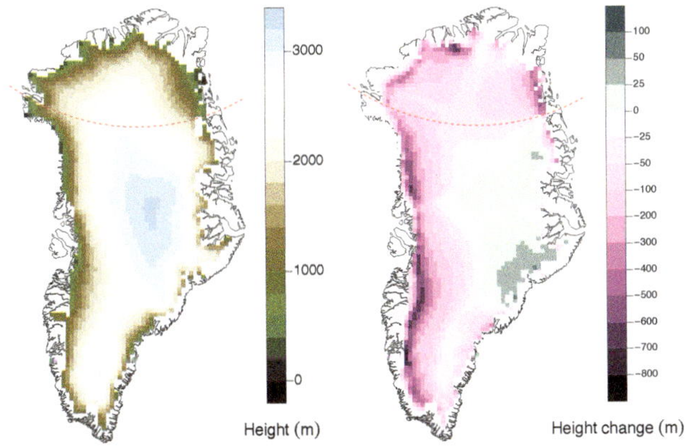

Fig. 1. Left: ice sheet surface elevation in the control experiments. Right: elevation change in the NonUn experiments (NonUn – control). Red dashed line is 77° N.

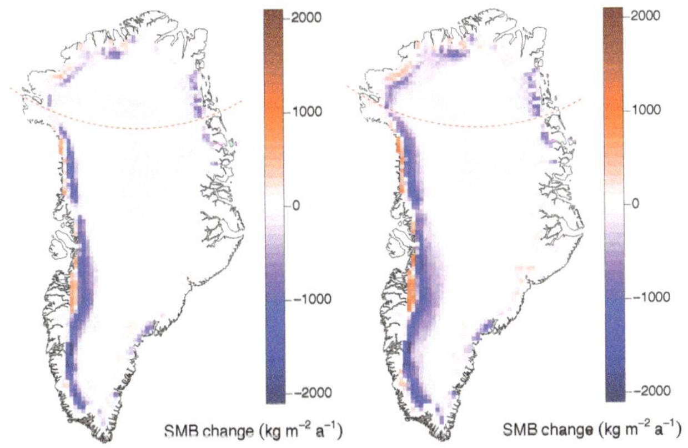

Fig. 2. Mean SMB change (perturbed minus control) in the NonUn experiments, 2000–2019 (left) and 2080–2099 (right). Red dashed line is 77° N.

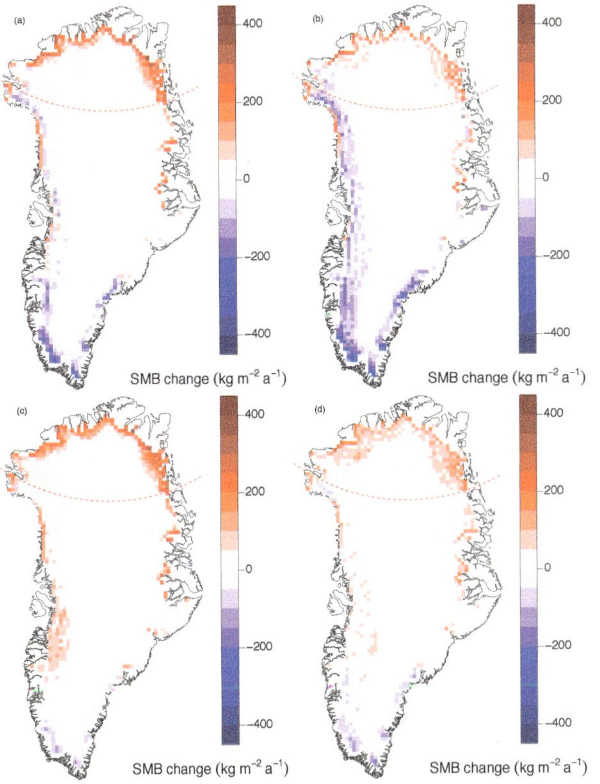

Fig. 3. Mean SMB change (perturbed minus control) in the −100 m (top row) and −50 m (bottom) experiments for the 2000–2019 (left column) and 2080–2099 (right) simulations. Red dashed line is 77° N.

experiments, there is an increase in SMB along the western ice sheet margin while there is a thinning of the ice sheet (Fig. 2). Here the lowering of the ice sheet surface dampens the "barrier wind" that brings warm air from the tundra along the ice sheet margin and enhances melting (van den Broeke and Gallée, 1996). In the uniform height change experiments, surface lowering can lead to either a decrease or increase in SMB (Fig. 3): a decrease in elevation exposes ice to warmer air temperatures, which can increase the moisture content of the air and enhance precipitation, but conversely an increase in elevation may cause air to rise and cool, also encouraging precipitation (Fettweis et al., 2005; Franco et al., 2012). These aspects show the importance of using a surface energy balance based RCM, rather than simpler models, to account for such phenomena. The consequences of this complexity for the parameterisation are discussed in Sect. 3.2.

Figure 4 shows SMB responses versus height changes for the two NonUn experiments, with arrows pointing from control to NonUn result, separated into regions north and south of latitude 77° N (this choice of latitude is explained later). The structure of the data is somewhat similar to that found by Helsen et al. (2011) for RACMO2/GR, with a broadly linear positive relationship below the equilibrium line altitude (ELA: the line at which SMB equals zero) and a negative, weaker relationship above the ELA. The behaviour is linear below the ELA within each time period because we use the simulation mean: in a constant climate, the average melting is approximately proportional to the average temperature, which is approximately proportional to elevation. The behaviour above the ELA, particularly south of 77° N (the majority of the ice sheet), is reminiscent of the complex relationship found between precipitation and height in MAR by Franco et al. (2012).

There is a clear offset between the beginning and end of the century. At a given height, particularly below the ELA, the SMB is lower in the warmer climate at the end of the century. This is partly due to the linear dependence of melting on local temperature in a constant climate (described above), but also to two mechanisms that accelerate the melting and runoff as the climate warms. The first is the positive ice albedo feedback. Bare ice appears each summer after the accumulated winter snowpack melts, and it has a lower

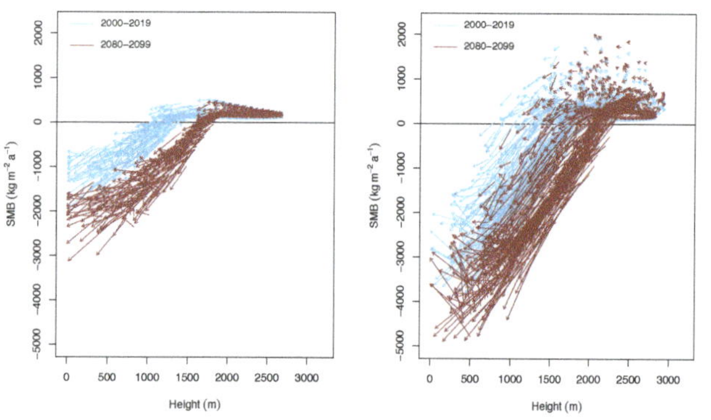

Fig. 4. Changes in SMB when perturbing the height of the MAR ice sheet from the control topography to the NonUn-based topography for grid cells north (left) and south (right) of 77° N. Arrows point from control to NonUn experiment. Data with height change $|\Delta h| < 25$ m are excluded.

albedo than snow. So in a warming climate the maximum area of bare ice (the ablation zone) increases, and a positive albedo feedback amplifies the warming. MAR has a more realistic, lower albedo for bare ice (around 0.45) than most RCMs and therefore a greater sensitivity to warming; Fettweis et al. (2013) estimate that surface melting increases exponentially with rising temperatures. The second mechanism is the type of precipitation falling on the ice sheet. In the latter part of the century most summer precipitation falls as rain rather than snow, and most of this runs off directly to the ocean rather than accumulating as ice. Both mechanisms accelerate the decrease in SMB as the A1B scenario progresses.

Figure 5 shows the SMB responses versus height changes for all of the perturbation experiments except NonUn t_2 (reserved as a test: Sect. 2.3), divided into four partitions of SMB (negative and positive) and region (north and south of 77° N). Each data point shows the SMB response ($\Delta S_i = S_i^{\mathrm{pert}} - S_i^{\mathrm{cont}}$) versus the height perturbation ($\Delta h_i = h_i^{\mathrm{pert}} - h_i^{\mathrm{cont}}$) for a given grid cell i, so each grid cell can appear up to five times. We exclude the 906 grid cells of the NonUn t_1 experiment that have $|\Delta h| < 25$ m (see Sect. 2.3). We also exclude cells in which the SMB crosses the ELA between the control and perturbed experiments (i.e. in which the perturbed and control SMB have opposite signs) to make distinct data sets for positive and negative SMB.

Most of the variation in Fig. 5 is from the NonUn simulation, because this has the widest range of height perturbations. The south has a steeper slope, a stronger relationship between ΔS and Δh, than the north.

The uniform perturbation experiments are the short vertical bands at $\Delta h = (-50\,\mathrm{m}, -100\,\mathrm{m})$. Most of the uniform experiment data in the south (bottom two subfigures of Fig. 5) show the behaviour we expect: when elevation decreases, SMB decreases (most points are in the bottom left quadrant, where both ΔS and Δh are negative; $\Delta S/\Delta h$ is therefore

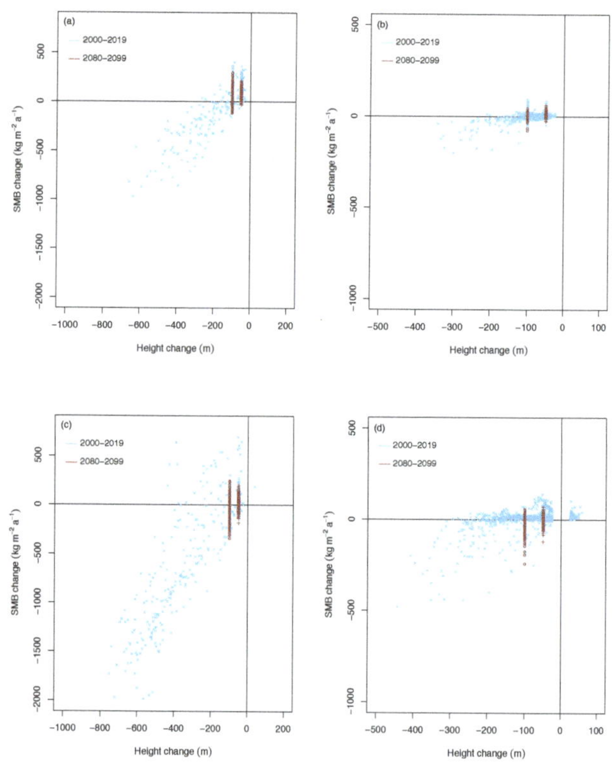

Fig. 5. Scatter plots of SMB change ΔS versus height change Δh for grid cells north (top row) and south (bottom) of 77° N, divided into grid cells with SMB in both the control and perturbed experiments less than zero (left column) and greater than or equal to zero (right). Data with height change $|\Delta h| < 25$ m are excluded.

positively signed). But in the north (top two subfigures), the response is the opposite: when elevation decreases, SMB increases (most points are in the top left quadrant, with positive ΔS and negative Δh; $\Delta S/\Delta h$ is negative). This change in response from north to south can be seen in the maps in Fig. 3, particularly for the -100 m experiment (top two subfigures) where the SMB change along the margin is positive (red) north of 77° N and mostly negative (blue) in the south. However, most of the uniform experiment data do lie within the range of the NonUn results.

2.2 Parameterisation structure

The parameterisation comprises four "SMB lapse rates", gradients that characterise a linear relationship between SMB change and surface elevation change. When testing the parameterisation, we use the gradients to adjust the control SMB using the NonUn height changes and compare with the actual NonUn SMB results. In a companion paper we use the parameterisation with several ice sheet models to dynamically adjust projections of future SMB as the GrIS shape evolves (Edwards et al., 2014). The four gradients correspond to the four possible combinations of the grid cell adjusted mean SMB over the past decade being positive or negative and the grid cell latitude being north or south of 77° N. We estimate these gradients from the ratios of SMB changes to height changes ($\Delta S/\Delta h$) in the surface elevation perturbation experiments. This parameterisation structure is determined by a combination of a priori choices and informal tests.

We choose the structure of our parameterisation with the following aims: to preserve as much of the SMB–elevation relationship in the MAR simulations as possible; to use as few assumptions as possible; to be applicable to any SMB forcing from MAR; and to be simple for the ice sheet modeller to implement. We test the ability of the parameterisation to reproduce the SMB field in the NonUn t_2 simulation when applied to the control t_2 simulation using the NonUn height changes.

We parameterise the relationship between elevation and mean SMB, using total SMB rather than its individual components (as in Franco et al., 2012) so it is easier to implement in ISMs and requires only one simulated variable as the input forcing. We also choose to parameterise changes in SMB as a function of changes in elevation (in common with Franco et al., 2012), rather than absolute values (as in Helsen et al., 2011). If we were to parameterise the relationship between absolute SMB S and absolute height h, using a linear model $S = a + bh$ (e.g. in Fig. 4), we would force the adjusted SMB to lie along a single line lying somewhere between the data from the two time periods t_1 and t_2, with large uncertainty in the intercept due to the climate dependence of SMB at a given height. Instead, we can parameterise the relationship between SMB changes and height changes, $\Delta S = b\Delta h$, estimating only the gradient b. This way SMB can be adjusted up or down the slope apparent in the data, rather than onto a single line with constant intercept. Eliminating the intercept in this way preserves the climate dependence of the SMB–elevation relationship in the MAR simulations, and removes half the unknown parameters. Working with anomalies rather than absolute values is also a standard approach in climate modelling, because the former are thought to be simulated more reliably than the latter.

The adjusted SMB is $S^{\mathrm{adj}} = S^{\mathrm{RCM}} + b\Delta h$, where S^{RCM} is the original SMB (kg m^{-2} a^{-1}), Δh the height change (m), and b the SMB–elevation gradient $b = \Delta S/\Delta h$ (kg m^{-3} a^{-1}). More specifically, for a given MAR grid cell in a year t a gradient b_t is used to adjust the control SMB S_t^{RCM} using the height difference between the NonUn and control experiments, $S_t^{\mathrm{adj}} = S_t^{\mathrm{RCM}} + b_t\,(h^{\mathrm{NonUn}} - h^{\mathrm{cont}})$. The gradient b_t is selected according to the "reference" SMB and latitude of the grid cell, where the reference is the mean of the adjusted SMB over the previous 10 yr (see Edwards et al., 2014, for more details). Using the adjusted SMB for the reference means the gradient selection evolves as part of the feedback, which helps to make the method more robust with changing climate.

Our height perturbation simulations allow us to derive the gradients directly from SMB responses to height changes for each grid cell, rather than the difference in SMB between grid cells in different locations on the ice sheet (as in Helsen et al., 2011). This is important because the SMB response may be determined by different physical processes due to local topography and atmospheric circulation patterns. Each grid cell i provides an estimate of the gradient $b = \Delta S/\Delta h$ from the SMB change (perturbed SMB minus control SMB, $\Delta S_i = S_i^{\mathrm{pert}} - S_i^{\mathrm{cont}}$) versus the elevation change (perturbed height minus control height, $\Delta h_i = h_i^{\mathrm{pert}} - h_i^{\mathrm{cont}}$). In Fig. 4 these correspond to the arrow slopes; in Fig. 5 they are the y axis values divided by the x axis values.

We choose not to make the gradients a function of grid cell location (Helsen et al., 2011; Franco et al., 2012), other than the north–south divide, to avoid dependence on the MAR grid resolution (Franco et al., 2012) and make the parameterisation as generic as possible. A spatially varying parameterisation would be tailored to the current shape of the ice sheet and the gradients would need to be interpolated for the ISM grid, which could lead to distorting edge effects at discontinuities such as the margin, ELA, and grid cell boundaries (e.g. Franco et al., 2012).

We do not make the gradients a function of climate or time (beginning versus end of the century), because this would restrict our ability to apply the parameterisation to other MAR simulations: for the missing years of the A1B scenario (2020–2079) we could interpolate or otherwise scale the results, but this would be less reliable or applicable for other emissions scenarios and simulations forced by other GCMs.

To guide our choices for other aspects of the parameterisation structure, we consider various methods of estimating and applying the gradients and quantify their relative success in reproducing SMB changes in one of the perturbation experiments. We estimate the gradients from the SMB responses in the two NonUn simulations, then use them to adjust the SMB in the control t_2 experiment according to the NonUn height changes. We quantify success by comparing the parameterised cumulative SMB change with the actual results in the NonUn t_2 simulation, in terms of both the root mean square error in the spatial pattern and the error in the GrIS total (not shown). We base our decisions on a combination of practical considerations (such as ease of implementation) and these informal sensitivity tests, rather than a systematic optimisation across all possible choices.

Our final gradients are a function of SMB sign (positive/negative) and region (north/south), because these divisions make substantial improvements to the parameterisation while not introducing much complexity when implementing in ISMs. The clear difference in SMB response above and below the ELA has already been discussed (Sect. 2.1). We also choose to divide by region because of the distinct regimes in Fig. 4 in which the north has a shallower gradient and larger intercept than the south. The uniform height change simulations also indicate that the the northern margin behaves differently (Fig. 3). We test the performance of north–south divisions in half degree intervals in the range 74–79° N, and also compare with using no division, and find that 77° N gives the best result.

We test two other functional dependencies for the gradients: eight divisions in SMB rather than two, and height dependence as well as SMB dependence. The improvements are not marked enough to justify the extra complexity.

We try three methods for estimating gradients: (a) a linear model of S versus h, (b) a linear model of ΔS versus Δh with zero intercept, and (c) a non-parametric method. We apply each to the four data sets (positive/negative SMB, north/south), and grid cells with $|\Delta h| < 25$ m are excluded. In method (a), a linear fit of S vs. h estimates the gradient b in Fig. 4; this is a similar approach to Helsen et al. (2011), except that we then make the SMB adjustment with our anomaly method rather than an intercept. In method (b), a linear fit of ΔS versus Δh estimates the gradient b in Fig. 5; a zero intercept reflects our expectation that mean SMB change is zero if there is no height change. In method (c), we use a non-parametric approach instead of a linear model. This takes the median of $\Delta S / \Delta h$ ratios (y/x in Fig. 5) as an estimate of b. Method (a) is the least successful, and (b) is the most successful. But we judge that (b) is not an appropriate method, because the fit residuals for grid cells above the ELA vary systematically as a function of height change. Part of this may be due to our constraint of a zero intercept, but the data also clearly have non-linear structure (Fig. 5). The non-parametric method avoids model assumptions such as normally distributed fit residuals, allowing us to capture all the aspects of the MAR response. Our final method is therefore based on (c), though we use the full distribution rather than the median (Sect. 2.3).

Our final parameterisation of the SMB–elevation feedback is therefore a set of four gradients, $\boldsymbol{b} = (b_{\mathrm{p}}^{\mathrm{N}}, b_{\mathrm{n}}^{\mathrm{N}}, b_{\mathrm{p}}^{\mathrm{S}}, b_{\mathrm{n}}^{\mathrm{S}})$, that are used to adjust SMB with a linear model of SMB change versus elevation change. A gradient is selected from the set of four according to whether the mean of the adjusted SMB in the previous decade is positive (p) or negative (n) and whether the grid cell is north or south of 77° N (N, S). The gradients are estimated from the ratios $\Delta S/\Delta h$ of grid cells in the MAR perturbation experiments.

The full SMB–elevation relationship is complex, but our aim is to create a parameterisation that is straightforward to implement: we have therefore partially linearised it, by partitioning the data four ways and by using a linear model of SMB adjustment. In the following section our aim is to account for this approximation with non-parametric assessment of uncertainties in those linear parameters. With this structure it is easy to implement the parameterisation, to use it in forcing an ice sheet model, and to assess the impacts of the parametric uncertainty (arising in large part from this linearisation) using additional ice sheet model simulations.

2.3 Parameter estimation

We now turn to formal statistical inference to obtain the final gradient values. We wish to assess the full uncertainty in the SMB–elevation relationship rather than using only tuned ("best estimate") values or performing ad hoc sensitivity tests. This is particularly important given there are opposite sign SMB responses to elevation changes in the simulations. So we estimate full probability distributions for each of the four gradients ($\boldsymbol{b} = b_{\mathrm{p}}^{\mathrm{N}}, b_{\mathrm{n}}^{\mathrm{N}}, b_{\mathrm{p}}^{\mathrm{S}}, b_{\mathrm{n}}^{\mathrm{S}}$) using a Bayesian approach. This also allows us to propagate the probabilistic SMB–elevation feedback uncertainty to predictions of the GrIS contribution to sea level (Edwards et al., 2014).

We derive initial ("prior") distributions for the four gradients using SMB responses from five of the six perturbation simulations: -100 m, t_1 and t_2; -50 m, t_1 and t_2; and NonUn t_1. These five experiments thus include parameter estimates under different climates (t_1 and t_2), different height changes (-50 m, -100 m, and NonUn), and different locations for a uniform height change (-50 m, -100 m).

We reserve the final simulation (NonUn t_2) as a test of the parameterisation, reweighting the prior distributions using the degree of success in reproducing the cumulative sea level change to obtain updated ("posterior") distributions. We choose NonUn t_2 because the NonUn height changes span a wider range and are closer in spatial pattern to those expected in a warmer climate than the uniform height changes, and because the SMB signal is larger for t_2 than for t_1; we are more concerned that the parameterisation is valid under a warmer climate than the present day.

This division of simulations allows us to try a wide range of candidates for parameter values but assign larger weights to those that match the target we wish to reproduce: the aggregate behaviour of the whole ice sheet.

We use histograms of the ratio of SMB changes to height changes, $\Delta S / \Delta h$ (Fig. 6), as a basis for our prior distributions for the four linear gradient values. These are the same data as in Fig. 5, which shows ΔS versus Δh. Our minimum threshold for the denominator, $|\Delta h| \geq 25$ m, removes extreme values from the tails of these distributions which stabilises estimation of the ratios. All four distributions show that SMB is sometimes positively correlated with height, sometimes negatively correlated. Above the ELA (Figs. 6b and d) the histograms for $b_{\mathrm{p}}^{\mathrm{N}}$ and $b_{\mathrm{p}}^{\mathrm{S}}$ are very narrow: the SMB responses for a given height change are small in magnitude with little variation. Below the ELA (Figs. 6a and c) the histograms for $b_{\mathrm{n}}^{\mathrm{N}}, b_{\mathrm{n}}^{\mathrm{S}}$ are much broader, showing the wide variation in response for different regions of the ice sheet. These histograms are dominated by the four uniform perturbation simulations.

Each of the four histograms has a different number of grid cells, so we take equally sized subsets of each to obtain a joint sample of the gradient set $\boldsymbol{b}$: for each histogram we order the values of $\Delta S / \Delta h$ and take the 0.5 % to 99.5 % quantile values in 0.5 % steps, giving 199 samples of the four gradients

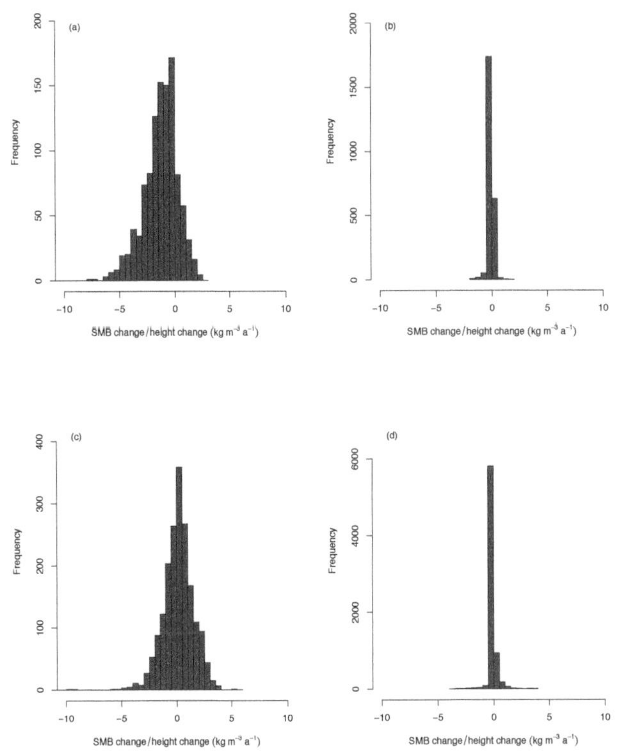

Fig. 6. Histograms of the ratio $\Delta S/\Delta h$ for grid cells north (top row) and south (bottom) of 77° N, divided into grid cells with SMB in both the control and perturbed experiments less than zero (left column) and greater than or equal to zero (right). Data with height change $|\Delta h| < 25$ m are excluded.

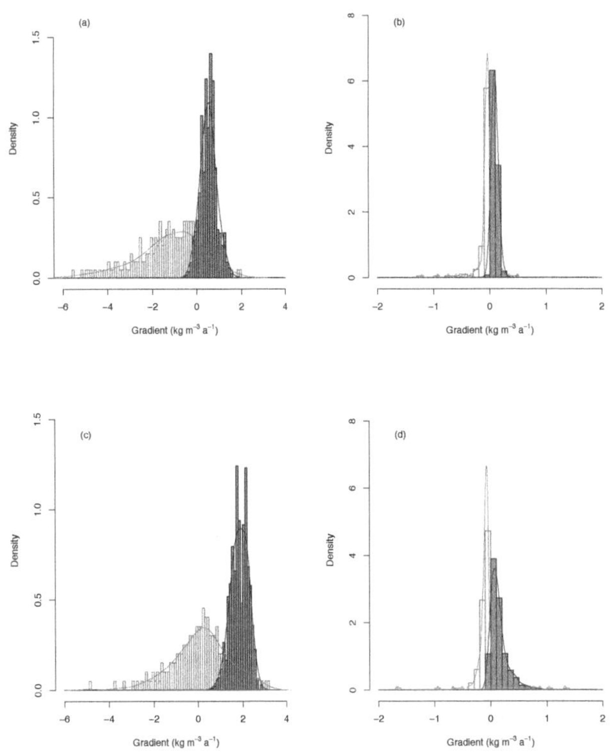

Fig. 7. Prior (light grey) and posterior (dark grey) distributions of the four value gradient set, $\boldsymbol{b} = (b_p^N, b_n^N, b_p^S, b_n^S)$ for regions north (b^N, top row) and south (b^S, bottom) of 77° N, and SMB less than zero (b_n, left column) and greater than or equal to zero (b_p, right).

$(b_p^N, b_n^N, b_p^S, b_n^S)$. These prior distributions are shown in light grey in Fig. 7.

We use each of these 199 prior estimates of the gradient set to adjust the control SMB in 2080–2099 according to the NonUn height change, and assess their success in reproducing the target NonUn t_2 experiment. Each gradient set is used to calculate a spatial pattern of cumulative SMB change and the corresponding total GrIS cumulative sea level contribution.

We simplify the statistical modelling by choosing comparisons so that the differences (discrepancies) between the adjusted and target SMB at each location are approximately i.i.d. (independent and identically distributed) in space. We make the comparisons approximately independent by "thinning" (Rougier and Beven, 2013), using only every 5th grid cell (125 km spacing). This spacing removes spatial correlation: an empirical variogram of the thinned discrepancies is flat for all lengths up to around 800 km (the width of the ice sheet). We assume the discrepancies are identically distributed in space, that is, that the model is equally likely to match the target at every location. We also assume that the discrepancies are normally distributed. In the absence of further information and as a first attempt to describe parameterisation uncertainty, these choices and assumptions allow us to avoid the difficult task of modelling the spatial correlation and variation of the discrepancies.

These assumptions translate to a simple metric for assessing the gradient estimates. The scoring, "likelihood", function is a multivariate (for multiple locations) independent Gaussian with constant variance; the exponent is the sum of squared differences between the adjusted SMB and the target SMB over the subsampled grid cells (independent: a product of Gaussians) divided by the "discrepancy variance" σ^2 (identically distributed: constant variance). The multiplicative constant is discarded due to normalisation later. So the score s_j for the jth of 199 samples of $\boldsymbol{b}$ is

$$s_j = \exp\left[\frac{-1}{2\sigma^2}\sum_i (f_i^j - z_i)^2\right], \qquad (1)$$

where f is the adjusted SMB, z the target SMB, and i the grid cell index. The discrepancy variance is a parameter that represents how closely we expect the parameterised SMB to match the target; our choice is discussed below.

The weight given to each gradient set is the normalised score, $w_j = s_j / \sum_j s_j$. Note that a single weight is calculated for each gradient set $\boldsymbol{b}$, rather than individual weights for each of the four components. The most successful

Table 1. The 2.5 % quantile, best estimate, and 97.5 % quantile estimates of the SMB–elevation gradients in $kg\,m^{-3}\,a^{-1}$, below (SMB < 0) and above (SMB ≥ 0) the ELA, for regions north and south of 77° N.

	Region	2.5 %	Best estimate	97.5 %
SMB < 0	North	−0.22	0.56	1.33
	South	1.03	1.91	2.61
SMB ≥ 0	North	−0.03	0.09	0.23
	South	−0.07	0.07	0.59

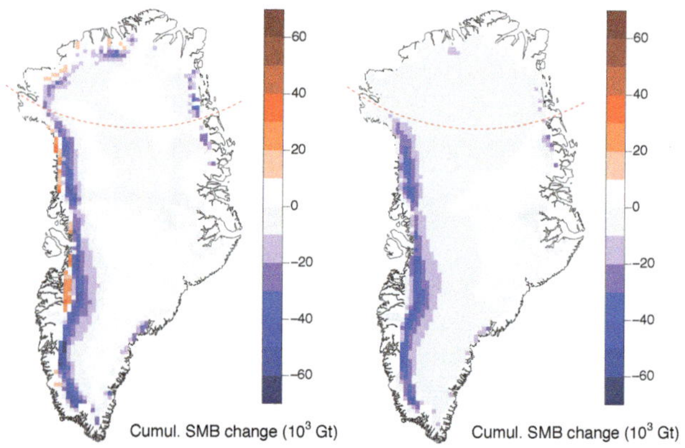

Fig. 8. Cumulative SMB change at the end of the NonUn 2080–2099 simulation: (left) target MAR simulation (perturbed minus control) and (right) result from maximum likelihood gradient set $\tilde{\boldsymbol{b}}$ applied to the NonUn height change (adjusted minus control). Red dashed line is 77° N.

("maximum likelihood") gradient set $\tilde{\boldsymbol{b}}$ has the smallest sum of squared differences and therefore the largest weight. We calculate posterior distributions for the four components of $\boldsymbol{b}$ by reweighting the prior distributions with the normalised weights. We estimate probability densities from the histograms with kernel density estimates and use these to estimate the modes of the posterior distributions, which are our best estimates of the gradients. As we are in a Bayesian framework, our uncertainties are expressed as "credibility intervals" rather than confidence intervals. We estimate 95 % credibility intervals with bootstrapping: we resample 100 000 times from the 199 gradient values (with replacement, using the normalised weights), smooth these with the same bandwidth, and estimate the 2.5 % and 97.5 % quantiles.

Our statistical framework requires minimal choices: the form of the likelihood function; the spacing for the subsampling; and a value for the discrepancy variance. We also choose to set the bandwidth (standard deviation of the smoothing) for the kernel density estimation because the automatically chosen value (Silverman, 1986) does not seem to adequately resolve the distribution shapes. We test various options and make our final choices with the following considerations: thinning so that the discrepancies appear approximately uncorrelated in space; the variance σ^2 chosen such that the weights are not concentrated on a small number of gradient estimates and most or all of the discrepancies for the maximum likelihood parameterisation ($\tilde{\boldsymbol{b}}$) are in the range $\pm 3\sigma$ (Pukelsheim, 1994); and the posterior distribution of total GrIS sea level contribution is close to the target. We choose the smoothing bandwidth so that the density profile captures the main features of the histogram. Our final choices are a Gaussian likelihood function; subsampling distance 5 grid cells (125 km); discrepancy variance $\sigma^2 = (20 \times 10^3\ \mathrm{Gt})^2$; and bandwidths $0.15\,kg\,m^{-3}\,a^{-1}$ for gradients below the ELA and $0.05\,kg\,m^{-3}\,a^{-1}$ above the ELA. Sensitivity tests for these choices are described in the next section.

2.4 Results

Figure 8 shows the adjusted cumulative SMB from the maximum likelihood parameterisation ($\tilde{\boldsymbol{b}}$) and the target. The maximum likelihood gradient set reproduces the target well in most areas, but cannot reproduce the SMB increases with decreasing elevation along the western and southeastern ice sheet margins. Figure 9 shows the discrepancies between the two for all grid cells and the subset used for the likelihood calculation. Most of the discrepancies are small over the ice sheet interior and larger at the margin.

Figure 7 shows the posterior distributions (dark grey) for the four gradients; Table 1 gives the best estimates and 95 % credibility intervals. The posterior distributions are mostly positive, with much larger gradients below the ELA, particularly in the south, than above. Most of the distributions are fairly symmetric, except the south above the ELA which has a low best estimate and a long tail of larger values. The weighting has a particularly strong effect for grid cells below the ELA, drastically narrowing the distributions: effectively the likelihood scoring gives high weights to the gradient estimates derived from the NonUn 2000–2019 experiment (large, positive values), rather than the uniform height change experiments (small, positive and negative values), because these are most successful in reproducing the patterns of change in the NonUn 2080–2099 experiment.

We can apply the same weights to the total GrIS cumulative sea level contributions for each sample of the gradient set (Fig. 10). The prior distribution is centred close to zero: the prior estimate of the elevation feedback is that it has no net effect. The update narrows and shifts the posterior distribution so that it is centred on the target, a positive contribution from the feedback.

We test the sensitivity of the results to the elevation threshold and statistical modelling choices. Varying the threshold (from the default 25 m) between 10 m and 50 m in 5 m intervals changes the results by no more than $0.02\,kg\,m^{-3}\,a^{-1}$, and in most cases $0.01\,kg\,m^{-3}\,a^{-1}$ or zero, for the majority of the gradient best estimates and CI bounds. The exceptions are the best estimates in the south (b_S^n, b_S^p) and the upper CI

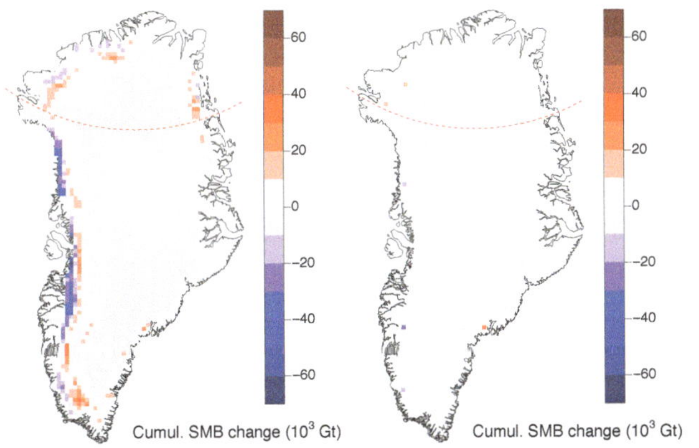

Fig. 9. Cumulative SMB change at the end of the NonUn 2080–2099 simulation: (left) error in the maximum likelihood gradient set $\tilde{b}$ applied to the NonUn height change (adjusted minus perturbed), and (right) the subset of those grid cells used in the calculation of the weights (discrepancies $f_i^j - z_i$ in Eq. 1). Red dashed line is 77° N.

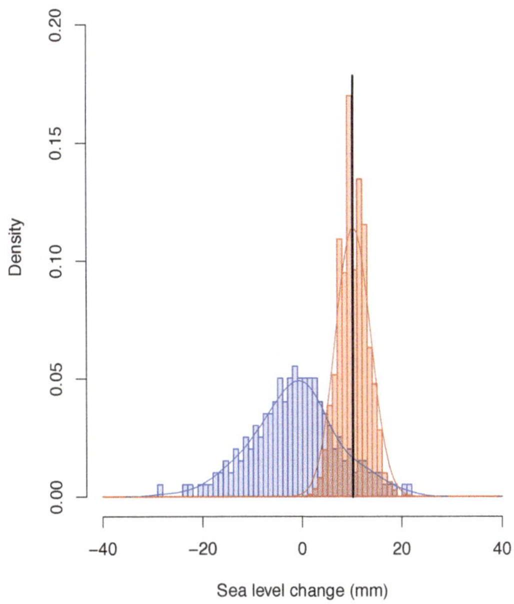

Fig. 10. Prior (blue) and posterior (red) distributions of cumulative change in sea level at 2099 using parameterised elevation feedback for height changes in the NonUn simulation. The target is the result from the NonUn 2080–2099 experiment (vertical black line).

bound for the latter (b_S^p), where the changes are in the range 0.06–0.12 kg m^{-3} a^{-1}. These correspond to a 5 % change for b_S^n; for the smaller gradient b_S^p the fractional changes are larger, but the changes plateau (i.e. the gradient estimates stabilise) above 25–30 m, as intended with the use of the threshold.

We try substituting the Gaussian likelihood with a Cauchy (Student's t distribution with one degree of freedom; very heavy tailed), scaled to match a Gaussian at the 25 % and 75 % quantiles. Our motivation is that the histogram of discrepancies for the maximum likelihood gradient set is fairly sharply peaked. The effect of a Cauchy likelihood is to distribute the weights over a much smaller number of gradient sets, which drastically narrows the posterior distributions. If we also reduce the bandwidths to match these narrower distributions (from 0.15 to 0.05 kg m^{-3} a^{-1} below the ELA and from 0.05 to 0.03 kg m^{-3} a^{-1} above), the CI widths decrease by 54–84 %, and the best estimates increase by 13–38 % for three of the gradients and 186 % (from 0.07 to 0.20 kg m^{-3} a^{-1}) for b_p^S. Because the weights are so concentrated, and we wish to be conservative with uncertainty estimates, we choose the Gaussian likelihood. An alternative approach would be to set a larger discrepancy variance for the ice sheet margin grid cells than the interior, though one might be less confident in assigning the value of two uncertain parameters rather than one.

The discrepancies for the maximum likelihood parameterisation are all within $\pm 1.5\sigma$, which indicates that our discrepancy variance is too large; on the other hand, reducing σ concentrates the weights on a smaller number of gradient estimates, leading to narrower posterior distributions and 95 % CIs. Changing σ from 20 to 15 or 25 Gt does not affect the best estimates much (0–11 %) except for the small-valued b_p^S (43 %). Increasing or decreasing σ by 5 Gt has the effect of increasing or decreasing the CI widths by 12–24 %. Decreasing σ to 15 or 10 Gt broadens the discrepancies to about 2σ, but concentrates the weights rather more. Again, we err on the side of conservatism in our choice.

In the grid cell sampling (required for independence), using different spacing does not have a monotonic effect on the results. Decreasing the spacing from 5 grid cells to 4 (100 km) or increasing it to 6 (150 km) both have the effect of decreasing most best estimates and CI widths. This indicates it is not a problem of using too short a correlation length (violating the independence assumption) but of sensitivity to the grid cell sampling, most likely at the ice sheet margin. Of these three choices, the 5 cell spacing produces the best match to the cumulative sea level change in Fig. 10; in other words, both the 4 and 6 cell spacings concentrate the weights on smaller gradients (smaller SMB adjustments), which match the target spatial pattern well for the particular sampled cells but perform poorly for the ice sheet total using all grid cells. We alter the offset of the sampling, which also has a non-monotonic effect. Shifting both the longitudinal and latitudinal offsets by −3 cells gives a small decrease in the best estimate (0 to −4 %), while offsets of −2, −1 and +1 all give higher best estimates (15–34 %, except b_p^S 71–100 %). The effect on CI width is also mixed; the largest effect is on b_p^S, up to 26 %.

Using a larger discrepancy variance for the margin than the interior would reduce the sensitivity of the results to sampling, because the margin grid cells would have less effect on the likelihood value.

Using automatically set bandwidths gives wider CIs, but oversmooths the distributions, for SMB < 0 and narrower CIs, but undersmooths the distributions, for SMB ≥ 0. Changing our bandwidths from 0.15 to 0.1 or 0.2 $\mathrm{kg\,m^{-3}\,a^{-1}}$ below the ELA and from 0.05 to 0.03 or 0.07 $\mathrm{kg\,m^{-3}\,a^{-1}}$ above affects the CI widths by small amounts ±0–6 %, except for $b_{\mathrm{p}}^{\mathrm{N}}$ (−23 to 27 %).

3 Discussion

3.1 Advantages and strengths

The main advantage of parameterising the GrIS SMB–elevation feedback is that it allows us to force ISMs with SMB simulated by the MAR RCM, which has a more physically realistic representation of the processes than simple schemes such as the PDD, while incorporating the feedback.

The parameterisation can also be used without ISMs to make a first order adjustment to SMB, improving projections such as Rae et al. (2012) and Fettweis et al. (2013) by incorporating the elevation feedback (in effect omitting only the dynamical ice response): for this, the SMB in a given year for each grid cell can be converted to an ice-equivalent height change. A third use is adjusting low resolution SMB fields to the observed surface elevation, for better comparisons with observations or inputs to ISMs (as do Franco et al., 2012).

We have confidence in our parameterisation due, for example, to the similarity in patterns between the maximum likelihood result and the target (Fig. 8), and the centring of the posterior sea level distribution on the target (Fig. 10), and have quantified the effects of the complex non-linear responses in MAR on the feedback uncertainty.

There are several advantages to our approach relative to the parameterisations by Helsen et al. (2011) and Franco et al. (2012). The first relate to our RCM simulations, in which the relationship between SMB and height appears to be more complex (e.g. Fig. 4). For Helsen et al. (2011), this may be partly due to the different schemes in MAR and RACMO2/GR, but in general it is due to our use of simulations in which both the surface elevation and climate boundary conditions are altered. Franco et al. (2012) alter the grid resolution, which produces local changes to elevation; Helsen et al. (2011) do not use altered topography to estimate their SMB gradients, though they do to assess their performance. Neither force the RCM with a global climate different to the present day. Altering the elevation means there is no need for a "space-for-time" substitution. This improves the relevance and robustness of the parameterisation because it is based on results from height changes at a given location, rather than height changes across different spatial locations. We found it is also important to use a wide range of height perturbations, that is, that 50–100 m changes are not sufficient to explore the relationship. It may also be important to apply height changes with the spatial pattern expected under climate change (NonUn) rather than a uniform or fractional (Hakuba et al., 2012) lowering, because the effects on local atmospheric circulation are potentially quite different. We see it is important, particularly for RCMs that include the albedo feedback, to assess the elevation feedback under different global climate conditions, rather than studying one climate era (Franco et al., 2012) or correcting the ice sheet elevation for other climates using a temperature lapse rate (Helsen et al., 2011).

The second set of advantages relate to our parameterisation structure. Using only a gradient (in common with Franco et al., 2012), rather than a gradient and intercept (Helsen et al., 2011), is more robust because it minimises the problem of the climate-dependent offset. In other words, parameterising the relationship between SMB changes and height changes, rather than absolute values, retains more information about the response. A further aspect of flexibility is our choice to estimate the gradients with a non-parametric method (no assumed functional form) rather than a linear model as both Helsen et al. (2011) and Franco et al. (2012) do. Furthermore our parameterisation is very flexible because, unlike the previous studies, it does not depend on spatial location (other than the north–south divide) so it does not depend on the RCM resolution or require interpolation to the ISM grid, and is easy to implement.

The third advantage relates to parameter assessment. We estimate the gradients within a formal probabilistic framework. This allows us to provide not only a best estimate parameterisation but the full probability distributions, so that ISMs can be used to explore the effect of this uncertainty on the GrIS contribution to sea level and express these as credibility intervals.

3.2 Limitations and further work

This is a parameterisation of the SMB-feedback response in an RCM, not of the real world. We have not attempted to estimate the parametric or structural uncertainty of MAR, and have not used an observational constraint. One approach to explore structural uncertainty would be to compare with parameterisations derived for other RCMs. Assessing MAR parametric uncertainty would require a perturbed parameter ensemble such as the 11-member HadRM3 ensemble of Murphy et al. (2009), which is very computationally expensive. We could incorporate observations into the elevation feedback by using them, rather than the target simulation, to calculate the likelihood. We would have to take care that a parameterisation based on observed SMB changes would give a coherent result when applied to RCM simulations. However, the effect of a first-order SMB adjustment on sea level is negligible for a present day ERA-INTERIM forced simulation (not shown), so observational constraints might in any case be of limited use.

We use MAR because it is the most successful of the three RCMs presented by Rae et al. (2012) at reproducing the current SMB of the GrIS. If we were to study additional RCMs, we would derive the parameterisation separately for each. Our preliminary assessment of another RCM used by Rae et al. (2012), HIRHAM, indicates that the SMB response to height is much more linear and less variable than in MAR, most likely because fewer processes are incorporated such as the albedo feedback. Uncertainty in SMB projections are generally thought to be dominated by the choice of GCM, rather than RCM (e.g. Rae et al., 2012); two GCMs are used for the projections presented by Edwards et al. (2014).

We note that the parameterisation may not be robust for changes in elevation and SMB much greater than those in the NonUn A1B 2080–2099 simulation.

We could parameterise each component of SMB separately (Franco et al., 2012) or make the parameterisation structure more complicated in other ways described, but this would have hindered our aim to test the results from the parameterisation in several ISMs (Edwards et al., 2014).

Our estimation of the gradients is non-parametric, but our adjustment of SMB with these gradients uses a linear model with zero intercept. This linear model is an approximation we use to reduce the complexity of the parameterisation. Figure 4 indicates that the relationship is not quite linear above or below the ELA: for example, the gradient in the north is slightly shallower at the lowest elevations. Figure 5 indicates that the intercept may be non-zero below the ELA. A more complex parameterisation could account for these departures from our model, though it would be harder to implement.

Figures 2 and 3 show that a surface lowering can often lead to an increase in SMB. This is particularly the case in the uniform elevation change simulations (Fig. 3) for the north and east, for small elevation changes, and for the beginning of the century. It is also apparent in the NonUn simulations (Fig. 2) along the western margin. This behaviour is likely to derive from the precipitation component of SMB, which has a complex, non-linear relationship with surface elevation (Franco et al., 2012). The maximum likelihood gradient set does not reproduce the SMB increases with decreasing height in the west and southeast (Fig. 8). The probability distributions do incorporate this variation by including the full range of responses: in other words, other samples from $\boldsymbol{b}$ give different correction patterns depending on whether the individual components are positive or negative (Edwards et al., 2014). But within an individual ISM simulation the four components of $\boldsymbol{b}$ are held constant. One way to represent this complex behaviour more fully would be with a stochastic parameterisation, in which the gradients are randomly sampled from the distributions through the simulation rather than held constant. This would incorporate the effect of both positive and negative values of each gradient within a single simulation rather than separate simulations (as in Edwards et al., 2014). However, this would require much more complex statistical modelling to describe the spatial and temporal correlation structure of the gradients, and more complex implementation.

We exclude grid cells with opposite sign SMB in the control and perturbation simulations when estimating the gradients, because of our division at the ELA. (This exclusion only applies to estimating the gradients, not to applying them: when adjusting the MAR SMB, a grid cell may be above the ELA before and below after). The arrows that cross the ELA in Fig. 4 indicate that this filtering may tend to remove smaller than average gradients from the below ELA sample and larger gradients from the above ELA sample. There is no significant change in albedo in grid cells that remain on one side of the ELA or the other, so these show a linear relationship between SMB and elevation. Non-linearity occurs mainly for grid cells that are above the ELA (where no bare ice appear in summer) in the control simulation and move below the ELA with a new elevation, or vice versa, but these are not included in the analysis. This exclusion might therefore lead to an underestimate of the uncertainty.

Boundaries can lead to unexpected edge effects. If the ice sheet were to retreat past 77° N then the parameterisation would shift to using only the south values (i.e. a larger response). However, as we have discussed, we would not recommend using this parameterisation for elevation and SMB changes far beyond our simulations.

The gradients change stepwise across the north–south boundary at 77° N. In principle, this could be smoothed out with a soft transition in a slightly more complex implementation. We include only one boundary rather than several to minimise edge effects. It might be useful to use a further regional division, west–east at around 40° W, because MAR projects different precipitation responses to a warming climate: along the eastern coast snowfall tends to increase, while along the western coast summer precipitation begins to fall as rain.

Our choice of a 25 m threshold was made to stabilise the gradient estimates. Most results are not sensitive to this choice, but we note that a different threshold would in particular alter the best estimate and CI upper bound of $b_{\mathrm{S}}^{\mathrm{p}}$.

Using the mean of each 20 yr simulation might lead to an underestimate of uncertainty by averaging over temporal variability. However, this variability is incorporated when the parameterisation is applied to an annual SMB time series, so care would have to be taken to avoid double counting.

Finally, the choices of the structure and parameter estimation depend on the approximations and the prioritisation of aspects described in Sects. 2.2 and 2.3. Different choices for the statistical modelling might be justifiable, and future work could explore this more thoroughly. As mentioned, an alternative choice might be to set a larger discrepancy variance for the ice sheet margin grid cells than the interior. The subset of data chosen to ensure independent discrepancies could also be selected by hand rather than regularly spaced, with the aim of choosing the most informative cells: for example, picking cells along the margin and a smaller number from the interior. Aggregation of data is another possible method for removing correlation (Rougier and Beven, 2013). Different statistical modelling choices would certainly be appropriate if parameterising the elevation feedback for a different RCM.

4 Conclusions

Surface mass balance of the GrIS can be modelled with the sophisticated, physically based energy balance schemes available in some RCMs, but this is usually at the expense of including the elevation feedback. To include the feedback requires coupling the RCM with an ISM, but this is computationally expensive and technically challenging, which effectively precludes the exploration of uncertainties in the structure and parameter values of the ISM, and in the elevation feedback in the RCM. The only way to incorporate the physical modelling of SMB processes and elevation feedback while also exploring these model uncertainties is with a parameterisation such as the one presented here.

We estimate the SMB–elevation feedback separately below and above the ELA and for regions north and south of 77° N from a set of MAR simulations in which we alter the ice sheet surface elevation. We make advances on previous parameterisations, in particular by including probabilistic assessment of the parameters. This demands statistical modelling choices that must balance adequate representation of the processes in MAR with ease of implementation for widespread use by ice sheet modellers. In general we approach this with linearisation of the processes (for simplicity of use) combined with Bayesian inference of parameter probability distributions (to capture the range of uncertainties arising from linearisation). The maximum likelihood parameterisation is successful in representing the cumulative sum, and most of the spatial pattern, of MAR SMB changes at the end of the century under the A1B scenario and non-uniform elevation changes. Best estimates and 95 % credibility intervals for the parameters provide ice sheet modellers with the opportunity to fully explore the non-linear behaviour. In Edwards et al. (2014) we propagate the probabilistic uncertainties presented here to future sea level projections for the GrIS.

Acknowledgements. This work was supported by funding from the ice2sea programme from the European Union 7th Framework Programme, grant number 226375. Ice2sea contribution number 120. S. Price, M. Hoffman, and M. Perego were supported by the US Department of Energy (DOE) Office of Science, Advanced Scientific Computing Research and Biological and Environmental Research programs. M. Hoffman was partially supported by the Center for Remote Sensing of Ice Sheets at the University of Kansas through US National Science Foundation grant ANT-0424589. Thanks to Jonty Rougier for helpful discussions. We thank Fuyuki Saito and two anonymous referees for their helpful reviews.

Edited by: E. Larour

References

Bamber, J. L., Layberry, R. L., and Gogenini, S. P.: A new ice thickness and bed data set for the Greenland ice sheet 1: measurement, data reduction, and errors, J. Geophys. Res., 106, 33773–33780, 2001.

Edwards, T. L., Fettweis, X., Gagliardini, O., Gillet-Chaulet, F., Goelzer, H., Gregory, J. M., Hoffman, M., Huybrechts, P., Payne, A. J., Perego, M., Price, S., Quiquet, A., and Ritz, C.: Effect of uncertainty in surface mass balance-elevation feedback on projections of the future sea level contribution of the Greenland ice sheet, The Cryosphere, 8, 195–208, doi:10.5194/tc-8-195-2014, 2014.

Ettema, J., van den Broeke, M. R., van Meijgaard, E., van de Berg, W. J., Bamber, J. L., Box, J. E., and Bales, R. C.: Higher surface mass balance of the Greenland ice sheet revealed by high-resolution climate modeling, Geophys. Res. Lett., 36, L12501, doi:10.1029/2009GL038110, 2009.

Fettweis, X.: Reconstruction of the 1979–2006 Greenland ice sheet surface mass balance using the regional climate model MAR, The Cryosphere, 1, 21–40, doi:10.5194/tc-1-21-2007, 2007.

Fettweis, X., Gallée, H., Lefebre, L., and van Ypersele, J.-P.: Greenland surface mass balance simulated by a regional climate model and comparison with satellite derived data in 1990–1991, Clim. Dynam., 24, 623–640, 2005.

Fettweis, X., Tedesco, M., van den Broeke, M., and Ettema, J.: Melting trends over the Greenland ice sheet (1958–2009) from spaceborne microwave data and regional climate models, The Cryosphere, 5, 359–375, doi:10.5194/tc-5-359-2011, 2011.

Fettweis, X., Franco, B., Tedesco, M., van Angelen, J. H., Lenaerts, J. T. M., van den Broeke, M. R., and Gallée, H.: Estimating the Greenland ice sheet surface mass balance contribution to future sea level rise using the regional atmospheric climate model MAR, The Cryosphere, 7, 469–489, doi:10.5194/tc-7-469-2013, 2013.

Franco, B., Fettweis, X., Lang, C., and Erpicum, M.: Impact of spatial resolution on the modelling of the Greenland ice sheet surface mass balance between 1990–2010, using the regional climate model MAR, The Cryosphere, 6, 695–711, doi:10.5194/tc-6-695-2012, 2012.

Goelzer, H., Huybrechts, P., Fürst, J. J., Nick, F. M., Andersen, M. L., Edwards, T. L., Fettweis, X., Payne, A. J., and Shannon, S. R.: Sensitivity of Greenland ice sheet projections to model formulations, J. Glaciol., 59, 733–749, doi:10.3189/2013JoG12J182, 2013.

Gordon, C., Cooper, C., Senior, C. A., Banks, H., Gregory, J. M., Johns, T. C., Mitchell, J., and Wood, R. A.: The simulation of SST, sea ice extents and ocean heat transports in a version of the Hadley Centre coupled model without flux adjustments, Clim. Dynam., 16, 147–168, 2000.

Hakuba, M. Z., Folini, D., Wild, M., and Schär, C.: Impact of Greenland's topographic height on precipitation and snow accumulation in idealized simulations, J. Geophys. Res, 117, D09107, doi:10.1029/2011JD017052, 2012.

Hanna, E., Huybrechts, P., Cappelen, J., Steffen, K., Bales, R. C., Burgess, E., McConnell, J. R., Peder Steffensen, J., van den Broeke, M., Wake, L., Bigg, G., Griffiths, M., and Deniz Savas, D.: Greenland Ice Sheet surface mass balance 1870 to 2010 based on Twentieth Century Reanalysis, and links with global climate forcing, J. Geophys. Res., 116, D24121, doi:10.1029/2011JD016387, 2011.

Helsen, M. M., van de Wal, R. S. W., van den Broeke, M. R., van de Berg, W. J., and Oerlemans, J.: Coupling of climate models and ice sheet models by surface mass balance gradients: application to the Greenland Ice Sheet, The Cryosphere, 6, 255–272, doi:10.5194/tc-6-255-2012, 2012.

Huybrechts, P. and de Wolde, J.: The dynamic response of the Greenland and Antarctic ice sheets to multiple-century climatic warming, J. Climate, 12, 2169–2188, 1999.

IPCC: Climate Change 2007: The Physical Science Basis. Contribution of Working Group I to the Fourth Assessment Report of the Intergovernmental Panel on Climate Change, edited by: Solomon, S., Qin, D., Manning, M., Chen, Z., Marquis, M., Avery, K. B., Tingor, M., and Miller, H. L., Cambridge University Press, Cambridge, United Kingdom and New York, NY, USA, 2007.

Janssens, I. and Huybrechts, P.: The treatment of meltwater retention in mass-balance parameterizations of the Greenland ice sheet, Ann. Glaciol., 31, 133–140, 2000.

Murphy, J. M., Sexton, D. M. H., Jenkins, G. J., Booth, B. B. B., Brown, C. C., Clark, R. T., Collins, M., Harris, G. R., Kendon, E. J., Betts, R. A., Brown, S. J., Humphrey, K. A., McCarthy, M. P., McDonald, R. E., Stephens, A., Wallace, C., Warren, R., Wilby, R., and Wood, R. A.: Climate change projections, in: UK Climate Projections, no. 2, Met Office Hadley Centre, Exeter, UK, 1–194, 2009.

Nakićenović, N., Alcamo, J., Davis, G., and de Vries, B.: Special report on emissions scenarios: a special report of Working Group III of the Intergovernmental Panel on Climate Change, Intergovernmental Panel on Climate Change, 2000.

Pukelsheim: The 3-sigma-rule, The American Statistician, 48, 88–91, 1994.

Rae, J. G. L., Aðalgeirsdóttir, G., Edwards, T. L., Fettweis, X., Gregory, J. M., Hewitt, H. T., Lowe, J. A., Lucas-Picher, P., Mottram, R. H., Payne, A. J., Ridley, J. K., Shannon, S. R., van de Berg, W. J., van de Wal, R. S. W., and van den Broeke, M. R.: Greenland ice sheet surface mass balance: evaluating simulations and making projections with regional climate models, The Cryosphere, 6, 1275–1294, doi:10.5194/tc-6-1275-2012, 2012.

Ridley, J., Huybrechts, P., Gregory, J., and Lowe, J.: Elimination of the Greenland Ice Sheet in a High CO_2 Climate, J. Clim., 18, 3409–3427, 2005.

Robinson, A., Calov, R., and Ganopolski, A.: An efficient regional energy-moisture balance model for simulation of the Greenland Ice Sheet response to climate change, The Cryosphere, 4, 129–144, doi:10.5194/tc-4-129-2010, 2010.

Roeckner, E., Bäuml, G., Bonaventura, L., Brokopf, R., Esch, M., Giorgetta, M., Hagemann, S., Kirchner, I., Kornblueh, L., Manzini, E., Rhodin, A., Schlese, U., Schulzweida, U., and Tompkins, A.: The atmospheric general circulation model ECHAM 5, Part I: Model description, Max-Planck-Institut für Meteorologie, Hamburg, Germany, 126 pp., 2003.

Rougier, J. C. and Beven, K. J.: Model limitations: the sources and implications of epistemic uncertainty, in: Risk and Uncertainty Assessment for Natural Hazards, edited by: Rougier, J. C., Sparks, R. S. J., and Hill, L. J., Cambridge University Press, Cambridge, 40–63, ISBN 978-1-107-00619-5, 2013.

Shepherd, A., Ivins, E. R., Geruo A., Barletta, V. R., Bentley, M. J., Bettadpur, S., Briggs, K. H., Bromwich, D. H., Forsberg, R., Galin, N., Horwath, M., Jacobs, S., Joughin, I., King, M. A., Lenaerts, J. T. M., Li, J., Ligtenberg, S. R. M., Luckman, A., Luthcke, S. B., McMillan, M., Meister, R., Milne, G., Mouginot, J., Muir, A., Nicolas, J. P., Paden, J., Payne, A.J., Pritchard, H., Rignot, E., Rott, H., Sandberg Sørensen, L., Scambos, T. A., Scheuchl, B., Schrama, E. J. O., Smith, B., Sundal, A. V., van Angelen, J. H., van de Berg, W. J., van den Broeke, M. R., Vaughan, D. G., Velicogna, I., Wahr, J., Whitehouse, P. L., Wingham, D. J., Yi, D., Young, D., and Zwally, H. J.: A Reconciled Estimate of Ice-Sheet Mass Balance, Science, 338, 1183–1189, 2012.

Silverman, B. W.: Density Estimation, Chapman and Hall, London, 1986.

Stone, E. J., Lunt, D. J., Rutt, I. C., and Hanna, E.: Investigating the sensitivity of numerical model simulations of the modern state of the Greenland ice-sheet and its future response to climate change, The Cryosphere, 4, 397–417, doi:10.5194/tc-4-397-2010, 2010.

Tarasov, L. and Peltier, W. R.: Greenland Glacial History and Local Geodynamic Consequences, Geophys. J. Int., 150, 198–229, 2002.

van de Berg, W. J., van den Broeke, M., Ettema, J., van Meijgaard, E., and Kaspar, F.: Significant contribution of insolation to Eemian melting of the Greenland ice sheet, Nat. Geosci., 4, 679–683, 2011.

van de Wal, R. S. W.: Mass-balance modelling of the Greenland ice sheet: a comparison of an energy-balance and a degree-day model, Ann. Glaciol., 23, 36–45, 1996.

van den Broeke, M. R. and Gallée, H.: Observation and simulation of barrier winds at the western margin of the Greenland ice sheet, Q. J. Roy. Meteor. Soc., 122, 1365–1383, 1996.

Vernon, C. L., Bamber, J. L., Box, J. E., van den Broeke, M. R., Fettweis, X., Hanna, E., and Huybrechts, P.: Surface mass balance model intercomparison for the Greenland ice sheet, The Cryosphere, 7, 599–614, doi:10.5194/tc-7-599-2013, 2013.

13

Modelling the transfer of supraglacial meltwater to the bed of Leverett Glacier, Southwest Greenland

C. C. Clason[1], D. W. F. Mair[2], P. W. Nienow[3], I. D. Bartholomew[3], A. Sole[4], S. Palmer[5], and W. Schwanghart[6]

[1]Department of Physical Geography and Quaternary Geology, Stockholm University, 106 91 Stockholm, Sweden
[2]Geography and Environment, University of Aberdeen, Aberdeen, AB24 3UF, UK
[3]School of Geosciences, University of Edinburgh, Edinburgh, EH8 9XP, UK
[4]Department of Geography, University of Sheffield, Sheffield, S10 2TN, UK
[5]Geography, College of Life and Environmental Sciences, University of Exeter, Exeter, EX4 4RJ, UK
[6]Institute of Earth and Environmental Science, University of Potsdam, 14476 Potsdam-Golm, Germany

Correspondence to: C. C. Clason (caroline.clason@natgeo.su.se)

Abstract. Meltwater delivered to the bed of the Greenland Ice Sheet is a driver of variable ice-motion through changes in effective pressure and enhanced basal lubrication. Ice surface velocities have been shown to respond rapidly both to meltwater production at the surface and to drainage of supraglacial lakes, suggesting efficient transfer of meltwater from the supraglacial to subglacial hydrological systems. Although considerable effort is currently being directed towards improved modelling of the controlling surface and basal processes, modelling the temporal and spatial evolution of the transfer of melt to the bed has received less attention. Here we present the results of spatially distributed modelling for prediction of moulins and lake drainages on the Leverett Glacier in Southwest Greenland. The model is run for the 2009 and 2010 ablation seasons, and for future increased melt scenarios. The temporal pattern of modelled lake drainages are qualitatively comparable with those documented from analyses of repeat satellite imagery. The modelled timings and locations of delivery of meltwater to the bed also match well with observed temporal and spatial patterns of ice surface speed-ups. This is particularly true for the lower catchment (< 1000 m a.s.l.) where both the model and observations indicate that the development of moulins is the main mechanism for the transfer of surface meltwater to the bed. At higher elevations (e.g. 1250–1500 m a.s.l.) the development and drainage of supraglacial lakes becomes increasingly important. At these higher elevations, the delay between modelled melt generation and subsequent delivery of melt to the bed matches the observed delay between the peak air temperatures and subsequent velocity speed-ups, while the instantaneous transfer of melt to the bed in a control simulation does not. Although both moulins and lake drainages are predicted to increase in number for future warmer climate scenarios, the lake drainages play an increasingly important role in both expanding the area over which melt accesses the bed and in enabling a greater proportion of surface melt to reach the bed.

1 Introduction

In the last decade it has been demonstrated that across large regions of the Greenland Ice Sheet (GrIS) surface meltwater is capable of penetrating through many hundreds of metres of cold ice via full-ice thickness crevasses, or moulins, and by the drainage of supraglacial lakes (e.g. Zwally et al., 2002; Das et al., 2008; Doyle et al., 2013). Evidence from remote sensing has shown the temporal and spatial patterns in lake formation and drainage during the melt seasons (e.g. McMillan et al., 2007; Sundal et al., 2009; Fitzpatrick et al., 2014) which indicates that the process is spatially extensive, with lake formation above 1800 m a.s.l. (Fitzpatrick et al., 2014). Once meltwater reaches the bed, the seasonal evolution of subglacial drainage system efficiency (e.g. Chandler et al., 2013), has been suggested to exert an important control on the dynamic response of the GrIS to surface meltwater inputs due to its modulation of the relationship between surface meltwater inputs and subglacial water pressure (Bartholomew et al., 2010, 2011a; Colgan et al., 2011; Hoffman et al., 2011; Sole et al., 2013). Consequently there has been renewed interest and significant progress in developing spatially distributed, coupled models of subglacial hydrology and ice flow at the glacier and ice sheet scale (e.g. Hewitt, 2013; de Fleurian et al., 2014).

There has, however, been less attention focused on the development of models which can simulate the delivery of surface runoff to the bed of the ice sheet, i.e. modelling the temporal and spatial evolution of surface-to-bed meltwater connections (Clason et al., 2012; Banwell et al., 2013). This is a significant limitation since it is increasingly clear that the dynamics of the overlying ice may be most sensitive to hydrology when and where there are transient changes in meltwater delivery to the bed (Schoof, 2010; Bartholomew et al., 2012), and where ice thickness and surface slope precludes the formation of stable channelized drainage (Meierbachtol et al., 2013; Doyle et al., 2013). Aside from the overall contribution to dynamics through basal sliding, modelling of surface-to-bed meltwater connections may also be important for glacier dynamics through ice deformation, due to a potential influence on cryo-hydrologic warming (Phillips et al., 2010; Colgan et al., 2011). Models of delivery of supraglacial meltwater to the ice sheet bed are thus essential if physically based coupling of models of surface meltwater generation, subglacial hydrology and ice sheet dynamics is envisaged.

Here we apply a simple model which simulates spatial and temporal patterns in the delivery of meltwater to the bed of an ice sheet to one catchment of the Southwest GrIS. The model requires spatially distributed inputs of surface elevation, ice surface velocities, accumulation and air temperature. The model is run for the ablation seasons of 2009 and 2010 for which contemporaneous investigations of meteorology, hydrology and ice dynamics have been undertaken and reported elsewhere (Bartholomew et al., 2011a, b). We investigate the sensitivity of the model to parameters controlling refreezing, surface runoff delay and spatial resolution, and the effect of enhanced atmospheric warming on temporal and spatial patterns of modelled ice–bed meltwater connections. In the absence of detailed direct observations of supraglacial drainage system evolution, we assess qualitatively the performance of the model through (1) the consistency between modeled and observed patterns of supraglacial lake drainages, and (2) a comparison between timings and locations of modelled delivery of meltwater to the subglacial drainage system and the measured dynamic responses of the ice sheet to changing meltwater inputs.

2 Study area

Our study is focused on Leverett Glacier, a land-terminating outlet glacier of the Southwest GrIS, with its terminus situated at 67.1° N, 50.1° W. The supraglacial hydrological catchment upstream of the main proglacial river was derived from a digital elevation model (DEM) of the ice surface produced from Interferometric Synthetic Aperture Radar (InSAR) data acquired in 1996 (Palmer et al., 2011). The catchment encompasses an ice-covered area of c. 1200 km^2 and extends to over 50 km inland of the margin, up to an elevation of c. 1550 m (Fig. 1). Meltwater leaves the catchment through a large subglacial conduit (Fig. 1, yellow star), feeding a proglacial river. We focus our modelling on the 2009 and 2010 melt seasons when peak discharge in the proglacial river was 317 and 398 m^3 s^{-1}, respectively (Bartholomew et al., 2011a).

3 Methods

The main components of the model, which has been applied in a previous version to the Croker Bay catchment of the Devon Ice Cap (Clason et al., 2012), comprise: (1) a degree-day model for meltwater generation; (2) an algorithm for routing meltwater across the ice surface (Schwanghart and Kuhn, 2010) and storing meltwater within supraglacial lakes; and (3) a model for calculating penetration depths of water-filled crevasses, after van der Veen (2007). The model, which is run here with a spatial resolution of 500 m and a temporal resolution of 1 day, is the first predictive (rather than prescriptive) model for moulin formation and the transfer of meltwater to the ice–bed interface applied to the Greenland ice sheet. Model outputs provide information on the location and timing of formation of surface-to-bed connections, the drainage of supraglacial lakes, the quantity of meltwater stored supraglacially and the quantity of meltwater delivered to the bed through each connection on each day.

3.1 Melt modelling and supraglacial meltwater retention

A lack of appropriate input data for energy balance modelling precludes its use here, so a degree-day model (Appendix A) was chosen for this application. Degree-day modelling is a simple approach for estimation of melting, but it has performed well in characterizing the relationship between melt and discharge in previous studies (Bartholomew et al., 2011b). Well calibrated degree-day factors (DDFs) for the catchment were calculated and calibrated for the Leverett glacier during 2009 (Appendix A). Meteorological data used for input to the degree-day model (Appendix A) were acquired at seven sites extending from the terminus of Leverett Glacier at 457 m (site 1, Fig. 1) into the ice sheet interior to 1716 m elevation (site 7, Fig. 1) (Bartholomew et al., 2011a). Daily accumulation was obtained from ultrasonic depth gauge measurements of surface height, and spring snowpack depth on 6 May 2009 was recorded at

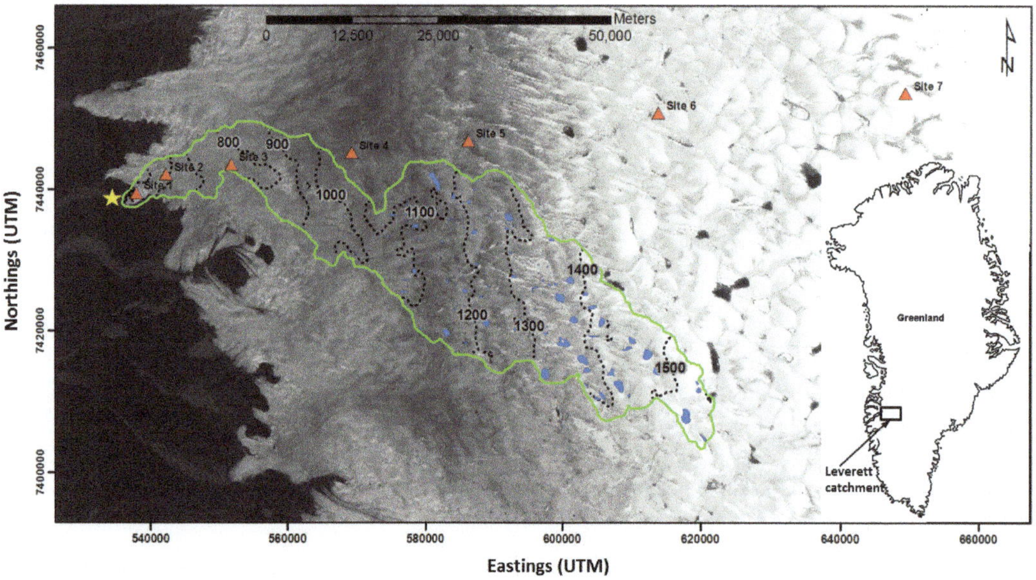

Figure 1. Leverett Glacier surface hydrological catchment (outlined in green). Contours show ice surface elevation (m a.s.l.); locations of meteorological data collection are depicted by red triangles; the location of proglacial discharge measurements is represented by the yellow star; and supraglacial lakes are highlighted in blue. The background image is from Landsat 7, band 2, captured on 5 August 2005.

each site (Fig. 1), resulting in an accumulation gradient of 256.6 mm w.e. per 1000 m ($R^2 = 0.76$).

Following application to the Croker Bay catchment (Clason et al., 2012) the model has been further developed to include both refreezing within the snowpack and a delay in meltwater routing across snow-covered cells. Model runs for the Leverett catchment without the inclusion of refreezing and runoff delay predicted lake drainages as early as May, which was not supported by observations, and studies such as Lefebre et al. (2002) and Box et al. (2006) demonstrate the considerable effect of refreezing on runoff. After Reeh (1991), meltwater retention due to refreezing in the snowpack was included by implementing the simple P_{max} coefficient, with a standard value of 0.6, supported by observations in the lower accumulation zone on west Greenland by Braithwaite et al. (1994). This coefficient is the fraction of the winter snowpack subject to refreezing over the course of a melt season, such that at the start of the model run P_{max} is applied to the spring snowpack to determine refreezing potential in each cell. At each time step meltwater is refrozen instantaneously until the refreezing potential in each cell is met, whereby future melting is allowed to runoff. Following Schuler et al. (2007) we do not differentiate between porewater refreezing and formation of superimposed ice.

To account for percolation and meltwater flow through the basal saturated layer a simple runoff delay, governed by local snow depth, was applied in all snow-covered cells. The length of the delay was based on flow rates for dye percolation through the snowpack and along the basal saturated layer of Haut Glacier d'Arolla (Campbell, 2007). The range of measured flow rates from Campbell (2007) give runoff delays ranging from 1 to 16 days for meltwater flow through 1 m deep snow and along a 500 m flow path (model spatial resolution). In our model we incorporate a moderately high meltwater routing delay of 10 days for 1 m deep snow, scaling this delay linearly with local snow depth, and thus assuming a constant density summer snowpack, such that there is no delay when there is no snow.

3.2 Meltwater routing and accumulation in supraglacial lakes

A single-flow direction algorithm was applied to route available surface meltwater across the ice surface based on surface elevation (Schwanghart and Kuhn, 2010; Schwanghart and Scherler, 2014), where the amount of meltwater in each cell weighted downstream flow accumulation. The 100 m Palmer et al. (2011) DEM was resampled to the standard model spatial resolution of 500 m. We did not define a threshold for discrete stream formation due to the spatial resolution of the DEM; instead meltwater was distributed across the ice surface by flow accumulation only. A total of 93 supraglacial lakes within the Leverett catchment were manually digitized in ArcGIS from lake extents visible on Landsat 7 ETM+ imagery acquired on 15 and 31 July 2009 (Fig. 1), which were assumed to be maximum lake extents. A fixed number of empty lakes were thus prescribed at the start of the season, rather than expanding up-glacier as the area experiencing melting becomes larger.

Prescription of lakes based on digitization from satellite imagery was chosen instead of automated DEM-based identification of lakes (e.g. Leeson et al., 2012; Arnold et al.,

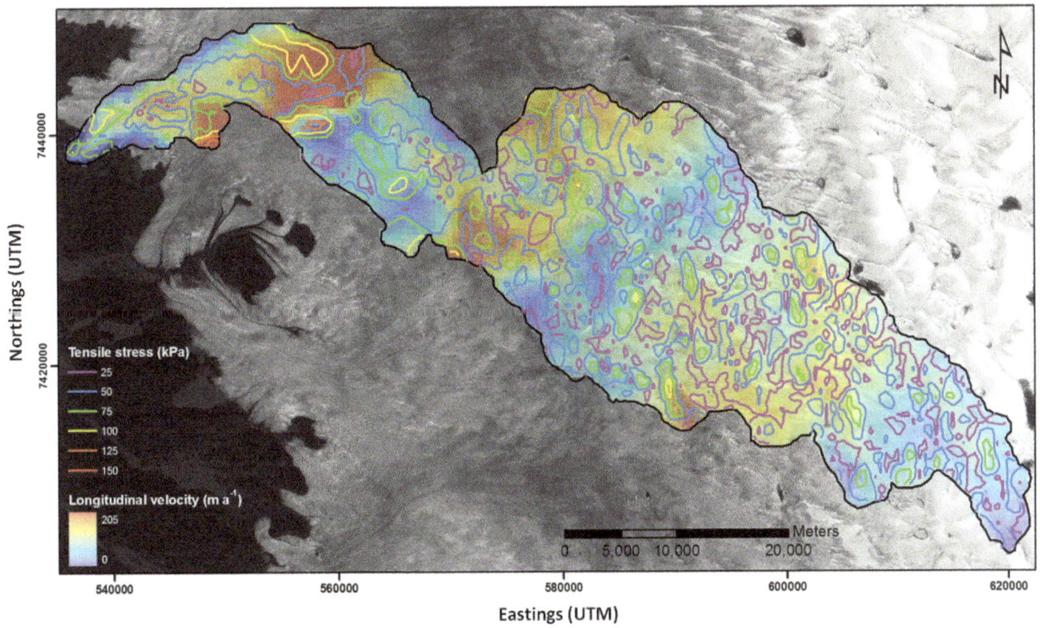

Figure 2. Longitudinally resolved (along-flow) ice surface velocities from InSAR data for the Leverett catchment (Joughin et al., 2010). Contours depict the ice surface tensile stress regime.

2014) to better capture the total number of lakes available for drainage. The 30 m resolution of Landsat imagery allows for higher accuracy than a 100 m resolution DEM in prescribing lake numbers and surface area, and furthermore, DEM-based models at best identify 78 % of lakes visible on remotely sensed imagery (Arnold et al., 2014). Modelled hydrofracture beneath lakes is very sensitive to meltwater volume, thus prescription of lakes from higher-resolution imagery was more appropriate for the purpose of predicting the timing of lake drainages and quantifying meltwater delivery to the bed. Given uncertainties associated with modelling lake volume based on depressions in DEMs, such as DEM vertical resolution, and since our model attempts only to predict when lakes drain, applying predictive tools to determine their location and maximum volume is beyond the requirements of this study.

Lake surface area was used to estimate lake volume based on a linear relationship derived between lake volume and surface area from data recorded by Box and Ski (2007) using MODIS for Southwest Greenland. There are two principal modes for supraglacial lake drainage: slow drainage events, where meltwater in lakes overtops and flows into downstream crevasses, moulins or other lakes (Hoffman et al., 2011; Tedesco et al., 2013); and fast drainage events, where large quantities of meltwater are delivered to the bed in a short period of time via hydrofracture, promoting a temporary ice dynamic response (e.g. Das et al., 2008; Doyle et al., 2013). Filling and overtopping of supraglacial lakes was accounted for within the flow accumulation routine (Clason et al., 2012) such that meltwater routed into a lake-containing cell will accumulate until reaching the prescribed lake volume. At this point the lake will overtop and contribute to downstream runoff, which may flow into downstream crevasses if the lake has not already drained locally through modelled hydrofracture (Appendix C). Supraglacial lakes in Southwest Greenland are more numerous, have a larger total area, and have a larger frequency of fast drainage than anywhere else on the ice sheet (Selmes et al., 2011), making them an important feature of the Leverett glacier catchment.

3.3 Modelling crevasse location and depth

Synthetic aperture radar data from RADARSAT (Joughin et al., 2010) provided annual mean ice surface velocity data for the Leverett catchment from which velocity components (Fig. 2) and surface stresses could be calculated. The von Mises criterion, σ_v, after Vaughan (1993) was applied for calculation of tensile stresses, and crevasse locations were predicted based upon a prescribed tensile strength (Appendix B). The depth of each crevasse is calculated using a model of water-filled crevasse penetration based on linear elastic fracture mechanics (van der Veen, 2007) driven by accumulated surface meltwater and the surface tensile stress regime (Fig. 2; Appendix C). The volume flux of meltwater to the ice–bed interface is calculated at the bottom of each full ice thickness crevasse. In addition to the propagation of surface crevasses, fracture beneath supraglacial lakes, and their consequent drainage, is also permitted when lake meltwater volume is large enough to drive a fracture through the ice thickness at a specific location according to Eq. (C1)

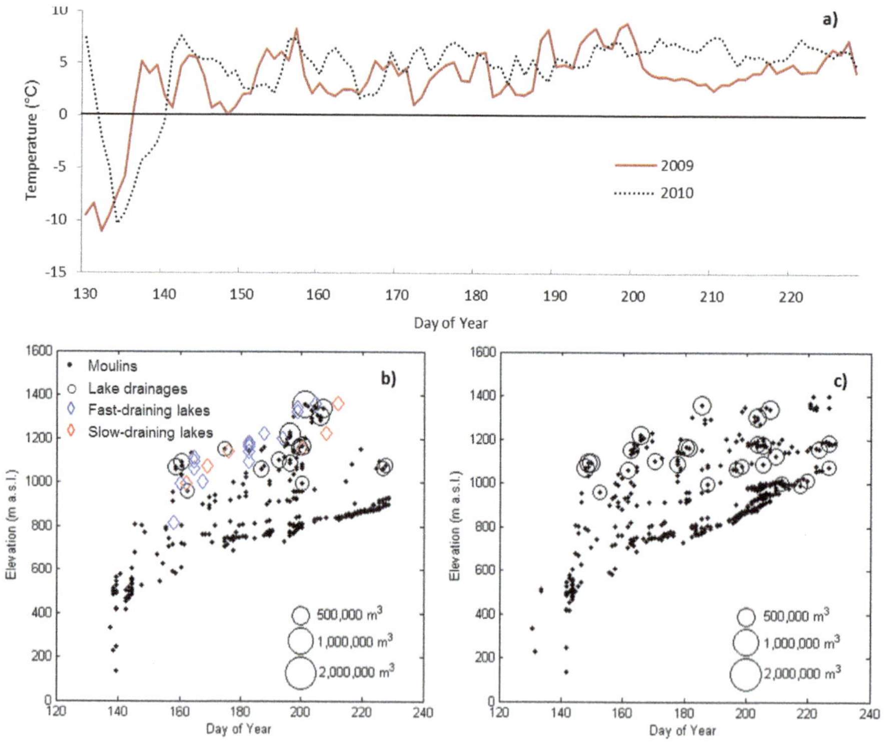

Figure 3. **(a)** Daily average air temperatures at site 1 (457 m a.s.l.) for 2009 and 2010, and moulin formation and lake drainage through the **(b)** 2009 and **(c)** 2010 melt seasons with elevation. Blue diamonds in **(b)** represent observed drainage of lakes in events between two MODIS images, and red diamonds represent lakes which drained over a period of several MODIS images; after Fig. 2a of Bartholomew et al. (2011a).

(Appendix C). Drainage of supraglacial lakes is permitted regardless of whether the tensile stress exceeds the prescribed ice tensile strength, since supraglacial lakes have been found to form in areas of low tensile or compressive surface stress (Catania et al., 2008).

4 Results

4.1 Application to Leverett 2009 and 2010 melt seasons

The model was first run for the 2009 melt season (run 1) with prescribed standard parameters of 75 kPa tensile strength, a 1 m depth-averaged crevasse width, an ice fracture toughness of 150 kPa $m^{1/2}$, P_{max} of 0.6, and a runoff delay of 10 days where snow is 1 m deep. In all subsequent runs these parameters remain the same unless otherwise stated. The timing of moulins first reaching the ice–bed interface in run 1 is depicted in Fig. 3b, where the number of moulins formed is shown to increase in elevation with time. This is due to expansion of the area experiencing melting, retreat of the snowline, increased meltwater delay with elevation, and also due to the thicker ice through which moulins at higher elevation must penetrate to reach the bed. Supraglacial lake drainages also occur at higher elevations over time, as supported by remote sensing observations in Southwest Greenland (Morriss et al., 2013).

The model was also run using meteorological data from the 2010 melt season (run 2), covering the same time period as 2009 (day 130 to day 228), allowing for an assessment of model response to increased meltwater production in the Leverett catchment. During this period daily average temperatures at site 1 were on average 1 °C higher than for 2009 (Fig. 3a). 2010 was characterized by high temperatures and significantly increased melt days across the GrIS, with temperatures highest in the west (Box et al., 2010). Melting occurred for up to 50 days longer than the 1979–2007 mean in areas of the western ice sheet, and during the month of May surface temperatures were as much as 5 °C higher than the 1971–2000 average according to Reanalysis data from NCEP/NCAR. Figure 3c illustrates the modelled temporal formation of surface-to-bed connections during the 2010 melt season, where moulins begin forming one week earlier in comparison to the cooler 2009 season.

In 2009 modelled surface-to-bed connections form up to c. 1400 m (Fig. 4a), delivering 76 % of surface-generated meltwater to the bed. Below 1000 m elevation there are large clusters of moulins, which are cells for which sufficient meltwater is produced to allow for full-thickness fracture propagation of a single crevasse without relying on inflow from upstream accumulated meltwater. The model sets the runoff ratio, or the proportion of meltwater transferred to the next downstream cell, to zero when routed meltwater

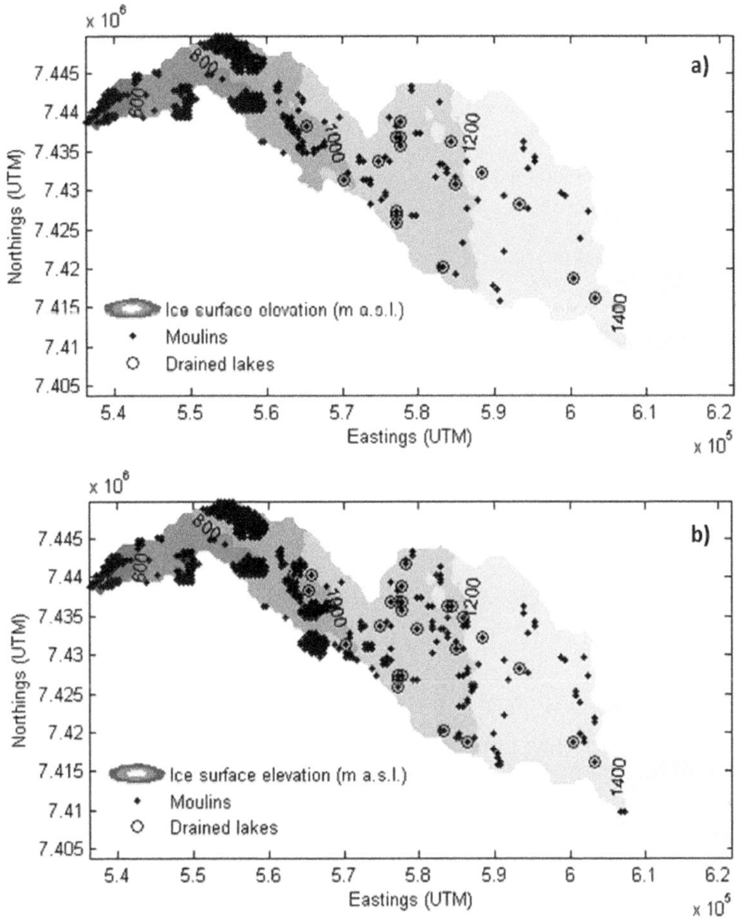

Figure 4. Spatial distribution of moulins and lake drainages for **(a)** 2009 and **(b)** 2010.

is captured by a crevasse. However, at low elevations melt rates are highest, enhanced by a smaller delay in meltwater transfer through cells with low spring snowpack depths. For the 2010 season the model predicts an increase in total moulin numbers of 44 % (Table 1) compared to 2009. Modelled lake drainages also increase in number from 17 in 2009 to 27 in 2010 (Table 1). Higher moulin numbers and lake drainages in 2010 causes the proportion of total meltwater that is (a) transferred to the bed to increase (by 9 %), and (b) stored supraglacially to decrease (by 5.7 %) (Table 1). In 2010 there is a notable increased clustering of moulins just below 1000 m and an increased number of lake drainages between 1100 and 1200 m elevation (Fig. 4b).

4.2 Sensitivity analysis

To investigate the influence of including refreezing within the model, P_{max} was changed to 0.4 and 0 in runs 3 and 4. Moulin numbers showed a modest increase of 3.9 and 8.6 % for runs 3 and 4, respectively (Table 1). Associated increase in meltwater transfer to the bed of 4 and 10 % was balanced by a near identical 4.3 and 11 % decrease in supraglacial meltwater storage, highlighting a strong control imposed by refreezing on meltwater availability for moulin formation.

The model was also tested for the upper and lower limits for runoff delay in runs 5 and 6, as derived from data by Campbell (2007). When a delay of only 1 day (at 1 m snow depth) was applied there was a small increase in moulin numbers of 5.2 %, due to the extended period during which melt is available to drive fracture propagation. Despite the increase in moulin numbers there was less than 1 % change in meltwater transfer to the bed and supraglacial storage (Table 1). Increasing the delay to 16 days for 1 m of snow had very little effect, with changes in meltwater transfer, storage and moulin numbers all less than 1 %. This is unsurprising as only the most upper reaches of the catchment are subject to the full meltwater transit delay, in an area receiving significantly less melt than in the lower elevation regions, where moulins are much less likely to form.

A limitation of the model is the control of spatial resolution on the number of crevasses with the potential to form connections to the bed. In runs 7 and 8 we thus ran the model at resolutions of 250 m and 1 km, respectively, excluding supraglacial lakes. Runs 7 and 8 produced a 50 %

Table 1. Total number of surface-to-bed connections formed, the percentage of surface-generated meltwater delivered to the bed and the percentage of surface-generated meltwater stored supraglacially via refreezing and percolation in the snowpack for each model run. Meltwater not accounted for by transfer to the bed or supraglacial storage is stored englacially inside crevasses which have not reached the bed.

Run name	Moulin numbers		Meltwater transfer		Supraglacial storage	
	Total moulins (number of lake drainages)	% change from initial run	% transfer from surface to bed	change from initial run	% of total generated meltwater	change from initial run
1 (2009)	327 (17)	n/a	76	n/a	17	n/a
2 (2010)	470 (27)	+44	85	+9.0	11	−5.7
3 ($P_{max} = 0.4$)	340 (17)	+3.9	80	+4.0	13	−4.3
4 ($P_{max} = 0$)	355 (17)	+8.6	86	+10	6.6	−11
5 (1-day runoff delay)	344 (16)	+5.2	77	+0.9	17	−0.5
6 16-day runoff delay)	329 (17)	+0.6	76	−0.2	17	+0.2
7 (250 m resolution)	489 (n/a)	+50	71	−5.1	17	0
8 (1 km resolution)	108 (n/a)	−67	81	+4.9	17	0
9 (A1B JJA min.)	479 (27)	+47	85	+8.8	12	−5.4
10 (A1B JJA mean)	549 (48)	+68	90	+14	8.0	−9.1
11 (A1B JJA max.)	685 (93)	+110	96	+20	3.7	−13

increase and 67 % decrease in moulin numbers respectively (Table 1), strongly controlled by the consequent changing number of surface crevasses. Although meltwater must be split between the available crevasses at each resolution, the available surface-produced meltwater is more than sufficient to drive many of these crevasses to the bed, resulting in only a small decrease in meltwater transfer of 5.1 % in run 7 and a small increase of 4.9 % in run 8. This relative insensitivity of meltwater transfer to changes in spatial resolution is encouraging for implementation within larger-scale ice sheet models. At such coarse spatial resolution prediction of the numbers of individual moulins is not yet possible, but prediction of areas where surface-to-bed meltwater transfer is an active process is important to simulate for subsequent forcing of subglacial hydrological models. There was no change in the amount of meltwater stored supraglacially in between runs 7 and 8 due to static controls on meltwater production and transport (Table 1). Instead, with an increase/decrease in crevasse numbers, the amount of water stored englacially in crevasses that do not reach the bed increases/decreases in runs 7 and 8, respectively. Since crevasse length is modified to equal cell width at each resolution, crevasse volume is also modified, resulting in a smaller quantity of meltwater necessary to produce the level of water-filling required to drive a crevasse to the bed.

Model sensitivity to tensile strength, fracture toughness and crevasse width was also tested for the Leverett domain, as described for application to the Croker Bay catchment on the Devon Ice Cap in Clason et al. (2012). Results of these tests illustrated the same model sensitivity to altering these parameters as was previously described: altering fracture toughness has no significant effect; altering tensile strength strongly influenced the total number of moulins due to controlling the number of surface crevasses; and while altering crevasse width has no impact on crevasse numbers, it does influence the number of moulins through altering the volume of the crevasse and thus how much water is necessary to drive it to the bed. In summary these tests show that the most important control on the spatial extent of moulins is the value of the tensile strength. Parameters which define crevasse geometry affect the rate at which water will fill a crevasse and are most important in determining the timing of the delivery of surface meltwater to the bed.

4.3 Moulin and lake density

The spatial densities of modelled moulins and drained lakes in different elevation bands were calculated to investigate how the model characterizes the change in the mechanism for delivery of meltwater to the bed with elevation (Fig. 5). During the 2009 melt season, the model predicts a marked reduction in moulin density above 1000 m. Lake drainages only occur above 750 m elevation, with the highest density of drainages occurring between 1000 and 1250 m, incorporating site 4 (1061 m) and site 5 (1229 m) (Fig. 1), which exhibit the largest velocity peaks of the four sites above 1000 m.

4.4 Sensitivity to atmospheric warming

To investigate the sensitivity of ice surface-to-bed meltwater connections across the catchment to enhanced atmospheric warming, the model was run with the 2009 Leverett meteorological data revised to reflect the IPCC (2007) A1B scenario June, July and August air temperature projections for the Arctic region. 2009 was an average melt season based on the 1981–2010 mean (Sole et al., 2013), from which the three A1B scenarios (minimum, mean and maximum) represent temperature rises of 1.2, 2.1, and 5.3 °C, respectively (IPCC,

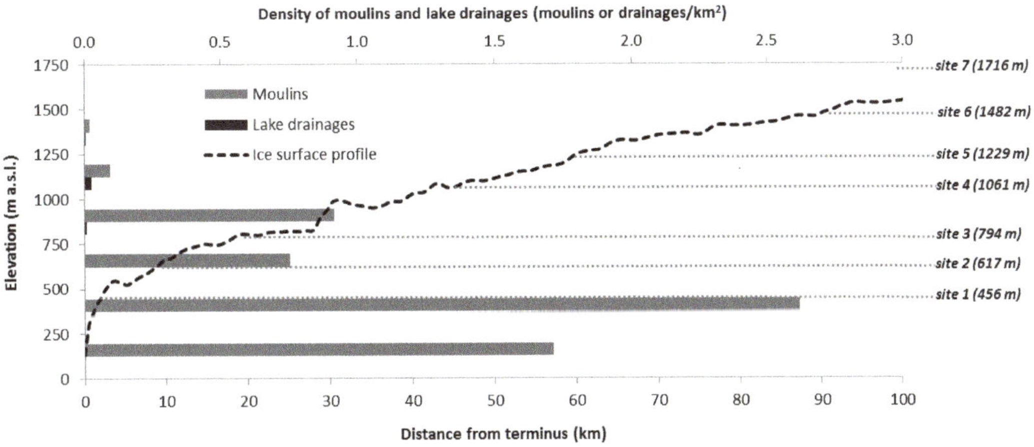

Figure 5. Density of moulins and lake drainages for 2009 within 250 m ice surface elevation bands. Sites of GPS velocity measurements (Fig. 1; Bartholomew et al., 2011b) are shown against the Leverett catchment ice surface profile. Note that only a very small area of the derived Leverett catchment lies below 500 m elevation, where outlet glaciers emerge at the margin of the ice sheet.

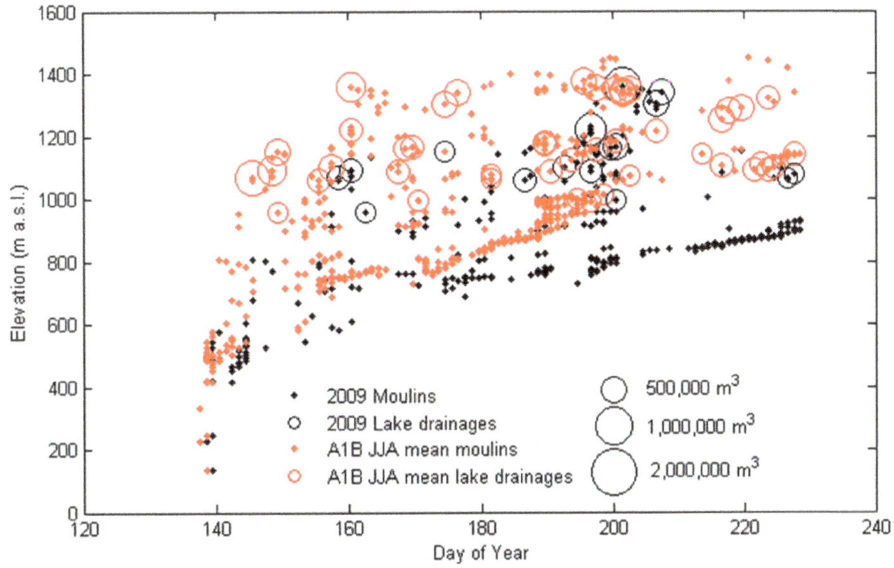

Figure 6. Spatial distribution of moulins and lake drainages for the 2009 melt season and the A1B mean June, July and August Arctic scenario of +2.1 °C (IPCC Fourth Assessment Report) applied to 2009 meteorological data.

2007), added uniformly to the 2009 temperature data. The results of running the model for increased future temperature scenarios (runs 9, 10 and 11; Table 1) show that in addition to an increase in moulin numbers (+47, 68 and 110 %) and much increased occurrence of lake drainages (+59, 182 and 447 %), applying these scenarios also resulted in increases of 8.8, 14 and 20 % in the proportion of surface-derived meltwater that is transferred to the bed, in comparison to model run 1.

Focusing on the mean scenario, below 750 m no change in moulin density is observed due to the smaller ice thicknesses and higher melt production resulting in all possible crevasses experiencing sufficient melt-filling to drive them to the bed. Although the melt-season starts just a few days earlier, a temporal shift in moulin formation is evident, with moulins at higher elevation forming much earlier than for the standard 2009 model run (Fig. 6), and with an additional increase in the density of moulins at elevations above 750 m (Fig. 7). Furthermore, there is an increase in occurrence of lake drainages at higher elevations, resulting in more widespread delivery of meltwater to the bed through large ice thicknesses, beginning earlier in the melt season (Figs. 6 and 7).

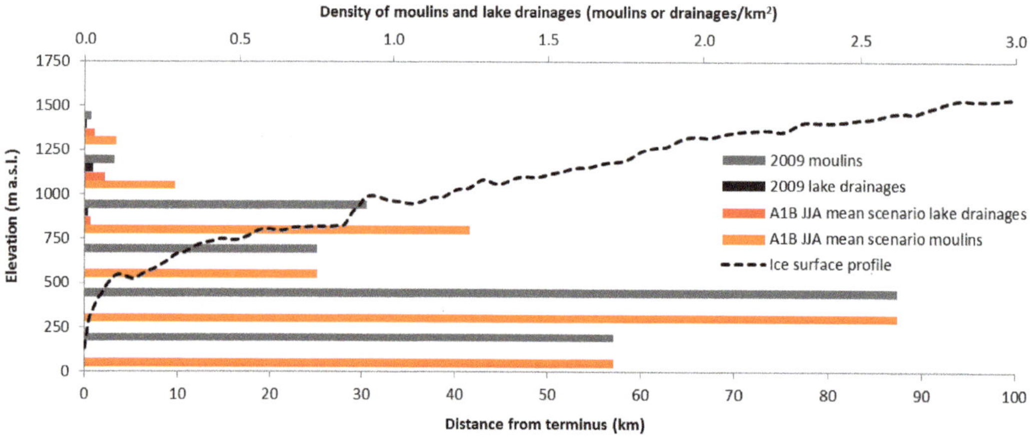

Figure 7. Density of moulins and lake drainages for A1B mean June, July and August Arctic scenario within 250 m ice surface elevation bands.

5 Assessment of model performance

5.1 Modelled and observed patterns of supraglacial lake drainage

A comparison between the modelled spatio-temporal pattern of lake drainages shown in Fig. 3a and a remote sensing-based assessment of lake drainage events in 2009 undertaken by Bartholomew et al. (2011a, their Fig. 2a) shows qualitative agreement. In both approaches: the majority of lakes drain between early June and mid-August; drainage of lakes mostly occurs between 1000 and 1400 m elevation, and there is a general trend showing an up-glacier progression in the timing of lake drainages of ~6–8 m elevation per day. In comparison of Fig. 4a with Fig. 1 of Bartholomew et al. (2011a) there is spatial clustering of drained lakes between ~1000 and ~1200 m elevation in both cases. The model also predicts relatively isolated drainages of lakes approaching 1400 m, as observed by Bartholomew et al. (2011a) on MODIS imagery. Despite lake locations, surface areas, and thus maximum volumes being prescribed in this study, identified lakes are not preconditioned to drain through hydrofracture, and require sufficient meltwater input and ice surface stress to drain to the bed. While the model is not trying to reproduce exact observations of lake drainage, of the 17 lakes predicted to drain during 2009, 7 are contemporaneous with lake drainage locations identified by Bartholomew et al. (2011a); this is not unreasonable given the assumptions of the model and of determining lake locations and drainages from satellite imagery. These comparisons demonstrate that the model can reproduce realistic spatial and temporal patterns of lake drainage behaviour.

5.2 Modelled meltwater delivery to bed and measured dynamic responses during 2009

We further assess the performance of the model through the consistency between modelled patterns in the delivery of meltwater to the subglacial drainage system and measured dynamic response of the ice sheet to changing meltwater inputs for the 2009 melt season. During the 2009 melt season horizontal ice surface velocities were measured at seven GPS units, sites 1 to 7 (Fig. 1), extending from the Leverett glacier at 456 m up onto the ice sheet at 1716 m elevation (Bartholomew et al., 2011b). The period of the melt season characterized by highest velocities began later at sites of increasing elevation, with initial acceleration recorded at sites 1 and 2 shortly after the onset of melting, while increased velocities at sites 5 and 6 were not recorded until much later in the season. This is due to retreat of the snowline and onset of melting at increasingly high elevation. Furthermore, the periods of enhanced velocity at sites 4, 5 and 6 (all above 1061 m) are not strongly associated with high positive degree days at these sites, in contrast to sites 1, 2 and 3 (all below 800 m).

Meltwater transferred to the bed each day within each elevation band was calculated to compare the timing of modelled meltwater discharge to the bed with the timings of significant speed-up events within each elevation band (Fig. 8). Between 0 and 499 m elevation, there is relatively little meltwater delivered to the bed through moulins, which reflects the very small area of the Leverett catchment below 500 m. Periods of increasing meltwater delivery to the bed between 500 and 999 m match well with periods of velocity increase early in the season (Fig. 8c and d). At the highest elevations within the catchment, above 1250 m, between ~day 200 and 210 there is also good agreement between the timing of meltwater delivery to the bed and the glacier speed-up. Between 1250 and 1499 m (Fig. 8), meltwater delivery to the bed is predicted in near-equal amounts from moulins and the

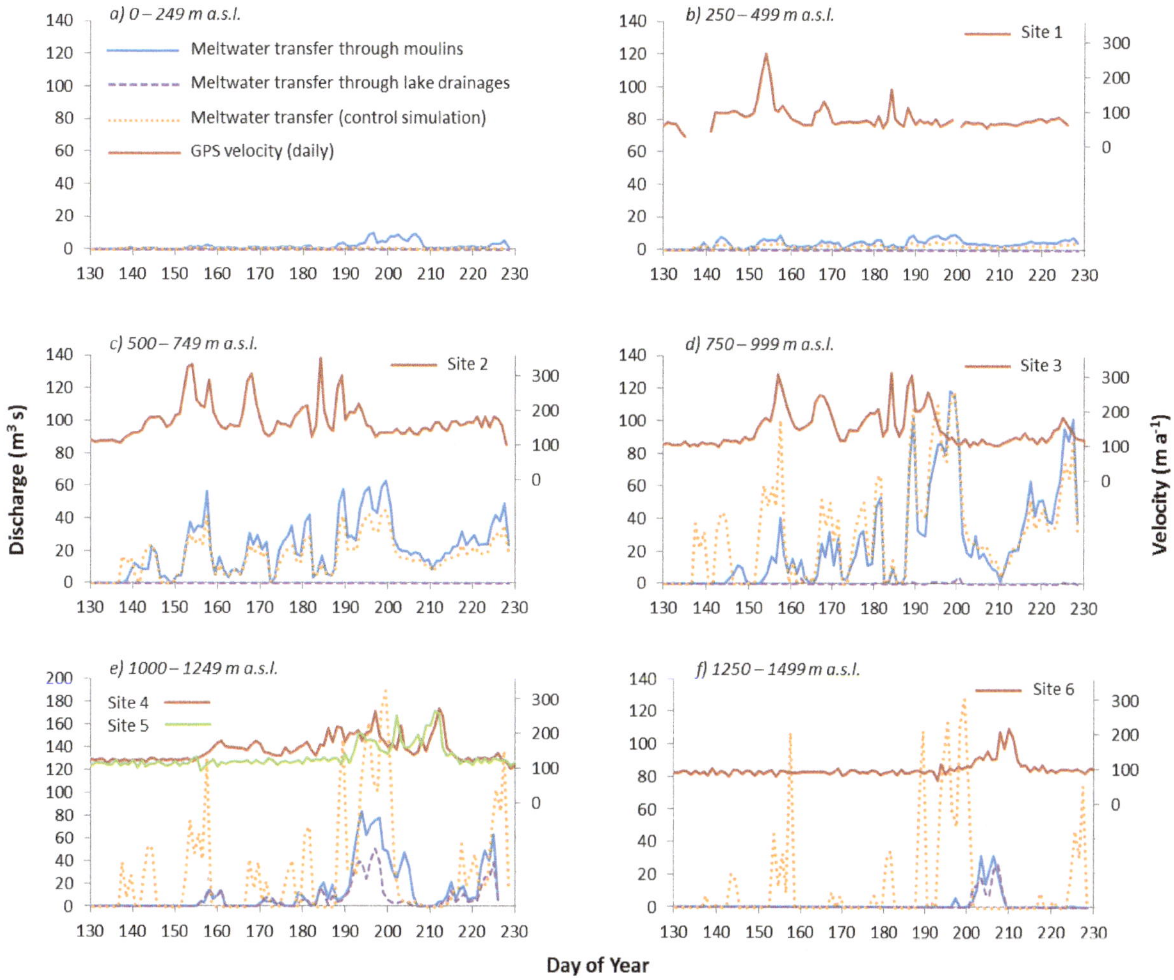

Figure 8. Supraglacial meltwater delivered to the bed each day through modelled lake drainages, moulins, and for the control simulation within ice surface elevation bands of 250 m during 2009. Ice surface velocities from GPS sites 1–6 are plotted within their respective elevation bands (after Bartholomew et al., 2011b). Note the extended y-axis in (**e**).

drainage of supraglacial lakes, highlighting the greater significance of lake drainages at high elevations.

To evaluate the necessity of predictive transfer of meltwater compared with routing all surface-generated meltwater to the bed (e.g. Shannon et al., 2013), a control simulation was run such that all meltwater was delivered to the bed locally and instantaneously, subject to storage and delay of meltwater through refreezing and percolation. The results of this control simulation (Fig. 8) reveal that without the additional modelling of surface meltwater runoff routing, hydrofracture through the ice, and the filling and drainage of supraglacial lakes, correspondence between the timing of increased meltwater transfer and increased ice surface velocities gets progressively worse with elevation. Between 750 and 999 m, meltwater transfer occurs early in the season, ~day 135, with no corresponding velocity increase (Fig. 8d). At 1000–1249 m, the correspondence between velocity and meltwater transfer for the control simulation continues to worsen in the early season, and breaks down completely for the whole season above 1250 m. These results highlight the importance of accounting for delay in meltwater transfer to the bed through storage in lakes, transport in supraglacial streams, and in meltwater delivery through moulins for which hydrofracture to the bed takes longer in areas of thicker ice.

6 Discussion

In light of future climate scenarios, incorporating the transfer of surface-derived meltwater to the bed is imperative if ice sheet models are to fully consider the behaviour and development of the subglacial drainage system, and the consequent ice velocity responses that drive ice sheet evolution and contribution to sea level change. This study has applied a model for prediction of moulin formation and lake drainages to data sets for the Leverett Glacier catchment in Southwest Greenland, simulating the delivery of meltwater from the ice surface to the bed. The model was run for the 2009 and 2010 melt seasons and predicts high spatial densities of moulins below 1000 m as the principal mechanism for rapid delivery of meltwater to the glacier bed, a finding that is consistent with interpretations from field measurements of surface melting and velocity. Bartholomew et al. (2011b) suggested that at lower elevations, ice surface velocities respond to supraglacial meltwater routed quickly to the ice–bed interface through moulins, while at higher elevations the lack of correlation between positive degree days and ice velocities may be indicative of a dynamic response to the delayed release of meltwater stored in supraglacial lakes. Our model results are consistent with this finding, showing a similar change in the mechanism for the delivery of meltwater to the bed with elevation such that moulins are more dominant below 1000 m and drained lakes of more importance above this (Fig. 5). Above c. 1000 m lake drainages play a much greater role in ensuring that meltwater reaches the bed through propagation of crevasses up to 1100 m deep (see Doyle et al., 2013), and into the ice sheet interior.

Many previous studies have demonstrated that the most likely cause of short-term ice surface speed-ups is the creation of areas of high water pressure at the bed of the ice sheet in response to high meltwater inputs to a drainage system that is not hydraulically efficient enough to accommodate transient high discharges at low pressure (Hoffman et al., 2011; Bartholomew et al., 2012; Sole et al., 2013). Across most of the catchment there is a strong association between periods when the model predicts rapid increases in meltwater delivery to the bed and episodes of ice surface speed-up. The model output is therefore consistent with previous process interpretations. At GPS sites 1, 2 and 3 the period when the modelled meltwater discharges to the bed rise to a peak are not associated with speed-ups (~ day 200). This is consistent with the proposition from interpretation of field evidence that in these regions of the ice sheet hydraulically efficient subglacial drainage channels eventually evolve which can accommodate high discharges at low pressures (e.g. Chandler et al., 2013).

The association between modelled meltwater delivery to the bed and observed ice sheet speed-ups is less obvious between 1000 and 1249 m a.s.l. (GPS sites 4 and 5). This may reflect model inadequacies or the effects of presenting modelled discharge as integrated values across an elevation band that covers a large horizontal extent. This elevation band is also likely to encompass the up-glacier limit in the extent to which efficient subglacial channels can evolve. Chandler et al. (2013) argued that channelized drainage could evolve up to 41 km from the ice sheet margin where the ice surface lies at a little over 1000 m a.s.l., i.e. at the lower range of this elevation band. However, this study also showed that inferred channels did not extend as far as 57 km where the ice sheet surface was 1230 m a.s.l. which is close to the upper range of the elevation band. Modelling of subglacial conduits by Meierbachtol et al. (2013) places an even lower limit of ~ 20 km on the up-ice extent of subglacial conduits, arguing that low surface slopes up-ice of the margin inhibit melting back of conduit walls. The conduits therefore cannot offset creep closure to accommodate increasing discharge. It is therefore likely that in the 1000–1249 m a.s.l. elevation band there is considerable spatial heterogeneity in subglacial drainage system evolution which would reduce the likelihood of observing a clear temporal association between spatially integrated modelled discharge and ice surface velocity. This assumption is supported by recent observations of hydraulic head and ice surface velocity in west Greenland by Andrews et al. (2014).

At the highest elevations within the catchment several processes combine to delay the delivery of meltwater to the bed: the vertical percolation and refreezing of melt in the snowpack, the slowing of horizontal surface runoff through the snowpack, and the accumulation of sufficient water in supraglacial lakes to initiate full-depth crevasse formation. The close agreement between the timing of modelled meltwater delivery to the bed and surface velocity speed-ups at the highest elevations in the catchment indicate that the model is able to characterize these processes effectively. This meltwater is delivered to the bed several days after the peak atmospheric temperatures during a relatively cool period between days 200 and 210 (see Figs. 3a and 8).

The comparison between 2009 and the warmer 2010 melt season and the testing of the sensitivity of the model results to atmospheric warming provides insight into how the catchment's hydrology may change under a warmer climate. The model shows the potential for an increased proportion of supraglacial meltwater to reach the bed, and that a larger area of the bed is directly affected by surface meltwater inputs, owing to the up-glacier expansion in the area affected by supraglacial lake drainages. This latter model outcome is supported by observations of an expansion in lake-covered area during warm years in the Russell Glacier catchment (Fitzpatrick et al., 2014). The modeled quantities of meltwater accessing the bed through lake drainage events shown here under the warmer climate scenarios are likely to give us a conservative view of what might be expected across the GrIS more generally, for two reasons. Firstly, the model uses a fixed, prescribed pattern of supraglacial lake cells which does not expand higher up-glacier as the melt extent increases. Secondly, the Leverett catchment only extends to c. 1550 m elevation and so cannot characterize the potential for a vast increase in the area where supraglacial lakes could form under a warmer climate. Both of these issues could be addressed by coupling this model with one that can predict the location of the formation of supraglacial lakes (e.g. Leeson et al., 2012). Nevertheless, the model clearly indicates that under a moderately warmer climate there will be an increase in the relative importance of supraglacial lake drainage in delivering melt to the bed of the ice sheet in the high-elevation areas of the ice sheet (above 1000 m elevation) despite ice thicknesses in excess of 1 km.

It is not clear what the long-term impact of more spatially extensive and more frequent lake drainages may be on longer-term ice dynamics across high-elevation areas of the ice sheet. Across the ablation area of the ice sheet it has been shown that there is no significant correlation between normalized surface melt and annual ice flow (Sole et al., 2013). It has been proposed that increased summer melting sustains large, widespread low-pressure subglacial channels which in turn promote more extensive and prolonged drainage of high pressure water from adjacent regions resulting in a greater drop in net basal water pressure and reduced displacement over the subsequent winter (Sole et al., 2013; Tedstone et al., 2013). This preconditioning of the ice–bed interface for reduced winter velocity limits the ice sheet's dynamic sensitivity to interannual variations in surface temperature and melt. However, a positive relationship between warmer summer air temperatures and annual velocities may be expected well above the ELA where the development of low-pressure channelized drainage is likely hindered by greater ice thicknesses and shallow surface slopes (Meierbachtol et al., 2013; Doyle et al., 2013). The long-term implications of increased melting during warmer years, such as that witnessed in 2010 and 2012 (Tedstone et al., 2013), on subglacial drainage configuration, basal water pressure, and consequently ice dynamics, are difficult to assess without coupling a model such as the one presented here to subglacial drainage and ice flow models (e.g. Hewitt, 2013; de Fleurian et al., 2014; Hoffman and Price, 2014).

7 Conclusions

A spatially distributed model for predicting the temporal and spatial patterns of moulin formation and lake drainages has been applied to the Leverett Glacier in Southwest Greenland. With minimal data requirements and a simple structure, the model is easily transferable to other areas, including those without supraglacial lakes. The model was run for the 2009 and 2010 ablation seasons, driven by in situ meteorological and melt observations, and assessed by comparison with independent interpretations of meltwater delivery to the bed based on analyses of ice dynamic response to atmospheric forcings. The response of the catchment's hydrology to future climate scenarios is also investigated, as is the model sensitivity to parametrization of refreezing, horizontal meltwater transit through surface snowpacks and the model's spatial resolution.

The model is successful in characterizing the spatial variation in the mechanisms for meltwater transfer from the surface to the bed. For the lower part of the catchment (< 1000 m a.s.l.) both the model and previous observations indicate that the development of moulins is the main mechanism for the transfer of surface meltwater to the bed. At the highest elevations (e.g. 1250–1500 m a.s.l.) the development and drainage of supraglacial lakes becomes increasingly important.

At the higher elevations, the delay between modelled melt generation and subsequent delivery of melt to the bed matches the observed delay between the peak air temperatures and subsequent velocity speed-ups. This indicates that the model effectively characterizes processes which delay the delivery of surface-generated melt to the ice sheet bed.

The temporal and spatial patterns of modelled lake drainages compare favourably with those seen from analyses of satellite imagery. The modelled timings and locations of delivery of meltwater to the bed match well with observed temporal and spatial patterns of ice surface speed-ups.

Results of modelling moulin formation and lake drainage for the warmer 2010 season, and particularly for future climate scenarios, indicate the potential for increased absolute and relative transfer of supraglacial meltwater to the bed during periods of increased surface melting. With atmospheric warming lake drainages play an increasingly important role in both expanding the area over which surface-derived melt accesses the bed and in enabling a greater proportion of surface melt to reach the bed. Model sensitivity testing demonstrates that the proportion of melt reaching the bed is relatively insensitive to refreezing thresholds, runoff delays and the spatial resolution of the model.

This work contributes to efforts to couple physically based models of surface meltwater generation, subglacial hydrology and ice sheet dynamics which will be required to fully understand past, contemporary and future sensitivity of ice sheet mass balance and dynamics to climate change.

Appendix A: Degree-day modelling

The model runs at a daily time step, and values of total melting each day, M_t, are determined by the application of a degree-day factor (DDF) for every day where mean temperature, T_t, equals or exceeds 0 °C:

$$M_t = (\mathrm{DDF} \cdot T_t) \quad T_t \geq 0°\mathrm{C} \tag{A1}$$

$$M_t = 0 \quad T_t < 0°\mathrm{C}. \tag{A2}$$

The sum of daily melt values occurring over N days thus gives total ablation, A:

$$A = \sum_{1}^{N} M_t. \tag{A3}$$

A DDF for snow ($\mathrm{DDF_s}$) is applied for snow-covered cells, with a DDF for ice ($\mathrm{DDF_i}$) applied when the cumulative melt exceeds the prescribed spring snowpack depth. Precipitation falling as snow is added to the snowpack depth, but rainfall, where air temperature is above 1 °C, is not included within the melt model due to the very small contribution it makes to total melt.

Temperature was recorded at 15 min intervals at each of seven sites in the Leverett catchment (Fig. 1) during the 2009 and 2010 melt seasons. The model is run for the contemporaneous period of data collection from 10 May (day 130) to 16 August (day 228) for each year. An air temperature lapse rate of 5.5 °C per 1000 m was calculated from the 2009 data ($R^2 = 0.96$). Degree-day factors for snow and ice ($\mathrm{DDF_s}$ and $\mathrm{DDF_i}$) of 5.81 and 7.79 mm w.e. $\mathrm{d^{-1}\,°C^{-1}}$, respectively, were determined based on calibration against ablation rates recorded by ultrasonic depth gauges during 2009.

Appendix B: Identification of areas of surface crevassing

Velocity data were first resolved into longitudinal and transverse components (Fig. 2), the directional derivatives of which were then used to calculate strain rates, $\dot{\varepsilon}_{ij}$. After Nye (1957), the constitutive relation was applied to convert strain rates to stresses, σ_{ij}:

$$\sigma_{ij} = B\dot{\varepsilon}_e^{(1-n)/n}\dot{\varepsilon}_{ij}, \tag{B1}$$

where $\dot{\varepsilon}_e$ is effective strain, and n is the flow law exponent with a value of 3. B is a viscosity parameter sensitive to ice temperature, and is related to the flow law as $B = A^{-1/n}$ (Vieli et al., 2006). For the Leverett catchment we apply an ice temperature of −5 °C, giving a flow law parameter, A, value of $9.3 \times 10^{-16}\,\mathrm{s^{-1}\,kPa^{-3}}$ (Cuffey and Paterson, 2010) and a viscosity parameter, B, value of $324\,\mathrm{kPa\,a^{1/3}}$.

For determining areas containing surface crevassing, ice surface tensile stresses, R_{ij}, were calculated based on the von Mises criterion, σ_v, after Vaughan (1993):

$$\sigma_v = (\sigma_1\sigma_1) + (\sigma_3\sigma_3) - (\sigma_1\sigma_3), \tag{B2}$$

where the maximum and minimum principal stresses, σ_1 and σ_3, are calculated from

$$\sigma_1 = \sigma_{max} = \frac{1}{2}\left(\sigma_{xx} + \sigma_{yy}\right) + \sqrt{\left[\frac{1}{2}\left(\sigma_{xx} - \sigma_{yy}\right)\right]^2 + \tau_{xy}^2} \tag{B3}$$

$$\sigma_3 = \sigma_{min} = \frac{1}{2}\left(\sigma_{xx} + \sigma_{yy}\right) - \sqrt{\left[\frac{1}{2}\left(\sigma_{xx} - \sigma_{yy}\right)\right]^2 + \tau_{xy}^2} \tag{B4}$$

and where σ_{xx}, σ_{yy} and τ_{xy} are the longitudinal, transverse and shear stresses respectively. The tensile stress is thus related to the von Mises criterion as

$$R_{ij} = \sqrt{\sigma_v}. \tag{B5}$$

Model cells containing surface crevassing were determined by prescribing a value of tensile strength, based on visual matching of the calculated surface tensile stresses (Fig. 2) with crevassing visible on Landsat 7 imagery. A tensile strength of 75 kPa was thus prescribed in the standard model parameters. This was the value that best represented spatial distribution of crevassing on imagery, without overprediction of crevasses in higher-elevation areas with numerous supraglacial lakes, which would have acted to impede lake filling through meltwater routing. This visual comparison approach was very simple, and based on tensile stresses calculated from annual mean velocities. For future applications we would recommend deriving tensile stresses from summer velocities where data exist, to ensure prescribing the most representative tensile strength for the ablation season. The tensile stresses are used both as an input to crevasse depth modelling and also for determining the runoff ratio of meltwater routed across the ice surface. The runoff ratio is 1 where cells no not contain crevasses, and 0 when tensile stresses exceed the prescribed tensile strength, such that upstream runoff is captured by surface crevasses, resetting downstream flow accumulation to zero.

Appendix C: Calculation of crevasse depths

The model uses accumulated surface meltwater and the surface tensile stress regime (Fig. 2) as inputs to a model of water-filled crevasse penetration to calculate crevasse depth, d, based on linear elastic fracture mechanics, after van der Veen (2007):

$$K_I = 1.12R_{xx}\sqrt{\pi d} - 0.683\rho_i g d^{1.5} + 0.683\rho_w g b^{1.5}. \tag{C1}$$

The net stress intensity factor, K_I, which describes elastic stresses incident on the tip of a crevasse, is found by summing the terms on the right which describe stress intensity factors relating to the tensile stress, the lithostatic stress of the ice, and the effect of water-filling within the crevasse. Acceleration due to gravity g, density of ice ρ_i and density of freshwater ρ_w are assigned the standard values of 9.81 m s^2, 918 and 1000 kg m^{-3} respectively. Surface tensile stresses R_{xx} derived from velocity data are used as input to the first term in the right-hand side. Meltwater accumulated in each cell determines the water level in a crevasse, b, in the third term using Q, the rate at which a crevasse is filled with water, and time, t, where

$$b = Qt. \qquad \text{(C2)}$$

The level of the meltwater, b, in a crevasse is also controlled by crevasse geometry. Accumulated daily surface meltwater is calculated as a depth of water equivalent generated across each 500 m × 500 m cell. This water then converges into the prescribed surface area of a crevasse when applied to Eq. (C1). The model assumes one crevasse per cell, and crevasse surface dimensions are prescribed as a depth-averaged width of 1 m and a length of 500 m (cell width) for the standard model runs. The width was prescribed at 1 m to represent a depth-average of observed crevasse widths in Greenland, ranging from the scale of metres at depth below the ice surface (e.g. Cook, 1956), to centimetres or decimetres as crevasses narrow towards the surface (e.g. Doyle et al., 2013).

The fracture toughness of ice, K_{IC}, is the critical stress at which a pre-existing flaw will begin to propagate, for which we prescribe a fracture toughness of 150 kPa m$^{1/2}$ as an average of values calculated by Fischer et al. (1995) and Rist et al. (1999). We prescribe an initial crevasse depth, or pre-existing flaw, of 1×10^{-7} m to ensure initiation of fracture propagation. Solving iteratively for depth, d, until K_I is less than the prescribed ice fracture toughness, K_{IC}, the model calculates the propagation depth of each crevasse. Crevasse propagation depths are calculated each day for cells where R_{xx} equals or exceeds the prescribed tensile strength, with depth increasing with time while propagation continues in response to daily accumulated surface meltwater.

The locations of moulins, delivering meltwater to the ice–bed interface, are predicted when crevasse depth equals ice thickness, which is based upon a 5 km ice thickness data set derived from ice-penetrating radar (Bamber et al., 2001). In this study we imply that a moulin is any connection where surface meltwater has forced propagation of a crevasse through the full ice thickness between the ice surface and the ice–bed interface, including crevasses beneath drained lakes. Intersection of supraglacial streams and surface crevasses can initiate the formation of traditional, circular moulins, although many of these connections will close within 1 year due to refreezing and due to creep closure of crevasses when the supply of meltwater is shut off (van der Veen, 2007). It is thus not assumed that the modelled surface-to-bed connections must take the form of traditional moulins, nor does the model account for perennial moulins reopened after the accumulation season.

The drainage of supraglacial lakes, identified by manual digitization of Landsat imagery, is accounted for within the model where it is assumed that a crevasse is present beneath each lake, regardless of the local tensile stress. The volume of meltwater stored in each lake is used to calculate the depth of meltwater within a crevasse, b, at each daily time step, converting stored meltwater in mm w.e. to crevasse water depth in m w.e., and adjusted for crevasse width and length. Drainage of lakes within one 24 h time step is supported by the sub-daily drainage of supraglacial lakes witnessed in Southwest Greenland by Das et al. (2008) and Doyle et al. (2013). Thus when Eq. (9) is solved for $K_I \geq K_{IC}$, where d is set to equal the ice thickness, lakes drain to the bed within one model time step since lake meltwater content has reached a level sufficient for crevasse propagation through the full ice thickness.

Author contributions. The modelling approach was developed by C. C. Clason, with meteorological and velocity data for Leverett glacier contributed by A. Sole and I. D. Bartholomew. The experimental concept was developed by C. C. Clason, D. W. F. Mair and P. W. Nienow. S. Palmer contributed the ice surface DEM, and W. Schwanghart developed the flow accumulation and lake filling subroutine. C. C. Clason wrote the original draft with contribution from D. W. F. Mair.

Acknowledgements. We acknowledge the College of Physical Sciences, University of Aberdeen, the Leverhulme Trust through a Study Abroad Studentship and the Swedish Radiation Safety Authority, for funding awarded to C. Clason. Data collection was supported by the UK Natural Environment Research Council (through a studentship to I. Bartholomew and grants to P. Nienow and D. Mair) and the Edinburgh University Moss Centenary Scholarship (I. Bartholomew). We acknowledge I. Joughin for providing InSAR velocity data, and the National Snow and Ice Data Centre, University of Colorado, and J. Bamber for ice thickness data. We thank two anonymous reviewers for their comments which helped to improve the paper.

Edited by: G. H. Gudmundsson

References

Andrews, L. C., Catania, G. A., Hoffman, M. J., Gulley, J. D., Lüthi, M. P., Ryserm C., Hawley, R. L., and Neumann, T. A.: Direct observations of evolving subglacial drainage beneath the Greenland Ice Sheet, Nature, 514, 80–83, doi:10.1038/nature13796, 2014.

Arnold, N. S., Banwell, A. F., and Willis, I. C.: High-resolution modelling of the seasonal evolution of surface water storage on the Greenland Ice Sheet, The Cryosphere, 8, 1149–1160, doi:10.5194/tc-8-1149-2014, 2014.

Bamber, J. L., Layberry, R. L., and Gogineni, S. P.: A new ice thickness and bed data set for the Greenland ice sheet: 1. Measurement, data reduction, and errors, J. Geophys. Res., 106, 33773–33780, 2001.

Banwell, A. F., Willis, I. C., and Arnold, N. S.: Modeling subglacial water routing at Paakitsoq, W Greenland, J. Geophys. Res.-Earth, 118, 1282–1295, 2013.

Bartholomew, I., Nienow, P., Mair, D., Hubbard, A., King, M., and Sole, A.: Seasonal evolution of subglacial drainage and acceleration in a Greenland outlet glacier, Nat. Geosci., 3, 408–411, 2010.

Bartholomew, I., Nienow, P., Sole, A., Mair, D., Cowton, T., Palmer, S., and Wadham, J.: Supraglacial forcing of subglacial drainage in the ablation zone of the Greenland ice sheet, Geophys. Res. Lett, 85, L08502, doi:10.1029/2011GL047063, 2011a.

Bartholomew, I. D., Nienow, P., Sole, A., Mair, D., Cowton, T., King, M. A., and Palmer, S.: Seasonal variations in Greenland Ice Sheet motion: Inland extent and behaviour at higher elevations, Earth Planet. Sc. Lett., 307, 271–278, 2011b.

Bartholomew, I., Nienow, P., Sole, A., Mair, D., Cowton, T., and King, M. A.: Short-term variability in Greenland Ice Sheet motion forced by time-varying meltwater drainage: Implications for the relationship between subglacial drainage system behaviour and ice velocity, J. Geophys. Res., 117, F03002, doi:10.1029/2011JF002220, 2012.

Box, J. E. and Ski, K.: Remote sounding of Greenland supraglacial melt lakes: implications for subglacial hydraulics, J. Glaciol., 53, 257–265, 2007.

Box, J. E., Bromwich, D., Veenhuis, B., Bai, L. S., Stroeve, J., Rogers, J., Steffen, K., Haran, T., and Wang, S. H.: Greenland Ice Sheet Surface Mass Balance Variability (1988–2004) from Calibrated Polar MM5 Output, J. Climate, 19, 2783–2800, 2006.

Box, J. E., Cappelen, J., Decker, D., Fettweis, X., Mote, T., Tedesco, M., and van de Wal, R. S. W.: Greenland, Arctic Report Card 2010, http://www.arctic.noaa.gov/reportcard (last access: 17 June 2014), 2010.

Braithwaite, R. J., Laternser, M., and Pfeffer, W. T.: Variations of near-surface firn density in the lower accumulation area of the Greenland ice sheet, Pakitsoq, West Greenland, J. Glaciol., 40, 477–485, 1994.

Campbell, F. M. A.: The role of supraglacial snowpack hydrology in mediating meltwater delivery to glacier systems, PhD thesis, University of Glasgow, Glasgow, UK, 227 pp., http://theses.gla.ac.uk/2871/ (last access: 17 June 2014), 2007.

Catania, G. A., Neumann, T. A., and Price, S. F.: Characterizing englacial drainage in the ablation zone of the Greenland ice sheet, J. Glaciol., 54, 567–578, 2008.

Chandler, D. M., Wadham, J. L., Lis, G. P., Cowton, T., Sole, A., Bartholomew, I., Telling, J., Nienow, P., Bagshaw, E. B., Mair, D., Vinen, S., and Hubbard, A.: Evolution of the subglacial drainage system beneath the Greenland Ice Sheet revealed by tracers, Nat. Geosci., 6, 195–198, 2013.

Clason, C. C., Mair, D. W. F., Burgess, D. O., and Nienow, P. W.: Modelling the delivery of supraglacial meltwater to the ice/bed interface: application to the southwest Devon Ice Cap, Nunavut, Canada, J. Glaciol., 58, 361–374, 2012.

Colgan, W., Steffen, K., Mclamb, W. S., Abdalati, W., Rajaram, H., Motyka, R., Phillips, T., and Anderson, R.: An increase in crevasse extent, West Greenland: Hydrologic implications, Geophys. Res. Lett., 38, L18502, doi:10.1029/2011GL048491, 2011.

Cook, J. C.: Some Observations in a Northwest Greenland Crevasse, Transactions, Am. Geophys. Union, 37, 715–718, 1956.

Cuffey, K. M. and Paterson, W. S. B.: The Physics of Glaciers, 4th Edn., Butterworth-Heinemann, Oxford, 2010.

Das, S. B., Joughin, I., Behn, M. D., Howat, I. M., King, M. A., Lizarralde, D., and Bhatia, M. P.: Fracture Propagation to the Base of the Greenland Ice Sheet During Supraglacial Lake Drainage, Science, 320, 778–781, 2008.

de Fleurian, B., Gagliardini, O., Zwinger, T., Durand, G., Le Meur, E., Mair, D., and Råback, P.: A double continuum hydrological model for glacier applications, The Cryosphere, 8, 137–153, doi:10.5194/tc-8-137-2014, 2014.

Doyle, S. H., Hubbard, A. L., Dow, C. F., Jones, G. A., Fitzpatrick, A., Gusmeroli, A., Kulessa, B., Lindback, K., Pettersson, R., and Box, J. E.: Ice tectonic deformation during the rapid in situ drainage of a supraglacial lake on the Greenland Ice Sheet, The Cryosphere, 7, 129–140, doi:10.5194/tc-7-129-2013, 2013.

Fischer, M. P., Alley, R. B., and Engelder, T.: Fracture toughness of ice and firn determined from the modified ring test, J. Glaciol., 41, 383–394, 1995.

Fitzpatrick, A. A. W., Hubbard, A. L., Box, J. E., Quincey, D. J., van As, D., Mikkelsen, A. P. B., Doyle, S. H., Dow, C. F., Hasholt,

B., and Jones, G. A.: A decade (2002–2012) of supraglacial lake volume estimates across Russell Glacier, West Greenland, The Cryosphere, 8, 107–121, doi:10.5194/tc-8-107-2014, 2014.

Hewitt, I. J.: Seasonal changes in ice sheet motion due to melt water lubrication, Earth Planet. Sc. Lett., 371–372, 16–25, 2013.

Hoffman, M. J. and Price, S.: Feedbacks between coupled subglacial hydrology and glacier dynamics, J. Geophys. Res.-Earth, 119, 414–436, 2014.

Hoffman, M. J., Catania, G. A., Neumann, T. A., Andrews, L. C., and Rumrill, J. A.: Links between acceleration, melting, and supraglacial lake drainage of the western Greenland Ice Sheet, J. Geophys. Res.-Earth, 116, F04035, doi:10.1029/2010JF001934, 2011.

IPCC: Climate Change 2007: The Physical Science Basis, in: Contribution of Working Group I to the Fourth Assessment Report of the Intergovernmental Panel on Climate Change, edited by: Solomon, S., Qin, D., Manning, M., Chen, Z., Marquis, M., Averyt, K. B., Tignor, M., and Miller, H. L., Cambridge University Press, Cambridge, UK, 2007.

Joughin, I., Smith, B. E., Howat, I. M., Scambos, T., and Moon, T.: Greenland flow variability from ice-sheet-wide velocity mapping, J. Glaciol., 56, 415–430, 2010.

Leeson, A. A., Shepherd, A., Palmer, S., Sundal, A., and Fettweis, X.: Simulating the growth of supraglacial lakes at the western margin of the Greenland ice sheet, The Cryosphere, 6, 1077–1086, doi:10.5194/tc-6-1077-2012, 2012.

Lefebre, F., Gallée, H., Van Ypersele, J., and Huybrechts, P.: Modelling of large-scale melt parameters with a regional climate model in south Greenland during the 1991 melt season, Ann. Glaciol., 35, 391–397, 2002.

McMillan, M., Nienow, P., Shepherd, A., Benham, T., and Sole, A.: Seasonal evolution of supra-glacial lakes on the Greenland Ice Sheet, Earth Planet. Sc. Lett., 262, 484–492, 2007.

Meierbachtol, T., Harper, J., and Humphrey, N.: Basal drainage system response to increasing surface melt on the Greenland ice sheet, Science, 341, 777–779, 2013.

Morriss, B. F., Hawley, R. L., Chipman, J. W., Andrews, L. C., Catania, G. A., Hoffman, M. J., Lüthi, M. P., and Neumann, T. A.: A ten-year record of supraglacial lake evolution and rapid drainage in West Greenland using an automated processing algorithm for multispectral imagery, The Cryosphere, 7, 1869–1877, doi:10.5194/tc-7-1869-2013, 2013.

Nye, J. F.: The distribution of stress and velocity in glaciers and ice sheets, P. Roy. Soc. Lond. A, 239, 113–133, 1957.

Palmer, S., Shepherd, A., Nienow, P., and Joughin, I.: Seasonal speedup of the Greenland Ice Sheet linked to routing of surface meltwater, Earth Planet. Sc. Lett., 302, 423–428, 2011.

Phillips, T., Rajaram, H., and Steffen, K.: Cryo-hydrologic warming: A potential mechanism for rapid thermal response of ice sheets, Geophys. Res. Lett., 37, L20503, doi:10.1029/2010GL044397, 2010.

Reeh, N.: Parameterization of melt rate and surface temperature on the Greenland ice sheet, Polarforschung, 59, 113–128, 1991.

Rist, M. A., Sammonds, P. R., Murrell, S. A. F., Meredith, P. G., Doake, C. S. M., Oerter, H., and Matsuki, K.: Experimental and theoretical fracture mechanics applied to Antarctic ice fracture and surface crevassing, J. Geophys. Res., 104, 2973–2986, 1999.

Schoof, C.: Ice sheet acceleration driven by melt supply variability, Nature, 468, 803–306, 2010.

Schuler, T. V., Loe, E., Taurisano, A., Eiken, T., Hagen, J. O., and Kohler, J.: Calibrating a surface mass-balance model for Austfonna ice cap, Svalbard, Ann. Glaciol., 46, 241–248, 2007.

Schwanghart, W. and Kuhn, N. J.: TopoToolbox: A set of Matlab functions for topographic analysis, Environ. Modell. Softw., 25, 770–781, 2010.

Schwanghart, W. and Scherler, D.: Short Communication: TopoToolbox 2 – MATLAB-based software for topographic analysis and modeling in Earth surface sciences, Earth Surf. Dynam., 2, 1–7, doi:10.5194/esurf-2-1-2014, 2014.

Selmes, N., Murray, T., and James, T. D.: Fast draining lakes on the Greenland Ice Sheet, Geophys. Res. Lett., 38, L15501, doi:10.1029/2011GL047872, 2011.

Shannon, S. R., Payne, A. J., Bartholomew, I. D., van den Broeke, M. R., Edwards, T. L., Fettweis, X., Gagliardini, O., Gillet-Chaulet, F., Goelzer, H., Hoffman, M. J., Huybrechts, P., Mair, D. W. F., Nienow, P. W., Perego, M., Price, S. F., Smeets, C. J. P. P., Sole, A. J., van de Wal, R. S. W., and Zwinger, T.: Enhanced basal lubrication and the contribution of the Greenland ice sheet to future sea-level rise, P. Natl. Acad. Sci. USA, 110, 14156–14161, doi:10.1073/pnas.1212647110, 2013.

Sole, A., Nienow, P., Bartholomew, I., Mair, D., Cowton, T., Tedstone, A., and King, M. A.: Winter motion mediates dynamic response of the Greenland Ice Sheet to warmer summers, Geophys. Res. Lett., 40, 3940–3944, 2013.

Sundal, A. V., Shepherd, A., Nienow, P., Hanna, E., Palmer, S., and Huybrechts, P.: Evolution of supra-glacial lakes across the Greenland Ice Sheet, Remote Sens. Environ., 113, 2164–2171, 2009.

Tedesco, M., Willis, I. C., Hoffman, M. J., Banwell, A. F., Alexander, P., and Arnold, N. S.: Ice dynamic response to two modes of surface lake drainage on the Greenland ice sheet, Environ. Res. Lett., 8, 034007, doi:10.1088/1748-9326/8/3/034007, 2013.

Tedstone, A. J., Nienow, P. W., Bartholomew, I. D., Sole, A. J., Cowton, T. R., Mair, D. W. F., and King, M. A.: Greenland ice sheet motion insensitive to exceptional meltwater forcing, P. Natl. Acad. Sci. USA, 110, 19719–19724, 2013.

van der Veen, C. J.: Fracture propagation as means of rapidly transferring surface meltwater to the base of glaciers, Geophys. Res. Lett., 34, L01501, doi:10.1029/2006GL028385, 2007.

Vaughan, D. G.: Relating the occurrence of crevasses to surface strain rates, J. Glaciol., 39, 255–266, 1993.

Vieli, A., Payne, A. J., Du, Z., and Shepherd, A.: Numerical modelling and data assimilation of the Larsen B ice shelf, Antarctic Peninsula, Philos. T. Roy. Soc. A, 364, 1815–1839, 2006.

Zwally, J. H., Abdalati, W., Herring, T., Larson, K., Saba, J., and Steffen, K.: Surface Melt-Induced Acceleration of Greenland Ice Sheet Flow, Science, 297, 218–222, 2002.

14

A decade (2002–2012) of supraglacial lake volume estimates across Russell Glacier, West Greenland

A. A. W. Fitzpatrick[1], A. L. Hubbard[1], J. E. Box[2], D. J. Quincey[3], D. van As[2], A. P. B. Mikkelsen[4], S. H. Doyle[1], C. F. Dow[5], B. Hasholt[4], and G. A. Jones[1]

[1]Institute of Geography and Earth Sciences, Aberystwyth University, Aberystwyth, UK
[2]Geological Survey of Denmark and Greenland (GEUS), Copenhagen, Denmark
[3]School of Geography, University of Leeds, Leeds, LS2 9JT, UK
[4]Department of Geography and Geology, University of Copenhagen, Denmark
[5]Glaciology Group, College of Science, Swansea University, Swansea, SA2 8PP, UK

Correspondence to: A. A. W. Fitzpatrick (aaf07@aber.ac.uk)

Abstract. Supraglacial lakes represent an ephemeral storage buffer for meltwater runoff and lead to significant, yet short-lived, episodes of ice-flow acceleration by decanting large meltwater and energy fluxes into the ice sheet's hydrological system. Here, a methodology for calculating lake volume is used to quantify storage and drainage across Russell Glacier, West Greenland, between 2002 and 2012. Using 502 MODIS scenes, water volume at ~ 200 seasonally occurring lakes was derived using a depth–reflectance relationship, which was independently calibrated and field validated against lake bathymetry. The inland expansion of lakes is strongly correlated with air temperature: during the record melt years of 2010 and 2012, lakes formed and drained earlier, attaining their maximum volume 38 and 20 days earlier than the 11 yr mean, as well as occupying a greater area and forming at higher elevations (> 1800 m) than previously. Despite occupying under 2 % of the study area, lakes delay the transmission of up to 7–13 % of the bulk meltwater discharged. Although the results are subject to an observational bias caused by periods of cloud cover, we estimate that across Russell Glacier, 28 % of supraglacial lakes drain rapidly (< 4 days). Clustering of such events in space and time suggests a synoptic trigger mechanism. Further, we find no evidence to support a unifying critical size or depth-dependent drainage threshold.

1 Introduction

Meltwater runoff from the Greenland Ice Sheet (GrIS) has increased since the 1960s (Hanna et al., 2005) and has been linked to seasonal ice-flow variability through basal hydrological forcing (Zwally et al., 2002; Joughin et al., 2008; van de Wal et al., 2008; Bartholomew et al., 2010). Horizontal and vertical ice surface motion lags peak daily melt and responds to rapid supraglacial lake (SGL) tapping (Shepherd et al., 2009; Das et al., 2008; Doyle et al., 2013). Furthermore, the spatial pattern of seasonal flow acceleration coincides with the development of ice sheet surface hydrology (Palmer et al., 2011; Fitzpatrick et al., 2013) where SGLs play a vital role via temporary storage and routing of surface meltwater. SGLs, which are prevalent across the entire GrIS (Selmes et al., 2011), also drain rapidly via hydraulic fracture propagation in ~ 2 h (Das et al., 2008; Doyle et al., 2013), forming moulins that provide a direct hydraulic connection to the ice sheet bed. Previous studies on SGLs have quantified spatial and temporal variations in SGL area across different sectors of the GrIS (e.g. Georgiou et al., 2009; Sundal et al., 2009; Selmes et al., 2011; Liang et al., 2012), and have estimated their volume (e.g. Box and Ski, 2007; Sneed and Hamilton, 2007), whilst others have attempted to model SGL evolution (e.g. Luthje et al., 2006; Banwell et al., 2012) and measure lake-bed melt rate (Tedesco et al., 2012). The extent to which temperatures correlate with melt volume and SGL storage remains unclear. Additionally, modelling studies (e.g.

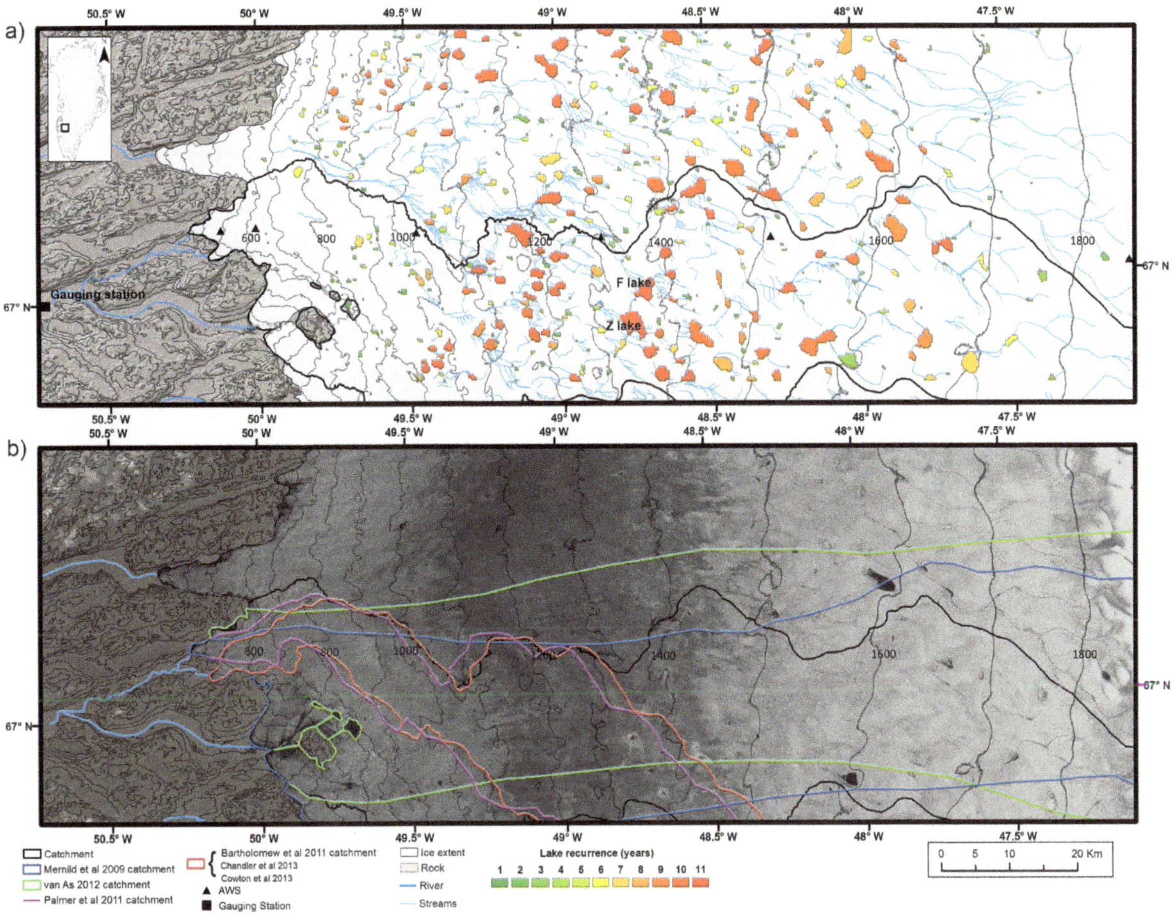

Fig. 1. **(a)** The location of all the SGLs within the study area between 2002–2012, coloured by their recurrence. The location of the AWS, gauging station and calibration lakes (lakes F and Z) are shown in addition to the supraglacial hydrological network, based on manual digitisation of LANDSAT and ASTER imagery between 2009–2012. The boundary of the defined Russell Glacier catchment is also shown; **(b)** a comparison of the catchment boundaries used in this study with those of previous studies (van As et al., 2012; Palmer et al., 2011; Mernild and Hasholt, 2009; Bartholomew et al., 2011; Chandler et al., 2013; Cowton et al., 2013), overlaid onto a Landsat 7 image from July 2012.

Clason et al., 2012; Banwell et al., 2012) infer the existence of a volumetric threshold beyond which rapid lake drainage is inevitable, however this has yet to be validated by field observations.

We investigate the seasonal evolution of SGLs over Russell Glacier using 11 yr (2002–2012) of quasi-daily lake volume estimates. This period includes extremely warm years when melt records were set (Nghiem et al., 2012; Tedesco et al., 2013), and therefore provides an opportunity to determine how the evolution of SGLs are modulated by regional climatic forcing. The year 2010 was the warmest on record (Cappelen et al., 2012; Box et al., 2010), 2.5 standard deviations above the 1973–2010 average in West Greenland (van As et al., 2012). Temperatures for April, June and July in 2012 also rank amongst the warmest on record (Cappelen et al., 2012), resulting in the second warmest summer (after 2010) and the largest spatial extent of melt as yet on record (Nghiem et al., 2012). We present a field calibrated algorithm for calculating changes in SGL volume during these years and compare them with longer term patterns derived over the period 2002–2012.

2 Study area

Russell Glacier is an ideal region to investigate SGL dynamics as this sector of the GrIS accounts for the highest areal extent of SGLs across the ice sheet (Selmes et al., 2011). The study area covers $\sim 6500\,km^2$ of the western GrIS between 66.8 and 67.3° N, centring on Russell Glacier and extends from the ice margin to ~ 2000 m elevation (~ 150 km inland, Fig. 1), and is hereafter referred to as the "Kangerlussuaq sector" of the GrIS. For hydrological impact assessment, this region benefits from a long-term meteorological record obtained from a transect of six automated weather stations extending well above the local equilibrium line altitude to 1830 m (van As et al., 2012; van de Wal et al., 2012) and bulk proglacial discharge gauged near a bridge across Watson River, Kangerlussuaq, since 2007 (Hasholt et al., 2013).

3 Methods

To investigate supraglacial lake dynamics, we extracted lake area, depth and volumes from 11 yr of daily-acquired MODIS imagery. The following sections detail the methods, limitations and uncertainties of this technique. The methods for deriving ancillary data (surface runoff data and proglacial discharge) and their spatial coverage are discussed.

3.1 Calculating lake extent

SGL areas were calculated using multitemporal Moderate Resolution Imaging Spectroradiometer (MODIS) imagery from the Terra satellite. We used the level 2, MOD09 product, which provides ungridded swaths of atmospherically corrected, calibrated and geolocated surface reflectance. MRTSwath version 2.2 (https://lpdaac.usgs.gov/tools) was used to convert the atmospherically corrected MOD09 land product from HDF swath format to a geographically referenced gridded image using the corresponding MODIS level 1B 1 km geolocation file for each image (MOD03). Although the region experiences 24 h daylight in summer, illumination conditions alter with changing solar zenith angle, therefore scenes used in this study were restricted to those captured at nadir. Scenes covering the entire melt season were obtained from 2002 to 2012 from day of year (DOY) 121 (1 May) to DOY 274 (30 September).

Previous studies classified supraglacial lakes in MODIS imagery using manual digitisation (McMillan et al., 2007; Lampkin and Vanderberg, 2011), band thresholds (Box and Ski, 2007; Liang et al., 2012), and object-orientated approaches (Sundal et al., 2009; Johansson et al., 2013). Here, we opted for a semi-automatic strategy to classify SGL extents, using a modified normalized difference water index (NDWI) (Huggel et al., 2002; Yang and Smith, 2013, Eq. 1) and freely available RSGISlib software (http://www.rsgislib.org). Before classification, MODIS bands within the visible spectrum (0.46 to 0.57 μm; bands 3 (blue) and 4 (green)) were sharpened from 500 m to 250 m resolution, using the ratio of bands 1 (red) at 250 m and 500 m resolution (L. Gumley et al., http://earthdata.nasa.gov/sites/default/files/field/document/MODIS_True_Color.pdf). The approach for identifying water cover was to isolate regions with a high reflectance in the blue spectral band (band 3), utilising their striking colour contrast against the surrounding ice. Thresholds for both red and blue bands were used in combination with the NDWI threshold to reduce misclassification of pixels with spectrally similar signals to SGLs. A cloud filter was also included in the classification using MODIS band 6 (1628–1652 nm) which is sensitive to solid and liquid phase water, and final SGL classification was subsequently carried out on pixels which had band 6 reflectance of less than 0.15 consistent with Box and Ski (2007).

$$\mathrm{NDWI} = \frac{(b3 - b1)}{(b3 + b1)} \quad (1)$$

3.2 Depth and volume estimation

Methods for calculating the volume of water contained in SGLs from satellite imagery have been established based on the principles of light transmission through water. Lake reflectance in MODIS band 1 reduces exponentially with depth allowing a depth–reflectance function to be extracted (Box and Ski, 2007). However, other studies have used a physically based retrieval to derive depth based on lake bottom albedo using the Bouger–Lambert–Beer law and applying the technique to ASTER scenes providing estimates of lake volume (Sneed and Hamilton, 2007). Numerically modelled depths using this technique were subsequently validated with in situ measurements (Sneed and Hamilton, 2011; Tedesco and Steiner, 2011).

In this study we opt to calculate lake volume using an empirically derived depth–reflectance relationship. This relationship was calibrated using bathymetry data of two SGLs surveyed in July 2010, referred to here as lake F (67.00° N, 48.71° W) and lake Z (66.96° N, 48.78° W) (Fig. 1a). On 26 June 2010, lake F was instrumented with water level sensors and on 29 June the lake perimeter was surveyed using a Leica System 500 Global Positioning System (GPS) receiver differentially corrected against a bedrock-mounted reference station. Following lake drainage on 30 June (Doyle et al., 2013), six elevation transects were surveyed over the lakebed using the GPS recording a location point every second. The combination of lake bathymetry and depth time series allowed for shoreline reconstruction (Doyle et al., 2013). The second, undrained lake was surveyed by mounting a depth sounder (Garmin X10) and GPS receiver (Leica System 500) on an inflatable raft, whilst the perimeter was surveyed using a similar GPS set-up. The depth sounder has a stated accuracy of ± 0.1 m. The GPS transect data and the extrapolated margins for both lakes were interpolated onto a 10 m $\times$ 10 m grid using kriging interpolation to yield lake bathymetry maps (Fig. 2). Lake perimeter measurements were overlain on MODIS imagery from corresponding days. MODIS pixels that contained more than 50 % water were accepted as lake pixels, constraining the minimum and maximum band thresholds for water. SGL bathymetries (Fig. 2) were interpolated to 250 m resolution to match the MODIS band 1 resolution, so that data could be extracted and plotted to reveal the relationship between depth and surface reflectance (Fig. 3a). Following Box and Ski (2007), we approximated the relationship between depth and surface reflectance using a least-squares fit, where D is depth and R is MODIS band 1

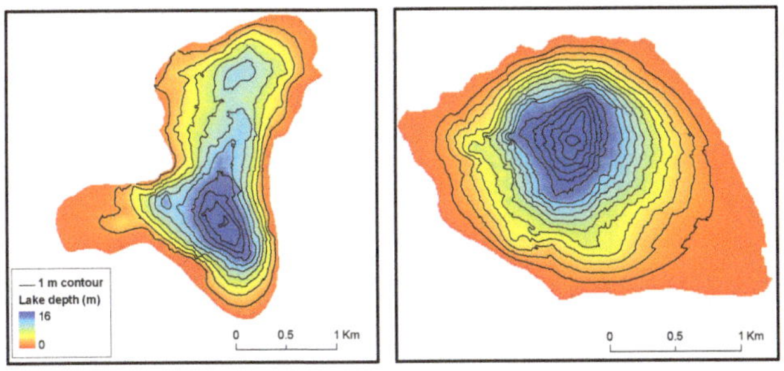

Fig. 2. Bathymetry maps of lake F and lake Z, data from which was used to calibrate the depth–reflectance curve.

reflectance:

$$D = \frac{0.65}{(0.035 + R)} - 0.4. \qquad (2)$$

The equation used in this study differs slightly to the one used by Box and Ski (2007) in that lake depth has a shallower upper and lower limit in relation to values of low and high reflectance respectively. This refinement is a result of our lake bathymetry measurements. Equation (2) was applied to all pixels classified as water across the study region to yield depth estimates, which were subsequently integrated over the SGL area to calculate water volume.

3.3 Surface melt, pro-glacial discharge and catchment delineation

Surface runoff data were derived from a surface energy balance model using AWS measurements and MODIS (MOD10A1 product) derived surface albedo as input (van As et al., 2012). The model iteratively searches the surface temperature for which the energy fluxes are in balance. When this fails due to a limiting melting-point temperature, the surplus energy yields snow or ice melt. Model input data were interpolated to 100 m elevation bins to determine the distributed melt patterns in the region. Model uncertainty for runoff totals was calculated to be between 6–14 % (van As et al., 2012).

Discharge measurements were recorded at a gauging station near the bridge crossing the Watson river at Kangerlussuaq (Fig. 1a). Discharge was derived using pressure transducers and a stage–discharge relationship, and at high stages with an acoustic doppler current profiler (Hasholt et al., 2013). The catchment described in Hasholt et al. (2013) has been refined in this study. The new catchment outline, hereafter referred to as the Russell Glacier catchment (Fig. 1a and b) was derived by manually digitising the locations of all major supraglacial streams using panchromatic (15 m resolution) and multispectral (30 m resolution) ASTER and LANDSAT images between 2009–2012. Multiple images acquired at different times in the melt season were used to map streams in each year to maximise coverage of the network. Using this map of the supraglacial hydrological network the

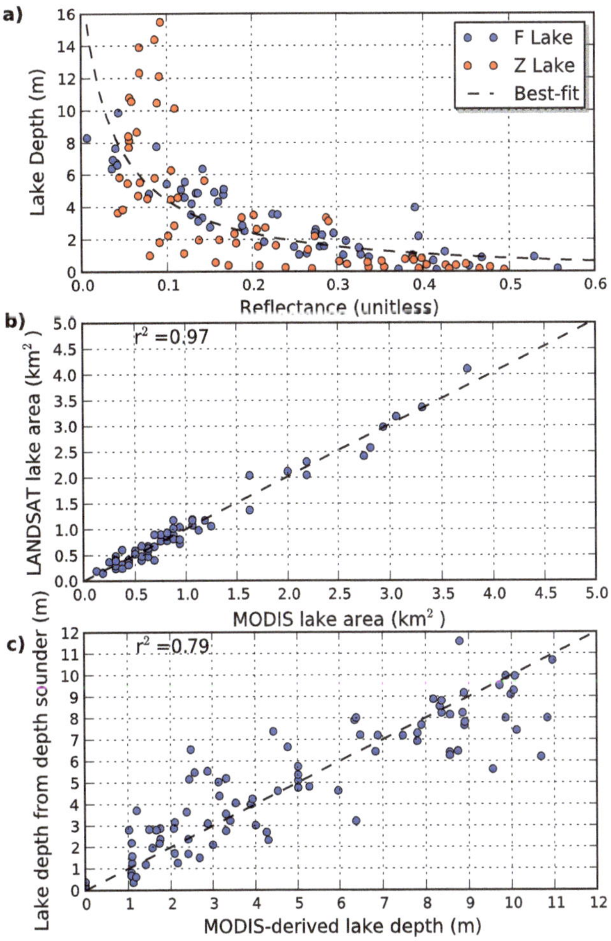

Fig. 3. (a) Lake F and Z MODIS band 1 reflectance vs. depth with best fit curve; **(b)** the relationship between lake area estimates derived from manually digitising 15 m resolution panchromatic LANDSAT 7 images and the automated MODIS classification; and **(c)** the relationship between independently measured lake depth for 3 SGLs and modelled lake depth using the MODIS depth–reflectance relationship.

Hasholt et al. (2013) catchment was edited, ensuring that the catchment boundary did not cross streams, and instead encompassed the stream network for each lake (Fig. 1a and b). For comparison, catchment boundaries from Palmer et al. (2011), van As et al. (2012), Mernild and Hasholt (2009), Cowton et al. (2013) and Chandler et al. (2013) are shown in Fig. 1b. Lake volume losses using data from the Kangerlussuaq sector of the GrIS and a subset of the lake database within the Russell Glacier catchment were compared to proglacial discharge.

3.4 Uncertainty estimation

3.4.1 Area

We evaluated the accuracy of the classification by comparing the area of individual SGLs extracted from MODIS data with a SGL area data set derived from finer spatial resolution (15 m) panchromatic LANDSAT ETM+ images. In total, 62 lakes (free of surface ice) were manually digitized from four panchromatic LANDSAT ETM+ images from different years and at different stages of the melt season (26 June 2008, 17 July 2004, 20 July 2005, 8 August 2006). The comparison shows a strong agreement between the two data sets (Fig. 3b) with strong correlation ($r^2 = 0.97$) and a root mean square deviation (RMSD) of $0.14\,\text{km}^2$, comparable with previous studies (e.g. $0.11\,\text{km}^2$, Selmes et al., 2011, and $0.22\,\text{km}^2$, Sundal et al., 2009). This RMSD value was multiplied by the number of SGLs occurring each day to provide error margins for total SGL area estimates.

The minimum area of any individual SGL identified from MODIS images is restricted by pixel size ($0.0625\,\text{km}^2$) thereby introducing potential bias in total lake area and its size distribution. Previous work using higher resolution products calculated that small lakes ($< 0.1\,\text{km}^2$) account for 12 % of the total lake area based on analysis of a single mid-season ASTER image (Sundal et al., 2009). In our study SGLs smaller than $0.0625\,\text{km}^2$ were identified from a LANDSAT ETM+ image (26 June 2008), indicating that they account for 7.5 % of the total SGL area at this time. From the same image, 11.1 % of SGLs were found to be smaller than $0.1\,\text{km}^2$ in line with the Sundal et al. (2009) study.

3.4.2 Volume

Depth surveys were undertaken on two SGLs located to the north of the study area on 14 and 15 August, 2005, referred to as Lakes A and B in Box and Ski (2007), and on Lake Disco on 22 July and 8–9 August, 2007. The results of these surveys were overlain on the corresponding MODIS derived depth maps enabling comparison (Fig. 3c). There is good agreement between the MODIS-derived depth and the raft-measured depth ($r^2 = 0.79$, RMSD = 1.47 m). To calculate percent volumetric error, the point measurements from Lake Disco surveys were gridded and interpolated onto a 10 m grid, with the lake margin delimited from a Landsat 7 image acquired on the 8 August 2007. The $9.5 \times 10^{-3}\,\text{km}^3$ calculated SGL volume agrees within 15 % of the volume from MODIS from 22 July ($11.3 \times 10^{-3}\,\text{km}^3$) and 8–9 August ($11.0 \times 10^{-3}\,\text{km}^3$) 2007. A further comparison between the recorded volumes of Lakes A ($7.1 \times 10^{-3}\,\text{km}^3$) and B ($7.8 \times 10^{-3}\,\text{km}^3$) from Box and Ski (2007), and the estimated volumes using the method described in this study ($5.9 \times 10^{-3}\,\text{km}^3$ and $6.7 \times 10^{-3}\,\text{km}^3$) reveals a volumetric difference of 15.6 % and 13.7 %, respectively.

3.5 Limitations

3.5.1 Cloud Cover

A key limitation to monitoring SGL evolution is the temporal and spatial continuity of available imagery. Cloud cover prevents the daily monitoring of SGLs and therefore the formation and drainage of some SGLs could have been missed. To limit the impact of clouds, SGL classification was only carried out on pixels that had a MODIS band 6 reflectance value of less than 0.15, consistent with Box and Ski (2007). Following automatic classification, images were manually inspected for cloud cover, and images which contained cloud that obscured the view of lakes were removed from the data set. This removal of cloudy images reduced the false positive identification of lakes due to cloud shadows, which in some cases have spectrally similar signatures to optically deep lakes. Although images included in the classification were selected when the Terra satellite was at or near nadir, the changing solar zenith throughout the course of the melt season meant that cloud shadows were present on some images. To eliminate this potential error each image was manually checked for the presence of clouds and their shadows. Furthermore, given that optically thin clouds reduce reflectance values in the red band – on which the depth reflectance function is applied – the removal of cloud-covered images is essential to accurately and consistently derive lake volumes. Cloud cover therefore limited the number of images used to evaluate SGL evolution each year (denoted as n in Fig. 4) and this constraint may introduce bias when identifying rapidly draining lakes. To identify such lakes we selected a four-day threshold as it provided the optimal interval for identifying fast draining lakes on a data set characterised by intermittent gaps due to cloud cover.

3.5.2 Prolonged blackout

Extended periods of cloud clover, termed blackouts, represent a problem for the analysis of lake drainage patterns as we are unable to quantify the amount of water stored in lakes which form and drain during these periods, therefore our results will likely underestimate lake storage and loss. During a prolonged blackout period with consistent cloud cover the degree of surface melting and its contribution to lake formation and expansion is unknown, an inherent limitation of this study. However, given that the average life cycle of a lake is ~ 10 days, we note only 13 blackout periods which extend beyond this time frame between the start of May and the end of August, over the entire 11 yr analysis. Even given a perfect data set, an uncertainty exists regarding lakes loss/gain during the time interval between satellite passes and hence results of this analysis must be considered minimum estimates of area and volume. Analysis of supraglacial lake evolution using optical satellite imagery will always face the problem of blackout periods due to cloud cover and is therefore a

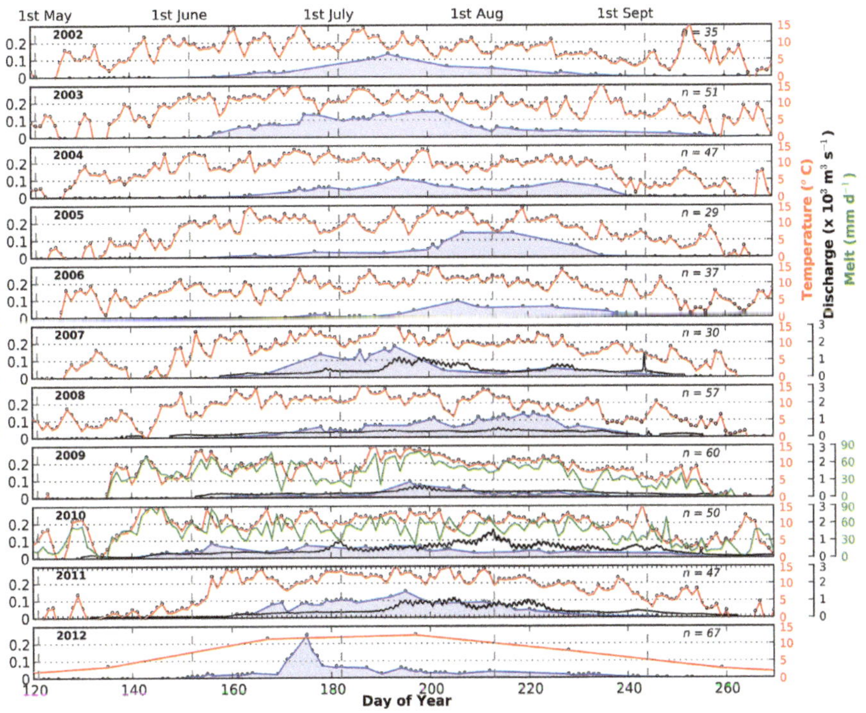

Fig. 4. Total lake volume within Kangerlussuaq sector of the GrIS (blue), Kangerlussuaq air temperature (red), surface melt (green) and proglacial Watson River discharge (black) between 2002–2012. Only monthly average air temperature data is currently available for 2012. Surface melt rate and Watson River discharge are only shown for years when data are available. The number of cloud-free observations of SGLs (denoted *n*) per year is given.

key limitation of the method. Although some studies have demonstrated the use of SAR data for monitoring lake evolution due to its ability to penetrate clouds, it is not on its own a replacement, due to its poor temporal coverage and can only be used to complement other visible–near infrared data sets. Despite the restrictions imposed by cloud cover and relatively low spatial resolution MODIS imagery remains the best source of data available for monitoring SGL evolution.

3.5.3 Classification

The classification we applied is sensitive to the presence of water and does not identify lakes covered by floating ice or snow. Therefore using our technique, lake area and volume will be underestimated in regions or seasons with a high proportion of ice/snow-covered lakes.

3.5.4 Depth–reflectance technique

The accuracy of the volume calculation relies on the calibrated lake depths being representative of the whole SGL population. Lakes deeper than our calibration lakes may exist elsewhere on the ice sheet: the maximum depths of lakes F, Z, A, B and Disco were 16 m, 16 m, 11.5 m, 10 m and 15.2 m, respectively. Box and Ski (2007) note that there is a cut-off to the depth–reflectance relationship, beyond which increased depth fails to reduce pixel reflectance. The depth–reflection function itself limits the accuracy of the volume results as it relies on accurate calculations of atmospherically corrected surface-reflection and bathymetry measurements. Previous remote sensing studies have presented lake area estimates (e.g. Selmes et al., 2011), but Liang et al. (2012) go a significant step further by using a first-order, conical lake area to volume approximation to derive lake volume estimates.

Tedesco and Steiner (2011) compared MODIS (band 3 and band 4) and LANDSAT (band 1 and 2) derived lake depths using the method of Sneed and Hamilton (2007) with in situ measurements. Tedesco and Steiner (2011) revealed that MODIS band 3 and 4 tends to overestimate and underestimate lake depth respectively when compared to in situ data. Here we used MODIS band 1 to derive lake depth because it is more sensitive to water depth than bands 3 and 4 (Box and Ski, 2007). To calibrate the relationship between reflectance and depth, we used bathymetry data from deeper lakes (up to 16 m, Fig. 2) than Tedesco and Steiner (2011), whose calibration lake had a maximum depth of 4.55 m. Tedesco and Steiner (2011) also found differences between the depth values derived from MODIS on board TERRA and AQUA, with values estimated by MODIS on TERRA being closest to those estimated by Landsat and measured in situ. On this basis we only used images from the TERRA satellite to derive depth estimates. We also acknowledge the limitations of

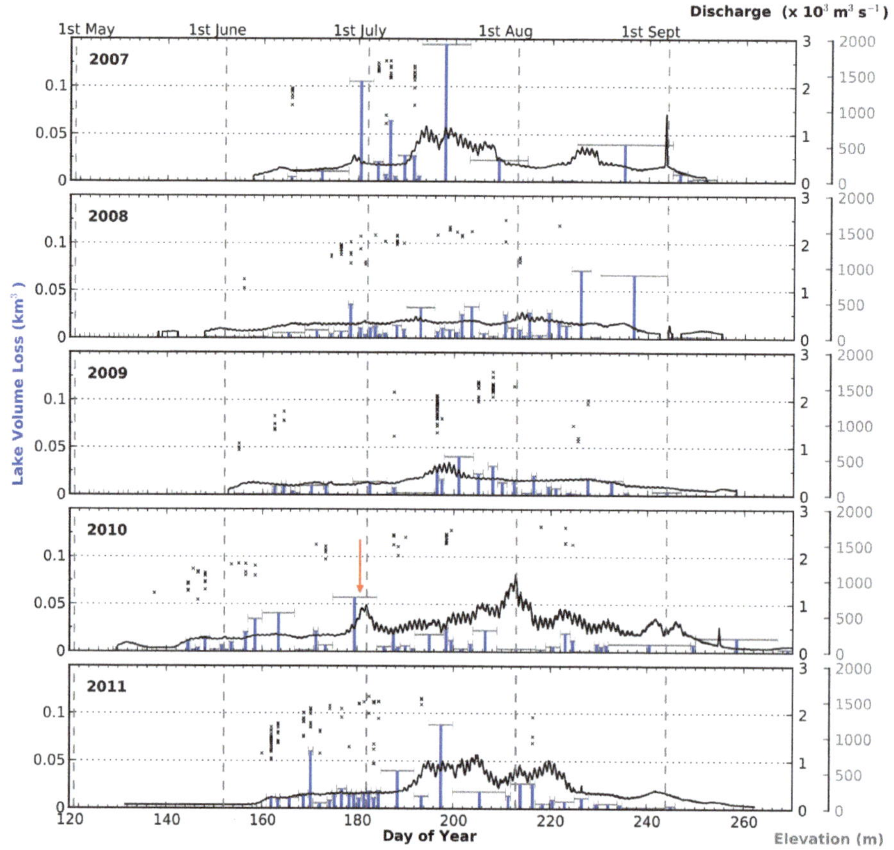

Fig. 5. Lake volume loss within the Kangerlussuaq sector of the GrIS and Watson River discharge between 2007 and 2011. The drainage date and elevation of rapidly draining SGLs are also shown (black crosses). The red arrow illustrates an example of lake drainage coinciding with a peak in proglacial discharge.

deriving volumes from mixed pixels, for example at the lake margins, where the presence of ice lowers the overall pixel reflectance leading to an underestimation of lake depth and hence volume.

4 Results

Our aggregate data set reveals that an average of 200 SGLs ($> 0.0625\,km^2$) are observed each year within the $6500\,km^2$ Kangerlussuaq sector of the GrIS (Fig. 1a). However, the total number of lakes varies from year to year (one standard deviation, ±20 lakes) and 21 % reoccurred annually (Fig. 1a). Mean area of an individual lake was $0.68\,km^2$ but they can grow as large as $8.2\,km^2$ (66.92° N, 48.15° W). SGLs formed 21 days earlier in 2010 compared to the 11 yr mean, with small lakes forming at low elevations by 9 May (Fig. 4 and Table 1). The maximum total volume of SGLs at any one time between 2002 and 2012 occurred on 23 June 2012 (DOY 175) when over $0.25\,km^3$ of water was stored across the Kangerlussuaq sector of the GrIS coincident with the warmest month in four decades (Cappelen et al., 2012). The largest individual SGL volume was $0.038\,km^3$ on 14 July 2011 (DOY 195), during the second warmest July on record.

Intra-annually, the timing of lake formation, periods of peak storage (Fig. 4) and lake-drainage events (Fig. 5) are highly variable, but it is clear that during high melt years (2010 and 2012), formation and drainage occurs much earlier in the season (regional maximum volume, Tv_{max}, was attained 38 and 20 days earlier than the 11 yr average) than in comparatively low melt years (e.g. 2002 and 2006; Figs. 4, 5 and 6a). The average maximum extent of lakes between 2002 and 2012 at each 200 m elevation interval illustrates that smaller lakes tend to form at lower elevations (< 1000 m, Fig. 6b). There is also large inter-annual variability in the fraction of rapidly draining lakes (defined here as those tapping within a 4 day period), with 28 % of lakes classified as such since 2002 (Fig. 7e). There is no obvious correlation with surface runoff, although the peak number of rapidly tapping lakes did occur in the peak melt year (2012). However, in record warm years, characterised by a greater number of cloud free MODIS images, it is possible that a sampling bias is introduced; that the greater number of rapid draining SGLs identified is a consequence of more complete temporal

Table 1. The date and day of year of maximum total lake volume (Tv_{max}) across the study region and maximum individual lake volume (Lv_{max}) between 2002 and 2012. The day the first SGL was observed, the number of days SGL were present and the elevation of the highest SGL (> 0.1 km^2) is also given.

Year	Day of Tv_{max}	Date of Tv_{max}	Tv_{max} (km^3)	Day of Lv_{max}	Date of Lv_{max}	Lv_{max} (km^3)	Day of first SGL	Number of days SGLs present	Highest SGL (> 0.1 km^2) elevation
2002	192	11 Jul	0.1369	192	11 Jul	0.0184	157	108	1661
2003	199	18 Jul	0.1496	203	22 Jul	0.0281	151	108	1660
2004	199	17 Jul	0.109	237	24 Aug	0.0231	151	90	1698
2005	207	26 Jul	0.144	217	5 Aug	0.0305	155	80	1706
2006	206	25 Jul	0.0958	236	24 Aug	0.0155	131	125	1652
2007	193	12 Jul	0.1844	191	10 Jul	0.0297	165	84	1757
2008	221	8 Aug	0.1451	221	8 Aug	0.0273	148	96	1697
2009	196	15 Jul	0.0938	197	16 Jul	0.0112	154	98	1689
2010	156	5 Jun	0.0896	192	11 Jul	0.0241	129	138	1790
2011	195	14 Jul	0.1546	195	14 Jul	0.038	159	87	1827
2012	174	22 Jun	0.2522	187	5 Jul	0.024	150	103	1826
Mean	194	13 Jul	0.142	206	25 Jul	0.0245	150	102	1724

coverage. For example in 2012, 67 images were used, 48 % greater than the 2002–2012 mean. In all years, SGLs tend to form and drain at progressively higher elevations throughout the melt season consistent with previous work (e.g. McMillan et al., 2007; Liang et al., 2012; Doyle et al., 2013). Using 2010 as an example (Fig. 8a), lakes can be seen storing and draining meltwater within five elevation bands. During 2010, 83 % of SGL volume was lost in lakes located between 1000 and 1600 m elevation (Fig. 8b) with only 9 % and 7 % of water lost from lakes below 1000 m and above 1600 m elevation, respectively.

Although most lakes form in the same locations each year, there is significant inter-annual variability in the maximum areal extent of SGLs within each 200 m elevation band (Fig. 9a). The maximum areal extent of lakes during relatively high temperature summers was 158 km^2 in 2003, 154 km^2 in 2007, 136 km^2 in 2010, 160 km^2 in 2011 and 160 km^2 in 2012, significantly larger than in other years. For example, SGL maximum areal extent was 40 % greater in the record melt year of 2012 compared with the low melt year of 2006. Furthermore, during 2012 lakes occupied a larger area (49 %) above 1400 m (96.6 km^2) than in previous years (11 yr average of 64.7 km^2) (Fig. 9a). The years 2003 and 2007 gave the second and third highest elevation lake extents above 1400 m a.s.l., 33 % and 25 % greater than the 11 yr mean, respectively. The maximum areal extent of lakes above 1400 m for each year in the period 2002–2012 was compared with mean monthly temperatures of June and July (Cappelen et al., 2012), revealing a positive correlation ($r^2 = 0.78$) supporting findings of SGLs forming at higher elevations in the highest temperature years (Fig. 9b). With the exception of the lower temperature 2002 and 2006 melt seasons, there is a significant temporal trend in both lake volume loss and the maximum elevation of lakes over the last decade (Fig. 7d, f).

Seasonal patterns of SGL meltwater storage closely correlate with melt rates derived from an energy balance model (van As et al., 2012) for 2009/10 (Fig. 4). Peaks in surface meltwater runoff and storage in lakes generally precede increases in discharge recorded from the Watson River (van As et al., 2012; Hasholt et al., 2013). Volume lost from lakes was compared to peaks in the river discharge record to determine if specific SGL-drainage events can be identified (Fig. 5). Our results indicate that SGL drainage does coincide with short-term perturbations in Watson River discharge, providing tentative evidence for the impact of changing supraglacial water storage on catchment-wide hydrology.

Lake-drainage maps (Fig. 10) illustrate the seasonal lake evolution, highlighting the spatial and temporal variations of rapidly (< 4 days) draining and relatively slow (> 4 days) draining lakes. In general, SGLs within the study area fit consistent drainage patterns each year, remaining rapid or slow draining and not switching between years. There is no obvious relationship between lake size (volume/area) and its drainage mode; fast or slow. In some parts of the study area, SGLs were observed to drain in clusters, i.e. lakes within close (< 6 km) proximity were observed to drain on the same day or within a couple of days of each other (Fig. 11), possibly impacting on the proglacial discharge hydrograph (Fig. 5).

5 Discussion

5.1 Inland expansion of supraglacial lakes

Using a field-calibrated and independently validated depth–reflectance function, we tracked the area and volume of 200 ± 20 individual SGLs in 11 melt seasons (2002–2012) across the Kangerlussuaq sector of the GrIS (Figs. 1a, 7a and

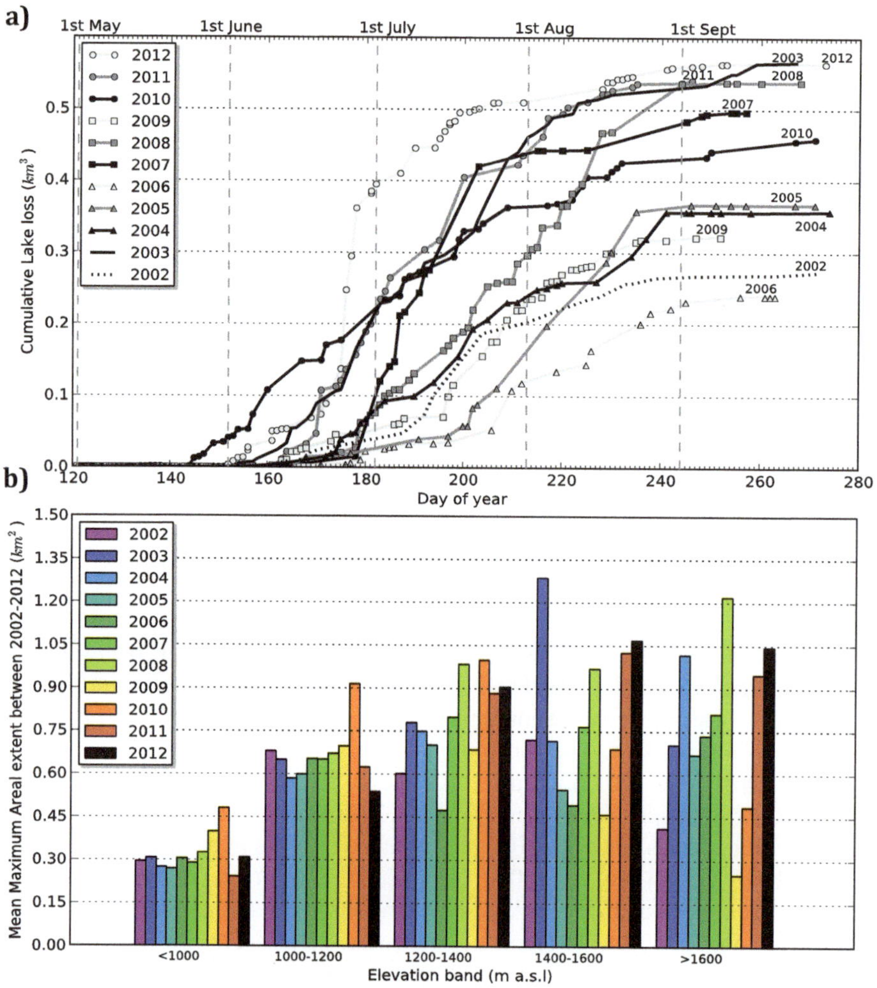

Fig. 6. (a) Cumulative lake volume loss within the Kangerlussuaq sector of the GrIS and **(b)** mean areal extent of SGLs binned at 200 m elevation intervals between 2002–2012.

10). The recurrence of SGLs in the same positions each year (Fig. 1a) is in agreement with previous work (Thomsen et al., 1988; Echelmeyer et al., 1991) and indicates that the lakes do not advect down-glacier with ice flow. Lampkin and Vanderberg (2011) and Sergienko (2013) demonstrated that ice surface topography is coupled to underlying basal topography, confirming that most SGLs form above the same subglacial basins each year. Within our study area, the persistent SGLs (21 % of the population) are preferentially located away from the fast-flow units as identified by Palmer et al. (2011).

The largest lake area observed within the study area was 8.25 km^2, coinciding with a maximum volume of 0.038 km^3, comparable to findings by Box and Ski (2007), although larger lakes have been observed in other studies (16.8 km^2 by Selmes et al. (2011)). On average, the maximum volume of any single lake (Lv_{max}) occurs on 25 July (DOY 206), $\sim$ 12 days after the regional maximum volume (Tv_{max}) (Table 1; Fig. 4) because larger (area) lakes tend to form at higher elevations, later in the season and also take longer to fill (Fig. 6b). The total number of SGLs within any year shows no relationship with Tv_{max} and Lv_{max} (Fig. 7b, c). For example in 2012, 200 lakes were observed, however Tv_{max} was 80 % greater than the 11 yr mean, and SGLs formed at higher elevation than in all previous years (Fig. 7d). Lake area varies with elevation, with smaller lakes forming at low (< 1000 m a.s.l.) and high elevations (> 1600 m a.s.l.), and the largest lakes forming in between (Fig. 6b). The majority (84 % in 2010) of water stored in lakes occurs within the 1000–1600 m elevation band (Fig. 8b), reflecting the effect of ice sheet hypsometry on lake morphology. We speculate that steeper slopes and crevassing associated with fast flow at lower elevations within this region (Palmer et al., 2011; Fitzpatrick et al., 2013) prevents lakes from growing as large before drainage occurs. At higher elevations (above 1600 m), lower temperatures and the shorter melt season limits meltwater availability and lake growth. However, SGL expansion in this upper melt zone is sensitive to positive temperature

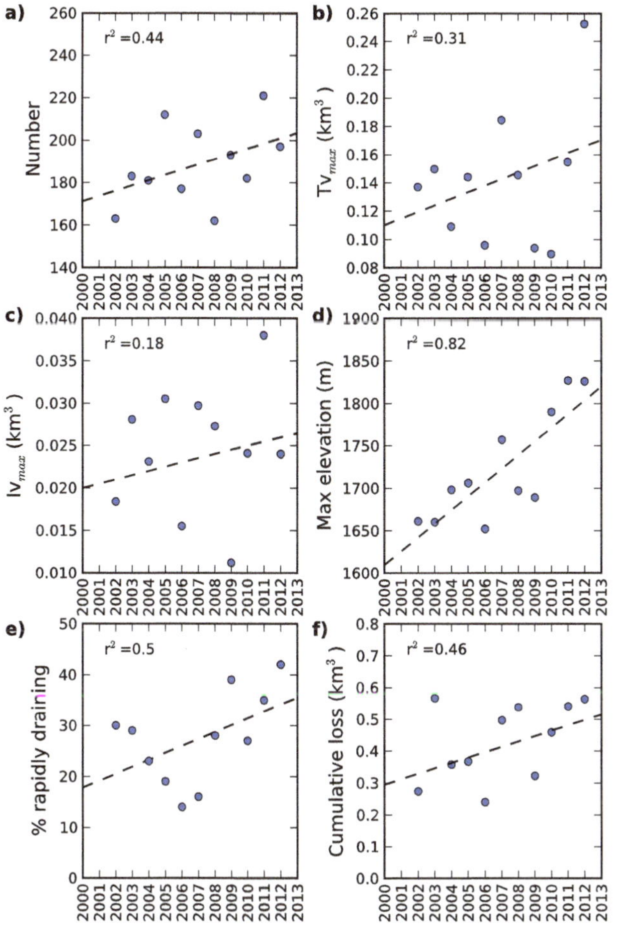

Fig. 7. Inter-annual differences in lake characteristics including **(a)** lake number, **(b)** regional maximum lake volume (Tv_{max}), **(c)** individual SGL volume (Lv_{max}), **(d)** maximum SGL elevation, **(e)** percentage of rapidly draining SGLs and **(f)** cumulative SGL volume loss.

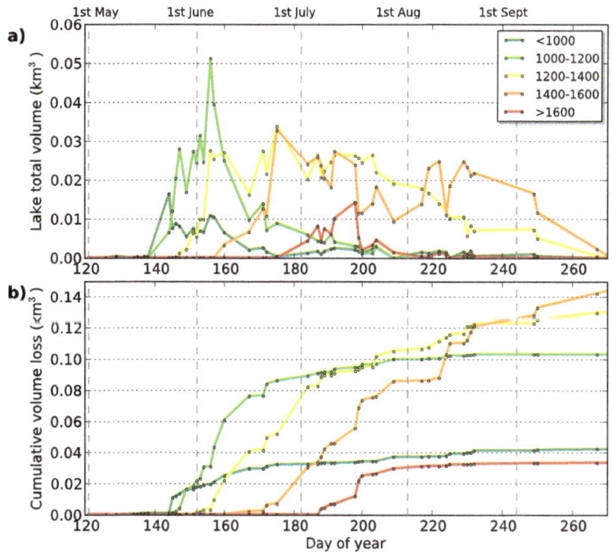

Fig. 8. **(a)** Total lake volume and **(b)** cumulative lake volume loss at 200 m elevation bands during 2010.

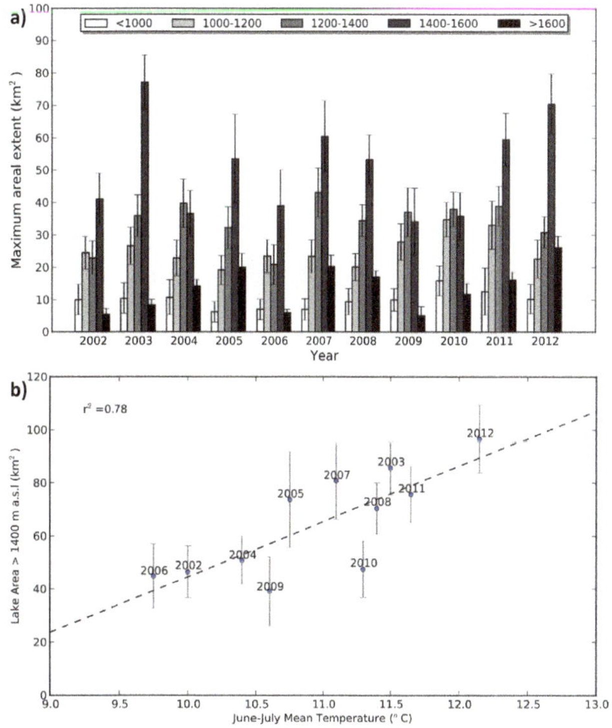

Fig. 9. **(a)** Maximum areal extent of SGLs within each 200 m elevation band and **(b)** the relationship between maximum areal extent of SGLs above 1400 m and mean peak summer temperatures (June/July) between 2002–2012.

and melt anomalies and may be used to indicate longer term climate trends (Howat et al., 2013).

Availability of surface runoff, governed in part by surface temperatures and winter snow pack, determines the initiation of lake formation (Fig. 4, Johansson et al., 2013). The timing and duration of individual lake-drainage events varies inter-annually (Fig. 10), explaining the large differences between the total volume of water stored in lakes in each year (Fig. 4). As previously stated, during record warm years (2010 and 2012) SGLs formed earlier and attained their Tv_{max} on DOY 156 and DOY 174, 38 and 20 days earlier than the 11 yr average, respectively (Fig. 4 and Table 1). The record melt of 2010 was exceeded in 2012 during which the largest melt extent within the satellite era (~ 97 % of the ice sheet) was observed (Nghiem et al., 2012). During 2012, the melt season lasted up to two months longer than the 1979–2011 mean (Tedesco et al., 2013), and although we recorded no increase in the number of lakes (Fig. 7a), the total volume of SGL (Tv_{max}) was ~ 80 % greater than the 11 yr mean and lakes were observed at higher elevations (> 1800 m) and occupied a greater area > 1400 m (96.6 km²) than in previous years (Figs. 7b, d and 9a, b).

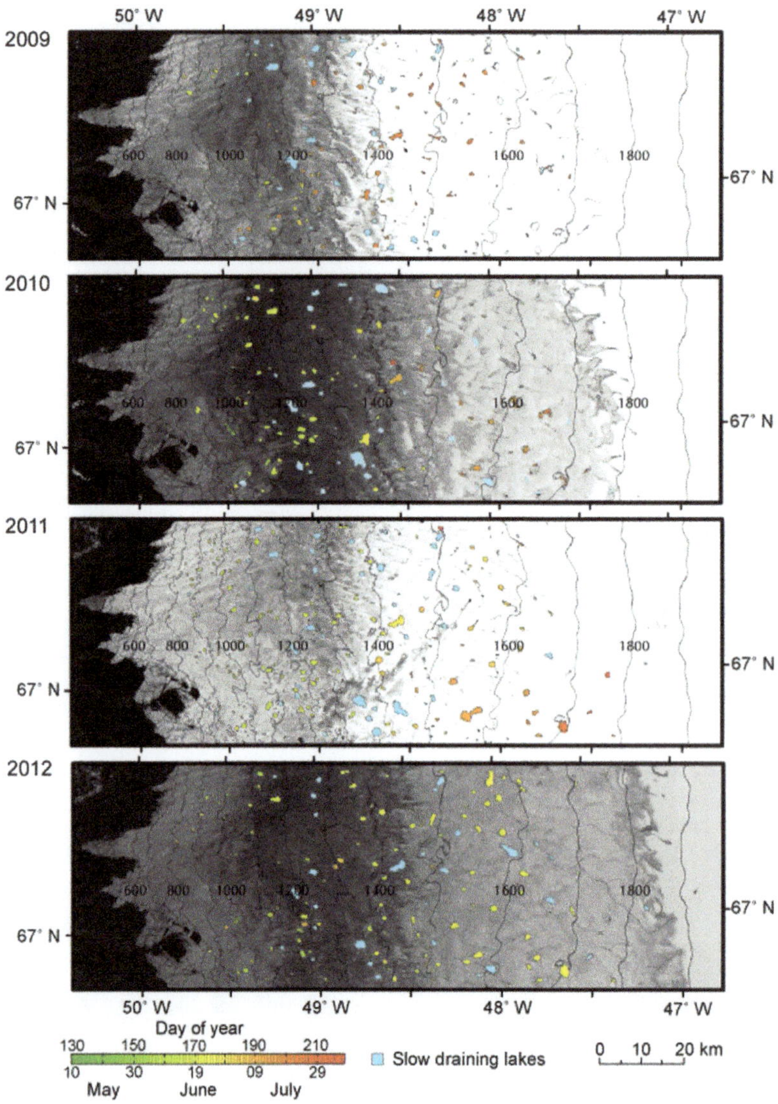

Fig. 10. Colour coded lake-drainage maps for 2009–2012 highlighting the inter-annual variability of lake drainage. Background images are Landsat 7 from 31 July 2009, 19 August 2010, 28 June 2011 and 16 July 2012.

Lake-drainage events have contributed directly to hydrograph anomalies in the discharge record from Watson River. For example, the rapid drainage of a number of large SGLs between DOY 175–184 (24 July–3 August) during 2010 (Fig. 5, Doyle et al., 2013) coincided with a peak in discharge without a corresponding peak in melt (Figs. 4 and 5).

Our data reveal considerable inter-annual variation in water storage and loss (Figs. 4 and 6a), with no two years following the same pattern. The rapid formation and drainage of lakes during the record melt year of 2012 strongly contrasts with the slow formation of lakes during the relatively cool year of 2006 (Fig. 6a). Assuming that all lakes within the study area encompassing the Kangerlussuaq sector of the GrIS region drain via the Watson River, a comparison between cumulative volume loss and discharge reveals that lakes temporarily store and release on average 13 % of total discharge (Table 2). However, if only lakes within the boundaries of the defined Russell Glacier catchment are considered to drain via the Watson River, the mean lake contribution to discharge is 7 % (max 11.5 %). Given the uncertainties associated with catchment delineation and deriving lake volumes, the estimates of lake storage are in line with modelling results over this region (Leeson et al., 2012), which have shown that supraglacial lakes stored a maximum of 12 % of total runoff during 2003. Using 2011 as an example, lake storage within the Palmer et al. (2011), van As et al. (2012) and Russell Glacier catchment was 3.5 %, 8.7 % and 7.6 %, respectively (Table 2). Previous studies note the uncertainties associated with defining a catchment for a single outlet from the GrIS (van As et al., 2012; Clason et al., 2012). It

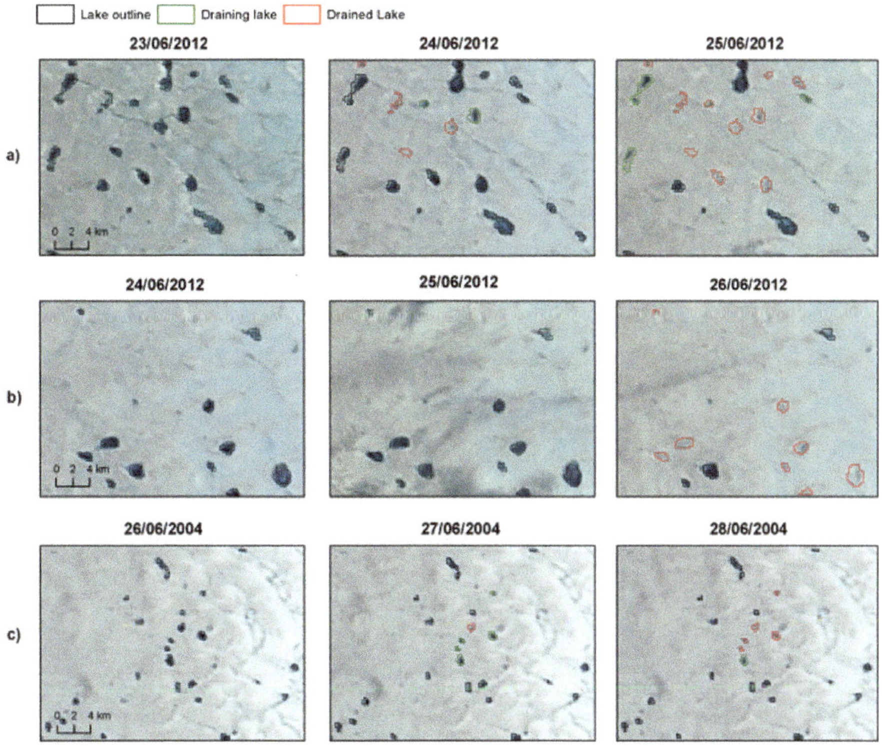

Fig. 11. Clustering of SGL-drainage events in three different regions of our study area: **(a)** 67.22° N, 48.01° W **(b)** 66.9° N, 47.9° W and **(c)** 67.02° N, 49.12° W. SGL outlines from the first day are shown in black, SGLs in the process of draining and those that have drained are shown in green and red, respectively.

Table 2. The contribution of lake volume to discharge from the Watson River between 2007–2011 for all lakes within the Kangerlussuaq sector of the GrIS and for only lakes within the boundaries of the defined catchment.

Year	Discharge (km^3)	Lake vol lost (km^3)	Lake storage for entire region (%)	Lake storage for defined catchment (%)
2007	3.67	0.4967	13.5	6.3
2008	2.86	0.5376	18.8	11.5
2009	2.57	0.3219	12.5	6.64
2010	5.46	0.4588	8.4	4.3
2011	3.97	0.54	13.6	7.63
Mean			13.3	7.3

is probable that water piracy occurs between catchments as catchment boundaries change depending upon the state of the supraglacial and subglacial systems.

Under a future warming climate (Meehl et al., 2007), SGLs are expected to expand further into the ice sheet interior as melt-season extent and duration increase (Howat et al., 2013). We present evidence supporting previous work (Liang et al., 2012; Howat et al., 2013), illustrating that in higher temperature years SGLs formed at higher elevations (Figs. 7d and 9b). Lakes occupied a significantly larger area at higher elevations (> 1400 m) during 2012 (Fig. 9a) than in previous years (49 % greater than the 11 yr average). This coincides with the highest mean June/July temperatures recorded in Kangerlussuaq (Fig. 9b) and the greatest Greenland melt extent on record (Nghiem et al., 2012). Furthermore, we find a positive correlation between the maximum areal extent of lakes above 1400 m a.s.l. and mean monthly temperatures of June and July (Fig. 9b). Preferential absorption of solar radiation by SGLs, causes expansion of their existing basins (Box and Ski, 2007; Tedesco et al., 2012), creating a positive albedo feedback and encouraging SGLs to reform in the same place in successive years. Despite the high melt season of 2010, the maximum extent of SGLs above 1400 m only matched the 11 yr mean (Fig. 9a). However, during 2010,

SGLs in each of the lower elevation bands were larger than all other years (Fig. 6b), indicating that the large volume of melt in 2010 primarily occurred across lower elevations. A similar pattern occurred in the warm summer of 2003, where SGLs occupied a below average extent above 1600 m (Fig. 9a), but the average maximum lake size between 1400–1600 m was 60 % greater than 11 yr mean (Fig. 6b).

5.2 Rapid lake-drainage events

Rapid SGL-drainage events provide a mechanism for surface water to penetrate through kilometre-thick ice, establishing hydraulic pathways which meltwater continues to drain through the remainder of the melt season (Das et al., 2008; Doyle et al., 2013). Over the 11 annual melt periods of this study, on average 28 % of the SGL population rapidly drained (Figs. 7e and 10). As the number of SGLs occurring higher up-glacier increases (Figs. 7d and 9a), more SGLs will rapidly drain, potentially expanding the region affected by melt-induced accelerated flow. We also note that at higher elevations, a number of SGLs appear to drain twice in higher melt years. For example, in 2012 a 5.2 km^2 SGL located at 1400 m drained completely on 23/24 June, subsequently refilled and by 5 July had completely drained again. Lake refilling has been observed in the field in automatic camera imagery (Brizgys and Box, 2005). Such observations indicate that the subglacial hydrological system may not readily adapt to surface meltwater inputs as has recently been postulated (e.g. Sundal et al., 2011). This suggests that basal hydrostatic creep–closure rates across the upper ablation zone where ice thickness is greater than 1 km exceed those processes which help to maintain an efficient hydrological system for the remainder of the melt season (Schoof, 2010; Sole et al., 2013).

Spatial and temporal clustering of SGL-drainage events (Figs. 10 and 11) provides evidence that neighboring SGLs dynamically trigger synoptic-scale tapping. Box and Ski (2007) calculate that a lake containing 0.0983 km^3 of water, upon drainage and without being laterally constrained will spread over a subglacial area of 97 km^2, assuming a void thickness of 1 m. We suggest that the drainage of neighbouring SGLs acts to mechanically trigger one another via short-term perturbations in the regional stress/strain regime transmitted over length scales on the order of tens of kilometres. Large surface rifts and fractures have been observed in SGLs that have drained rapidly (Das et al., 2008; Doyle et al., 2013), and are the primary mode of drainage identified in these studies. Many SGLs across the GrIS appear to drain in this manner each year and we find no common critical threshold of lake volume or depth which appears to control drainage initiation. Based on our results, we hypothesise that the rapid drainage trigger mechanism is site specific, governed by the interplay of seasonal routing and delivery of meltwater moderated by the regional topographic setting and stress/strain regimen together with the inherent mechanical resilience of the ice surface in the locality.

6 Conclusions

Using 502 cloud-free atmospherically corrected MODIS images, we derived water volumes from $\sim$ 200 seasonally occurring SGLs between 2002 and 2012 using a field-calibrated and validated depth–reflectance relationship. We find smaller lakes form at lower elevations (< 1000 m) and larger area lakes form at higher elevations later in the season, taking longer to fill. Although SGLs only occupied a relatively small region of our study area (2 %), they store a disproportionate volume of bulk runoff (7–13 %) and have important implications for ice dynamics through the release of surface meltwater into the subglacial hydrological system via rapid in situ drainage or through overflow into moulins. We find evidence of SGLs draining in clusters, causing short-term perturbations in discharge gauged at the Watson River, and infer that one drainage event dynamically triggers rapid tapping in neighbouring lakes. We find that lake size does not influence its drainage mechanism, and that there is no evidence for a critical lake depth or volume threshold to initiate rapid drainage.

In years of high summer temperatures, SGLs form and drain earlier in the season (e.g. 38 and 20 days earlier than the 11 yr mean during 2010 and 2012, respectively), and cover a larger surface area (e.g. 40 % greater in the record melt year of 2012 compared with the cooler year of 2006). Furthermore, inland expansion of SGLs is strongly correlated with air temperature ($r^2 = 0.78$), with lakes occupying a greater area within the upper ablation zone during warmer years. For example, in 2012, lake area was 49 % greater above 1400 m compared with the 11 yr mean and extended further inland (> 1800 m) than previously recorded. In a warming climate, spatial and temporal expansion of SGLs with concomitant surface to bed coupling may impact strongly on inland ice sheet flow dynamics and interior mass drawdown.

Acknowledgements. This research was funded by UK Natural Environmental Research Council grant NE/G005796 and the Greenland Analogue Project (Sub-project A). We also acknowledge startup support from the Aberystwyth University Research Fund, The Royal Geographical Society and the Gilchrist Trust without which this study would not have been initiated. MODIS data are distributed by the Land Processes Distributed Active Archive Center (LP DAAC), located at the USGS EROS Center (lpdaac.usgs.gov). J. Schmaltz is thanked for advice on MODIS data processing. Landsat data are distributed by the US Geological Survey (glovis.usgs.gov). A. A. W. Fitzpatrick kindly acknowledges a combined NERC/AU studentship bursary. The two anonymous reviewers and the authors of short comments are thanked for constructive comments.

Edited by: E. Larour

References

Banwell, A., Arnold, N., Willis, I., Tedesco, M., and Ahlstrøm, A.: Modeling supraglacial water routing and lake filling on the Greenland Ice Sheet, J. Geophys. Res., 117, F04012, doi:10.1029/2012JF002393, 2012.

Bartholomew, I., Nienow, P., Mair, D., Hubbard, A., King, M., and Sole, A.: Seasonal evolution of subglacial drainage and acceleration in a Greenland outlet glacier, Nat. Geosci., 3, 408–411, doi:10.1038/NGEO863, 2010.

Bartholomew, I., Nienow, P., Sole, A., Mair, D., Cowton, T., King, M., and Palmer, S.: Greenland; GPS; ice dynamics; supraglacial lakes; subglacial hydrology, Earth Planet. Sc. Lett., 307, 271–278, 2011.

Box, J. and Ski, K.: Remote sounding of Greenland supraglacial melt lakes: implications for subglacial hydraulics, J. Glaciol., 53, 257–265, doi:10.3189/172756507782202883, 2007.

Box, J., Cappelen, J., Decker, D., Fettweis, X., Mote, T., Tedesco, M., and van de Wal, R. S. W.: Greenland, in: Arctic Report Card 2010, edited by: Richer-Menge, J., 55–64, 2010.

Brizgys, G. and Box, J. E.: West Greenland ice sheet melt lake observations and modeling, Eos Transactions, American Geophysical Union (AGU), 86, Fall Meeting Supplement, Abstract C41A-07, 2005.

Clason, C. Mair, D. W. F., Burgess, D. O., and Nienow, P. W.: Modelling the delivery of supraglacial meltwater to the ice-bed interface: application to southwest Devon Ice Cap, Nunavut, Canada, J. Glaciol., 58, 361–374, 2012.

Cappelen, J., Laursen, E., Kern-Hansen, C., Boas, L., Wang, P., Jørgensen, B., and Carstensen, L.: Weather and Climate Data from Greenland 1958–2011, in: Technical Report 12–15, Danish Meterological Institute, Copenhagen, 2012.

Chandler, D. M., Wadham, J. L., Lis, G. P., Cowton, T., Sole, A., Bartholomew, I., Telling, J., Nienow, P., Bagshaw, E. B., Mair, D., Vinen, S., and Hubbard, A.: Evolution of the subglacial drainage system beneath the Greenland Ice Sheet revealed by tracers, Nat. Geosci., 6, 195–198, 2013.

Cowton, T., Nienow, P., Sole, A., Wadham, J. L., Lis, G. P., Bartholomew, I., Mair, D., and Chandler, D. M.: Evolution of drainage system morphology at a land-terminating Greenlandic outlet glacier, J. Geophys. Res.-Earth, 118, 29–41, 2013.

Das, S., Joughin, I., Behn, M., Howat, I., King, M., Lizarralde, D., and Bhatia, M.: Fracture propagation to the base of the Greenland ice sheet during supraglacial lake drainage, Science, 320, 778–781, doi:10.1126/science.1153360, 2008.

Doyle, S. H., Hubbard, A. L., Dow, C. F., Jones, G. A., Fitzpatrick, A., Gusmeroli, A., Kulessa, B., Lindback, K., Pettersson, R., and Box, J. E.: Ice tectonic deformation during the rapid in situ drainage of a supraglacial lake on the Greenland Ice Sheet, The Cryosphere, 7, 129–140, doi:10.5194/tc-7-129-2013, 2013.

Echelmeyer, K., Clarke, T., and Harrison, W.: Surficial glaciology of Jakobshavns Isbrae, West Greenland, J. Glaciol., 37, 368–382, 1991.

Fitzpatrick, A., Hubbard, A., Joughin, I., Quincey, D., van As, D., Mikkelsen, A., Doyle, S., Hasholt, B., and Jones, G.: Ice flow dynamics and surface meltwater flux at a land-terminating sector of the Greenland Ice Sheet, J. Glaciol., 59, doi:10.3189/2013JoG12J143, 2013.

Georgiou, S., Shepherd, A., McMillan, M., and Nienow, P.: Seasonal evolution of surpaglacial lake volume from ASTER imagery, Ann. Glaciol., 50, 95–100, 2009.

Hanna, E., Huybrechts, I., Janssens, J., Cappelen, K., Steffen, K., and Stephens, A.: Runoff and mass balance of the Greenland ice sheet: 1958–2003, J. Geophys. Res.-Atmos., 110, D13108, doi:10.1029/2004JD005641, 2005.

Hasholt, B., Mikkelsen, A., Nielsen, M., and Larsen, M.: Observations of runoff and sediment and dissolved loads from the Greenland ice sheet at Kangerlussuaq, West Greenland, 2007 to 2010, Z. Geomorphol., 57 (Suppl. 2), 3–27, doi:10.1127/0372-8854/2012/S-00121, 2013.

Howat, I. M., de la Peña, S., van Angelen, J. H., Lenaerts, J. T. M., and van den Broeke, M. R.: *Brief Communication* "Expansion of meltwater lakes on the Greenland Ice Sheet", The Cryosphere, 7, 201–204, doi:10.5194/tc-7-201-2013, 2013.

Huggel, C., Kaab, A., Haeberli, W., Teysseire, P., and Paul, F.: Remote sensing based assessment of hazards from glacier lake outbursts: a case study in the Swiss Alps, Can. Geotech. J., 39, 316–330, 2002.

Johansson, A., Jansson, P., and Brown, I.: Spatial and temporal variations in lakes on the Greenland Ice Sheet, J. Hydrol., 476, 314–320, 2013.

Joughin, I., Das, S., King, M., Smith, B., Howat, I., and Moon, T.: Seasonal speedup along the western flank of the Greenland Ice Sheet, Science, 320, 781–783, doi:10.1126/science.1153288, 2008.

Lampkin, D. J. and Vanderberg, J.: A preliminary investigation of the influence of basal and surface topography on supraglacial lake distribution near Jakobshavn Isbræ, Hydrol. Process., 25, 3347–3355, doi:10.1002/hyp.8170, 2011.

Leeson, A. A., Shepherd, A., Palmer, S., Sundal, A., and Fettweis, X.: Simulating the growth of supraglacial lakes at the western margin of the Greenland ice sheet, The Cryosphere, 6, 1077–1086, doi:10.5194/tc-6-1077-2012, 2012.

Liang, Y., Colgan, W., Qin, L., Steffen, K., Abdalati, W., Stroeve, J., Gallaherb, D., and Bayoub, N.: A decadal investigation of supraglacial lakes in West Greenland using a fully automatic detection and tracking algorithm, Remote Sens. Environ., 123, 127–138, 2012.

Luthje, M., Pedersen, L., Reeh, N., and Greuell, W.: Modelling the evolution of supraglacial lakes on the West Greenland ice-sheet margin, J. Glaciol., 52, 608–618, 2006.

McMillan, M., Nienow, P., Shepherd, A., Benham, T., and Sole, A.: Seasonal evolution of supraglacial lakes on the Greenland Ice Sheet, Earth Planet. Sc. Lett., 262, 484–492, 2007.

Meehl, G., Stocker, T., Collins, W., Friedlingstein, P., Gaye, A., Gregory, J., Kitoh, A., Knutti, R., Murphy, J., Noda, A., Raper, S., Watterson, I., Weaver, A., and Zhao, Z.: Global climate projections, in: Climate Change 2007: The Physical Science Basis, Contribution of Working Group I to the Fourth Assessment Report of the Intergovernmental Panel on Climate Change, Cambridge University Press, Cambridge, UK and New York, NY, USA, 2007.

Mernild, S. H. and Hasholt, B.: Observed runoff, jokulhlaups and suspended sediment load from the Greenland ice sheet at Kangerlussuaq, West Greenland, 2007 and 2008, J. Glaciol., 55, 855–858, 2009.

Nghiem, S. V., Hall, D. K., Mote, T. L., Tedesco, M., Albert, M. R., Keegan, K., Shuman, C. A., DiGirolamo, N. E., and Neuman, G.: The extreme melt across the Greenland Ice Sheet in 2012, Geophys. Res. Lett., 39, 1–6, doi:10.1029/2012GL053611, 2012.

Palmer, S., Shepherd, A., Nienow, P., and Joughin, I.: Seasonal speedup of the Greenland Ice Sheet linked to routing of surface water, Earth Planet. Sc. Lett., 302, 423–428, doi:10.1016/j.epsl.2010.12.037, 2011.

Schoof, C.: Ice-sheet acceleration driven by melt supply variability, Nature, 468, 803–806, doi:10.1038/nature09618, 2010.

Selmes, N., Murray, T., and James, T. D.: Fast draining lakes on the Greenland Ice Sheet, Geophys. Res. Lett., 38, L15501, doi:10.1029/2011GL047872, 2011.

Sergienko, O.: Glaciological twins: basally controlled subglacial and supraglacial lakes, J. Glaciol., 59, 3–8, 2013.

Shepherd, A., Hubbard, A., Nienow, P., King, M., McMillan, M., and Joughin, I.: Greenland ice sheet motion coupled with daily melting in late summer, Geophys. Res. Lett., 36, L01501, doi:10.1029/2008GL035758, 2009.

Sneed, W. and Hamilton, G.: Evolution of melt pond volume of the surface of the Greenland Ice Sheet, Geophys. Res. Lett., 34, L03501, doi:10.1029/2006GL028697, 2007.

Sneed, W. and Hamilton, G.: Validation of a method for determining the depth of glacial melt ponds using satellite imagery, Ann. Glaciol., 52, 15–22, doi:10.3189/172756411799096240, 2011.

Sole, A., Nienow, P., Bartholomew, I., Mair, I., Cowton, T., Tedstone, A. and King, M.: Winter motion mediates dynamic response of the Greenland Ice Sheet to warmer summers, Geophys. Res. Lett., 15, 3940–3944, doi:40, 10.1002/grl.50764, 2013.

Sundal, A., Shepherd, A., Nienow, P., Hanna, E., Palmer, S., and Huybrechts, P.: Evolution of supra-glacial lakes across the Greenland Ice Sheet, Remote Sens. Environ., 113, 2164–2171, doi:10.1016/j.rse.2009.05.018, 2009.

Sundal, A., Shepherd, A., Nienow, P., Hanna, E., Palmer, S., and Huybrechts, P.: Melt-induced speed-up of Greenland ice sheet offset by efficient subglacial drainage, Nature, 469, 521–524, doi:10.1038/nature09740, 2011.

Tedesco, M. and Steiner, N.: In-situ multispectral and bathymetric measurements over a supraglacial lake in western Greenland using a remotely controlled watercraft, The Cryosphere, 5, 445–452, doi:10.5194/tc-5-445-2011, 2011.

Tedesco, M., Lüthje, M., Steffen, K., Steiner, N., Fettweis, X., Willis, I., Bayou, N., and Banwell, A.: Measurement and modeling of ablation of the bottom of supraglacial lakes in western Greenland, Geophys. Res. Lett., 39, L02502, doi:10.1029/2011GL049882, 2012.

Tedesco, M., Fettweis, X., Mote, T., Wahr, J., Alexander, P., Box, J. E., and Wouters, B.: Evidence and analysis of 2012 Greenland records from spaceborne observations, a regional climate model and reanalysis data, The Cryosphere, 7, 615–630, doi:10.5194/tc-7-615-2013, 2013.

Thomsen, H., Thorning, L., and Braithwaite, R. J.: Glacier-hydrological conditions on the Inland Ice north-east of Jakobshavn/Ilulissat, West Greenland, Rapport Grønlands Geologiske Undersøgelse, 138, Cophenhagen, Denmark, 1988.

van As, D., Hubbard, A. L., Hasholt, B., Mikkelsen, A. B., van den Broeke, M. R., and Fausto, R. S.: Large surface meltwater discharge from the Kangerlussuaq sector of the Greenland ice sheet during the record-warm year 2010 explained by detailed energy balance observations, The Cryosphere, 6, 199–209, doi:10.5194/tc-6-199-2012, 2012.

van de Wal, R. S. W., Boot, W., van den Broeke, M. R., Smeets, C. J. P. P., Reijmer, C. H., Donker, J. J. A., and Oerlemans, J.: Large and rapid melt-induced velocity changes in the ablation zone of the Greenland Ice Sheet, Science, 321, 111–113, doi:10.1126/science.1158540, 2008.

van de Wal, R. S. W., Boot, W., Smeets, C. J. P. P., Snellen, H., van den Broeke, M. R., and Oerlemans, J.: Twenty-one years of mass balance observations along the K-transect, West Greenland, Earth Syst. Sci. Data, 4, 31–35, doi:10.5194/essd-4-31-2012, 2012.

Yang, K. and Smith, L. C.: Supraglacial streams on the Greenland Ice Sheet delineated from combined spectral- shape information in high resolution satellite imagery, IEEE Geosci. Remote S., 10, 801–805, 2013

Zwally, H., Abdalati, W., Herring, T., Larson, K., Saba, J., and Steffen, K.: Surface melt-induced acceleration of Greenland ice-sheet flow, Science, 297, 218–22, doi:10.1126/science.1072708, 2002.

15

Monitoring water accumulation in a glacier using magnetic resonance imaging

A. Legchenko[1], C. Vincent[2], J. M. Baltassat[3], J. F. Girard[3], E. Thibert[4,7], O. Gagliardini[2,6], M. Descloitres[1], A. Gilbert[2], S. Garambois[5], A. Chevalier[1], and H. Guyard[1]

[1]IRD, UJF-Grenoble 1/CNRS/G-INP, LTHE UMR5564, BP 53, Grenoble Cedex 9, 38041, France
[2]Laboratoire de Glaciologie et Géophysique de l'Environnement and CNRS – LGGE, 38041 Grenoble Cedex 9, France
[3]BRGM, BP 6009, 45060, Orléans Cedex 2, France
[4]IRSTEA, UR ETGR, Erosion torrentielle neige et avalanches, 2 rue de la Papeterie-BP 76, 38402 Saint Martin d'Hères, France
[5]Institut des Sciences de la terre (ISTerre), CNRS/ Université Joseph Fourier, Grenoble, France
[6]Institut Universitaire de France, IUF, Paris, France
[7]Université Grenoble Alpes, 38041 Grenoble, France

Correspondence to: A. Legchenko (anatoli.legtchenko@ird.fr)

Abstract. Tête Rousse is a small polythermal glacier located in the Mont Blanc area (French Alps) at an altitude of 3100 to 3300 m. In 1892, an outburst flood from this glacier released about 200 000 m^3 of water mixed with ice, causing much damage. A new accumulation of melt water in the glacier was not excluded. The uncertainty related to such glacier conditions initiated an extensive geophysical study for evaluating the hazard. Using three-dimensional surface nuclear magnetic resonance imaging (3-D-SNMR), we showed that the temperate part of the Tête Rousse glacier contains two separate water-filled caverns (central and upper caverns). In 2009, the central cavern contained about 55 000 m^3 of water. Since 2010, the cavern is drained every year. We monitored the changes caused by this pumping in the water distribution within the glacier body. Twice a year, we carried out magnetic resonance imaging of the entire glacier and estimated the volume of water accumulated in the central cavern. Our results show changes in cavern geometry and recharge rate: in two years, the central cavern lost about 73 % of its initial volume, but 65 % was lost in one year after the first pumping. We also observed that, after being drained, the cavern was recharged at an average rate of 20 to 25 $m^3\,d^{-1}$ during the winter months and 120 to 180 $m^3\,d^{-1}$ in summer. These observations illustrate how ice, water and air may refill englacial volume being emptied by artificial draining. Comparison of the 3-D-SNMR results with those obtained by drilling and pumping showed a very good correspondence, confirming the high reliability of 3-D-SNMR imaging.

1 Introduction

Water circulation in a glacier is an important factor that determines ice dynamics, runoff characteristics, and water quality. The recent, growing, concern over the response of glaciers to future-climate scenarios necessitates understanding of the hydrological processes in ice. A significant proportion of glaciers have a polythermal regime, where ice masses are composed of temperate ice (at the pressure-melting point) and cold ice (Irvine-Fynn et al., 2011). In a polythermal glacier, the coexistence of temperate and cold ice increases the potential for water storage within the glacier's drainage system. Observations at Austre Brøggerbreen in 1998 and 2000 showed that a water volume of approximately 8000 m^3 was retained in a single englacial channel (Irvine-Fynn et al., 2011). At Hansbreen, the annual water volumes in englacial conduits were estimated to be about $1.3 \times 10^6\,m^3$ (Benn et al., 2009). However, the total volume of accumulated water will depend upon channel density and dimensions, glacier size, and the rate at which summer-season outflow is curtailed. Water trapped within a glacier can lead to extreme discharge events (Haeberli, 1983; Jansson et al., 2003). In densely populated mountain areas, glaciers may affect a number of hazard-relevant issues, such as abrupt floods (Haeberli et al., 1989) or slope instability (Fischer et al., 2005). A better understanding of the hydrological processes in polythermal glaciers and the design of suitable protection measures against new hazards requires regular observations and detailed knowledge of the internal glacier structure, which can be obtained using surface geophysical methods.

In recent years, ground-penetrating radar (GPR) has been increasingly used for the detection of subglacial and englacial environments (Irvine-Fynn et al., 2006; Bælum and Benn, 2011). Using calibrations and additional temperature data, GPR helped estimate the water content in the ice (Hamran et al., 1996; Macheret and Glazovsky, 2000), and more recent GPR research examined temporal changes in glaciers and investigation of 2-D and 3-D structures (Grasmueck et al., 2004; Gross et al., 2004). Study of ice properties by combining surface and borehole radar measurements improves the accuracy of the results (Murray et al., 2000). Other non-invasive methods, such as seismic techniques (Lanz et al., 1998; Musil et al., 2002; Maurer and Hauck, 2007), frequency- and time-domain electromagnetic methods (FDEM and TDEM, respectively) (Hoekstra, 1978; Hauck et al., 2001) and electrical resistivity tomography (ERT) (Hauck and Vonder Mühll, 2003; Marescot et al., 2003; Kneisel, 2004) provide both redundant and complementary information about glacier features. Joint interpretation of data obtained with different methods provides more reliable and accurate results (Endres et al., 2009; Kim et al., 2010; Merz et al., 2012). However, variations in the physical parameters of the investigated formations observed by these traditional geophysical methods do not allow a unique interpretation in terms of water presence, and hence an estimate of the water volume accumulated in a glacier body can only be qualitative.

In this paper, we report our results of using the surface nuclear magnetic resonance method (SNMR), also known as the magnetic resonance sounding (MRS), for investigating hydrological processes in a glacier. SNMR is an emerging geophysical technique, specifically developed for hydrogeological investigations (Legchenko and Valla, 2002; Hertrich, 2008; Knight et al., 2012). Selective sensitivity to groundwater is the main advantage of SNMR in comparison with other geophysical tools, and the magnetic resonance phenomenon can also be used for investigating brine and water inclusions in ice (Callaghan et al., 1999). Note that cold ice with negligible interstitial water does not produce an SNMR response and will be interpreted as a dry material. However, temperate ice may contain liquid water and SNMR can quantify this water content. In its 2-D and 3-D implementation, SNMR was reported to be an efficient method for investigating subglacial formations (Lehmann-Horn et al., 2011) and for locating water-filled caverns and channels in a karst environment (Vouillamoz et al., 2003; Boucher et al., 2006; Girard et al., 2007; Legchenko et al., 2008). For 2-D and 3-D investigations, SNMR can be used with separate transmitting (Tx) and receiving (Rx) loops (Hertrich et al., 2009), or with a coincident Tx/Rx loop (Legchenko et al., 2011). We have studied the Tête Rousse glacier located in the Mont Blanc area (French Alps) using the latter configuration as being better adapted to high mountain conditions.

2 The 3-D-SNMR method

The SNMR field setup consists of a wire loop laid out on the surface, usually in a square with sides between 40 and 150 m. The loop is then energized by a pulse of alternating current $i(t) = I_0 \cos(\omega_0 t)$. The current frequency is set equal to the Larmor frequency of the protons ω_0 in the geomagnetic field $B_0(\omega_0 = \gamma B_0$ with γ being the gyromagnetic ratio). The pulse causes precession around the geomagnetic field of the spin magnetization of the protons in groundwater, which creates an alternating magnetic field that can be detected by the same loop after the pulse is terminated (the free-induction decay method). Oscillating at the Larmor frequency, the SNMR signal is measured by varying the pulse moment $q = I_0 \tau$, with I_0 and τ being the amplitude and duration of the pulse, respectively. The distribution of the water content in the subsurface can be derived from inversion of the SNMR signal.

The signal induced in the coincident transmitting/receiving (Tx/Rx) loop is proportional to the sum of the flux of all precessing magnetic moments and can be computed with the volume integral (Valla and Legchenko, 2002):

$$e_0(q) = I_0^{-1} \omega_0 \int_V B_1 M_\perp w(\boldsymbol{r}) \mathrm{d}V, \tag{1}$$

where $M_\perp = \sin(\gamma B_1 I_0^{-1} q/2)$ is the transverse component of spin magnetization, γ is the gyromagnetic ratio, B_1 is transmitted by the surface-loop magnetic-field component perpendicular to the geomagnetic field, I_0 is the current amplitude in the loop, q is the pulse moment, ω_0 is the Larmor frequency for protons in the geomagnetic field, $0 \le w(\boldsymbol{r}) \le 1$ is the water content, and $\boldsymbol{r} = r(x, y, z)$ is the coordinate vector.

The 3-D field setup is composed of overlapped measuring loops covering the area investigated with SNMR. The SNMR signal is independently measured in each receiving loop while varying the pulse moment in the transmitting loop. All individual soundings are incorporated into one data set for the 3-D inversion (Legchenko et al., 2011). For the inversion, the linear Eq. (1) is approximated by a matrix equation

$$\mathbf{Aw} = \mathbf{e}, \tag{2}$$

where $\mathbf{A} = [\tilde{a}_{i,j}]$ is a rectangular matrix with the elements representing the amplitude of the magnetic-resonance signal generated by water in corresponding cells. The experimental data set is $\mathbf{e} = (\tilde{e}_1, \tilde{e}_2, ..., \tilde{e}_i, \tilde{e}_I)^T$, the water content in the corresponding cell is $\mathbf{w} = (w_1, w_2, ..., w_j, ..., w_J)^T$, and the symbol T denotes the transposition. For simplicity, we assume that cell size is constant throughout the investigated volume.

For analyzing the resolution of matrix Eq. (2) we use the singular value decomposition, or SVD (Aster et al., 2005). For that, the matrix $\mathbf{A}$ is presented as

$$\mathbf{A} = \mathbf{USV}^T, \tag{3}$$

where $\mathbf{U}$ is an $I \times I$ orthogonal matrix representing the data space, $\mathbf{V}$ is an $J \times J$ orthogonal matrix representing the model space, and $\mathbf{S}$ is an $I \times J$ diagonal matrix with non-negative diagonal elements called singular values. The singular values of $\mathbf{S}$ are arranged in decreasing magnitude. The first p values are nonzero and others may be zero. Very small singular values represent numerical instabilities, which is why we consider only significant singular values and partition $\mathbf{S}$ as

$$\mathbf{S} = \begin{bmatrix} \mathbf{S}_p & 0 \\ 0 & 0 \end{bmatrix}, \tag{4}$$

where $\mathbf{S}_p$ is a $p \times p$ diagonal matrix composed of the significant singular values. Thus, we can write a reduced-rank approximation as

$$\mathbf{A} = \mathbf{U}_p \mathbf{S}_p \mathbf{V}_p^T. \tag{5}$$

The matrix $\mathbf{U}_p$ forms an orthonormal basis for the data space and the matrix $\mathbf{V}_p$ for the model space.

The resolution of the inversion can be estimated using the model resolution matrix $\mathbf{R}_m$, which is a symmetric matrix describing how the inversion represents the original model and can be computed as

$$\mathbf{R}_m = \mathbf{V}_p \mathbf{V}_p^T. \tag{6}$$

If $rank(\mathbf{A}) = p = J$, then the model will be perfectly recovered by the inversion. If $p = J$, then the resolution will not be perfect.

The uncertainty of the inversion can be estimated assuming independent and identically distributed normal data errors σ^2. For the inversion result $\mathbf{w}$, we define the 95 % confidence intervals as

$$\frac{1}{\sigma\sqrt{2\pi}} \int_{-1.96\sigma}^{1.96\sigma} \exp\left(-\frac{w^2}{2\sigma^2}\right) \mathrm{d}w \approx 0.95. \tag{7}$$

Then we use the covariance for the model

$$Cov(\mathbf{w}_{L_2}) = \sigma^2 \mathbf{V}_p \mathbf{S}_p^{-2} \mathbf{V}_p^T. \tag{8}$$

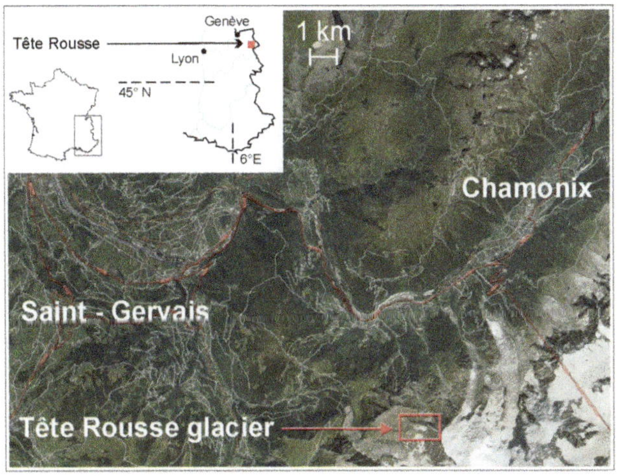

Fig. 1. Location of the Tête Rousse glacier.

The 95 % confidence intervals $\mathbf{w}_{0.95}$ for a model parameter having a normal distribution around the mean $\mathbf{w}_{L_2}$ is given as

$$\mathbf{w} = \mathbf{w}_{L_2} \pm \mathbf{w}_{0.95}, \tag{9}$$

where

$$\mathbf{w}_{0.95} = 1.96 \times \sqrt{diag\left(Cov(\mathbf{w}_{L_2})\right)}. \tag{10}$$

Inversion of the 3-D-SNMR data is ill-posed. Different methods for resolving ill-conditioned inverse problems can be found in the literature (e.g., Tarantola, 1987). For our study, the inversion was conducted according to Tikhonov's regularization method (Tikhonov and Arsenin, 1977). To find an approximate solution to the matrix Eq. (2), this method supposes minimization of Tikhonov's functional (Legchenko and Shushakov, 1998):

$$TF(\eta) = \| \mathbf{Aw} - \mathbf{e} \|_{L_2} + S_{x,y,z} = \min, \tag{11}$$

where $S_{x,y,z} = \eta_x \times \left(\frac{\partial}{\partial x}\mathbf{w}\right)^2 + \eta_y \times \left(\frac{\partial}{\partial y}\mathbf{w}\right)^2 + \eta_z \times \left(\frac{\partial}{\partial z}\mathbf{w}\right)^2$ is an estimate of the smoothness of the solution and $\eta_x, \eta_y, \eta_z > 0$ are the smoothing factors in each direction. For the optimization itself, we used the conjugate gradient method (Stoer and Bulirsch, 1980).

3 Investigated area

The area investigated with 3-D-SNMR is located in the French Alps (Fig. 1). Tête Rousse is a small glacier, whose surface area in 2007 was about 0.08 km^2 (Fig. 2). A temperate accumulation area and a predominantly cold ablation area characterize its thermal regime. Reported results of numerical simulations of snow cover and englacial temperatures confirmed by the temperature measurements in boreholes show that the glacier was almost entirely cold before

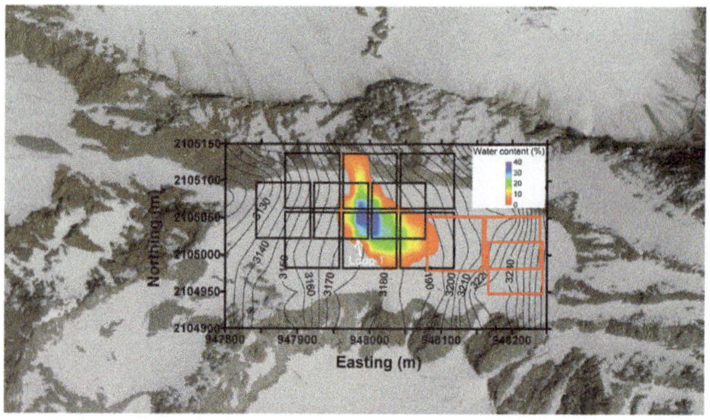

Fig. 2. Aerial photo of the Tête Rousse glacier with an approximate projection of the SNMR measuring loops (black squares) and water-filled cavern detected in 2009 (in the Lambert-II co-ordinate system). Additional SNMR loops used in 2012 are shown as red squares. Isoclines show the elevation of the glacier surface in 2009.

1870, but then gradually warmed up until 1920 (Gilbert et al., 2012). In that year, the glacier was almost entirely temperate except for its central deeper part that remained cold. Between 1920 and 1950, the glacier started freezing again, but the temperate part remained large. Nowadays, except for the first ten meters close to the surface and influenced by seasonal temperature variations, the glacier consists of temperate and cold ice. In its western part, below 3160 m elevation, the glacier body is composed of cold ice with a temperature below −2 °C. Above 3180 m, the basal ice is temperate or near temperate in the first 10 to 20 m above bedrock. In the central part, between 3160 and 3180 m, the ice is temperate from approximately 30 m below the glacier surface down to bedrock (Vincent et al., 2012a).

Both topology and thermal regime are favorable for water accumulation in the Tête Rousse glacier. In 1892, the outburst flood from this glacier released about 200 000 m^3 of water mixed with ice and caused much damage (Vallot and Delebecque, 1892; Mougin and Bernard, 1905). Between the catastrophe and 1904, three drainage tunnels were constructed for preventing water accumulation in the glacier. One of the tunnels was maintained until 2009, though no water was drained through this tunnel over the last 106 yr. The uncertainty relative to the glacier conditions initiated an extensive glaciological study aiming at evaluating the hazard. In 2007, mapping of the glacier with ground-penetrating radar (GPR) revealed an anomaly that was interpreted as a possible accumulation of water (Vincent at al., 2010). This result initiated a 3-D-SNMR survey in September 2009 that discovered a large water-filled cavern containing about 55 000 m^3 of water (Vincent et al., 2012b). The 3-D-SNMR data were interpreted alone, though using additional information on the glacier thickness derived from GPR results.

The SNMR results guided a drilling program of more than 20 boreholes, which confirmed the existence of a water-filled cavern and allowed defining the cavern size and location. Drainage of the cavern revealed a water volume of approximately 48 000 m^3. This cavern was also investigated with borehole sonar, but its complex shape disturbed the sonar measurements and it was not possible to obtain an entire 3-D model of the cavern. The volume of 24 500 m^3 obtained from sonar was less than the 55 000 m^3 estimated with the SNMR method and 48 000 m^3 confirmed by pumping.

Whatever the mechanism of water penetration into the glacier, this water reached the bottom and was trapped by the thermal barrier formed by cold basal ice. Today, about 3000 people are exposed to the Tête Rousse hazard and the water in the cavity has therefore been pumped every fall since 2010. For monitoring the water accumulation, we used regular 3-D-SNMR surveys. GPR and sonar data were found to be less efficient for monitoring and we did not use these methods for water-volume estimates.

4 Numerical results

During our study, we used half-overlapping square loops ($80 \times 80\,m^2$) that covered most of the glacier (Fig. 2). Investigation depth with this loop was about 80 m. For investigating the resolution of the inverse problem we discretized the subsurface into equal cells with sizes of $20\,m \times 20\,m \times 5\,m$ and $40\,m \times 40\,m \times 10\,m$. The resolution matrix $\mathbf{R}_m$ for these two cell sizes is shown in Fig. 3, which shows that the main diagonal of the resolution matrix clearly dominates in both cases. For the discretization with cells of $20\,m \times 20\,m \times 5\,m$ we obtained $rank(\mathbf{R}_m)/J = 0.86$, and using cells of $40\,m \times 40\,m \times 10\,m$ we obtained $rank(\mathbf{R}_m)/J = 0.9$. Hence, when $\mathbf{R}_m \approx \mathbf{I}$ we obtain a good theoretical resolution for both cell sizes. However, assuming a 1 nV noise we can estimate the resolution for a more realistic case. For that, we computed the 95 % confidence intervals using Eq. (10). Figure 4 shows the 95 % confidence intervals for each cell; when using discretization with $20\,m \times 20\,m \times 5\,m$ cells we obtained large 95 % confidence intervals, indicating a poor resolution. Inversion with $40\,m \times 40\,m \times 10\,m$ cells will give a better resolution. Thus we estimated the resolution of 3-D-SNMR inversion to be not better than the $40\,m \times 40\,m \times 10\,m$ target size. Smaller targets cannot be accurately resolved. Note that resolution also depends on the noise level.

When water drains through a glacier, it may accumulate not only in large caverns but also in small volumes that cannot be resolved by 3-D-SNMR inversion. Let us see how the sensitivity of the method is distributed within the area investigated with one SNMR loop. For this, we computed the amplitude of the signal generated by one $20\,m \times 20\,m \times 5\,m$ cell for different cell positions (Fig. 5). Figure 5a shows a horizontal slice corresponding to a depth of 20 m (top of the cell at 20 m) below the loop, which is shown by a black dashed line. The white-line square corresponds to a $40\,m \times 40\,m \times 10\,m$ cell and the numbers

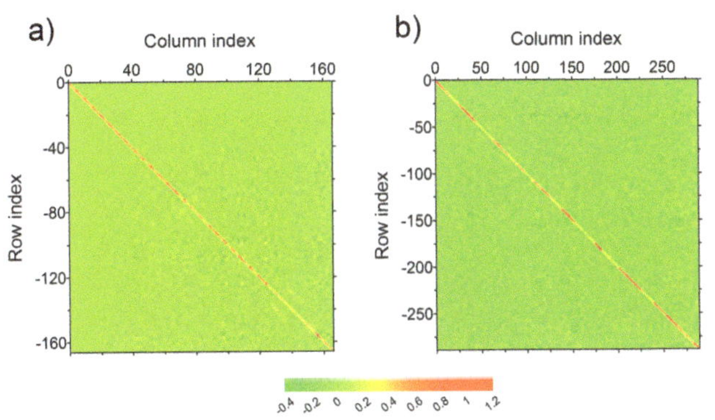

Fig. 3. The model resolution matrix $\mathbf{R}_m$ computed considering the discretization of the matrix **A**: **(a)** cells of 40 m × 40 m × 10 m; **(b)** cells of 20 m × 20 m × 5 m.

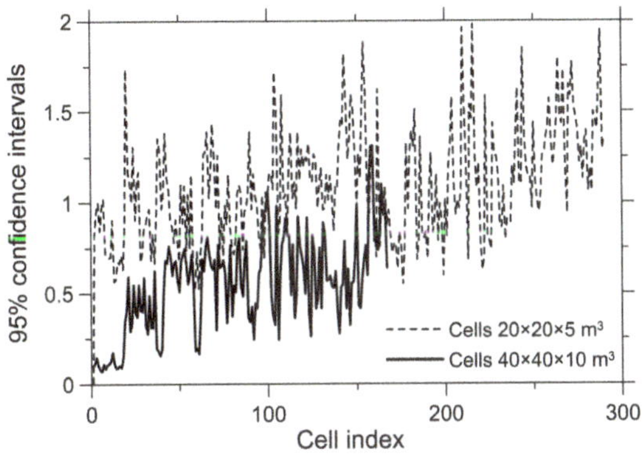

Fig. 4. The 95 % confidence intervals computed considering the discretization of the matrix **A**: **(a)** cells of 40 m × 40 m × 10 m; **(b)** cells of 20 m × 20 m × 5 m.

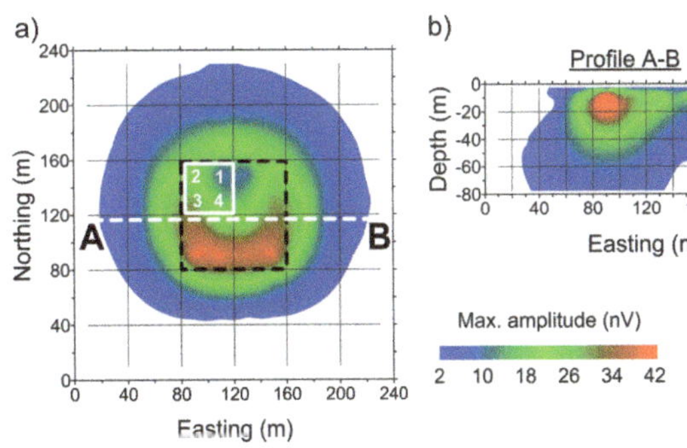

Fig. 5. 3-D sensibility of the SNMR loop computed for 20 m × 20 m × 10 m cells: **(a)** the amplitude of SNMR signal as a function of the cell location within the loop-affected area computed for the depth of 20 m; **(b)** the amplitude of SNMR signal as a function of the cell location shown as a cross-section along profile A–B. The SNMR loop is shown with a dashed black line and cell size is shown with a white line. Four cells of 20 m × 20 m × 5 m are shown by numbers 1 to 4 within one 40 m × 40 m × 10 m cell.

show four 20 m × 20 m × 5 m cells. Figure 5b shows a cross-section corresponding to profile A–B in Fig. 5a: depending on the location within the loop area, the same volume of water may generate four different amplitudes. This heterogeneous and asymmetric sensitivity of the SNMR loop is due to the inclination of the geomagnetic field and to the elliptical polarization of the loop's magnetic field (Weichmann et al., 2000; Girard et al., 2007; Hertrich, 2008). Thus, modeling results show that small targets cannot be resolved with 3-D-SNMR inversion and consequently small water volumes cannot be accurately measured with one SNMR loop. However, if we use a 3-D field setup consisting of several overlapping loops, then areas with a different sensitivity for different loops will overlap and the overall sensitivity within the 3-D setup will be much more homogeneous.

Water within a glacier occurs in caverns and fractures of different size and location. Each small reservoir has 100 % of the water content, but as they are located in ice the water content averaged over the volume that can be resolved by 3-D-SNMR inversion may be less than 100 %. For example, if an ice volume of 40 m × 40 m × 10 m contains a 20 m × 20 m × 5 m water-filled cavern, the water content estimated with 3-D-SNMR in the ice volume will be 12.5 % instead of 100 % expected in the target volume and 0 % in ice. If, additionally, the location of the target is not accurately known, then the SNMR estimate of the volume will be biased by the inhomogeneity of the sensitivity.

3-D-SNMR inversion uses regularization based on the smoothness constraint, which allows obtaining a stable solution, but may not be justified for the strongly heterogeneous water distribution in a glacier. Our numerical experiments consisted of the inversion of 3-D targets of different size and location; this provided the approximate location of the initial model, but in small targets the water volume may be overestimated up to twofold. We explain this by the regularization effect, as the inversion result is an image that results from a trade-off between the misfit and the smoothness constraint. Misfit is directly linked to water volume through the integral equation, but the smoothness constraint represents an assumption on the solution shape and physically is not linked to water volume. For example, if a cell produces a small signal, then the volume of water in this cell will have a small effect on the misfit. However, the smoothness constraint will attribute to this cell a water content equal to that in the neighboring cells, thus increasing the volume estimate, which is not justified by the data.

To cope with this heterogeneous SNMR sensitivity in the measuring setup and a limited resolution of the 3-D-SNMR inversion, we developed an interpretation approach based on a learning procedure. As a learning procedure, we simulated numerical models of different size and location using measuring conditions corresponding to a field survey. Interpretation of the synthetic and experimental data was carried out in two steps: a smooth inversion with the regularization followed by forward modeling. First, smooth 3-D inversion provided an approximate volume and location of water in the

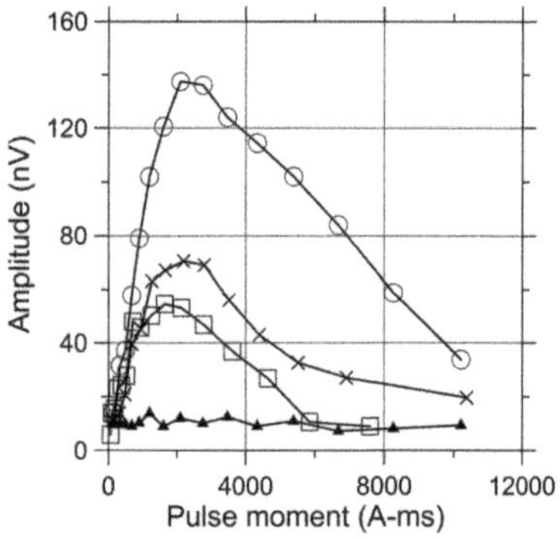

Fig. 6. Amplitude of the SNMR signal vs. pulse moment measured with loop no. 1. Circles show the amplitude measured in 2009 when the cavern was undisturbed by pumping ($V = 53\,500\,\mathrm{m}^3$), crosses correspond to 2011 data ($V = 18\,500\,\mathrm{m}^3$) and squares are 2012 measurements ($V = 11\,200\,\mathrm{m}^3$). Triangles show measurements of the ambient electromagnetic noise.

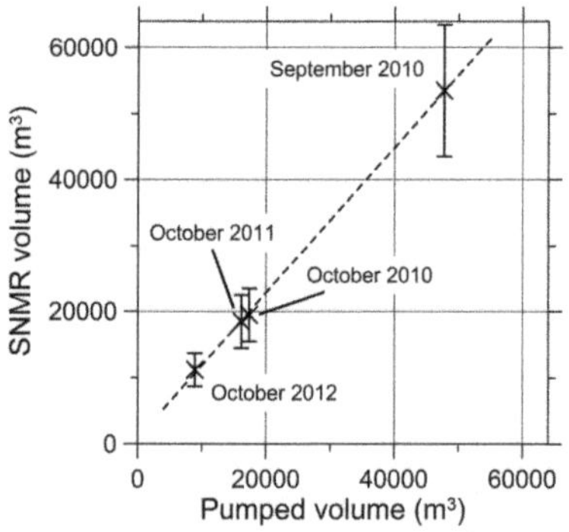

Fig. 7. 3-D-SNMR water volume estimated before each pumping against the water volume extracted by pumping.

glacier without any prior knowledge of the solution. After that, the obtained image of the water distribution was approximated by a model consisting of several boxes. The number, location and water content in each box were selected so that the computed theoretical signal would fit the experimental data. For estimating the minimum and maximum possible amounts of water that can explain the measured signals, we used models with water located in areas with high and low sensitivities of the SNMR loops. Below, we discuss that this expert approach produced good results when applied to investigations of the Tête Rousse glacier.

5 Experimental results

Nine loops were used between 2009 and 2011 (Fig. 2; black squares) and three additional loops were added for surveys carried out in 2012 (red squares). We used NUMIS^{plus} equipment manufactured by IRIS Instruments (France). Data processing and 3-D inversion were done with the SAMOVAR-11x4 software package developed at the Institut de Recherche pour le Développement (IRD, France).

5.1 3-D-SNMR measurements

Figure 2 shows the location of the central cavern as a horizontal slice at an elevation of 3124 m, corresponding to a depth of 46 m below the glacier surface. Note that the average water content given by 3-D-SNMR inversion is 40 % instead of the 100 % expected for bulk water in the cavern. This discrepancy can be explained by the limited resolution of magnetic resonance imaging, which shows an area larger than the true cavern with a water content averaged over this larger area. The water reservoir may also have a complex geometry unresolved by the inversion.

Figure 6 shows the amplitude of the SNMR signal vs. pulse moment, measured with loop no. 1 located over the central cavern (see Fig. 2 for location). The measurements were carried out at different periods, when the cavern contained different volumes of accumulated water. In all three cases, the measured SNMR signal was significantly larger than the ambient electromagnetic noise. The smaller signal thus corresponds to smaller volumes of water, which demonstrates the sensitivity of the method to water-volume variations.

5.2 Water volume estimate

Between 2010 and 2012, the central cavern was drained three times, and once we conducted 3-D-SNMR measurements during such pumping. A 3-D-SNMR survey carried out in June 2010 did not detect any changes in the water distribution between September 2009 and June 2010, and we assume that the 3-D-SNMR results obtained in 2009 can be compared with the pumping results obtained in 2010. Thus, we have four points for comparing 3-D-SNMR estimates of the water volume and the volume extracted by pumping from this cavern. A summary of the measurements of the water volume accumulated in the central cavern using 3-D-SNMR and pumping is presented in Table 1. Figure 7 shows a good correlation (dashed line) between the water volume estimated by 3-D-SNMR and that extracted by pumping.

Further comparison with the pumping data shows that 3-D-SNMR consistently overestimated the volume of water relative to that extracted by pumping. The average discrepancy between 3-D-SNMR and pumping was calculated as approximately $3100\,\mathrm{m}^3$. At least some part of this volume can be attributed to residual water remaining in the glacier after pumping, because technically it is impossible to extract all the water. Moreover, water in hydraulically disconnected caverns and channels contributes to the SNMR signal, but cannot be pumped out. It is also known that the water content in temperate ice usually varies between 0.7 and 2.5 % (Lliboutry, 1976; Bradford et al., 2009) and may be as high

Table 1. Summary of measurements of the water volume accumulated in the central cavern using 3-D-SNMR and pumping.

Action	Schedule									
	5–16/09/2009	26/08–15/10/2010	25–28/09/2010	28/09–15/10/2010	3–9/06/2011	22–24/09/2011	26/09–9/10/2011	22–24/06/2012	14–22/08/2012	24/09–8/10/2012
3-D-SNMR volume (m^3)	53 500		19 500		7000	18 500		8500	11 200	
Pumped volume (m^3)		47 728		17 362			16 162			8904
Discrepancy (m^3)		5572		2138			2338			2296
Discrepancy (%)		10.8		11			12.6			20.5

as 9 % (Pettersson et al., 2004). This water will increase the SNMR signal, but only part of it can be pumped out.

The location of the temperate ice in the Tête Rousse glacier was approximately known (Gilbert et al., 2012; Vincent et al., 2012b) and we estimated its potential water volume by performing SNMR measurements outside the cavern location. Our estimate showed not more than 2 % of water in temperate ice. More accurate estimates were limited by the insufficient signal-to-noise ratio for small water contents. To this end, for areas occupied by temperate ice we added 2 % water content to the model of the water-filled cavern in the glacier body. Modeling results showed that up to 2000 m^3 of the water volume measured by 3-D-SNMR may represent the contribution of water in temperate ice. However, the pumping operations took 44 days in 2010, 13 days in 2011 and 17 days in 2012, during which the cavern was continuously recharged. The recharge rate was unknown, but as snow melt on the Tête Rousse glacier in September/October is minimal, the recharge can be assumed to be slightly higher than in winter (20 to 30 $m^3\,d^{-1}$). Therefore, an additional water volume of 250 to 1200 m^3 was expected in the cavern; this contributed to the pumping results, but could not be taken into account by the 3-D-SNMR estimate. All these processes are difficult to quantify, but they cause additional inaccuracy of the volume estimates made by both 3-D-SNMR and pumping.

5.3 Monitoring

Twice a year, we carried out magnetic resonance measurements that allowed imaging about 75 % of the glacier body. Each 3-D-SNMR campaign provided a water distribution within the glacier and an estimate of the water volume. Measurements were made in early June, after winter recharge, and in September after the summer recharge.

The central cavern, discovered in 2009, was the starting point of our monitoring. The 3-D image of the cavern presented in Fig. 8 consists of horizontal slices of the water content at different altitudes, showing the cavern location within the glacier. The repeated 3-D-SNMR measurements provided time-lapse images of the cavern. For example, Fig. 8 shows the undisturbed cavern observed in 2009 and Fig. 9 shows the same cavern imaged in 2011 after one draining-and-refilling cycle. Comparison of these images shows that the cavern remained in the same place, but that the draining caused changes in the cavern geometry and in the water-content distribution. In 2011, the maximum water content was estimated at 15 % instead of the 40 % observed in 2009, pointing to a smaller cavity and a smaller volume of water in the cavern in 2011.

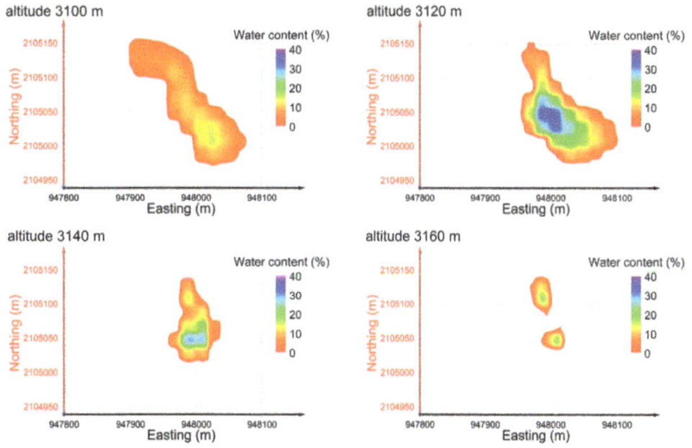

Fig. 8. 3-D-SNMR image of the water-filled cavern observed in September 2009 and shown as slices at different altitudes.

To compare SNMR and pumping results, a 3-D-SNMR image can be transformed into a 1-D plot of the water volume per unit depth for different elevations. For this, we performed integration in the x and y directions of the 3-D water distribution $w(\boldsymbol{r})$ provided by inversion. The 3-D-SNMR and pumping results obtained between 2009 and 2012 are shown in Figs. 10 and 11. In Fig. 10, the volume of accumulated water derived from the 3-D-SNMR results is plotted vs. depth and is compared with the volume of water extracted by pumping. We observe a generally good correspondence between SNMR and pumping data for the depth interval corresponding to the main cavern volume. However, at the bottom of the cavern, the correlation is not as good as for the shallow part, because of a more complex geometry of the deeper

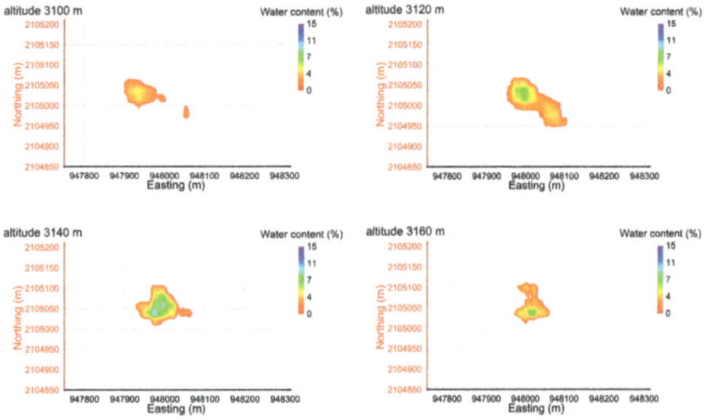

Fig. 9. 3-D-SNMR image of the water-filled cavern observed in September 2011 and shown as slices at different altitudes.

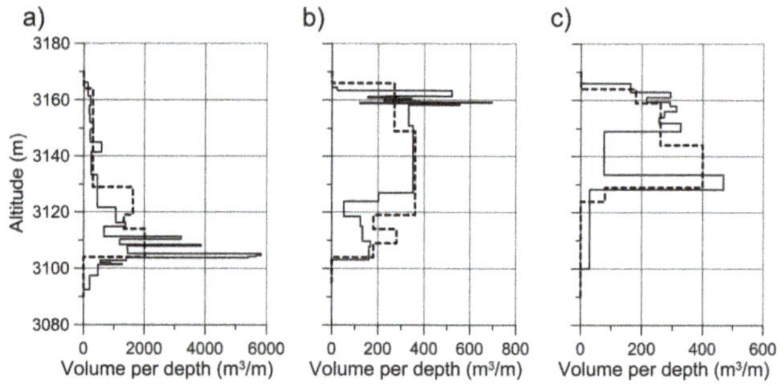

Fig. 11. Water volume per unit depth extracted from different depth intervals (solid line) and water volume per unit depth estimated with the 3-D-SNMR (dashed line): **(a)** August–October 2010; **(b)** September–October 2011; **(c)** August 2012.

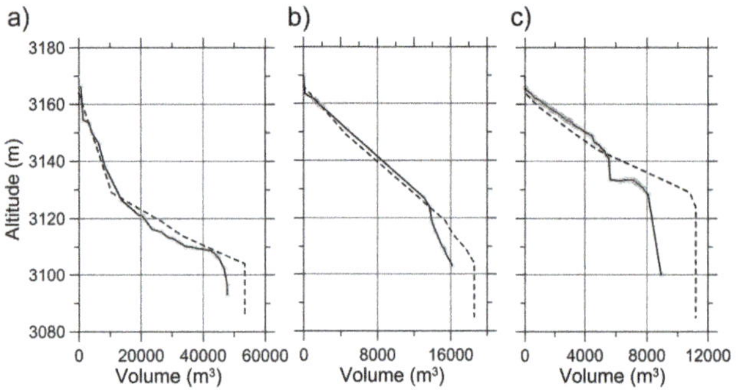

Fig. 10. Accumulated water volume vs. depth extracted by pumping (solid line) and water volume estimated with 3-D-SNMR (dashed line): **(a)** August–October 2010; **(b)** September–October 2011; **(c)** August 2012.

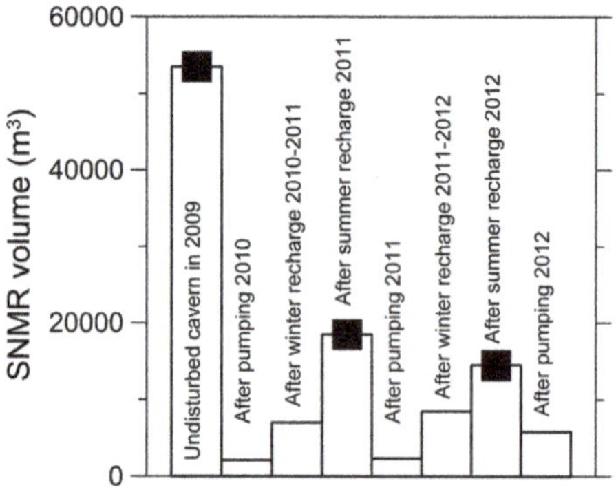

Fig. 12. Monitoring of the water volume accumulated in the central part of the Tête Rousse glacier between 2009 and 2012. Black squares show the volume after full recharge.

part of the cavern where 3-D-SNMR tends to overestimate the water volume due to processes discussed in Sect. 5.2. When the cavern was larger (2009 to 2011), the correlation was better than in 2012, probably because the cavern geometry became more complex after the first drainage. Figure 11 shows the water volume per unit depth vs. depth derived from 3-D-SNMR and pumping results, which are derivatives of the functions shown in Fig. 10. The correlation is generally good, but the rapid variations in extracted volume, observed by pumping, were not resolved by the 3-D-SNMR inversion. Figure 11 clearly shows that the main reservoir discovered in 2009 was located below 3125 m and that it became significantly smaller after the first draining–refilling cycle. In 2012, water was only observed above 3125 m. A summary of the water-volume monitoring is shown in Table 2.

5.4 Recharge of the central cavern

During our monitoring, the cavern was continuously recharged. The results of water-volume measurements are shown in Fig. 12, indicating that the water volume in the fully recharged cavern was reduced after each draining–refilling cycle. In one year, the cavern lost about 65 % of its volume, and after two years it was reduced by about 73 %. Note that in Tables 1 and 2 the water volume observed in 2012 (11 200 m^3) corresponds only to water in the central cavern that was drained, thus allowing comparison between 3-D-SNMR and pumping. In Fig. 12, however, we consider all water observed in the glacier (14 500 m^3), including 3300 m^3 in a small cavern, which was imaged but not affected by pumping.

For estimating the recharge rate, we used the 3-D-SNMR estimates of water volume. Regular magnetic resonance measurements were made in early June for estimating the winter recharge rate, and in August or September for estimating the summer recharge rate. The average recharge rate was estimated as

$$Q_n = (V_{n+1} - V_n)/(t_{n+1} - t_n), \tag{12}$$

where V_n is the water volume in the cavern corresponding to the nth measurement, t_n is time, and $n = 0, 1, 2, .., N$.

The average recharge rate observed between 2010 and 2012 is shown in Fig. 13. The summer recharge rate was estimated to be between 120 and 180 m^3 d^{-1}, and recharge continued throughout the winter at an average rate of 20 to

Table 2. Summary of the 3-D-SNMR and pumping estimates of the changes in water volume in the central cavern caused by drainage. Evolution of the volume was estimated relative to the initial volume observed in 2010.

Method	Depth interval (m)	2010 (m^3)	2011 (m^3)	2012 (m^3)	Evolution 2010–2011 (%)	Evolution 2011–2012 (%)
Pumping	3170–3125	13 617	13 663	8464	0.3	−38.2
	3125–3095	34 111	2499	440	−92.7	−6.0
	3170–3095	47 728	16 162	8904	−66.1	−15.2
3-D-SNMR	3170–3125	10 600	11 700	10 800	10.4	−8.5
	3125–3095	42 900	6800	400	−84.1	−14.9
	3170–3095	53 500	18 500	11 200	−65.4	−13.6

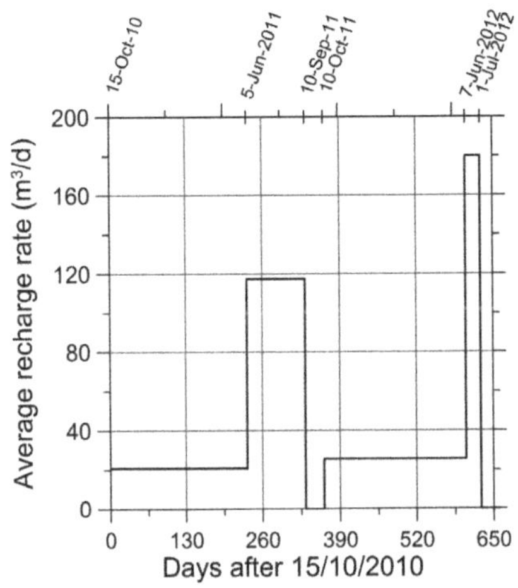

Fig. 13. Estimated average recharge rate of the cavern in the Tête Rousse glacier for the period between 15 October 2010 and 1 August 2012.

$25\,m^3\,d^{-1}$. We calculated the latter by considering the volume of residual water left in the cavern after pumping, which was estimated as the difference between the water volume given by 3-D-SNMR before pumping and the volume of water extracted by pumping (Table 1).

To explain such winter recharge, when no melting occurs at the glacier surface and only solid precipitation exists at this altitude, we looked for a water reservoir located at a higher elevation than the central cavern where water was not expected before. To that end, the measuring setup was extended towards the eastern part of the glacier (see Fig. 2 for location). The 3-D-SNMR image of the glacier obtained in 2012 using the additional loops (Fig. 14) in fact showed two distant reservoirs. Figure 14 shows water distribution starting from 3120 m because no water was observed below. The central reservoir is the cavern discovered in 2009; this became smaller and shallower in comparison with previous years, and its maximum water content was only 5 % instead of the 15 % observed in 2011 and the 40 % in 2009. The glacier surface was about 3173 m high at 948 000° E, thus showing that the central cavern was extended to the glacier surface (confirmed by boreholes). The second (upper) cavern is located at about 150–200 m east of the central cavern, at a higher elevation and in a dangerous area with difficult access. For this reason, accurate 3-D-SNMR imaging of this cavern was not possible in 2012. The 3-D-SNMR estimate of the water volume in the upper cavern suggests over $5000\,m^3$, but this preliminary result requires further investigation. To verify the existence of a possible high-yield hydraulic link between the two caverns (tunnel or large channel), we took magnetic resonance measurements over the upper cavern before and immediately after draining of the central cavern, which showed that the 3-D-SNMR signal generated by water in the upper cavern did not change. This observation confirms that, if the pumping affected the upper cavern, this was below the detection threshold of the instrument.

6 Discussion

The annual drainage of a large cavern in the central part of the Tête Rousse glacier caused evolutionary rather than revolutionary changes in ice-cavern geometry, providing us with a rare opportunity of studying the glacier's transitional regime imposed by pumping. Tête Rousse thus is a real-scale physical model of a polythermal glacier, providing useful data that contribute to a better understanding of the different processes taking place in this type of glacier.

The use of 3-D-SNMR allowed quantifying the temporal and spatial changes in water distribution caused by the annual draining of a water-filled cavern. During our experiments, we confirmed that the 3-D-SNMR method is reliable and cost-effective compared to drilling and pumping for investigating water in a glacier. Unambiguous identification of liquid water in ice, localization of water-storage areas, and the possibility of estimating the water volume stored in a glacier are three advantages of 3-D-SNMR compared with traditional geophysical methods. In a highly heterogeneous environment, non-invasive imaging of a large area allows visualizing the entire glacier, which is a competitive advantage of 3-D-SNMR over drilling. Indeed, not all boreholes

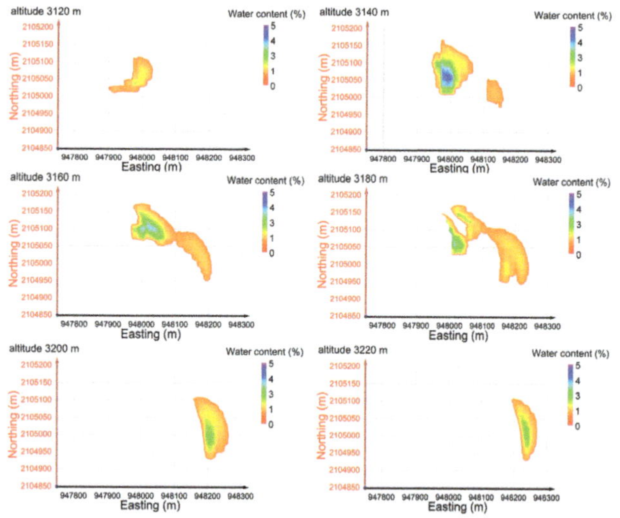

Fig. 14. 3-D-SNMR image of the water-filled cavern observed in October 2012 and shown as slices at different altitudes.

may be representative, and the need to drill many holes renders the fieldwork longer and more expensive (Gordon et al., 2001). Applied under glacier conditions, pumping may cause changes in the internal structure of the investigated glacier, but 3-D-SNMR is a non-destructive method. However, 3-D-SNMR and pumping are complementary techniques that provide close, but different, visions of glacier hydrology. For example, 3-D-SNMR detects and visualizes water, regardless of whether or not a hydraulic link exists between different areas, but pumping may provide information on the hydraulic connection of different reservoirs.

Figure 14 shows that water in the Tête Rousse glacier accumulated in two separate reservoirs (central and upper caverns). The upper cavern would be preferentially recharged by melt water, but the central cavern may accumulate both surface and englacial water with corresponding refilling rates. Our results show that a large tunnel does not directly connect these two caverns, but that water may flow from the upper cavern through a subglacial drainage system and probably through subglacial sediment with a relatively low hydraulic conductivity. This slow drainage would explain the observed continuous recharge of the central cavern during the winter months. In summer, higher water fluxes may cause a higher hydraulic conductivity of the englacial channels (Raymond et al., 1995; Hubbard et al., 1995), as well as more water flow on the ice surface. Such increased water circulation may explain the higher summer-recharge rate observed during our monitoring. However, the subglacial drainage system in the Tête Rousse glacier is still not well known, and different drainage mechanisms typical for a polythermal glacier may coexist (Fountain and Walder, 1998).

The possibility of cavern-roof collapse during pumping was not excluded (Gagliardini et al., 2011). For security reasons, the glacier surface was under direct observation and small movements were continuously monitored with a differential GPS. In July 2012, a large crevasse appeared and ice deformation became visible on the surface over the cavern location, indicating that the latter caused this deformation. Monitoring of water accumulated in the glacier showed that, before 2012, each pumping event caused the volume of water in the central cavern to diminish. Both 3-D-SNMR and pumping results reveal that in one year (2010–2011) the cavern lost about 90 % of the volume in its deeper part below 3125 m (Table 2). The relatively rapid rate of temperate ice deformation may explain these observations. For example, Reynaud (1987) observed a 60 % reduction in inactive, englacial conduit at depths of about 100 m over a 20-day period. In addition, it cannot be excluded that part of the reduced volume of accumulated water may be attributed to air pockets in drained and refilled voids within the ice body.

Despite the new knowledge about the Tête Rousse glacier that has been acquired over the past three years, we still have an incomplete understanding of the glacial-hydrological processes and of glacier-ice dynamics. For example, while a cold-ice barrier may explain the formation of the central cavern, the existence of the upper cavern was not expected. We hope to continue our monitoring, leading to the precise location and size of the upper cavern that will complete our understanding of the glacier structure, and of the ice-deformation processes taking place in the glacier.

7 Conclusions

Imaging of the water distribution in the Tête Rousse glacier showed two main water-storage areas and allowed estimating the volume of accumulated water. Continuous cavern recharge occurs both in summer and during the winter months. The summer recharge rate was estimated to be six to eight times higher than the winter one. Our monitoring results revealed significant changes in the internal glacier structure, caused by the artificial drainage. In one year, the cavern lost about 65 % (30 000 m^3) of its initial volume, which was explained by creep deformation of the surrounding ice after draining and the probably formation of air pockets within the ice. These observations provide useful information to decision makers for developing protection measures minimizing the risk of a catastrophic flood from the glacier.

For this study, we used a newly developed 3-D-SNMR imaging method that allowed visualizing the englacial water and estimating the volume of water accumulated within the Tête Rousse glacier. The 3-D-SNMR results were found to be in a good agreement with pumping results and other observations, thus confirming the high reliability and efficiency of 3-D-SNMR imaging. As far as we know, this is the first report of applying large-scale magnetic resonance imaging to monitoring water accumulation in a glacier.

Acknowledgements. The town of Saint-Gervais (France), *Service de Restauration des Terrains en Montagne* (RTM), and the *Préfecture de la Haute Savoie* (France) funded this study. We thank N. Karr, F. Charles and V. Tairraz for useful advice and for helping with the field measurements. We also thank all the people who assisted us during this work. We acknowledge reviewers A. D. Booth and J. West for their critical comments that contributed to improving the presentation of our results.

Edited by: J. O. Hagen

References

Aster, R. C., Borchers, B., and Thurber, C. H.: Parameter Estimation and Inverse Problems, Elsevier Academic Press, New York, 2005.

Bælum, K. and Benn, D. I.: Thermal structure and drainage system of a small valley glacier (Tellbreen, Svalbard), investigated by ground penetrating radar, The Cryosphere, 5, 139–149, doi:10.5194/tc-5-139-2011, 2011.

Benn, D., Gulley, J., Luckman, A., Adamek, A., and Glowacki, P. S.: Englacial drainage system formed by hydrologically driven crevasse propagation, J. Glaciol., 55, 513–523, doi:10.3189/002214309788816669, 2009.

Boucher, M., Girard, J. F., Legchenko, A., Baltassat, J. M., Dörfliger, N., and Chalikakis, K.: Using magnetic resonance soundings to locate a water-filled karst conduit, J. Hydrol., 330, 413–421, doi:10.1016/j.jhydrol.2006.03.034, 2006.

Bradford, J. H., Nichols, J., Mikesell, T. D., and Harper, J. T.: Continuous profiles of electromagnetic wave velocity and water content in glaciers: an example from Bench Glacier, Alaska, USA, Ann. Glaciol., 50, 1–9, doi:10.3189/172756409789097540, 2009.

Callaghan, P. T., Dykstra, R., Eccles, C. D., Haskell, T. G., and Seymour, J. D.: A nuclear magnetic resonance study of Antarctic sea ice brine diffusivity, Cold Reg. Sci. Technol., 29, 153–171, doi:10.1016/S0165-232X(99)00024-5, 1999.

Endres, A. L., Murray, T., Booth, A. D., and West, L. J.: A new framework for estimating englacial water content and pore geometry using combined radar and seismic wave velocities, Geophys. Res. Lett., 36, L04501, doi:10.1029/2008GL036876, 2009.

Fischer, L., Kääb, A., Huggel, C. and Nötzli, J.: Geology, glacier changes, permafrost and related slope instabilities in a high-mountain rock face: Monte Rosa east face, Italian Alps, Geophys. Res. Abstr., 7, 00518, doi:10.1029/2008GL036876, 2005.

Fountain, A. G. and Walder, J. S.: Water flow through temperate glaciers, Rev. Geophys., 36, 299–328, doi:10.1029/97RG03579, 1998.

Gagliardini, O., Gillet-Chaulet, F., Durand, G., Vincent, C., and Duval, P.: Estimating the risk of glacier cavity collapse during artificial drainage: The case of Tête Rousse Glacier, Geophys. Res. Lett., 38, L10505, doi:10.1029/2011GL047536, 2011.

Gilbert, A., Vincent, C., Wagnon, P., Thibert, E., and Rabatel, A.: The influence of snow cover thickness on the thermal regime of Tête Rousse Glacier (Mont Blanc range, 3200 m a.s.l.): Consequences for outburst flood hazards and glacier response to climate change, J. Geophys. Res., 117, F04018, doi:10.1029/2011JF002258, 2012.

Girard, J. F., Boucher, M., Legchenko, A., and Baltassat, J. M.: 2-D magnetic resonance tomography applied to karstic conduit imaging, J. Appl. Geophys., 63, 103–116, doi:10.1016/j.jappgeo.2007.08.001, 2007.

Gordon, S., Sharp, M., Hubbard, B., Willis, I., Smart, C., Copland, L., Harbor, J., and Ketterling, B.: Borehole drainage and its implications for the investigation of glacier hydrology: experiences from Haut Glacier d'Arolla, Switzerland, Hydrol. Process., 15, 797–813, doi:10.1002/hyp.184, 2001.

Grasmueck, M., Weger, R., and Horstmeyer, H.: Three-dimensional ground-penetrating radar imaging of sedimentary structures, fractures, and archaeological features at submeter resolution, Geology, 32, 933–936, doi:10.1130/G20776.1, 2004.

Gross, R., Green, A. G., Horstmeyer, H., and Begg, J. H.: Location and geometry of the Wellington Fault (New Zealand) defined by detailed three-dimensional georadar data, Geophys. Res. Lett., 109, B05401, doi:10.1029/2003JB002615, 2004.

Hamran, S. E., Aarholt, E., Hagen., J. O., and Mo, P.: Estimation of relative water content in a sub-polar glacier using surface-penetration radar, J. Glaciol., 42, 533–537, 1996.

Haeberli, W.: Frequency and characteristics of glacier floods in the Swiss Alps, Ann. Glaciol., 4, 85–90, 1983.

Haeberli, W., Alean, J. C., Müller, P., and Funk, M.: Assessing risks from glacier hazards in high mountain regions: some experiences in the Swiss Alps, Ann. Glaciol., 13, 96–102, 1989.

Hauck, C. and Vonder Mühll, D.: Using DC resistivity tomography to detect and characterise mountain permafrost, Geophys. Prospect., 51, 273–284, doi:10.1046/j.1365-2478.2003.00375.x, 2003.

Hauck, C., Guglielmin, M., Isaksen, K., and Vonder Mühll, D.: Applicability of frequency domain and time-domain electromagnetic methods, Permafrost Periglac., 12, 39–52, doi:10.1002/ppp.383, 2001.

Hertrich, M.: Imaging of groundwater with nuclear magnetic resonance, Prog. Nucl. Mag. Res. Sp., 53, 227–248, doi:10.1016/j.pnmrs.2008.01.002, 2008.

Hertrich, M., Green, A. G., Braun, M., and Yaramanci, U.: High-resolution surface NMR tomography of shallow aquifers based on multioffset measurements, Geophysics, 74, G47–G59, doi:10.1190/1.3258342, 2009.

Hoekstra, P.: Electromagnetic methods for mapping shallow permafrost, Geophysics, 43, 782–787, doi:10.1190/1.1440853, 1978.

Hubbard, B. P., Sharp, M. J., Willis, I. C., Nielsen, M. K., and Smart, C. C.: Borehole water-level variations and the structure of the subglacial hydrological system of Haut Glacier d'Arolla, Valais, Switzerland, J. Glaciol., 41, 572–583, 1995.

Irvine-Fynn, T. D. L., Moorman, B. J., Williams, J. L. M., and Walter, F. S. A.: Seasonal changes in groundpenetrating radar signature observed at a polythermal glacier, Bylot Island, Canada,

Earth Surf. Proc. Land., 31, 892–909, doi:10.1002/esp.1299, 2006.

Irvine-Fynn, T. D. L., Hodson, A. J., Moorman, B. J., Vatne, G., and Hubbard, A. L.: Polythermal Glacier Hydrology: A review, Rev. Geophys., 49, RG4002, doi:10.1029/2010RG000350, 2011.

Jansson, P., Hock, R., and Schneider, T.: The concept of glacier storage: A review, J. Hydrol., 282, 116–129, doi:10.1016/S0022-1694(03)00258-0, 2003.

Kim, K. Y., Lee, J., Hong, M. H., Hong, J. K., Jin, Y. K., and Shon, H.: Seismic and radar investigations of Fourcade Glacier on King George Island, Antarctica, Polar Res., 29, 298–310, doi:10.1111/j.1751-8369.2010.00174.x, 2010.

Kneisel, C.: New insights into mountain permafrost occurrence and characteristics in glacier forefields at high altitude through the application of 2-D resistivity imaging, Permafrost Periglac., 15, 221–227, doi:10.1002/ppp.495, 2004.

Knight, R., Grunewald, E., Irons, T., Dlubac, K., Song, Y., Bachman, H. N., Grau, B., Walsh, D., Abraham, J. D., and Cannia, J.: Field experiment provides ground truth for surface nuclear magnetic resonance measurement, Geophys. Res. Lett., 39, L03304, doi:10.1029/2011GL050167, 2012.

Lanz, E., Maurer, H. R., Green, A. G., and Ansorge, J.: Refraction tomography over a buried waste disposal site, Geophysics, 63, 1414–1433, doi:10.1190/1.1444443, 1998.

Legchenko, A. V. and Shushakov, O. A.: Inversion of surface NMR data, Geophysics, 63, 75–84, doi:10.1190/1.1444329, 1998.

Legchenko, A. and Valla, P.: A review of the basic principles for proton magnetic resonance sounding measurements, J. Appl. Geophys., 50, 3–19, doi:10.1016/S0926-9851(02)00127-1, 2002.

Legchenko, A., Ezersky, M., Camerlink, C., Al-Zoubi, A., Chalikakis, K., and Girard, J. F.: Locating water-filled karst caverns and estimating their volume using magnetic resonance soundings, Geophysics, 73, 51–61, doi:10.1190/1.2958007, 2008.

Legchenko, A., Descloitres, M., Vincent, C., Guyard, H., Garambois, S., Chalikakis, K., and Ezerski, M.: Three-dimensional magnetic resonance imaging for groundwater, New J. Phys., 13, 025022, doi:10.1088/1367-2630/13/2/025022, 2011.

Lehmann-Horn, A., Walbrecker, J. O., Hertrich, M., Langston, G., McClymont, A. F., and Green, A. G.: Imaging groundwater beneath a rugged proglacial moraine, Geophysics, 76, B165–B172, doi:10.1190/geo2011-0095.1, 2011.

Lliboutry, L.: Physical processes in temperate glaciers, J. Glaciol., 16, 151–158, 1976.

Macheret, Y. and Glazovsky, A. F.: Estimation of absolute water content in Spitsbergen glaciers from radar sounding data, Polar Res., 19, 205–216, 2000.

Marescot, L., Loke, M. H., Chapellier, D., Delaloye, R., Lambiel, C., and Reynard, E.: Assessing reliability of 2D resistivity imaging in permafrost and rock glaciers using the depth of investigation index method, Near Surf. Geophys., 1, 57–67, doi:10.3997/1873-0604.2002007, 2003.

Maurer, H. and Hauck, C.: Instruments and Methods Geophysical imaging of alpine rock glaciers, J. Glaciol., 53, 110–120, doi:10.3189/172756507781833893, 2007.

Merz, K. J., Maurer, H. R., Rabenstein, L., Buchli, T., and Springman, S. M.: Geophysical characterization of unstable permafrost in the Turtmann Valley, Switzerland, Geophys. Res. Abstr., 14, EGU2012–2107, 2012.

Mougin, P. and Bernard, C.: Études exécutées au glacier de Tête Rousse, in: Annales de l'Observatoire météorologique, physique et glaciaire du Mont Blanc (altitude 4,358 métres), edited by: Vallot, J., Tome VI, Paris, 1905.

Murray, T., Stuart, G. W., Fry, M., Gamble, N. H., and Crabtree, D. M.: Englacial water distribution in a temperate glacier from surface and borehole radar velocity analysis, J. Glaciol., 46, 389–398, doi:10.3189/172756500781833188, 2000.

Musil, M., Maurer, H., Green, A. G., Horstmeyer, H., Nitsche, F. O., Vonder Mühll, D., and Springman, S.: Shallow seismic surveying of an Alpine rock glacier, Geophysics, 67, 1701–1710, doi:10.1190/1.1527071, 2002.

Pettersson, R., Jansson, P., and Blatter, H.: Spatial variability in water content at the cold-temperate transition surface of the polythermal Storglaciären, Sweden, J. Geophys. Res., 109, F02009, doi:10.1029/2003JF000110, 2004.

Raymond, C. F., Benedict, R., Harrison, W., Echelmeyer, K., and Sturm, M.: Hydrological discharges and motion of Fels and Black Rapids Glaciers, Alaska, USA: Implications for the structure of their drainage systems, J. Glaciol., 41, 290–304, 1995.

Reynaud, L.: The November 1986 survey of the Grand Moulin on the Mer de Glace, Mont Blanc Massif, France, J. Glaciol., 33, 130–131, 1987.

Stoer, J. and Bulirsch, R.: Introduction to numerical analysis, Springer-Verlag Berlin, 1980.

Tarantola, A.: Inverse problem theory. Methods for data fitting and model parameter estimation, Elsevier Science Publ.Co., Inc., 1987.

Tikhonov, A. and Arsenin, V.: Solution of ill-posed problems: John Wiley & Sons, Inc., 1977.

Valla, P. and Legchenko, A.: One-dimensional modelling for proton magnetic resonance sounding measurements over an electrically conductive medium, J. Appl. Geophys., 50, 217–229, doi:10.1016/S0926-9851(02)00141-6, 2002.

Vallot, J. and Delebecque, A.: Sur les causes de la catastrophe survenue à Saint-Gervais (Haute Savoie), C. R. Acad. Sci., 115, 264–266, 1892.

Vincent, C., Garambois, S., Thibert, E., Lefèbvre, E., Le Meur, E., and Six, D.: Origin of the outburst flood from Tête Rousse glacier in 1892 (Mont-Blanc area, France), J. Glaciol., 56, 688–698, doi:10.3189/002214310793146188, 2010.

Vincent, C., Descloitres, M., Garambois, S., Legchenko, A., Guyard, H., Lefebvre, E., and Gilbert, A.: Detection of a subglacial lake in Glacier de Tete Rousse (Mont Blanc area, France), J. Glaciol., 58, 866–878, doi:10.3189/2012JoG11J179, 2012a.

Vincent, C., Descloitres, M., Garambois, S., Legchenko, A., Guyard, H., Thibert, E., Gilbert, A., Karr, N., and Tairraz, V.: Détection d'une poche d'eau au glacier de Tête Rousse en 2010 et mesures préventives pour éviter une catastrophe, La Houille Blanche, 2, 34–41, doi:10.1051/lhb/2012013, 2012b.

Vouillamoz, J. M., Legchenko, A., Albouy, Y., Bakalowicz, M., Baltassat, J. M., and Al-Fares, W.: Localization of saturated karst aquifer with magnetic resonance sounding and resistivity imagery, Ground Water, 41, 578–587, doi:10.1111/j.1745-6584.2003.tb02396.x, 2003.

Weichmann, P. B., Lavely, E. M., and Ritzwoller, M.: Theory of surface nuclear magnetic resonancewith applications to geophysical imaging problems, Phys. Rev. E, 62, 1290–1312, doi:10.1103/PhysRevE.62.1290, 2000.

16

Heat sources within the Greenland Ice Sheet: dissipation, temperate paleo-firn and cryo-hydrologic warming

M. P. Lüthi[1,*], C. Ryser[1], L. C. Andrews[2,3], G. A. Catania[2,3], M. Funk[1], R. L. Hawley[4], M. J. Hoffman[5], and T. A. Neumann[6]

[1]Versuchsanstalt für Wasserbau, Hydrologie und Glaziologie (VAW), ETH Zurich, 8093 Zurich, Switzerland
[2]Institute for Geophysics, The University of Texas at Austin, Austin, Texas, 78758, USA
[3]Dept. of Geological Sciences, The University of Texas at Austin, Austin, Texas, 78713, USA
[4]Dept. of Earth Sciences, Dartmouth College, Hanover, New Hampshire, 03755, USA
[5]Fluid Dynamics and Solid Mechanics Group, Los Alamos National Laboratory, Los Alamos, New Mexico, 87545, USA
[6]NASA Goddard Space Flight Center, Code 615, Greenbelt, Maryland, 20770, USA
[*]now at: Geographical Institute, University of Zurich, 8057 Zurich, Switzerland

Correspondence to: M. P. Lüthi (martin.luethi@geo.uzh.ch)

Abstract. Ice temperature profiles from the Greenland Ice Sheet contain information on the deformation history, past climates and recent warming. We present full-depth temperature profiles from two drill sites on a flow line passing through Swiss Camp, West Greenland. Numerical modeling reveals that ice temperatures are considerably higher than would be expected from heat diffusion and dissipation alone. The possible causes for this extra heat are evaluated using a Lagrangian heat flow model. The model results reveal that the observations can be explained with a combination of different processes: enhanced dissipation (strain heating) in ice-age ice, temperate paleo-firn, and cryo-hydrologic warming in deep crevasses.

1 Introduction

Vertical ice temperature profiles of the Greenland Ice Sheet (GrIS) carry information on past upstream surface temperature and accumulation rates. This information is modified by vertical and horizontal stretching of the ice during flow, smoothed by diffusion, and altered by subglacial and englacial heat sources.

Characteristic ideal profiles at different locations in the ice sheet have been obtained with numerical advection-diffusion models (e.g., Budd et al., 1982; Letreguilly et al., 1991; Funk et al., 1994; Greve, 1997). Full-depth ice temperature profiles from the ablation zone of the GrIS have been published for only nine drill sites: five in the ablation zone of the Paakitsoq area, downstream of Swiss Camp (Thomsen et al., 1991) and four in Jakobshavn Isbræ (Iken et al., 1993; Lüthi et al., 2002). Despite the limited spatial extent of these observations, comparison between modeled and observed temperature profiles allow us to explore the complexity of heat sources in the ablation zone.

Previous work from Jakobshavn Isbrae demonstrates that comparison of modeled and measured ice temperature profiles provides a means to infer ice-dynamical characteristics such as vertical stretching of basal ice as it enters Jakobshavn Isbræ (Iken et al., 1993; Funk et al., 1994). In this case, the validity of the assumed initial and boundary conditions was assessed by an almost perfect match of measured and modeled ice temperatures at drill site DUCK at the margins of Jakobshavn Isbræ (Fig. 1) (Lüthi et al., 2002). It is noteworthy that the temperature profile at 500 km from Funk et al. (1994) is similar to the profile at Swiss Camp (Thomsen et al., 1991) whereas all other temperature profiles from the Paakitsoq area (Thomsen et al., 1991) are considerably warmer. These warmer temperatures must be caused by heat sources not accounted for in the model. Such heat sources are likely flowing water in moulins and englacial pathways or freezing water in crevasses, processes recently referred to

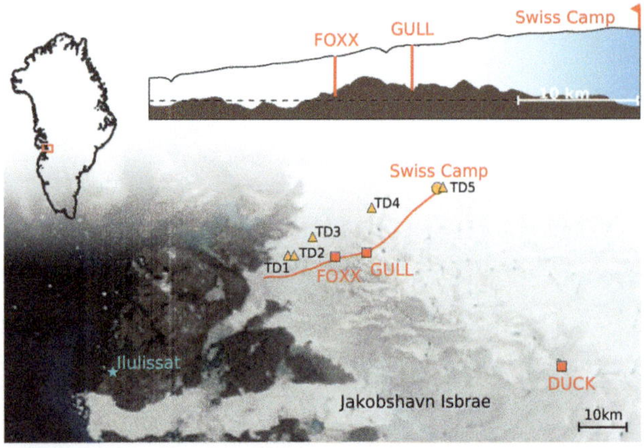

Figure 1. The study site indicated with red area within Greenland outline is illustrated with a MODIS satellite image provided by NASA/GSFC (2010). The drill sites FOXX and GULL are located on a flow line downstream of Swiss Camp, and north of Jakovshavn Isbræ. Site DUCK is from 1995 (Lüthi et al., 2002). Drill sites by Thomsen are marked with yellow triangles and indicated by TD1-5. The inset shows bed and surface topographies along the flow line at 5-fold vertical exaggeration (data from DTU Space and Remote Sensing, 2005; Gogineni, 2012).

as "cryo-hydrologic warming" (Phillips et al., 2010, 2013). Future increases in melt area extent might therefore lead to more cryo-hydrologic warming, a mechanism considered to cause rapid future warming of the ice sheet (Phillips et al., 2013).

In this study we present full-thickness temperature profiles from four new boreholes. Using a numerical heat flow model we calculate expected ice temperature profiles for our study sites. Comparison of modeled profiles with measurements shows warmer temperatures of the ice in the ablation area. Possible sources of this extra heat are investigated and discussed.

2 Data and methods

2.1 Drill sites

The location of the three drill sites in the western ablation zone of the GrIS is indicated on the map in Fig. 1. Sites FOXX and GULL were instrumented in summer 2011 with sensors for ice deformation, subglacial pressure, and ice temperature (Ryser et al., 2014a, b; Andrews et al., 2014). Deep drilling at Swiss Camp and installation of thermistor string TD5 in 1990 was performed by Danish Technological University (DTU; Thomsen and Thorning, 1992; Ahlstrom, 2007).

2.2 Temperature measurements

At sites FOXX and GULL temperature was measured with the custom-built digital borehole sensor system DIBOSS (described in Ryser et al., 2014b) at depths below 300 m. These deep sensors are complemented with analog thermistors closer to the surface. In each DIBOSS sensing unit an IST TSic-716 semiconductor temperature sensor with integrated 14 Bit AD-converter (resolution 4 mK, ±70 mK absolute) was mounted in direct contact with the enclosing aluminum housing. These sensors were logged in 10 min intervals for 2 years with a Campbell CR-1000 data logger. Depending on ambient ice temperature, the readings of temperature equilibrated to their undisturbed values after one to three months.

For measurements close to the surface, sets of two thermistor strings were deployed in two boreholes at site FOXX, and in one borehole at site GULL. Each string consisted of an 18-core cable with 9 NTC thermistors (Fenwal 135-103FAG-J01) that were shielded from pressure and moisture, and for which individual calibration curves were determined in a calibration bath at six to eight reference temperatures in the range of −17.5 to 10 °C with a digital multi-meter. Reference thermistors, calibrated to an absolute accuracy of 20 mK by the Swiss Federal Office of Metrology, were used to determine the bath temperature. The maximum difference between reference temperatures and the calibration curve was 40 mK. In the field, resistances were measured with a full bridge circuit logged by a Campbell CR-1000 data logger, and occasional readings with a digital multi-meter over the course of 2 years. With the setup described above, the absolute accuracy of all measured temperatures is estimated to be better than 70 mK.

Figure 2 shows ice temperature measured at sites FOXX, GULL and TD5 (Swiss Camp; Thomsen and Thorning, 1992). Ice temperature increases from the highest site towards the margin. A notable exception is profile FOXX2 which was recorded in 86 m distance from FOXX1, and which is even colder than GULL.

2.3 Heat sources

Heat sources within the ice are due to dissipation (strain heating), or related to flowing or freezing water. Here we calculate order of magnitude estimates of these effects. For this purpose we assume ice with a density of $\rho_i = 900\,\mathrm{kg\,m^{-3}}$, heat capacity $C_i = 2093\,\mathrm{J\,kg^{-1}}$ and latent heat of freezing $L = 333.5\,\mathrm{kJ\,kg^{-1}}$.

2.3.1 Freezing water

The heat energy from freezing water needed to raise the temperature of 1 m^3 of ice by $\Delta T = 1$ K is

$$W = \rho_i C_i \Delta T = 1.88\,\mathrm{MJ\,m^{-3}}. \quad (1)$$

To produce this amount of heat by freezing, 5.6 kg of water are required. Such water is readily available at the glacier surface during the ablation season, and also in the lower reaches of the accumulation area where permanent storage of water within the firn was recently discovered, although 250 km south of our sites (e.g., Harper et al., 2012; Forster et al., 2013). Water seeping through cracks can freeze and provide an extensive heat source down to the bottom of such cracks.

2.3.2 Strain heating

The volumetric heat production rate P due to dissipation (strain heating) under shear deformation is calculated under the assumption of the shallow ice approximation and $n = 3$ as

$$P = \dot{\varepsilon}_{ij}\sigma_{ij} \sim 2\dot{\varepsilon}_{xz}\sigma_{xz} = 2A(T)\sigma_{xz}^4, \quad (2)$$

where the shear stress $\sigma_{xz} = \rho_i\, g\, h \tan\alpha$ is calculated from the density ρ_i, gravity g, surface slope α and depth h below the surface. The value of P therefore varies with current surface slope, and depends on depth and current temperature. The temperature dependent flow rate factor $A(T)$ is taken from Cuffey and Paterson (2010) which agrees well with measurements of ice deformation at sites FOXX and GULL (Ryser et al., 2014b). At a shear stress of 0.1 MPa and with $A\,(-5\,°\mathrm{C}) = 29.3\,\mathrm{MPa^{-3}\,a^{-1}}$ the heating power is $P = 0.006\,\mathrm{MJ\,m^{-3}\,a^{-1}}$, equivalent to a heating rate of 0.31 K per century.

2.4 Heat flow model

The heat flow model is adapted to model the lower accumulation and upper ablation areas in our region of study. The model tracks a flowline with origin at the ice sheet center (0 km; Fig. 7b in Funk et al., 1994). Figure 3 shows our area of interest, located on this flow line between coordinate 450 and 530 km. In this coordinate system, the 1990 drill site TD5 (Thomsen et al., 1991) at Swiss Camp is located at coordinate 498 km at the equilibrium line. The two 2011 drill sites FOXX and GULL are located at coordinates 520 and 530 km.

Heat flow is modeled with the finite element method in two dimensions. The model implements a block of ice in Lagrangian description. The finite element library Libmesh (Kirk et al., 2006) was used to solve the transient heat diffusion equation in a Lagrangian reference frame (i.e., the mesh nodes follow the deforming material; e.g., Hutter and Jöhnk, 2004)

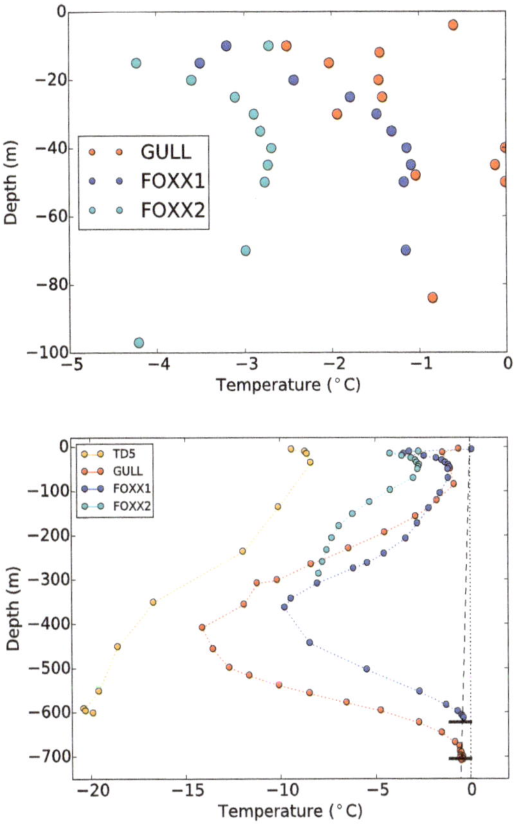

Figure 2. Measured ice temperatures. **(a)** shows near-surface temperatures at sites GULL and FOXX. Holes FOXX1 and FOXX2 are 86 m apart. **(b)** shows measured ice temperatures in four boreholes at the drill sites TD5 (Swiss Camp; data from Thomsen et al., 1991), GULL and FOXX. The dashed line indicates the pressure melting temperature.

$$\rho_i C_i \frac{\partial T}{\partial t} = \nabla(k\nabla T) + P, \quad (3)$$

with temperature T, density ρ_i, specific heat capacity C_i, heat conductivity k and volumetric heat production rate P. The model domain consists of a block discretized with rectangular Quad4 elements with Galerkin weighting (linear approximation of temperature). Implicit time stepping was implemented with a standard Crank–Nicolson scheme.

Each model run consisted of the deformation of the block of ice (the model domain) during horizontal motion over the distance of 100 km (from 450 to 550 km; Fig. 3). The horizontal velocity was taken from Funk et al. (1994), resulting in a travel time from Swiss Camp to FOXX of ~900 years. During motion along the flow line the model domain (the block) is horizontally and vertically stretched or compressed to comply with the local ice thickness. Vertical shearing was neglected (i.e., $\dot{\varepsilon}_{xz} = 0$) so that the block moved in plug-like flow, whereas dissipative heat production according to Eq. (2) was included. Stretching was applied homogeneously throughout the ice body honoring mass continuity

Table 1. Designation of model runs with different heat sources. These sources are combined in different model runs, e.g., run mEC.

mC	refreezing of water-filled crevasses
mD	dissipation from strain heating/reference run
mE	dissipation from strain heating with enhancement
mF	temperate firn above equilibrium line

with $\dot{\varepsilon}_{zz} = -\dot{\varepsilon}_{xx}$ and $\dot{\varepsilon}_{yy} = 0$ (with coordinates x in flow direction, y across, z vertical). Nodes were moved according to their position with respect to the origin of the coordinate system of the Lagrangian model domain. Heat (i.e., temperature at the model nodes as the solution from an earlier time step) is therefore advected with the mesh in horizontal and vertical direction.

Mass accumulation at the upper surface was simulated by adding new elements on top of the domain, in which temperature was set to the surface temperature. Surface temperature and mass balance are taken from Fig. 4 in Funk et al. (1994), which are similar to the 1900–2000 values of a more recent reconstruction (Box, 2013). Similarly, mass removal by surface melt was implemented by removing elements from the top of the domain. In the Libmesh implementation this was achieved in a simple manner by solving the heat flow equation in a subdomain of a larger mesh with constant topology. With this approach, only the extent of the active subdomain was adjusted in every time step. To obtain a high resolution of the grid close to the surface the mesh was refined to a resolution of 0.5 m in the top 150 m of the active subdomain, using the adaptive mesh refinement with hanging nodes implemented in Libmesh, as illustrated in Fig. 4. The unrefined mesh consisted of 100 elements in the vertical, corresponding to a vertical mesh size of 20 to 7 m, depending on total ice column height. A constant time step size of 1 year was used, which together with the vertical mesh size of 0.5 m at the surface gives a numerically stable solution. This model setup closely reproduced analytical solutions for two setups: Sinusoidal forcing of the surface temperature, and a moving boundary solution for constant surface ablation (Vorkauf, 2014).

The initial dimensions of the model domain were 1000 m length and 2000 m thickness. The mesh was refined around the vertical coordinate 1670 m corresponding to the vertical position of the initial surface. Elements above the current surface were flagged to belong to the inactive subdomain. The heat flow Eq. (3) was solved in the active subdomain below the surface, with Dirichlet boundary conditions (prescribed temperature) at the uppermost active element, and melting point temperature at the bottom. On the upstream and downstream faces a zero-flux boundary condition was prescribed. Figure 4 illustrates the progressive deformation of the mesh, the refinement of elements around the surface, and the varying number and extent of active elements. In all model runs the mesh was considerably finer than shown in Fig. 4; 200 elements in the vertical, 20 in the horizontal, and the mesh was refined around the surface in 5 steps instead of the 3 steps displayed in the figure for illustration.

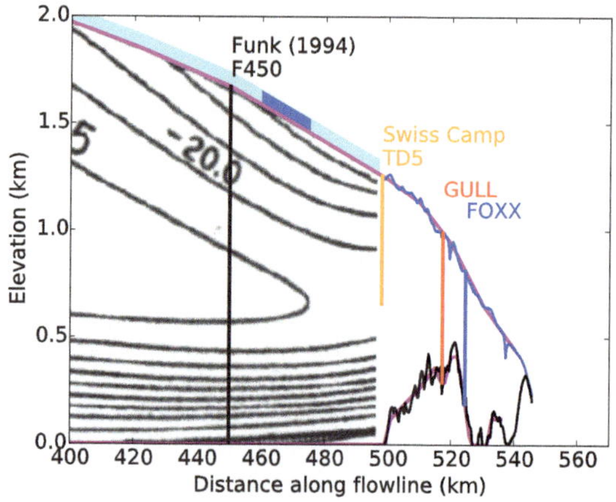

Figure 3. The locations and depths of temperature profiles at sites TD5, GULL and FOXX are shown as colored vertical lines on a background plot of the modeled temperature field from Fig. 7b in Funk et al. (1994). The temperature profile F450 at horizontal coordinate 450 km is used as model input. Bedrock (black) and surface (blue) was obtained from radar data (Gogineni, 2012). Smoothed versions, indicated with purple lines, were used as model input. The light blue area indicates cold firn conditions, and the dark blue area between 460 and 475 km the area within which the temperate firn that emerges at TD5 was accumulated between 1570 and 1730.

The initial temperature profile F450, located 450 km along the flow line, is taken from Funk et al. (1994, Fig. 8b). This profile was obtained by solving the advection-diffusion equation with a discrete phase-change boundary along the flow line from the ice sheet center, with prescribed surface and basal velocity. In our model, the block is assumed to move at prescribed horizontal velocities, also taken from Fig. 4 in Funk et al. (1994). These velocities are similar to those from recent satellite imagery feature tracking (Joughin et al., 2008).

3 Model results

In the following we compare model runs of the heat flow model with borehole data. The model runs are driven with different internal heat sources, and a modified surface boundary condition. These assumptions are independent from each other, and are combined to investigate the origin of measured temperatures profiles. Designations of model runs implementing these assumptions are given in Table 1. The relative importance of these contributions is evaluated by comparison of modeled to measured vertical temperature profiles from Swiss Camp (TD5), GULL and FOXX (both profiles).

Figures 5–7 show comparisons of measured temperature profiles to model results from the reference run mD, and for runs with additional heat sources. The reference model

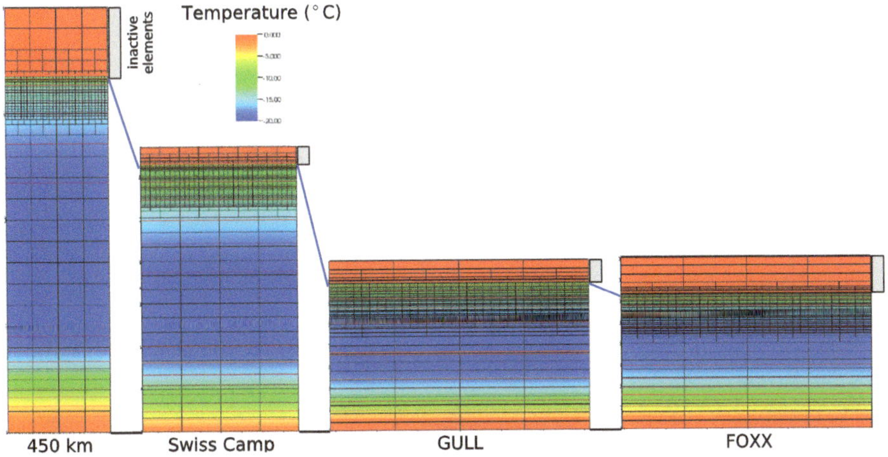

Figure 4. The gradual deformation of the computing mesh along the flow line to accommodate for the varying ice thickness. The ice surface is marked with blue lines between meshes, and inactive elements are indicated with gray bars (red elements on top). The precise position of the surface with respect to the mesh is tracked with adaptive mesh refinement. Modeled ice temperatures are indicated with colors. The greenish elements close to the surface are accumulated during flow (thickest at Swiss Camp at the equilibrium line) and slowly removed from the top due to ablation (sites GULL and FOXX). Note that computational mesh shown is for illustration only, actual model runs were performed on considerably finer meshes.

run mD (dissipation only) reveals that measured ice temperatures are considerably higher than modeled throughout the ice column (Fig. 5). The maximum extra heat energy of 27 MJ m^{-3} corresponds to a temperature difference of up to 14 K. Following Eq. (1), correcting the temperature difference between the modeled (mD) and observed temperature profiles in the upper part of the model domain would require additional heat from refreezing water equivalent to 9 % by mass.

Adding enhanced dissipation through enhanced shearing ice deformation in the Wisconsin ice (model run mE; Fig. 5) explains the observed extra heat between 500 and 700 m depth. The vertical extent of Wisconsin ice was taken from Ryser et al. (2014b), but the enhancement factor had to be set to 5 instead of the measured (and commonly assumed) value of 3. Since the whole model domain moves at the same velocity, the basal ice moves faster than in reality, and therefore has less time to warm. A model run with half surface velocity (not displayed) shows that enough heat is produced at an enhancement factor of 3 to explain measured temperature in the lowest 200 m.

The effect of accumulation of temperate firn (run mF) is shown in Fig. 6. In this model run, the surface temperature between 460–475 km was set to 0 °C, which corresponds to the 160-year time span 1570–1730 C.E. This seemingly arbitrary horizontal extent of temperate firn conditions coincides with reconstructed warm temperatures at the deep drill sites GRIP and Dye3 (Dahl-Jensen et al., 1998) (see Sect. 4).

4 Discussion

Any shape of temperature profile within the ice can be obtained by carefully positioning heat sources during certain time spans within the ice body. The aim of this study is not to perfectly match the observed ice temperature distribution but to investigate how several simple source mechanisms might contribute to the observed temperature profiles. The lower panels of Figs. 5–7 indicate the amount of extra heat per volume needed to match the measured temperatures. This heat is likely provided by several heat sources: dissipation (strain heating), temperate paleo-firn, and cryo-hydrologic warming.

For the ice near the base, strain heating is a crucial process to provide the required 10 MJ m^{-3} of heat (Fig. 5). Calculations with Eq. (2) show that for the considered ~900 years of ice flow the warming is of the order 3 K which is sufficient to explain the measured temperature of the lowest 200 m.

The relatively warm ice between 100 and 300 m depth at site Swiss Camp (profile TD5; Fig. 5) can be explained by temperate firn conditions in the past. Setting the surface temperature of the accumulated firn between horizontal coordinate 460 and 475 km to melting temperature (Fig. 3) yields the temperature profile shown in Fig. 6. This time span coincides with reconstructed warm temperatures in 1570–1730 (Dahl-Jensen et al., 1998). We assume that during this warm period surface melt events were frequent in the accumulation zone, and therefore the firn was considerably warmed, or even at melting temperature throughout the year. Such conditions with perennial water within the firn have recently been found in the accumulation area in southern Greenland (Humphrey et al., 2012; Harper et al., 2012). Currently, firn

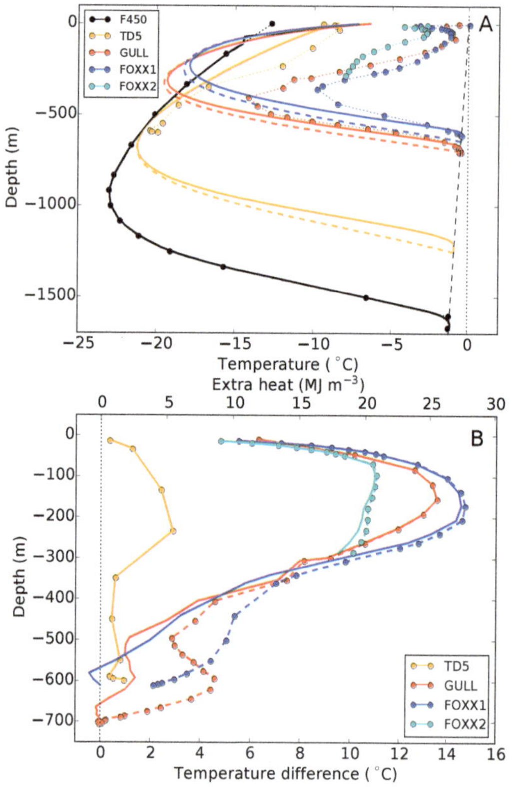

Figure 5. Measured and modeled temperatures at the three sites Swiss Camp (TD5), GULL and FOXX. Symbols connected by dotted lines indicate measurements. Dashed lines refer to the reference model run mD (dissipation only). Solid lines refer to run mE (additional heat sources from enhanced shear straining in ice-age ice). The scale on top of the lower panel shows the extra heat per volume with respect to model run mD (dissipation only).

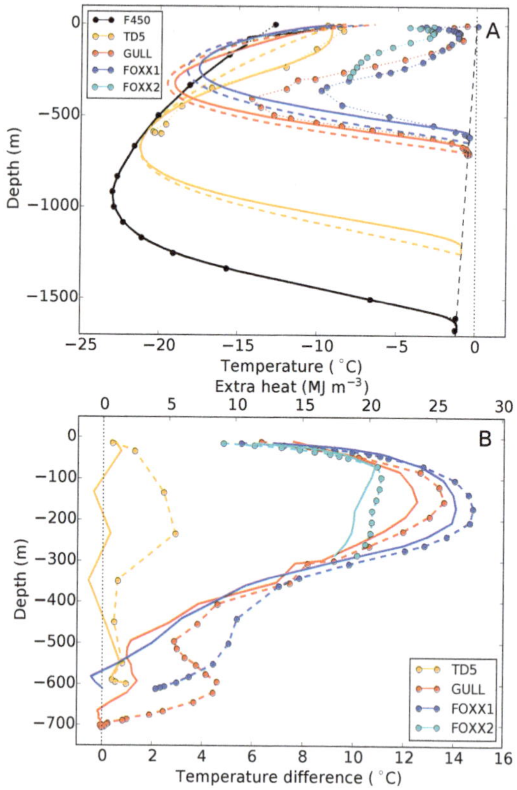

Figure 6. Same as Fig. 5 but for model run mEF (enhanced shear straining and temperate paleo-firn).

temperatures are again warming in the accumulation zone upstream of the study area (Polashenski et al., 2014), and will leave their imprint in the thermal state of the ice sheet. Warm paleo-firn only affects the upper ablation area, and explains only measured temperatures of borehole TD5, but is not sufficient to reproduce warm temperatures at depth at sites GULL and FOXX.

The temperate paleo-firn and ice accumulated in the lower accumulation area is presently emerging as relatively warm ice in the upper ablation area. Further downstream, ice accumulated under cold conditions in the dry-snow accumulation area emerges at the surface. Both thermal regimes are clearly visible at our drill sites: at GULL a flat, dirty and slushy ice surface is reminiscent of temperate Alpine glaciers. At FOXX the surface is bright white with many deep cryoconite holes, which is usually found when cold ice emerges in the ablation area (e.g., Ryser et al., 2013). Our near-surface temperature profiles (Fig. 2a) support this notion, with temperatures above $-2\,°C$ at GULL, but colder near-surface temperatures at downstream drill site FOXX. It is likely this difference in surface temperature, and therefore the distribution of dust (dirty ice vs. cryoconite holes) that leads to the dark band visible in the upper ablation zone of the western GrIS (Wientjes and Oerlemans, 2010; Wientjes et al., 2012).

For the central ice body an important heat source is needed to provide the $25\,MJ\,m^{-3}$ heat, corresponding to the observed 12–14 K temperature difference at GULL and FOXX (Fig. 5). Strain heating due to horizontal stress gradients is one possible heat source. Caterpillar-like horizontal extension and compression on diurnal and longer time scales has been observed in the study area (Ryser et al., 2014a). This heat source is, however, not strong enough: a quick calculation with Eq. (2), A $(-10\,°C)$ and a continuous, very high horizontal–longitudinal stress gradient of 0.1 MPa yields a heat production of only $0.0011\,MJ\,m^{-3}\,a^{-1}$ which is one order of magnitude lower than the shear strain-induced heating of basal ice discussed above.

To explain the high temperature at GULL and FOXX, the only conceivable heat source at depth is advection of heat through flowing, or ponding and freezing water between Swiss Camp and GULL. A major crevasse zone halfway between the sites, and moulins draining water from the surface have been mapped (Thomsen et al., 1988; Phillips et al., 2011). The extent and amount of such heat sources was investigated with several model runs with simple source geometries and durations. Assuming a single crevasse advected

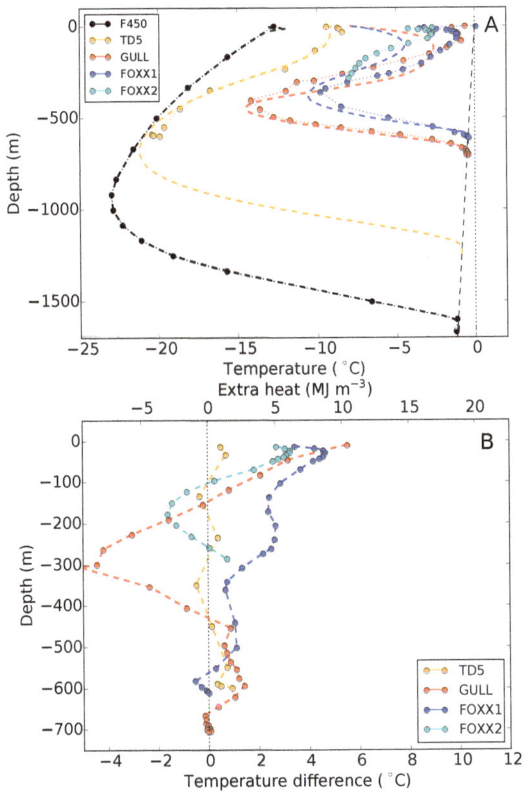

Figure 7. Same as Fig. 5 but for model run mCEF (crevasses, enhanced shear straining, and temperate paleo-firn).

to the drill site, its depth would have to be of the order of 400 m, and it would need to provide a temperature of 0 °C for 100 years. Figure 7 shows results from a model run assuming a series of crevasses of 400 m depth and 100 m spacing which provide a temperature of 0 °C for 50 years.

Very deep crevasses (300–400 m) with tens of years of activity are only possible if they are water-filled (Van der Veen, 1998), but do not penetrate to the bottom. While there is hydrofracturing as a mechanism to create a deep crevasse, there must be a mechanism to stop their depth penetration. Limited water supply for hydrofracturing is one possible cause, which is very likely within the small upstream surface area within a crevasse field. Another possible limiting factor for crevasse penetration to the bed is tougher ice at the bottom. The critical crack-tip loading rate at −20 °C is orders of magnitude lower than at temperatures approaching the melting point (Schulson and Duval, 2009, Table 9.1). We therefore suggest that water-driven crevasses stop their downwards growth once they reach warmer ice. If such very deep, water-filled crevasses indeed exist is unknown, although some observational evidence of strong englacial reflectors was detected in the study area (Catania et al., 2008; Catania and Neumann, 2010). These strong reflectors are likely due to water stored within the glacier ice.

An alternative explanation to crevasses are moulins, vertical shaft systems draining water from the surface. Moulins are stationary drainage features on an undulating surface, and are only active for a few years until they move out of surface depressions (Catania and Neumann, 2010). Their spacing along a flow line amounts to the duration of their activity times the flow velocity, and thus hundreds of meters in the area of our drill sites (Thomsen et al., 1988; Phillips et al., 2011). A series of model runs was performed that implement moulins as vertical shafts at melting temperature during their activity (Vorkauf, 2014). As line sources they provide little warming to the surrounding ice. Modeled temperature difference at the site of a moulin is below 1 K after 10 years of inactivity (Vorkauf, 2014). Furthermore, modeled vertical patterns of ice temperature around moulins do not agree with the observed temperature profiles. Therefore we conclude that very deep water-filled crevasses, rather than moulins, are the main heat sources in the ablation area.

An earlier modeling study of thermal properties in the same area (Phillips et al., 2013) concluded that ice warmed by cryo-hydrologic features has an important influence on ice deformation rates, and is causing the increasing flow velocities observed at the surface. The main difference in modeling approaches is their assumption of continuous heat sources spread over the entire ice thickness. Neither our data nor the interpretation with the heat flow model support this assumption. The ice in the lowest 200 m of the ice column shows no signs of warming through cryo-hydrologic features, as the temperatures there can be explained by diffusion and dissipation alone, as shown in Fig. 5. Since vertical shear deformation is highest at depth (Ryser et al., 2014a, b), temperature changes in the central ice body will have a minor influence on ice deformation rates, and therefore surface velocities.

A noteworthy feature of the presented ice temperature profiles is the difference between the two profiles at site FOXX (Fig. 2). The temperature difference between the two boreholes which are 86 m apart amounts to 4 K in 200–300 m depth. Furthermore, the profiles intersect below 300 m depth. Only a combination of very localized heat sources can produce these different shapes of the temperature profiles in such close vicinity. No simple heat source patterns, nor the influence of active vertical moulins or horizontal conduits produce similar temperature profiles (Vorkauf, 2014). The observation of very different temperature patterns in neighboring boreholes cautions against interpretation of single boreholes in areas where englacial heat sources strongly affect the thermal structure of the ice.

5 Conclusions

We showed that the thermal structure of the ice in the ablation area of the GrIS is dominated by strong heat sources providing extra heat, as compared to modeled heat diffusion alone. Three types of heat sources were identified with a modeling study: dissipation, temperate paleo-firn, and cryo-hydrologic features, represented by very deep water-filled crevasses.

An important conclusion of the modeling study is that temperate firn was accumulated in the now cold accumulation area during a warm period of about 160 years, during 1570–1730 C.E. (Dahl-Jensen et al., 1998). This temperate paleo-firn is presently emerging as relatively warm ice in the upper ablation area, and is likely the cause for the observed dark band in the ablation area of the western GrIS (Wientjes and Oerlemans, 2010; Wientjes et al., 2012).

Future warming of the firn areas of the GrIS by extreme melt events, such as in summer 2012 (McGrath et al., 2013), are likely in a warming Arctic. Extended areas of surface melt might lead to increasing firn temperatures, affecting large parts of the ice sheet. Such recent warming has already been detected in the study area (Polashenski et al., 2014).

The central ice body at our study sites is 10–15 K warmer than modeled with heat diffusion alone. To heat this ice to observed temperatures, a continuous heat source with melting point temperature is required which penetrates down to 300–400 m below the surface. The most likely sources are very deep, water-filled crevasses which persist for decades as they move through the extensional stress regime of a crevasse zone.

All of these heat sources were inferred from comparison of models with measurements. How important localized sources, such as crevasses and moulins, are for the thermal regime of the ice is still a matter of debate. Sampling ice temperature at sites suited for drilling (un-crevassed depressions with surface streams) yields a skewed picture of the full variability encountered in nature. Without a systematic sampling of ice temperatures to great depth, the information content extracted from individual holes is to be taken with care. This argument is impressively illustrated by the two temperature profiles from site FOXX which are 86 m apart, but show large differences in the upper 300 m, and even cross below.

Acknowledgements. We thank several people who were essential in this project: Cornelius Senn, Edi Imhof, Thomas Wyder, Andreas Bauder, Christian Birchler, Michael Meier, Blaine Moriss, and Fabian Walter. We acknowledge the constructive reviews by Martin Truffer and two anonymous referees.

This project was supported by Swiss National Science Foundation Grant 200021_127197, US-NSF Grants OPP 0908156, OPP 0909454 and ANT-0424589 (to CReSIS), NASA Cryospheric Sciences, and Climate Modeling Programs within the US Department of Energy Office of Science. Logistical support was provided by CH2MHill Polar Services.

Edited by: I. M. Howat

References

Ahlstrom, A. P.: Previous glaciological activities related to hydropower at Paakitsoq, Ilulissat, West Greenland, Tech. Rep., 25, Danmarks og Grønlands Geologiske Undersøgelse, 2007.

Andrews, L. C., Catania, G. A., Hoffman, M. J., Gulley, J. D., Lüthi, M. P., Ryser, C., Hawley, R. L., and Neumann, T. A.: Direct observations of evolving subglacial drainage beneath the Greenland Ice Sheet, Nature, 514, 80–83, doi:10.1038/nature13796, 2014.

Box, J. E.: Greenland Ice Sheet mass balance reconstruction, Part II: Surface mass balance (1840–2010), J. Climate, 26, 6974–6989, doi:10.1175/JCLI-D-12-00518.1, 2013

Budd, W., Jacka, T., Jenssen, D., Radok, U., and Young, N.: Derived physical characteristics of the Greenland Ice Sheet, Meteor. Dept. Pub. no. 23, University of Melbourne, Melbourne, 1982.

Catania, G. A. and Neumann, T. A.: Persistent englacial drainage features in the Greenland Ice Sheet, Geophys. Res. Lett., 37, L02501, doi:10.1029/2009GL041108, 2010.

Catania, G. A., Neumann, T. A., and Price, S. F.: Characterizing englacial drainage in the ablation zone of the Greenland ice sheet, J. Glaciol., 54, 567–578, doi:10.3189/002214308786570854, 2008.

Cuffey, K. and Paterson, W.: The Physics of Glaciers, Elsevier, Burlington, MA, USA, 2010.

Dahl-Jensen, D., Mosegaard, K., Gundestrup, N., Clow, G. D., Johnsen, S. J., Hansen, A. W., and Balling, N.: Past temperatures directly from the Greenland Ice Sheet, Science, 282, 268–271, doi:10.1126/science.282.5387.268, 1998.

DTU Space, M. and Remote Sensing, Lyngby, D.: Ice radar data, unpublished data, 2005.

Forster, R. R., Box, J. E., van den Broeke, M. R., Miège, C., Burgess, E. W., van Angelen, J. H., Lenaerts, J. T. M., Koenig, L. S., Paden, J., Lewis, C., Gogineni, S. P., Leuschen, C., and McConnell, J. R.: Extensive liquid meltwater storage in firn within the Greenland ice sheet, Nat. Geosci., 7, 418–422, doi:10.1038/NGEO2043, 2013.

Funk, M., Echelmeyer, K., and Iken, A.: Mechanisms of fast flow in Jakobshavns Isbrae, Greenland; Part II: Modeling of englacial temperatures, J. Glaciol., 40, 569–585, 1994.

Gogineni, P.: CReSIS Greenland Radar Data, Lawrence, Kansas, USA, Digital Media, available at: http://data.cresis.ku.edu/ (last access: 1 October 2013), 2012.

Greve, R.: Application of a polythermal three-dimensional ice sheet model to the Greenland ice sheet: response to a steady-state and transient climate scenarios, J. Climate, 10, 901–918, 1997.

Harper, J., Humphrey, N., Pfeffer, W., Brown, J., and Fettweis, X.: Greenland ice-sheet contribution to sea-level rise buffered by meltwater storage in firn, Nature, 491, 240–243, doi:10.1038/nature11566, 2012.

Humphrey, N. F., Harper, J. T., and Pfeffer, W. T.: Thermal tracking of meltwater retention in Greenland's accumulation area, J. Geophys. Res., 117, F01010, doi:10.1029/2011JF002083, 2012.

Hutter, K. and Jöhnk, K.: Continuum Methods of Physical Modeling: Continuum Mechanics, Dimensional Analysis, Turbulence, Springer, Berlin, Germany, 2004.

Iken, A., Echelmeyer, K., Harrison, W. D., and Funk, M.: Mechanisms of fast flow in Jakobshavns Isbrae, Greenland, Part I: Measurements of temperature and water level in deep boreholes, J. Glaciol., 39, 15–25, 1993.

Joughin, I., Das, S. B., King, M. A., Smith, B. E., Howat, I. M., and Moon, T.: Seasonal speedup along the western flank of the Greenland Ice Sheet, Science, 320, 781–783, 2008.

Kirk, B., Peterson, J. W., Stogner, R. H., and Carey, G. F.: libMesh: a C++ library for parallel adaptive mesh refinement/coarsening simulations, Eng. Comput., 22, 237–254, doi:10.1007/s00366-006-0049-3, 2006.

Letreguilly, A., Reeh, N., and Huybrechts, P.: The Greenland ice sheet through the last glacial-interglacial cycle, Palaeogeogr. Palaeocl., 90, 385–394, 1991.

Lüthi, M. P., Funk, M., Iken, A., Gogineni, S., and Truffer, M.: Mechanisms of fast flow in Jakobshavns Isbrae, Greenland; Part III: Measurements of ice deformation, temperature and cross-borehole conductivity in boreholes to the bedrock, J. Glaciol., 48, 369–385, doi:10.3189/172756502781831322, 2002.

McGrath, D., Colgan, W., Bayou, N., and Steffen, K.: Recent warming at Summit, Greenland: global context and implications, Geophys. Res. Lett., 40, 2091–2096, doi:10.1002/grl.50456, 2013.

Phillips, T., Rajaram, H., and Steffen, K.: Cryo-hydrologic warming: a potential mechanism for the thermal response of ice sheets, Geophys. Res. Lett., 37, L20503, doi:10.1029/2010GL044397, 2010.

Phillips, T., Leyk, S., Rajaram, H., Colgan, W., Abdalati, W., McGrath, D., and Steffen, K.: Modeling moulin distribution on Sermeq Avannarleq glacier using ASTER and WorldView imagery and fuzzy set theory, Remote Sens. Environ., 115, 2292–2301, doi:10.1016/j.rse.2011.04.029, 2011.

Phillips, T., Rajaram, H., Colgan, W., Steffen, K., and Abdalati, W.: Evaluation of cryo-hydrologic warming as an explanation for increased ice velocities in the wet snow zone, Sermeq Avannarleq, West Greenland, J. Geophys. Res.-Earth, 118, 1241–1256, doi:10.1002/jgrf.20079, 2013.

Polashenski, C., Courville, Z., Benson, C., Wagner, A., Chen, J., Wong, G., Hawley, R., and Hall, D.: Observations of pronounced Greenland ice sheet firn warming and implications for runoff production, Geophys. Res. Lett., 41, 4238–4246, doi:10.1002/2014GL059806, 2014.

Ryser, C., Lüthi, M., Blindow, N., Suckro, S., Funk, M., and Bauder, A.: Cold ice in the ablation zone: its relation to glacier hydrology and ice water content, J. Geophys. Res., 118, 693–705, doi:10.1029/2012JF002526, 2013.

Ryser, C., Lüthi, M. P., Andrews, L. C., Hoffman, M. J., Catania, G. A., Hawley, R. L., and Neumann, T. A.: Caterpillar-like ice motion in the ablation zone of the Greenland Ice Sheet, J. Geophys. Res.-Earth, 119, 2258–2271, doi:10.1002/2013JF003067, 2014a.

Ryser, C., Lüthi, M., Andrews, L., Hoffman, M., Catania, G., Hawley, R., Neumann, T., and Kristensen, S. S.: Sustained high basal motion of the Greenland Ice Sheet revealed by borehole deformation, J. Glaciol., 60, 647–660, doi:10.3189/2014JoG13J196, 2014b.

Schulson, E. and Duval, P.: Creep and Fracture of Ice, Cambridge University Press, 2009.

Thomsen, H. H. and Thorning, L.: Ice Temperature Profiles for Western Greenland, Tech. rep., Groenl. Geol. Unders., Copenhagen, 1992.

Thomsen, H. H., Thorning, L., and Braithwaite, R. J.: Glacier-Hydrological conditions on the inland ice North-East of Jakobshavn/Ilusissat, West Greenland, Report 138, Grønlands Geologiske Undersøgelse, Copenhagen, Denmark, 1988.

Thomsen, H. H., Olesen, O., Braithwaite, R. J., and Boggild, C.: Ice drilling and mass balance at Pakitsoq, Jakobshavn, central West Greenland, Report 152, Grønlands Geologiske Undersøgelse, Copenhagen, Denmark, 1991.

Van der Veen, C. J.: Fracture mechanics approach to penetration of surface crevasses on glaciers, Cold Reg. Sci. Technol., 27, 31–47, 1998.

Vorkauf, M.: Modeling the influence of moulins and temperate firn on the vertical temperature profile of the Greenland Ice Sheet, Master's thesis, VAW, ETH Zürich, Switzerland, 2014.

Wientjes, I. G. M. and Oerlemans, J.: An explanation for the dark region in the western melt zone of the Greenland ice sheet , The Cryosphere, 4, 261–268, doi:10.5194/tc-4-261-2010, 2010.

Wientjes, I., van de Wal, R., Schwikowski, M., Zapf, A., Fahrni, S., and Wacker, L.: Carbonaceous particles reveal that Late Holocene dust causes the dark region in the western ablation zone of the Greenland ice sheet, J. Glaciol., 58, 787–794, doi:10.3189/2012JoG11J165, 2012.

17

Enthalpy benchmark experiments for numerical ice sheet models

T. Kleiner, M. Rückamp, J. H. Bondzio, and A. Humbert

Alfred Wegener Institute, Helmholtz Centre for Polar and Marine Research, Am Alten Hafen 26, 27568 Bremerhaven, Germany

Correspondence to: T. Kleiner (thomas.kleiner@awi.de)

Abstract. We present benchmark experiments to test the implementation of enthalpy and the corresponding boundary conditions in numerical ice sheet models. Since we impose several assumptions on the experiment design, analytical solutions can be formulated for the proposed numerical experiments. The first experiment tests the functionality of the boundary condition scheme and the basal melt rate calculation during transient simulations. The second experiment addresses the steady-state enthalpy profile and the resulting position of the cold–temperate transition surface (CTS). For both experiments we assume ice flow in a parallel-sided slab decoupled from the thermal regime.

We compare simulation results achieved by three different ice flow-models with these analytical solutions. The models agree well to the analytical solutions, if the change in conductivity between cold and temperate ice is properly considered in the model. In particular, the enthalpy gradient on the cold side of the CTS goes to zero in the limit of vanishing temperate-ice conductivity, as required from the physical jump conditions at the CTS.

1 Introduction

Ice sheets and glaciers can be distinguished by their thermal structure into cold, temperate and polythermal ice masses. While in cold ice the temperature is below the pressure melting point, in temperate ice the pressure melting point is reached. In temperate ice the heat generated by viscous deformation cannot give rise to temperature changes, but will be used for melting (Fowler, 1984; Blatter and Hutter, 1991). Thus temperate ice may contain a liquid water content (moisture). Polythermal ice masses contain both cold ice and temperate ice, separated by the cold–temperate transition surface (CTS, Greve, 1997a, b). The large ice sheets in Greenland and Antarctica show the polythermal structure of a Canadian-type glacier, which are mostly cold except for a temperate layer at the base (Aschwanden et al., 2012, and references therein). The liquid water inclusion in temperate ice makes this ice considerably softer than cold ice, resulting in a strong relationship between viscosity and moisture content (Duval, 1977; Lliboutry and Duval, 1985). The importance of this feature for the ice dynamics is obvious especially for temperate ice at the base where stresses are highest.

The enthalpy scheme presented in Aschwanden and Blatter (2009) and Aschwanden et al. (2012) describes temperature and water content in a consistent and energy-conserving formulation. Changes in the enthalpy are caused by changes of temperature in the cold ice part and by changes of the water content in the temperate ice part. The CTS position is implicitly given as the level-set of the pressure melting point and can be derived from the enthalpy field. Therefore, no restriction to the topology and shape of the CTS exists and there is no need to track it as in front-tracking models (e.g. Hutter et al., 1988; Blatter and Hutter, 1991; Greve, 1997a, b). Compared to the front-tracking models neither jump conditions nor kinematic conditions are required at the CTS.

The enthalpy scheme has already been used in model studies for the Greenland Ice Sheet. In the "reference-implementation" of Aschwanden et al. (2012) the enthalpy scheme was compared to a cold ice scheme. A simplified version of the enthalpy scheme (regarding basal boundary conditions and ice rheology) was used to assess the effect of the initial thermal regime on century-scale simulations (Seroussi et al., 2013). Thus far we are lacking analytical solutions for thermo-mechanically coupled polythermal ice flow to test the enthalpy implementations in ice sheet models.

Here, two numerical experiments for the enthalpy field are presented for which analytical solutions exist. Similar to other studies on ice sheet modelling (Huybrechts et al., 1996; Bueler et al., 2005; Pattyn et al., 2012) we aim to verify the enthalpy method by comparing numerical solutions to analytical solutions under simplified boundary conditions. While artificially constructed exact solutions require additional compensatory terms to be incorporated in the numerical model (e.g. Bueler et al., 2005, 2007), the proposed experiments are chosen in a way that numerical models should be able to perform them with no or only minor modifications of their source codes. Therefore it is ensured that the models run through the same model components and execute the same code for the proposed numerical experiments and for real-world simulations.

2 Theory

2.1 Governing equations

Compared to thermodynamics usage, the enthalpy described in Aschwanden et al. (2012) is the specific internal energy. The work associated with changing the volume is not considered, since ice is assumed to be incompressible. The use of the name "enthalpy" is made to match other cryospheric applications (e.g. Notz and Worster, 2006). In the enthalpy approach temperature T and moisture ω are diagnostically computed from the modelled enthalpy field E (units: $\mathrm{J\,kg^{-1}}$). The following transfer rules are used:

$$E(T,\omega,p)=\begin{cases} c_i(T-T_{\mathrm{ref}}), & \text{if } E<E_{\mathrm{pmp}} \\ E_{\mathrm{pmp}}+\omega L, & \text{if } E\geq E_{\mathrm{pmp}}, \end{cases} \tag{1}$$

where p is the pressure, T_{ref} is a reference temperature (to have positive values for the enthalpy for typical temperatures in glaciers), and L the latent heat of fusion. The enthalpy of the solid ice at the pressure melting point is defined as $E_{\mathrm{pmp}}=E_s(p)=c_i(T_{\mathrm{pmp}}(p)-T_{\mathrm{ref}})$, where $T_{\mathrm{pmp}}(p)=T_0-\beta p$ is the pressure melting point temperature, β is the Clausius–Clapeyron constant and T_0 is the melting point at standard pressure (see Table A1 for parameter values).

The enthalpy field equation of the ice–water mixture depends on whether the mixture is cold ($E<E_{\mathrm{pmp}}$) or temperate ($E\geq E_{\mathrm{pmp}}$):

$$\rho_i\left(\frac{\partial E}{\partial t}+\boldsymbol{v}\cdot\nabla E\right)=-\nabla\cdot\boldsymbol{q}_i+\Psi, \tag{2}$$

with the ice density ρ_i, the ice velocity vector $\boldsymbol{v}=(v_x, v_y, v_z)$, the conductive flux $\boldsymbol{q}_i$, and the heat source by internal deformation Ψ. The conductive flux in cold ice is represented by Fourier's law in enthalpy form with the conductivity $K_c=k_i/c_i$. In temperate ice the conductive flux is composed of the sensible heat flux (caused by variations in the pressure melting point) and latent heat flux, thus

$$\boldsymbol{q}_i=-\begin{cases} K_c\nabla E, & \text{if } E<E_{\mathrm{pmp}} \\ k_i\nabla T_{\mathrm{pmp}}(p)+K_0\nabla E, & \text{if } E\geq E_{\mathrm{pmp}}. \end{cases} \tag{3}$$

At the present state, K_0 is poorly constrained. To test the sensitivity of the models on this parameter, different values have been used according to Table A1.

2.2 Boundary conditions

At the upper ice surface the enthalpy is prescribed from the surface temperature with zero moisture content corresponding to a Canadian-type polythermal glacier (see Blatter and Hutter, 1991). In the following description of the basal conditions $T'(p)=T-T_{\mathrm{pmp}}(p)+T_0=T+\beta p$ is the temperature relative to the melting point, H_w is the basal water layer thickness and $\boldsymbol{n}_b$ is the outward pointing normal vector. The type of basal boundary condition (Neumann or Dirichlet) is time dependent. The decision chart for local conditions given in Aschwanden et al. (2012, Fig. 5) need to be evaluated at every time step. The chart encompasses four different situations:

- *Cold base (dry)*: if the glacier is cold at the base and without a basal water layer (i.e. $E<E_{\mathrm{pmp}}$ and $H_w=0$), then

  $$K_c\nabla E\cdot\boldsymbol{n}_b=q_{\mathrm{geo}}. \tag{4}$$

  The geothermal flux is the only source of heat as basal sliding and therefore frictional heating is forbidden for ice with temperatures below the pressure melting point. The geothermal flux is assumed to be constant, thus changes of the heat storage in the underlying bedrock cannot affect the basal heat budget of the ice.

- *Temperate base*: if the glacier is temperate at the base without an overlying temperate ice layer, but with melting conditions at the base (i.e. $E\geq E_{\mathrm{pmp}}$, $H_w>0$ and $\nabla T'\cdot\boldsymbol{n}_b<\beta/K_c$), then

  $$E=E_{\mathrm{pmp}}. \tag{5}$$

  This condition applies to basal ice maintained at the enthalpy of the pressure melting point, when the geothermal flux and frictional heating (caused by sliding) exceed the heat flux away from the base into the ice. In this case the remaining heat is used for melting.

- *Temperate ice at base*: if the glacier is temperate at the base with an overlying temperate ice layer (i.e. $E\geq E_{\mathrm{pmp}}$, $H_w>0$ and $\nabla T'\cdot\boldsymbol{n}_b=\beta/K_c$), we let

  $$K_0\nabla E\cdot\boldsymbol{n}_b=0. \tag{6}$$

  The proposed insulating Neumann boundary condition suppresses the diffusive enthalpy flux into the temperate ice layer even in the case of $K_0\neq 0$. In this case the energy at the base is balanced by the basal melt rate calculation.

- *Cold base (wet)*: if the glacier is cold, but has a liquid water layer at the base that is refreezing (i.e. $E < E_{\text{pmp}}$ and $H_{\text{w}} > 0$), then

$$E = E_{\text{pmp}}. \tag{7}$$

It is assumed here that the subglacial water is at the pressure melting point and the heat stored in the water layer does not allow the basal enthalpy to be below the pressure melting point (continuity of temperature). As a consequence, the refrozen ice has a zero water content.

Note that, in addition to the temperate base condition, $E \geq E_{\text{pmp}}$, it is necessary to check if there is a temperate layer of ice above, $\nabla T' \cdot \boldsymbol{n}_{\text{b}} = \beta / K_{\text{c}}$.

Since we are dealing with polythermal glaciers, melting of ice or refreezing of liquid water at the base plays a role. The calculated melting/refreezing rate, a_{b} (units: m s^{-1} ice equivalent), obeys

$$a_{\text{b}} = \frac{F_{\text{b}} - (\boldsymbol{q}_{\text{i}} - \boldsymbol{q}_{\text{geo}}) \cdot \boldsymbol{n}_{\text{b}}}{L \rho_{\text{i}}}, \tag{8}$$

with the frictional heating F_{b} due to basal sliding, the heat flux in the ice $\boldsymbol{q}_{\text{i}}$, and the geothermal flux $\boldsymbol{q}_{\text{geo}}$ entering the ice at the base.

Although explicit boundary conditions for CTS are not required in the enthalpy scheme, they are used to evaluate the numerical results and to derive analytical solutions later in the text. According to Greve (1997b) melting, freezing and parallel flow conditions must be distinguished depending on the CTS velocity. The enthalpy method allows for all three conditions in general. However the basal boundary conditions used in Aschwanden et al. (2012) only permit melting conditions, as Eq. (6) inhibits the increase of enthalpy towards the CTS. Further, the numerical models applied here do not allow a discontinuous enthalpy solution in the case of freezing or parallel flow conditions ($\omega > 0$ at the CTS).

In the case of melting conditions at the CTS, the total enthalpy flux (advective and diffusive) at both sides of the CTS must be equal:

$$\rho \boldsymbol{v} E^{+} + K_{\text{c}} \nabla E^{+} \cdot \boldsymbol{n}_{\text{CTS}}^{+} = \rho \boldsymbol{v} E^{-} - K_0 \nabla E^{-} \cdot \boldsymbol{n}_{\text{CTS}}^{-}, \tag{9}$$

where the superscripts "+" and "−" denote the cold and the temperate side of the interface, respectively, and $\boldsymbol{n}_{\text{CTS}}^{+}$ and $\boldsymbol{n}_{\text{CTS}}^{-}$ are the normal vectors pointing toward the CTS. This is based on the general assumption that the total heat flux leaving a representative volume through a particular face must be identical to the flux entering the next representative volume through the same face.

At the CTS, ice at its pressure melting point and without any moisture flows into the temperate layer. Hence, the enthalpy is continuous at the CTS:

$$E^{+} = E^{-}. \tag{10}$$

Assuming a horizontal CTS and according to Eqs. (9) and (10), the enthalpy derivative at the CTS is discontinuous in the given case of $K_{\text{c}} \neq K_0$:

$$K_{\text{c}} \left. \frac{\partial E}{\partial z} \right|^{+} = K_0 \left. \frac{\partial E}{\partial z} \right|^{-}. \tag{11}$$

The condition further implies that for $K_0 \rightarrow 0$ the enthalpy gradient on the cold side of the CTS (+) vanishes.

3 Numerical models

The numerical models used here are all three-dimensional flow models including a thermal component for ice. They all allow the evolution of the ice thickness, although this is not applied here.

3.1 TIM-FD^3 (finite differences)

In the Thermocoupled Ice-flow Model (TIM-FD^3, Kleiner and Humbert, 2014) the relevant equations are discretized using finite differences in terrain-following (sigma) coordinates. For the advective terms in Eq. (2) the hybrid difference scheme of Spalding (1972) is used. This scheme switches between the second-order central-difference scheme and the first-order upwind-difference scheme according to the local cell Péclet number. It allows stable numerical solutions for the advection-dominated transport in the temperate ice layer.

The conductive terms in Eq. (2) are discretized using the second-order central-difference scheme for the second derivative, where the conductivities are evaluated midway between the grid nodes (e.g. Greve and Blatter, 2009, chap. 5.7.3). The transport due to sensible heat flux in the temperate layer $\Gamma = \nabla \cdot (k_{\text{i}} \nabla T_{\text{pmp}}(p)) = -\beta \nabla \cdot (k_{\text{i}} \nabla p)$ is assumed to be small and considered as a source term in the model. The time stepping is performed using a semi-implicit Crank–Nicolson scheme with a constant time step.

Special attention is required for the diffusion term, since the conductivity is discontinuous at the CTS. The most straightforward procedure for obtaining the interface conductivity would be to assume a linear variation of the conductivity between nodes (arithmetic mean). However, this approach cannot handle the abrupt changes of conductivity at the CTS. We use the harmonic mean of the conductivities, as suggested by Patankar (1980, chap. 4.2.3) not only at the CTS but for all conductivities evaluated between grid nodes.

3.2 ISSM (finite element)

The open source Ice Sheet System Model (ISSM; https://issm.jpl.nasa.gov/) is applied here. A detailed model description can be found e.g. in Larour et al. (2012). It now implements the entire set of field equations and boundary conditions of the enthalpy formulation presented by Aschwanden et al. (2012). Since Seroussi et al. (2013), the implementation has been completed by adding the basal boundary condition and basal melting rate scheme as described in Aschwanden et al. (2012, Fig. 5).

The enthalpy field equation is discretized using a finite-element method with linear elements. The steady-state equation and implicit time stepping scheme, respectively, give rise to a nonlinear system. It is solved using a parallelized solver. The numerical scheme can be stabilized using artificial diffusion or streamline upwind diffusion. For best comparison to the respective analytical solutions no numerical stabilization has been used here. Jumps in heat conductivity at the CTS are being accounted for by taking a volume-weighted harmonic mean of the heat conductivities over the element, see Patankar (1980).

3.3 COMice (finite element)

Numerical solutions are obtained using the COMice model (Rückamp et al., 2010) that is based on the commercial finite-element software COMSOL Multiphysics© (www.comsol.com). The domain is approximated by a structured triangular mesh with vertical equidistant layers. Enthalpy (Eq. 2) is solved with first-order Lagrange elements stabilized with streamline diffusion. The time derivatives are discretized using the implicit backward Euler scheme. An adaptive time stepping method according to Hindmarsh et al. (2005) controls the chosen time step with respect to a given tolerance. We apply Newton's method to solve the resulting system of nonlinear algebraic equations at each time step.

The step of the conductivity from $K_c(E)$ to K_0 at the CTS is implemented using Comsol's built-in operator `circumcenter(expr)`:

$$K(E) = \begin{cases} K_c, & \text{if } \texttt{circumcenter}(E) < E_{pmp} \\ K_0, & \text{else} \end{cases}. \qquad (12)$$

The operator interpolates the enthalpy solution to the circumcentre of the mesh element to which the point belongs. In doing so, `circumcenter(E)` is constant on each triangle and discontinuous along the edges. Therefore the conductivity jump is located on a mesh edge. This implementation shows better and faster convergence compared to other tested methods like a Heaviside function or a smoothed Heaviside function as used in the COMSOL implementation of Aschwanden and Blatter (2009) to compute the conductivity jump at the CTS. For post-processing, the CTS position is linearly interpolated between nodes.

4 Experiment description

4.1 Experiment A: parallel-sided slab (transient)

The simulation set-up is designed to test the implementation of the basal decision chart for boundary conditions and melting rates (Aschwanden et al., 2012, Fig. 5). Depending on the different thermal situations that occur at the base, the numerical code may have to switch between Neumann and Dirichlet boundary conditions for the enthalpy and the corresponding basal melt rate calculation. The main idea of this set-up is to test the reversibility during transient simulations. The conservation of water volume is also addressed here. An initially cold ice body that runs through a warmer period with an associated build-up of a liquid water layer at the base must be able to return to its initial steady state. This requires refreezing of the liquid water at the base. To test this behaviour we assume a simple heat conducting block of ice.

A parallel-sided slab of ice of constant thickness H is considered. The surface is parallel to the bed and has a constant inclination $\gamma = 0°$ to guarantee $|\boldsymbol{v}| = 0$ and $\Psi = 0$. To make the set-up basically vertical 1-D, in order to be able to consider only vertical heat transport, we impose periodic boundary conditions at the sides of the block. Hence the horizontal extension does not play a role. The geothermal flux $\boldsymbol{q}_{geo}$ at the base is constant. All parameters and their values are listed in Table A1.

The model run is as follows:

Initial phase (I): starting under cold conditions with an imposed surface temperature of $T_s = T_{s,c} = -30\,°C$ and an isothermal initial temperature field $T(0, z) = T_{s,c}$ the simulation is run for 100 ka.

Warming phase (II): the surface temperature is switched to $T_s = T_{s,w} = -10\,°C$ and the simulation is continued for another 50 ka.

Cooling phase (III): the surface temperature is switched back to the initial value of $T_s = T_{s,c}$ and the simulation is continued for a further 150 ka.

Since Ψ is zero, a temperate layer of ice at the base will not form and cold ice conditions hold everywhere inside the ice. The ice thickness and vertical alignment of the block is held constant over time although a significant water layer can be built up during the warming phase. Further, the water is stored at the base and no restriction of the maximum water layer thickness is applied.

4.2 Experiment B: polythermal parallel-sided slab (steady state)

To test the numerical solution for enthalpy in a vertical ice column with ice advection, we apply the "Slab with Melting Conditions at the CTS" set-up with a known analytical solution for $K_0 = 0\,\mathrm{kg\,m^{-1}\,s^{-1}}$ (Greve and Blatter, 2009, chap. 9.3.6). However, the knowledge about latent heat flux in temperate ice is poorly constrained as laboratory experiments and field observations are scarce. We vary the values of K_0 to highlight the effect on the resulting polythermal structure.

Similar to Experiment A, a parallel-sided slab of constant ice thickness H and a constant surface and bed inclination

γ in the x-direction is considered (Table A1). Ice flow is decoupled from the thermal quantities by using a constant flow rate factor A. The velocity throughout the ice column is prescribed as

$$v_x(z) = \frac{A(\rho g \sin\gamma)^3}{2}\left(H^4 - (H-z)^4\right), \quad (13)$$

$$v_y(z) = 0, \quad (14)$$

$$v_z(z) = -a_s^\perp = \text{const.} \quad (15)$$

Note that this set-up is not mass conservative, as there is no process considered that balances the accumulation rate required for a constant ice thickness. For simplicity we do not account for ice thickness evolution. The geothermal flux $\boldsymbol{q}_{\text{geo}}$ is set to zero and basal sliding is neglected ($F_b = 0$). Strain heating $\Psi = 4\mu\dot{\varepsilon}_{\text{eff}}^2$ is the only source of heat, where μ and $\dot{\varepsilon}_{\text{eff}}$ are the viscosity and the effective strain rate. The Glen–Steinemann power-law rheology (Glen, 1955; Steinemann, 1954) for the deformation of ice is used, thus

$$\mu = \frac{1}{2}A^{-1/3}\dot{\varepsilon}_{\text{eff}}^{-2/3}, \quad (16)$$

$$\dot{\varepsilon}_{\text{eff}} = \frac{1}{2}\frac{\partial v_x}{\partial z} = A(\rho g \sin\gamma)^3(H-z)^3. \quad (17)$$

The strain heating is largest at the base and reaches $\sim 2.6 \times 10^{-3}\,\text{W m}^{-3}$.

According to the assumptions in Greve and Blatter (2009, p. 246) the enthalpy conductivity K_0 in the temperate ice is zero, and the enthalpy flux at the cold site of the CTS (Eq. 11) must vanish. The CTS in this experiment is uniquely determined because the vertical velocity is downward. At the ice surface ($z = H$) the enthalpy is prescribed corresponding to the surface temperature $T_s = -3\,°\text{C}$ and zero water content. At the ice base ($z = 0$) one of the boundary conditions given in Eqs. (4)–(7) holds depending on the basal thermal conditions. All simulations start from a constant enthalpy corresponding to a temperature of $-1.5\,°\text{C}$ and zero water content. An analytical solution for the steady-state enthalpy profile based on the solution of Greve and Blatter (2009) is given in Appendix A2. The solution leads to a CTS position of approx. 19 m above the bed. The conductivity ratio $\text{CR} = K_0/K_c$ varies from $\text{CR} = 10^{-1}$ to 10^{-5} for TIM-FD^3 and COMice and to 0 for ISSM, respectively for this set-up. The simulations are performed on vertically equidistant layers using different vertical resolutions $\Delta z = (10.0, 5.0, 2.0, 0.5)$ m.

Note that in both experiments outlined above no frictional heating at the base occurs. Drainage of moisture that exceeds a certain limit to the base needs to be considered, when a coupling of moisture to the ice viscosity is used, but is also ignored in this study. The implementation of a basal hydrology model is beyond the scope of this study, hence basal water is accumulated at the place of origin with no restriction to the water layer thickness.

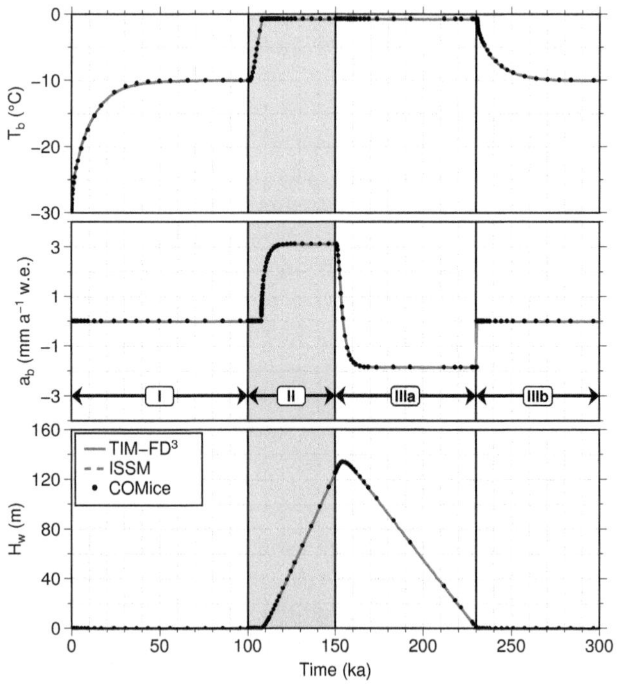

Figure 1. Results for Experiment A simulated with TIM-FD^3 (blue), ISSM (red) and COMice (black) overlay each other. Phases I to III are described in the main text. The warming phase II is shaded in grey.

5 Results

5.1 Experiment A

The set-up does not allow for a temperate ice layer and therefore enthalpy variations are given only by temperature variations. The simulated basal temperatures, basal melt rates and the basal water layer thicknesses over time are shown in Fig. 1.

As heat conduction is the only process of heat transfer, the vertical enthalpy profiles are linear in the steady states, which are reached at the end of each phase. At the steady states of the initial (I) and the cooling (III) phase the total vertical temperature gradient is given by the geothermal flux at the base and Eq. (4). This leads to the basal temperature of $T_b^{(\text{I,III})} = T_{s,c} + H\,q_{\text{geo}}/k_i = -10\,°\text{C}$ and zero melting at the base, revealed by all three models ($|\Delta T| < 5 \times 10^{-2}\,°\text{C}$).

In the warming phase (II) the basal temperature reaches the pressure melting point after a few thousand years and a basal water layer develops based on the basal melt rates. At the end of this phase temperatures reach the steady state ($|\Delta T| < 5 \times 10^{-2}\,°\text{C}$) and the basal melt rates can be calculated based on the steady-state temperature gradient between the surface and the base according to Eq. (8) as

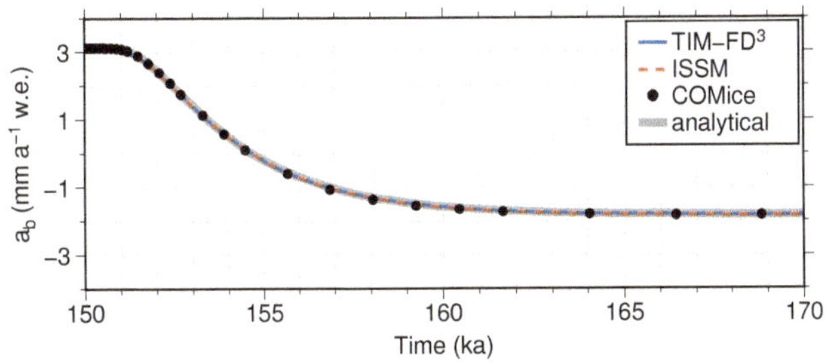

Figure 2. Simulation results compared to the analytical solution (thick solid grey line) for phase IIIa in Experiment A. TIM-FD^3 as blue solid line, ISSM as red dashed line, and COMice as black filled circles.

$$a_b^{(II)} = \frac{1}{\rho_w L}\left(q_{geo} + k_i \frac{T_{s,w} - T_{pmp}}{H}\right). \quad (18)$$

For this setting the basal melt rate is $a_b^{(II)} = 3.1 \times 10^{-3}\,\mathrm{m\,a^{-1}}$ water equivalent (w.e.). The models agree well with Eq. (18) as shown in Fig. 1 ($|\Delta a_b^{(II)}| < 10^{-5}\,\mathrm{m\,a^{-1}}$ w.e.).

Phase III can be separated into two different parts: phase IIIa where the base is temperate because of the remaining basal water layer from phase II, and phase IIIb, where all subglacial water is refrozen and the base returns to cold conditions. As long as a basal water layer exists, the basal temperature is kept at pressure melting point independent of the applied surface temperature and temperature profile according to Eq. (7). At the end of phase IIIa, the basal melt rates can therefore be found by replacing $T_{s,w}$ with $T_{s,c}$ in Eq. (18). Due to the low surface temperature, refreezing conditions arise and reach steady-state values of $a_b^{(IIIa)} = -1.84 \times 10^{-3}\,\mathrm{m\,a^{-1}}$ w.e. at the end of this phase as shown by the model solutions ($|\Delta a_b^{(IIIa)}| < 10^{-5}\,\mathrm{m\,a^{-1}}$ w.e.).

Since we have not included either a hydrology model or a reasonable upper limit for the subglacial water layer thickness, it is free to reach arbitrary thicknesses. That, in turn, is an advantage of the set-up, as we want to observe the system behaviour over longer time periods. The simulations lead to a maximum water layer thickness of ~130 m that occurs a few thousand years after the end of the warming phase (II). A realistic liquid water layer thickness of about 2 m would vanish in a few time steps and would not allow for steady-state considerations at the end of IIIa.

We have chosen phase IIIa to compare not only the quasi-steady-state solutions of the models at the end of each phase, but also the transient behaviour of the models compared to the analytical solution. For the comparison we use the basal melt rate instead of the temperature profile, since the correct melt rate requires a correct temperature profile and is easier to compare. In Fig. 2 the simulated basal melt rates for the first 20 ka of phase IIIa are compared to the analytical solution given in Appendix A1.

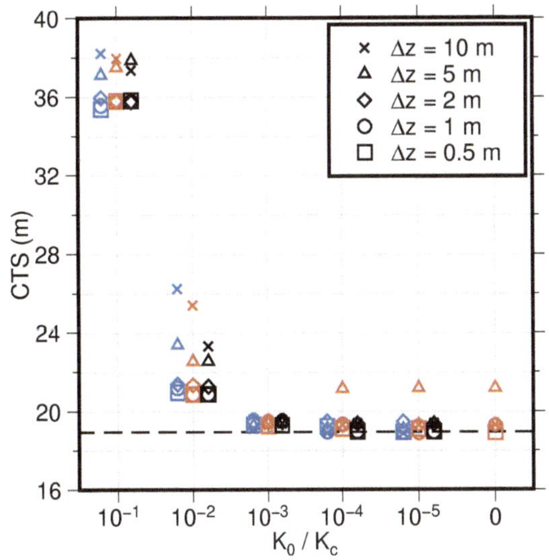

Figure 3. Comparison of simulated steady-state CTS positions for different values of the temperate ice conductivity in Experiment B. The different models are shown as: TIM-FD^3 (blue), ISSM (red) and COMice (black). Results of different models are slightly shifted on the x-axis to not overlay each other. The dashed black line indicates the CTS position of the analytical solution derived for $K_0 = 0$.

After ~1000 years the cold signal from the surface reaches the base and melting starts to decrease until the temperature gradient in the overlying ice does not allow for further melting and refreezing sets in. All models agree well with the analytical solution. The COMice solution is sometimes slightly below the analytical solution because of the very large time steps. The transition between melting and freezing occurs after ~4684.7 years in the analytical solution. Model simulations show this transition at a comparable modelled time.

All model results clearly reveal reversibility: after the whole simulation period of 300 ka, the models return to the initial steady state at the end of phase I.

5.2 Experiment B

Here, model results of the steady-state simulations of experiment B are compared to the analytical solution given in Appendix A2. For TIM-FD^3 and COMice the steady state is assumed after 1000 model years, while in ISSM a thermal steady-state solver is applied. The final steady-state CTS positions for all simulations are shown in Fig. 3.

For the maximum value of temperate ice conductivity ($\mathrm{CR} = 10^{-1}$) and the highest vertical resolution ($\Delta z = 0.5$ m) the models result in CTS position slightly below 36 m. In these simulations the thickness of the temperate ice layer is almost doubled compared to the results achieved by using the smallest value of temperate ice conductivity ($\mathrm{CR} = 10^{-5}$) with the same vertical resolution. The CTS positions decrease with decreasing CR and converge to the analytical solution. The models have approximately the same spread for the different vertical resolutions. The spread of the CTS position is smallest for $CR = 10^{-3}$ independent of the applied model. Compared to ISSM, TIM-FD^3 and COMice implementations do not allow for solving the case $K_0 = 0$ as in the analytical solution.

The steady-state enthalpy profiles and the corresponding temperature and moisture profiles are shown in Fig. 4 together with the analytical solution given in Appendix A2. The profiles are shown for the lowest (10 m) and highest (0.5 m) vertical resolution and the lowest conductivity ratio $CR = 10^{-5}$ used by all models. The results of all models agree well with the analytical solution for high resolutions. At coarser resolutions the simulated enthalpy profiles differ noticeably from the analytical solution. In the following we compare enthalpy differences as $\Delta E = E_{\text{analytic}} - E_{\text{simulated}}$.

In the ISSM simulation with the coarsest resolution ($\Delta z = 10$ m), the enthalpy differs from the analytical solution by $\sim 1720\,\text{J kg}^{-1}$ close to the CTS. This results in a temperature difference of $\sim 0.9\,°\text{C}$ in the cold ice part. TIM-FD^3 and COMice reveal also a lower enthalpy at the cold side of the CTS compared to the analytical solution, but only to a minor extent (TIM-FD^3: $\sim 0.2\,°\text{C}$, COMice: $\sim 0.1\,°\text{C}$). Note that the analytical solution only holds for $K_0 = 0$, thus small differences are expected here.

As the method chosen for interpolating heat conductivities in ISSM strongly favours the lower value, a quasi-isolating layer thicker than in the analytical solution is artificially created. Thus, heat flux into the upper cold ice column decreases, and that column cools. Vice versa, excess heat is accumulated in the lower temperate ice column, such that this part of the ice column heats up. The result is a negative temperature and positive water fraction offset. It scales with vertical mesh resolution, but stays detectable even on the highest mesh resolution tested here.

In the TIM-FD^3 simulation with the coarsest resolution ($\Delta z = 10$ m), the enthalpy difference ΔE is largest at the base ($\sim 2530\,\text{J kg}^{-1}$). As the base is temperate, this difference in the enthalpy corresponds to a difference in the basal water content of $\sim 0.8\,\%$. With this resolution the temperate ice layer needs to be resolved within the lowermost three grid points. The slope in the profile is caused by second-order one-sided discretization (e.g. Payne and Dongelmans, 1997) of the basal boundary condition (Eq. 6) in TIM-FD^3. Compared to the FE models neither strain heating nor transport of heat is considered for basal grid nodes.

With increasing vertical resolution the maximum deviation from the analytical solution decreases for all models. For the highest resolution ($\Delta z = 0.5$ m) and $CR = 10^{-5}$ the maximum differences are $\sim 150\,\text{J kg}^{-1}$, $\sim 100\,\text{J kg}^{-1}$ and $\sim 10\,\text{J kg}^{-1}$ for TIM-FD^3, ISSM and COMice, respectively. The differences remain positive, thus the enthalpy is slightly underestimated. Only ISSM is able to perform the experiment with $K_0 = 0$ as in the analytical solution, but the maximum enthalpy difference does not further decrease. As expected from Eq. (11) all models show small enthalpy gradients at the cold side of the CTS.

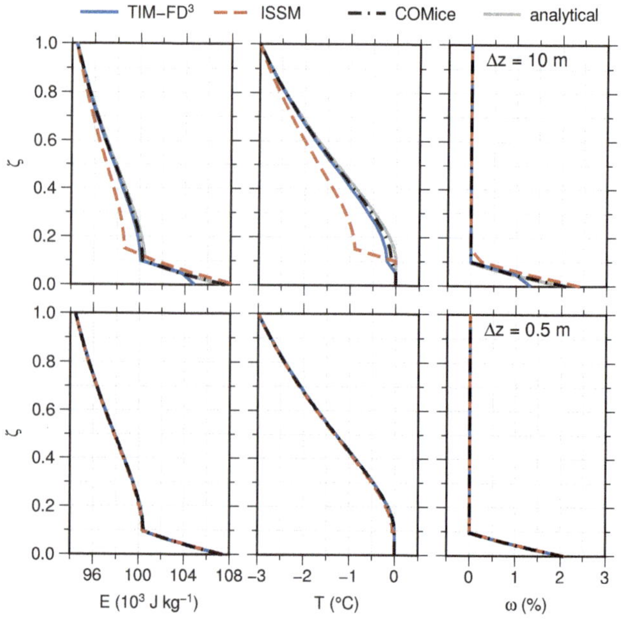

Figure 4. Simulated steady-state profiles of the enthalpy E, the temperature T and the water content ω for TIM-FD^3 (blue), ISSM (red) and COMice (black) compared to the analytical solution (grey). $\zeta = z/H$ is the normalized vertical coordinate. The vertical resolution is $\Delta z = 10$ m (upper row) and $\Delta z = 0.5$ m (lower row), $CR = K_0/K_c = 10^{-5}$. In the lower row the model results overlay the analytical solution.

The observed order of accuracy measured as the root-mean-square deviation (RMSD) to the analytical solution has been obtained in the series of vertical mesh refinements from $\Delta z = 10$ to 0.5 m and is shown in Fig. 5. The models TIM-FD^3 and ISSM show approximately first-order convergence as $\Delta z \rightarrow 0$, while in COMice the RMSD drops only for Δz below 2 m. The finite difference discretization scheme in TIM-FD^3 is formally second-order accurate in space (and time) and the finite element models ISSM and COMice use linear basis functions, thus one would expect second-order convergence as $\Delta z \rightarrow 0$ for smooth problems. However, this is not the case here, since the observed order of accuracy depends on the strength of discontinuities (conductivity ratio between cold and temperate ice) and on the CTS implementation details.

6 Discussion

All three models are able to run the time-dependent experiment A and agree with the analytical solutions in terms of absolute values, timing and reversibility. However, not all types of basal boundary conditions have been tested here. Since the absence of strain heating suppresses the formation of a tem-

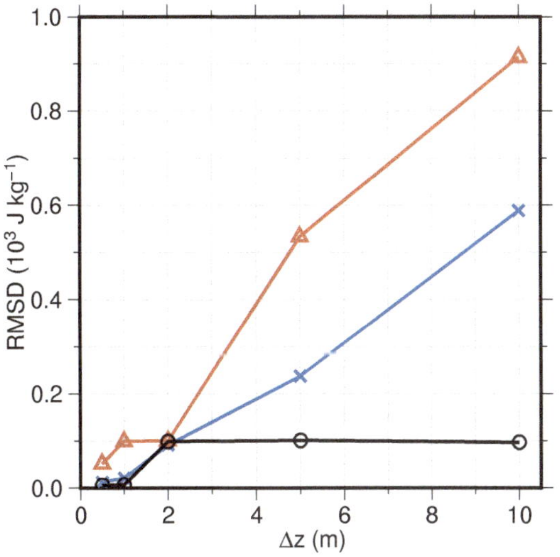

Figure 5. The root-mean-square deviation (RMSD) of the model results to the analytical solution (Experiment B) for different vertical grid resolutions Δz. Model results for TIM-FD^3 (blue crosses), ISSM (red triangles) and COMice (black circles) are obtained for the lowest conductivity ratio $\mathrm{CR} = K_0/K_\mathrm{c} = 10^{-5}$ applied to all models.

perate ice layer at the base, the insulating boundary condition (Eq. 6) could not be tested.

Beside the test of the implementation of the boundary conditions, this experiment addresses the importance of a basal water layer. Although the surface temperature changes, the basal temperature is kept at pressure melting point as long as a basal water layer exists. The amount of water at the base is crucial for the temperatures in the ice, because it acts as an energy buffer. It slows down the response of basal temperatures to surface cooling. The water layer thicknesses simulated here are unrealistically high compared to conditions under real ice masses. More realistic simulations would require a subglacial hydrology model, but this is beyond the scope of this paper.

Experiment B addresses whether the models are able to reproduce the steady-state analytical solution for certain polythermal conditions including advection, diffusion and strain heating. The models agree well with the analytical solution for $K_0 \to 0$, if the vertical resolution is high. All models meet the transition conditions for the melting CTS, although no explicit boundary conditions are implemented. An adequate treatment of the abrupt change of conductivity at the CTS in the numerical discretization scheme is required to achieve this behaviour. The usage of an arithmetic mean (TIM-FD^3) or a Heaviside as well as a smoothed Heaviside function (COMice) for the conductivity jump leads to oscillations in the enthalpy solution that are visible e.g. in a time-varying CTS position. Consequently, no steady-state solution is reached under these conditions. The harmonic mean approach of Patankar (1980, chap. 4.2.3) for the conductivity (TIM-FD^3 and ISSM) leads to a continuous heat flux at the CTS and violates the condition of Eq. (11) (non-continuous). Nevertheless, the harmonic mean strongly favours the lower conductivity K_0 for small ratios $\mathrm{CR} = K_0/K_\mathrm{c}$ and this leads to the apparent jump in $\partial E/\partial z$.

TIM-FD^3 tends to underestimate the water content at the base of a temperate ice layer. This would result in stiffer ice at the base. In typical applications of the model the vertical layers are not equidistant as in this study, but refined towards the base. We therefore expect only a minor influence on the velocity field. ISSM simulations underestimate the temperature in the cold part accompanied by an overestimation of the water content in the basal temperate layer at coarse resolution. Implications for the overall stiffness are hard to obtain. Ice would deform more in the temperate part at the base, but less in the cold part above.

The understanding of moisture transport in the temperate ice is poor. If the latent heat flux can be represented as in Aschwanden et al. (2012), then it is crucial to consider the assumption made on the chosen value of K_0. Simulations with a relatively high value of K_0 would lead to a much thicker temperate ice layer in contrast to simulations where $K_0 \approx 0$. Stable numerical solutions could be obtained for temperate ice diffusivities in the chosen range of $K_0 \approx 10^{-4}$ to $10^{-8}\,\mathrm{kg\,m^{-1}\,s^{-1}}$ and 0 for ISSM. The lower bound is therefore several magnitudes lower than $K_0 = 10^{-4}\,\mathrm{kg\,m^{-1}\,s^{-1}}$ as the lowest value possible for a stable solution in Aschwanden and Blatter (2009). If one assumes a vanishing latent heat flux in the temperate part of a glacier, we would recommend using a value of $K_0 \approx 10^{-6}\,\mathrm{kg\,m^{-1}\,s^{-1}}$ ($\mathrm{CR} = 10^{-3}$). For this value the CTS positions of all models are close to the analytical solution and show the smallest spread with varying vertical resolutions (Fig. 3).

The evolution equation for the enthalpy field is similar to the temperature evolution equation already implemented in thermomechanically coupled ice sheet models. Therefore an enthalpy scheme allows one to convert so-called "cold ice" models into polythermal ice models with only minor modifications, but with the restriction of melting conditions at the CTS. The question of whether exclusive melting conditions at the CTS are valid in an ice sheet is not conclusive. At least simulations of the Greenland Ice Sheet based on a two-layer front-tracking scheme performed with the polythermal ice model SICOPOLIS indicate that freezing conditions are relatively rare (Greve, 1997a, b). In the most recent version of SICOPOLIS the enthalpy scheme of Aschwanden et al. (2012) and a modified version of this scheme have been implemented as conventional one-layer enthalpy scheme and one-layer melting CTS scheme, respectively (Blatter and Greve, 2014). Thus comparisons to the two-layer front-tracking scheme can be performed for continental-scale ice sheets in the future.

The dynamics of glaciers, ice caps and ice sheets are strongly linked to the description of the rheology of temperate ice and its uncertainties. Besides the limited knowledge on the rheology of temperate ice, the current experimentally based relationship for the flow rate factor is only valid for water contents up to 1 % (Duval, 1977; Lliboutry and Duval, 1985). However, actual water contents found in temperate and polythermal glaciers are sometimes substantially larger (up to 5 %, Bradford and Harper, 2005). The advantage of deriving the water content by solving numerically for the enthalpy is limited by the use of a flow rate factor with a restricted validity range. Consequently, deformation experiments with temperate ice are urgently needed.

7 Conclusions

The proposed numerical experiments provide tests for the enthalpy implementation in numerical ice sheet models. All models applied here (TIM-FD^3, ISSM, COMice) are able to perform these experiments successfully and agree to the analytical solutions. The enthalpy scheme determines the cold–temperate transition surface (CTS) and the vertical enthalpy profile in a polythermal glacier correctly without the need of tracking the CTS explicitly and applying additional conditions at this internal boundary. This is in particular the case for high vertical resolution for all three models. TIM-FD^3 and COMice also perform well for low vertical resolution, while the ISSM solution shows a significant enthalpy difference to the analytical solution although the analytical CTS position is met. There is a clear need for an empirical determination of the temperate ice conductivity K_0 and an improved description of the temperate ice rheology.

Table A1. Constants and model parameters used.

Quantity	Value	Units
Seconds per year, spy	31 556 926	$\mathrm{s\,a^{-1}}$
Gravitational acceleration, g	9.81	$\mathrm{m\,s^{-2}}$
Density of ice, ρ_i	910	$\mathrm{kg\,m^{-3}}$
Density of water, ρ_w	1000	$\mathrm{kg\,m^{-3}}$
Reference temperature, T_{ref}	223.15	K
Melting point at standard pressure, T_0	273.15	K
Specific heat capacity, c_i	2009.0	$\mathrm{J\,kg^{-1}\,K^{-1}}$
Thermal conductivity, k_i	2.1	$\mathrm{W\,m{-}1\,K^{-1}}$
Experiment A[a]:		
Ice thickness, H	1000	m
Geothermal flux, q_{geo}	0.042	$\mathrm{W\,m^{-2}}$
Latent heat of fusion, L	3.34×10^5	$\mathrm{J\,kg^{-1}}$
Clausius–Clapeyron constant, β	7.9×10^{-8}	$\mathrm{K\,Pa^{-1}}$
Moisture mass diffusivity, K_0	$k_i/c_i \times 10^{-1}$	$\mathrm{kg\,m^{-1}\,s^{-1}}$
Experiment B[b]:		
Ice thickness, H	200	m
Geothermal flux, q_{geo}	0.0	$\mathrm{W\,m^{-2}}$
Latent heat of fusion, L	3.35×10^5	$\mathrm{J\,kg^{-1}}$
Clausius–Clapeyron constant, β	0.0	$\mathrm{K\,Pa^{-1}}$
Rate-factor, A	5.3×10^{-24}	$\mathrm{Pa^{-3}\,s^{-1}}$
Moisture diffusion coefficient, K_0	$k_i/c_i \times 10^{-1}$	
	⋮	
	$k_i/c_i \times 10^{-5}$	$\mathrm{kg\,m^{-1}\,s^{-1}}$

[a] Aschwanden et al. (2012); [b] Greve and Blatter (2009).

Appendix A: Analytical solutions

A1 Basal melt rate in Experiment A

To derive the basal melt rate for phase IIIa of Experiment A it is assumed that the temperature is in steady state at the end of the warming phase II. For this set-up Eq. (2) simplifies to the one-dimensional form

$$\rho_i \frac{\partial E}{\partial t} = \frac{\partial}{\partial z}\left(K_c \frac{\partial E}{\partial z}\right). \tag{A1}$$

We have only cold ice conditions in the interior of the ice body and K_c as well as ρ_i are constants. Based on the transfer rules in Eq. (1), Eq. (A1) can be written as an evolution equation for the temperature:

$$\frac{\partial T}{\partial t} = \kappa \frac{\partial^2 T}{\partial z^2} \text{ and } \kappa = \frac{k_i}{\rho_i c_i}. \tag{A2}$$

We determine the evolution of $T(z,t)$ starting from the initial condition (steady-state temperature profile of phase II)

$$T(z,0) = T_0(z) = T_{pmp} + \left(T_{s,w} - T_{pmp}\right) z/H \tag{A3}$$

and Dirichlet conditions at the upper and lower surface

$$T(H,t) = T_{s,w} \text{ and } T(0,t) = T_{pmp}. \tag{A4}$$

The basal temperature is kept at pressure melting point by the basal water layer (Eq. 7). Solutions of the heat equations can be found by separation of variables and Fourier analysis and require homogeneous boundary conditions. Therefore, the temperature deviation Θ is used instead of T, thus

$$T(z,t) = T_{eq}(z) + \Theta(z,t), \tag{A5}$$

where the steady-state profile for this set-up is again a linear

$$T_{eq}(z) = T_{pmp} + \left(T_{s,c} - T_{pmp}\right) z/H. \tag{A6}$$

Substitution of Eq. (A5) into Eq. (A2) and application of the steady-state solution $T_{eq}(z)$ implies that $\Theta(z,t)$ satisfies the homogeneous heat equation

$$\frac{\partial \Theta}{\partial t} = \kappa \frac{\partial^2 \Theta}{\partial z^2} \tag{A7}$$

with homogeneous Dirichlet boundary conditions,

$$\Theta(0,t) = \Theta(H,t) = 0 \text{ for } t > 0 \tag{A8}$$

and the initial condition

$$\Theta(z,0) = T_0(z) - T_{eq}(z) \text{ for } 0 \le z \le H. \tag{A9}$$

The solution of Eqs. (A7)–(A9) for Θ can be obtained using the method of separation of variables and leads to (e.g. Dubin, 2003)

$$\Theta(z,t) = \sum_{n=1}^{\infty} A_n e^{\lambda_n t} \sin\frac{n\pi z}{H}, \quad \text{where } \lambda_n = -\kappa\left(\frac{n\pi}{H}\right)^2. \tag{A10}$$

Setting $t = 0$ the Fourier coefficients A_n can be found by matching the initial condition Eq. (A9):

$$\Theta(z,0) = \sum_{n=1}^{\infty} A_n \sin\frac{n\pi z}{H} = T_0(z) - T_{eq}(z). \tag{A11}$$

Thus for the Fourier sine series, the coefficients A_n are determined as

$$A_n = \frac{1}{H}\int_0^H \left(T_0(z) - T_{eq}(z)\right) \sin\frac{n\pi z}{H} \mathrm{d}z. \tag{A12}$$

Inserting the initial condition (Eq. A3) and the steady-state profile (Eq. A6) into Eq. (A12) leads to

$$A_n = (-1)^{n+1} \frac{2\left(T_{s,w} - T_{s,c}\right)}{n\pi}. \tag{A13}$$

Based on the analytical solution of the temperature profile (Eq. A10) the basal melt rate (Eq. 8) is

$$a_\mathrm{b} = \frac{q_\mathrm{geo} - q_\mathrm{i}}{\rho L} = \frac{1}{\rho L}\left(q_\mathrm{geo} + k\frac{\partial T}{\partial z}|_{z=0}\right), \tag{A14}$$

where

$$\begin{aligned}\frac{\partial T}{\partial z}|_{z=0} &= \frac{\partial T_\mathrm{eq}(z)}{\partial z}|_{z=0} + \frac{\partial \Theta(z,t)}{\partial z}|_{z=0} \\ &= \frac{T_\mathrm{s,c} - T_\mathrm{pmp}}{H} + \sum_{n=1}^{\infty}\frac{n\pi}{H}A_n e^{\lambda_n t}.\end{aligned} \tag{A15}$$

The sum is evaluated up to $n = 25$ to produce the analytical solution shown in Fig. 2.

A2 Analytical solution Experiment B

The following derivation of the analytical solution to experiment is a modification of the derivation of the "parallel-sided polythermal slab" provided by Greve and Blatter (2009). Under the assumptions given there and the variable transform $z = \zeta H$, the enthalpy field Eq. (2) reduces to

$$\mathcal{D}\frac{\partial^2 E}{\partial \zeta^2} + \mathcal{M}\frac{\partial E}{\partial \zeta} = -\mathcal{K}(1-\zeta)^4, \text{ if } E < E_\mathrm{pmp} \tag{A16}$$

$$\mathcal{M}\frac{\partial E}{\partial \zeta} = -\mathcal{K}(1-\zeta)^4, \text{ else.} \tag{A17}$$

Here,

$$\mathcal{D} = \frac{K_\mathrm{i}}{\rho},\ \mathcal{M} = H a_\mathrm{s}^\perp,\ \mathcal{K} = \frac{2A}{\rho}(\rho g \sin\gamma)^4 H^6. \tag{A18}$$

Let E^+ be a solution of Eq. (A16) and E^- a solution to Eq. (A17). Then the enthalpy solution for the entire ice column is given by $E = E^- \mathbb{I}_{[0,\zeta_\mathrm{m})} + E^+ \mathbb{I}_{[\zeta_\mathrm{m},\,1]}$, where ζ_m is the position of the CTS.

At the CTS the continuity condition for the enthalpy Eq. (10) holds and due to the neglect of water conductivity in temperate ice the right-hand side of Eq. (11) is zero. A solution E^+ to Eq. (A16) is given by a solution to the homogeneous differential equation E_h associated with Eq. (A16) and a particular solution E_p:

$$E^+ = E_\mathrm{h} + E_\mathrm{p},\ \text{with} \tag{A19}$$

$$E_\mathrm{h}(\zeta) = c_1 e^{-\mathcal{M}\zeta/\mathcal{D}} + c_2,\ \text{and} \tag{A20}$$

$$E_\mathrm{p}(\zeta) = \sum_{k=1}^{5} a_k \zeta^k. \tag{A21}$$

The coefficients $a_1, \ldots, a_5$ of E_p can be found by balancing powers in Eq. (A16) (see Greve and Blatter, 2009). The three remaining unknowns, c_1, c_2 and ζ_m, can now be derived from the conditions at the CTS (Eqs. 10 and 11) and the given surface enthalpy. Inserting E^+ yields

$$E_\mathrm{s} = c_1 e^{-\mathcal{M}/\mathcal{D}} + c_2 + \sum_{k=1}^{5} a_k \tag{A22}$$

$$E_\mathrm{pmp} = c_1 e^{-\mathcal{M}\zeta_\mathrm{m}/\mathcal{D}} + c_2 + \sum_{k=1}^{5} a_k \zeta_\mathrm{m}^k \tag{A23}$$

$$0 = -c_1\frac{\mathcal{M}}{\mathcal{D}} e^{-\mathcal{M}\zeta_\mathrm{m}/\mathcal{D}} + \sum_{k=1}^{5} k a_k \zeta_\mathrm{m}^{k-1}. \tag{A24}$$

With c_1 from Eq. (A24) and c_2 from Eq. (A22), then Eq. (A23) becomes an implicit definition for ζ_m, whose root can be determined using a numerical solver. Then c_1 and c_2 follow accordingly.

A solution E^- for the temperate ice part can be found by integrating the temperate version of Eq. (A17) directly. E^- is then fully determined by Eq. (11):

$$E^-(\zeta) = E_\mathrm{pmp} + \frac{\mathcal{K}}{5\mathcal{M}}\left((1-\zeta)^5 - (1-\zeta_\mathrm{m})^5\right). \tag{A25}$$

Acknowledgements. The authors would like to thank H. Blatter, R. Greve, F. J. Navarro, J. Otero and F. Ziemen for fruitful discussions on enthalpy and ice sheet models. We acknowledge the support of the ISSM developer team (E. Larour, H. Seroussi and M. Morlighem) for the implementation of the enthalpy formulation in ISSM.

Edited by: F. Pattyn

References

Aschwanden, A. and Blatter, H.: Mathematical modeling and numerical simulation of polythermal glaciers, J. Geophys. Res., 114, F01027, doi:10.1029/2008JF001028, 2009.

Aschwanden, A., Bueler, E., Khroulev, C., and Blatter, H.: An enthalpy formulation for glaciers and ice sheets, J. Glaciol., 58, 441–457, doi:10.3189/2012JoG11J088, 2012.

Blatter, H. and Greve, R.: Comparison and verification of enthalpy schemes for polythermal glaciers and ice sheets with a one-dimensional model, Polar Science, submitted, preprint at ArXiv E-Prints, arXiv:1410.6251, 2014.

Blatter, H. and Hutter, K.: Polythermal conditions in Arctic glaciers, J. Glaciol., 37, 261–269, 1991.

Bradford, J. H. and Harper, J. T.: Wave field migration as a tool for estimating spatially continuous radar velocity and water content in glaciers, Geophys. Res. Lett., 32, L08502, doi:10.1029/2004GL021770, 2005.

Bueler, E., Lingle, C. S., Kallen-Brown, J. A., Covey, D. N., and Bowman, L. N.: Exact solutions and the verification of numerical models for isothermal ice sheets, J. Glaciol., 51, 291–306, 2005.

Bueler, E., Brown, J., and Lingle, C. S.: Exact solutions to the thermomechanically coupled shallow-ice approximation: effective tools for verification, J. Glaciol., 53, 499–516, doi:10.3189/002214307783258396, 2007.

Dubin, D.: Numerical and Analytical Methods for Scientists and Engineers Using Mathematica, Wiley, Hoboken, New Jersey, USA, 656 pp., 2003.

Duval, P.: The role of the water content on the creep rate of polycrystalline ice, International Association of Hydrological Sciences Publication 118, Symposium on Isotopes and Impurities in Snow and Ice, Grenoble, 29–33, 1977.

Fowler, A. C.: On the transport of moisture in polythermal glaciers, Geophys. Astro. Fluid, 28, 99–140, doi:10.1080/03091928408222846, 1984.

Glen, J. W.: The Creep of Polycrystalline Ice, P. Roy. Soc. Lond. A, 228, 519–538, 1955.

Greve, R.: Application of a polythermal three-dimensional ice sheet model to the Greenland Ice Sheet: response to steady-state and transient climate scenarios, J. Climate, 10, 901–918, 1997a.

Greve, R.: A continuum-mechanical formulation for shallow polythermal ice sheets, Philos. T. Roy. Soc. Lond., 355, 921–974, 1997b.

Greve, R. and Blatter, H.: Dynamics of ice sheets and glaciers, in: Advances in Geophysical and Environmental Mechanics and Mathematics, Springer, Berlin, Heidelberg, 287 pp., doi:10.1007/978-3-642-03415-2, 2009.

Hindmarsh, A. C., Brown, P. N., Grant, K. E., Lee, S. L., Serban, R., Shumaker, D. E., and Woodward, C. S.: SUNDIALS: Suite of Nonlinear and Differential/Algebraic Equation Solvers, ACM T. Math. Software, 31, 363–396, doi:10.1145/1089014.1089020, 2005.

Hutter, K., Blatter, H., and Funk, M.: A model computation of moisture content in polythermal glaciers, J. Geophys. Res.-Sol. Ea., 93, 12205–12214, doi:10.1029/JB093iB10p12205, 1988.

Huybrechts, P., Payne, T., Abe-Ouchi, A., Calov, R., Fabre, A., Fastook, J. L., Greve, R., Hindmarsh, R. C. A., Hoydal, O., Johannesson, T., MacAyeal, D. R., Marsiat, I., Ritz, C., Verbitsky, M. Y., Waddington, E. D., and Warner, R.: The EISMINT benchmarks for testing ice-sheet models, Ann. Glaciol., 23, 1–12, 1996.

Kleiner, T. and Humbert, A.: Numerical simulations of major ice streams in western Dronning Maud Land, Antarctica, under wet and dry basal conditions, J. Glaciol., 60, 215–232, doi:10.3189/2014JoG13J006, 2014.

Larour, E., Seroussi, H., Morlighem, M., and Rignot, E.: Continental scale, high order, high spatial resolution, ice sheet modeling using the Ice Sheet System Model (ISSM), J. Geophys. Res., 117, F01022, doi:10.1029/2011JF002140, 2012.

Lliboutry, L. A. and Duval, P.: Various isotropic and anisotropic ices found in glaciers and polar ice caps and their corresponding rheologies, Ann. Geophys.-Italy, 3, 207–224, 1985.

Notz, D. and Worster, M. G.: A one-dimensional enthalpy model of sea ice, Ann. Glaciol., 44, 123–128, doi:10.3189/172756406781811196, 2006.

Patankar, S. V.: Numerical Heat Transfer and Fluid Flow, McGraw-Hill, New York, 1980.

Pattyn, F., Schoof, C., Perichon, L., Hindmarsh, R. C. A., Bueler, E., de Fleurian, B., Durand, G., Gagliardini, O., Gladstone, R., Goldberg, D., Gudmundsson, G. H., Huybrechts, P., Lee, V., Nick, F. M., Payne, A. J., Pollard, D., Rybak, O., Saito, F., and Vieli, A.: Results of the Marine Ice Sheet Model Intercomparison Project, MISMIP, The Cryosphere, 6, 573–588, doi:10.5194/tc-6-573-2012, 2012.

Payne, A. J. and Dongelmans, P. W.: Self-organization in the thermomechanical flow of ice sheets, J. Geophys. Res., 102, 12219–12233, doi:10.1029/97JB00513, 1997.

Rückamp, M., Blindow, N., Suckro, S., Braun, M., and Humbert, A.: Dynamics of the ice cap on King George Island, Antarctica: field measurements and numerical simulations, Ann. Glaciol., 51, 80–90, doi:10.3189/172756410791392817, 2010.

Seroussi, H., Morlighem, M., Rignot, E., Khazendar, A., Larour, E., and Mouginot, J.: Dependence of century-scale projections of the Greenland ice sheet on its thermal regime, J. Glaciol., 59, 1024–1034, doi:10.3189/2013JoG13J054, 2013.

Spalding, D. B.: A novel finite difference formulation for differential expressions involving both first and second derivatives, Int. J. Numer. Meth. Eng., 4, 551–559, doi:10.1002/nme.1620040409, 1972.

Steinemann, S.: Results of preliminary experiments on the plasticity of ice crystals, J. Glaciol., 2, 404–413, 1954.

18

Simulating the Greenland ice sheet under present-day and palaeo constraints including a new discharge parameterization

R. Calov[1], A. Robinson[2,3,1], M. Perrette[1], and A. Ganopolski[1]

[1]Potsdam Institute for Climate Impact Research, Potsdam, Germany
[2]Universidad Complutense Madrid, 28040 Madrid, Spain
[3]Instituto de Geociencias, UCM-CSIC, 28040 Madrid, Spain

Correspondence to: R. Calov (calov@pik-potsdam.de)

Abstract. In this paper, we propose a new sub-grid scale parameterization for the ice discharge into the ocean through outlet glaciers and inspect the role of different observational and palaeo constraints for the choice of an optimal set of model parameters. This parameterization was introduced into the polythermal ice-sheet model SICOPOLIS, which is coupled to the regional climate model of intermediate complexity REMBO. Using the coupled model, we performed large ensemble simulations over the last two glacial cycles by varying two major parameters: a melt parameter in the surface melt scheme of REMBO and a discharge scaling parameter in our parameterization of ice discharge. Our empirical constraints are the present-day Greenland ice sheet surface elevation, the surface mass balance partition (ratio between total ice discharge and total precipitation) and the Eemian interglacial elevation drop relative to present day in the vicinity of the NEEM ice core. We show that the ice discharge parameterization enables us to simulate both the correct ice-sheet shape and mass balance partition at the same time without explicitly resolving the Greenland outlet glaciers. For model verification, we compare the simulated total and sectoral ice discharge with other estimates. For the model versions that are consistent with the range of observational and palaeo constraints, our simulated Greenland ice sheet contribution to Eemian sea-level rise relative to present-day amounts to 1.4 m on average (in the range of 0.6 and 2.5 m).

1 Introduction

Modelling the response of the Greenland ice sheet (GrIS) to anthropogenic warming has already been undertaken for more than 2 decades (Huybrechts et al., 1991; van de Wal and Oerlemans, 1997; Huybrechts and de Wolde, 1999; Greve, 2000) and attracted considerable attention in recent years (Vizcaíno et al., 2010; Goelzer et al., 2011; Graversen et al., 2011; Applegate et al., 2012; Lipscomb et al., 2013; Stone et al., 2013), including higher-order and full Stokes modelling approaches (Price et al., 2011; Seddik et al., 2012; Goelzer et al., 2013). The recent SeaRISE ice sheet modelling project (Nowicki et al., 2013) highlighted the importance of treatment of processes at the ice-ocean interface for the response of models of the Greenland ice sheet to future climate change.

Observational data indicate that during the past decade mass loss by the GrIS, both through surface melt and enhanced ice discharge, has contributed appreciably to global sea level rise (Shepherd et al., 2012). The latest projections suggest that the GrIS will contribute notably to sea level rise during the next century (Hanna et al., 2013). In the longer-term perspective, the GrIS can become even more important, because even if the global temperature is stabilized at the level of 2 °C above preindustrial, the GrIS would still continue to melt and, in the long-term perspective, can lose a significant fraction of its mass even for moderate warming (Ridley et al., 2005; Robinson et al., 2012).

Models of the GrIS contain a number of parameters that can be used for tuning the model using observational constraints. Present-day extent and surface elevation of the GrIS are accurately known and it is natural to use them as such constraints (e.g. Stone et al., 2010; Greve et al., 2011; Lipscomb et al., 2013). At the same time, it is known that coarse-resolution ice-sheet models have problems simulating the correct margins of the GrIS and they systematically overestimate its volume. One reason is that under present-day conditions, most ice discharge into the ocean occurs through relatively narrow outlet glaciers. As a result, although ice discharge into the ocean currently accounts for more than half of surface accumulation, the majority of Greenland the ice sheet margin is located several tens of kilometres away from the ocean. Since current ice-sheet models do not resolve outlet glaciers and their interaction with the ocean in the fjords, the modelled GrIS needs too much contact with the ocean to produce realistic discharge. This leads to systematic overestimation of the ice area and volume and makes observational "geometrical" constraints difficult to apply.

However, the observed shape of the GrIS is not the only characteristic that can serve as a constraint for the GrIS models. Recently, Robinson et al. (2011) introduced the mass balance partition (defined as the ratio between total ice discharge into the ocean and total precipitation over the GrIS, MBP) as another constraint on the GrIS. The MBP is an important characteristic for short-term as well as for long-term (future) behaviour of the GrIS, because it determines the GrIS mass balance sensitivity to climate change. In particular, the present-day MBP is related to the long-term stability properties of the GrIS (Robinson et al., 2012), i.e. for low MBP values (large surface melt) the modelled GrIS is more susceptible to warming than for high ones. For the long-term stability of the GrIS, the MBP is a more important characteristic than the present-day shape of the GrIS. However for short-term (centennial time scale) future global warming simulations, such an ice sheet would be an unfavourable initial condition. This is because during a considerable portion of time of such a future simulation, the modelled ice would melt in areas where in reality no ice exists (Goelzer et al., 2013).

Since standard coarse-resolution GrIS ice sheet models cannot simulate a realistic present-day surface orography of the GrIS and at the same time have the correct mass balance partition, we developed a novel approach, which allows us to circumvent this problem without resolving individual ice streams and outlet glaciers – and without an increase in computational cost. This approach is in the spirit of our previous modelling work (Robinson et al., 2010, 2011, 2012) and is based on a rather simple semi-empirical parameterization of ice discharge through the outlet glaciers. We propose the usage of this approach until a more complete representation of fast processes is available. Considering the above concerns, this approach is feasible for short-term as well as for long-term simulations.

In addition to the present-day constraints on the ice-sheet shape and the MBP, we use the Eemian as a palaeo constraint. Eemian conditions have already been recognized earlier as an important palaeo constraint for GrIS model parameters (Tarasov and Peltier, 2003) and have been applied more recently in several studies (Robinson et al., 2011; Born and Nisancioglu, 2012; Stone et al., 2013). While the Greenland Summit position might be located too far in the interior of the GrIS to serve as a strong palaeo constraint, the position of the new NEEM ice core appears more promising, because it is located rather near the ice margin, a location which is very sensitive to the climate changes during the Eemian (Born and Nisancioglu, 2012). In the present paper, we make use of the recently published estimate of the Eemian elevation drop at a position of about 200 km upstream of NEEM (NEEM community members, 2013), where the borehole ice sampled today was deposited during the Eemian.

The paper is summarized as follows. First, we give a short description of the ice-sheet model SICOPOLIS and the regional energy-moisture balance model REMBO (Sect. 2). Our new discharge parameterization is comprehensively explained in Sect. 3. Different metrics of the ice-sheet shape are introduced in Sect. 4 and the the model setup is given in Section 5. We discuss the behaviour of the shape metrics in Sect. 6, while we present constraints on our model parameters in Section 7. Further simulations inspect Eemian climate and GrIS stability (Sect. 8) and compare our findings with those of others (Sect. 9). We close with a discussion and finally our conclusions.

2 Model description

For this study, we used the three-dimensional polythermal ice-sheet model SICOPOLIS (version 2.9) coupled to the regional energy-moisture balance model (REMBO). SICOPOLIS treats the evolution of ice thickness, ice temperature and water content (Greve, 1997) based on the shallow ice approximation (Hutter, 1983). The dependence of the ice velocities on the ice temperature and water is introduced via the rate factor. SICOPOLIS enables a free and easy choice of several parameters including resolution. In our paper, Greenland is mapped onto a stereographic plane with 76×141 grid points (20 km grid spacing) using the topographic data set by Bamber et al. (2001). The vertical is resolved by 90 layers with decreasing layers thickness towards the bed of the ice sheet. A 10-layer thermal rock bed is coupled to the overlying ice sheet via heat fluxes. The geothermal heat flux is prescribed at the lower border of the thermal bedrock. We use the Weertman-type sliding law with powers as described in Robinson et al. (2011). The bedrock adjusts to the load caused by the ice sheet's weight using a local lithosphere relaxing asthenosphere model with a time delay of 3000 years.

The regional energy-moisture balance model REMBO is a climate model of intermediate complexity and it is described in detail by Robinson et al. (2010). REMBO uses diffusion-type equations for surface air temperature and atmospheric water content. For temperature, the well-known Budyko-

Sellers energy balance approach is implemented. Planetary albedo is related to surface albedo via a linear parameterization based on empirical data. The lateral boundary conditions for temperature and relative humidity are taken from climatology (Uppala et al., 2005) for the 1958–2001 period, which in this paper is referred to as "present day". REMBO includes a 1-layer snowpack model with a simple parameterization of refreezing. Surface albedo depends on snow thickness and the melt rate. Surface melt is computed using a simple parameterization of van den Berg et al. (2008) and depends on both temperature and insolation. The formula for surface melt contains a free parameter c_{m} (melt parameter), which is one of the major parameters determining the sensitivity of the ice sheet to climate change (Robinson et al., 2010). It reads

$$m = \frac{1}{\rho_{\mathrm{w}} L_{\mathrm{m}}} \left[\tau_{\mathrm{s}} (1 - \alpha_{\mathrm{s}}) S + c_{\mathrm{m}} + \lambda T \right], \quad (1)$$

which relates the surface melt m with the free melt parameter c_{m}. The remaining variables ρ_{w}, L_{m}, τ_{s}, α_{s}, S, λ and T, are the density of water, the latent heat of ice melting, the total transmissivity, the surface albedo, the insolation at the top of the atmosphere, the long wave radiation coefficient and the surface temperature, respectively. It is important to note that REMBO resolves the seasonal cycle of temperature and insolation, which are both important for properly modelling surface melt (van de Berg et al., 2011). Please, refer to Robinson et al. (2010) for more details.

The coupling between the models is bi-directional, i.e. SICOPOLIS provides the climate model with information about surface elevation and spatial extent of the ice sheet. In turn, REMBO provides SICOPOLIS with surface mass balance and mean annual surface temperature.

3 Ice discharge parameterization

Most ice discharge of the GrIS is brought into the ocean via fast-flowing narrow outlet glaciers (Rignot and Mouginot, 2012), most of which cannot be resolved by the coarse-resolution GrIS models used for palaeoclimate simulations and long-term sea level change projections. This motivates us to include this portion of ice discharge in the mass balance of the ice sheet, which reads

$$\frac{\partial h}{\partial t} = -\nabla \cdot \boldsymbol{q} - d + b + b^*, \quad (2)$$

where h is the ice thickness, $\nabla \cdot \boldsymbol{q}$ is the divergence of large-scale lateral ice flow explicitly resolved by the model, d is the divergence of the ice flow associated with non-resolved ice streams/outlet glaciers which is parameterized (see below), b is the surface mass balance and b^* is the basal mass balance, which is rather small. Note that, although b and d have the same units ($\mathrm{m\,s^{-1}}$) and are treated as surface fluxes, they represent completely different physical processes. The

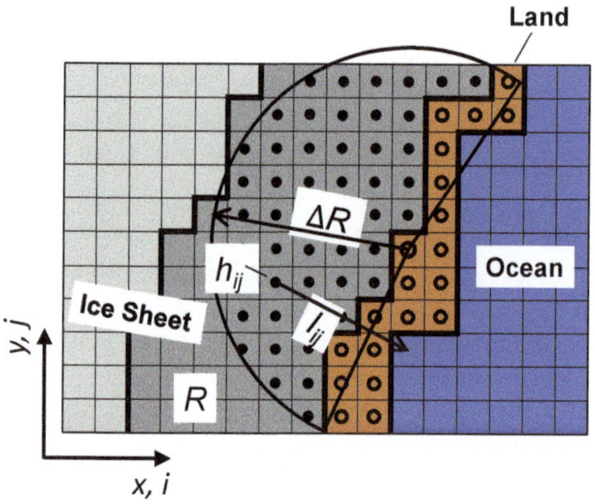

Figure 1. Principle sketch of the discharge parameterization over a part of the horizontal computational domain. The gray shading shows the ice-covered cells, while the dark gray shaded area R indicates the region over the ice sheet where the discharge parameterization applies. The (half) circle illustrates how that band R with the width of about ΔR is determined in our scheme. Namely, the centre of such a circle is applied to every land point (open circles over the brown area). The smallest distance to the ocean l_{ij} is depicted here for one example ice grid point. It is determined for every grid point inside the band R. For the discharge parameterization, the ice thickness h_{ij} and the smallest distance to the ocean l_{ij} are evaluated at grid points (i, j), see Eq. (3).

integral of d over the entire area of the GrIS represents the total solid ice discharge into the ocean through outlet glaciers. Total solid ice discharge into the ocean is equal to the sum of parameterized ice discharge and the lateral ice discharge into the ocean explicitly computed by SICOPOLIS. Under present day climate conditions the latter is much smaller than the parameterized solid discharge. This is explained by the fact that simulated modern GrIS has rather limited direct contact with the ocean. Note that Eq. (2) per definition conserves mass.

The divergence of the explicitly non-resolved fast lateral ice flow d in each grid cell (i, j) is parameterized in a simple heuristic statistical approach via the local thickness of ice h and the distance from the actual grid cell to the nearest ocean grid cell l as

$$d = \begin{cases} c\frac{h^p}{l^q} & \text{inside area } R, \\ 0 & \text{outside area } R, \end{cases} \quad (3)$$

where c, p and q are constant model parameters. R is defined as the area which consists of grid points located within a distance not more than ΔR from the nearest ice-free land surface point or the nearest ocean point in case if the ice sheet has direct contact with the ocean (see Fig. 1). The rationale for this parameterization is the following. The ice thickness near the ice margins represents the amount of ice available to the outlet glaciers, while the distance to the coast can be regarded as a statistical measure of the outlet glacier density:

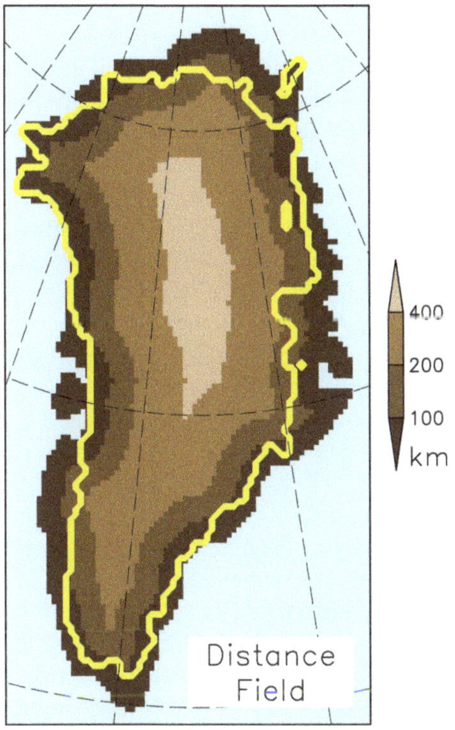

Figure 2. Distance field over the entire Greenland area in km. It is determined by the minimal distance of every land grid point (ice-free and ice-covered ones) to the coast (first ocean grid point, see Fig. 1). It defines the length l in the discharge parameterization (Eq. 3). The yellow line indicates the ice margin of the present-day Greenland ice sheet.

if the ice margin is far away from the coast, it is very unlikely that any outlet glacier has contact with the ocean and there is only minor calving flux into the ocean, while one would expect a large calving flux for small distances to the coast. It is assumed that significant influence of outlet glaciers does not penetrate inside of the ice sheet (distance more than ΔR).

Additionally, we assume that ice discharge into the ocean via fast moving ice streams only occurs from the area where ice surface is descending toward the coast. This is enforced by setting a maximum value of $\alpha_0 = 60°$ to the angle between the gradients of surface elevation ∇z_s and distance field ∇l (see Fig. 2 for the characteristics of the field l). Using the definition of the scalar product for the angle between above two gradients, this reads

$$\frac{\nabla z_s \cdot \nabla l}{|\nabla z_s||\nabla l|} \geq \cos(\alpha_0), \text{ evaluated at } (i, j). \quad (4)$$

This prevents our parameterization from simulating ice discharge into the ocean from an ice margin, which is oriented towards the interior of the Greenland island, which could happen when the GrIS is retreating under warm climates, and when only a few small ice sheets remain; see the small land bridge between the two ice caps in Fig. 3b for an example.

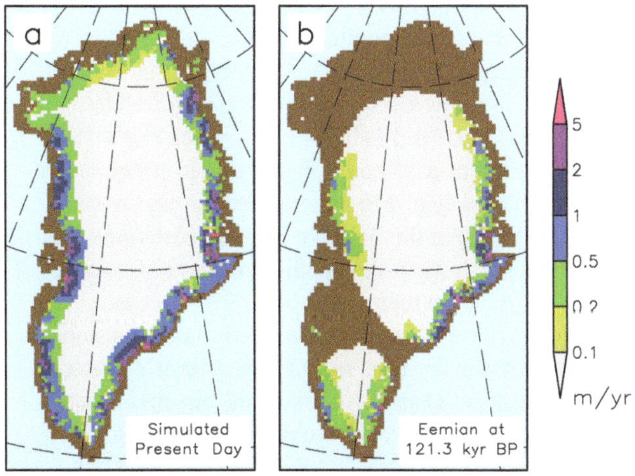

Figure 3. Simulated parameterized divergence of fast ice flow in m yr^{-1} (Eq. 3) at **(a)** present day and, **(b)** at Eemian from the model version with the most reduced Eemian ice sheet.

The discharge parameterization is applied only to the ice-covered grid cells that are located not more than ΔR (here $\Delta R = 120$ km) from the ice margin (see Fig. 1). The resulting belt encloses the regions of ice with rather high velocities as found by satellite measurements (Joughin et al., 2010; Rignot and Mouginot, 2012). This reflects that our parameterization primarily accounts for the ice discharge via the near margin fast-flow and outlet glaciers.

The value of parameter c depends strongly on the choice of the powers p and q to maintain total ice discharge known form observational data. For convenience, we normalize c as

$$c = c_0 c_d, \quad (5)$$

i.e. for any fixed value of p and q we selected c_0 such that c_d has a value of about one. In practice, after selecting p and q we chose c_0 so that the parameterization applied to observed Greenland elevation matched observed total discharge, for which we used 350 Gt yr^{-1}. The latter value is just about the average of the totals of ice discharge as found by Reeh (1994) and Rignot and Kanagaratnam (2006). Although the discharge for the modelled present-day GrIS will not be precisely 350 Gt yr^{-1}, such an approach guarantees that all valid values of the parameter c_d keep the order of magnitude of about one for any power p and q. We thus have three free parameters c_d, p and q in our ice discharge parameterization (Eqs. 3 and 5).

As seen in Fig. 2, the minimal distance to the coast l can reach up to about 400 km in the centre of Greenland. In our parameterization, the inverse dependence of the ice discharge on the distance to the coast results in high ice discharge over regions very near to the coast (dark brown colour) and low discharge further inland (lighter brown colours). For a distance of around 400 km, the parameterized divergence of fast flow would nearly vanish. In our discharge parameterization, this is only relevant for an ice sheet under warmer climates, which can be much smaller than the present-day GrIS. In fact, because our parameterization only applies for $\Delta R < 120$ km around the ice margins, most parameterized divergence of fast flow occurs at the near-margin ice area for present-day Greenland (Fig. 3a). These are the regions where a high present-day ice discharge due to many outlet glaciers is observed (e.g. Moon et al., 2012), e.g. over the north-western region of the GrIS. For regions with fewer marine outlet glaciers, e.g. in the south-west, the observed ice margin resides rather in regions further from the coast.

In the model, the ice-marginal ring ΔR and the distance field are computed every time step. The former is necessary, because the ice sheet changes its shape in time and because the coastline can potentially change its shape due to sea level change.

4 Measures of geometrical characteristics of an ice sheet

There are numerous possibilities to define a measure of the performance of a model based on the comparison of simulated geometrical characteristics of an ice sheet with observational data. The simplest is arguably to use the error in simulated total ice area and ice volume, which we define as

$$\mathrm{err}(A) = \frac{|A_{\mathrm{mod}} - A_{\mathrm{obs}}|}{A_{\mathrm{obs}}} \cdot 100,$$
$$\mathrm{err}(V) = \frac{|V_{\mathrm{mod}} - V_{\mathrm{obs}}|}{V_{\mathrm{obs}}} \cdot 100, \quad (6)$$

where A_{obs}, A_{mod}, V_{obs} and V_{mod} are the observed ice area, the modelled ice area, the observed ice volume and the modelled ice volume, respectively. These errors can approach zero in principle, but this does not guarantee accurate simulation of the ice-sheet geometry, since regional errors can compensate each other. Therefore, we choose a stronger constraint based on the error in ice thickness expressed relatively to the total ice thickness. This reads

$$\mathrm{err}(H) = \frac{\sum_{ij} |H_{ij}^{\mathrm{mod}} - H_{ij}^{\mathrm{obs}}|}{\sum_{ij} H_{ij}^{\mathrm{obs}}} \cdot 100, \quad (7)$$

where H_{ij}^{obs} and H_{ij}^{mod} are the observed and modelled ice thickness at the horizontal grid position (i, j), respectively. The indices i, j run over the entire domain of the computational area and assume that the ice thickness is zero outside the ice-covered area. This error only approaches zero when the ice-sheet thickness is correctly simulated in each grid cell. Simulations assessed with the different error measures are presented in Sect. 6.

5 Simulation setup

Following Robinson et al. (2011), we run the coupled REMBO-SICOPOLIS model through two glacial cycles starting at 250 kyr BP. These simulations serve a dual purpose: to perform a model spin-up necessary to simulate the present-day state of the GrIS and to apply palaeoclimate constraints (see below) to additionally reduce the range of model parameters. To drive the model through two glacial cycles, we apply variations in insolation due to changes in orbital parameters, equivalent CO_2 concentration and regional temperature anomaly obtained from the CLIMBER-2 model (Petoukhov et al., 2000; Calov et al., 2005; Ganopolski et al., 2010). We took these anomalies from the standard simulation as in Ganopolski and Calov (2011). The applied forcing is illustrated in Fig. 1 by Robinson et al. (2011). To generate an ensemble of model realisations, we vary two parameters: the discharge parameter c_{d} (Sect. 3) and the melt parameter c_{m} (Sect. 2). The discharge parameter c_d is varied in steps of 0.2 and the melt parameter c_m in steps of $5\,\mathrm{W\,m^{-2}}$. The geothermal heat flux is set to $50\,\mathrm{m\,W\,m^{-2}}$ and the sliding coefficient to $15\,\mathrm{m\,(yr\,Pa)^{-2}}$. All other parameters are the same as in Robinson et al. (2010).

We fix the powers p and q by minimizing the relative error in present-day ice thickness (see Sect. 4). Based on ensemble simulations over the parameter space (c_{d}, c_{m}) for different powers p and q (not explicitly displayed here), we found that decreasing p and increasing q reduce the error in ice thickness err(H). For simplicity, we chose the integers $p = 1$ and $q = 3$. To normalize the parameter c_{d} (see Sect. 3 and Eq. (5)), we set the dependent parameter $c_0 = 2.61 \times 10^4\,\mathrm{m^3\,s^{-1}}$. All ensemble simulations presented in this paper are performed with these values for the powers and for the parameter c_0 in the ice discharge parameterization.

6 Simulations of the GrIS in the entire parameter space

Figure 4 compares the three error measures introduced in Sect. 4 for the ice-sheet shape in the phase space of the melt, c_{m}, and discharge, c_{d}, parameters. At first sight, the error fields in the three panels look similar. The smallest errors appear approximately along the descending diagonal, which is characterized by decreasing values of c_{m} and increasing values of c_{d}. One can also see that the parameter combinations with small errors are not limited to our sampled space: these regions expand in the direction of the descending diagonal. This underlines the need for more constraints (Sect. 7).

Figure 4 illustrates that our discharge parameterization allows us to reduce the errors in total area and volume practically to zero. However, we found no parameter combination for which the error in ice thickness was much lower then 20 %. Still, these are considerable improvements compared

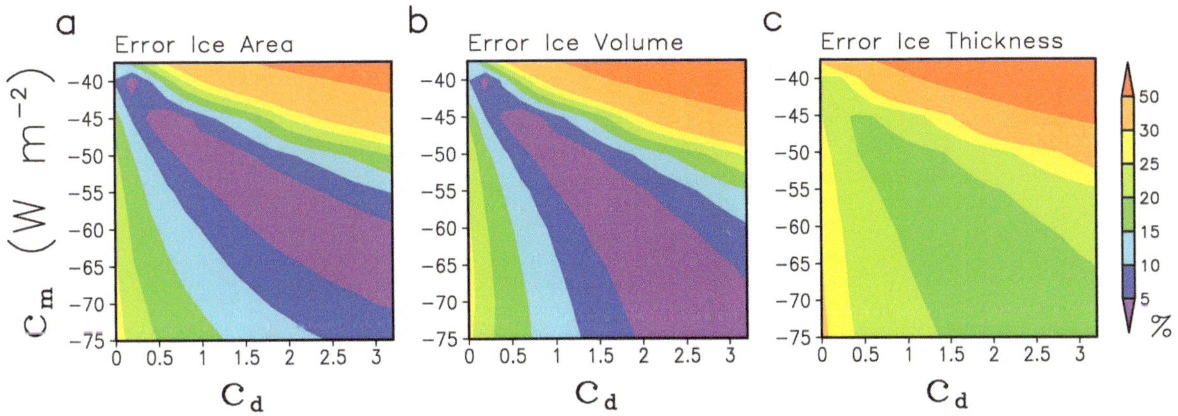

Figure 4. Error measures for a modelled ice sheet in the $\{c_d\} \times \{c_m\}$ parameter space. Relative errors in **(a)** ice area, **(b)** ice volume and **(c)** ice thickness, all in percent.

to the standard version of the model without ice discharge parameterization constrained only by the mass balance partition (Robinson et al., 2011), which overestimates total ice volume and area by ca. 20 % and has a relative thickness error of ca. 30 %. When the mass balance partition constraint is ignored, one can improve model performance in simulation of ice sheet shape by increasing surface melt. By choosing $c_m = -40\,\mathrm{W\,m^{-2}}$, one can make all three errors comparable with those of the best model version with ice discharge parameterization. However, such a high melt factor practically eliminates ice discharge into the ocean and, as shown below, drastically affects the GrIS stability. In fact, it causes the GrIS to be unstable even under near-present-day climate conditions.

7 Constraining the model parameters

In addition to the relative error in present-day ice thickness, we use the following as further empirical constraints on the ensemble of the model realizations: the present-day surface mass balance partition and the Eemian drop in surface elevation relative to present day at the upstream position of the NEEM ice borehole. Figure 5a–c illustrates all constraints used in this paper. We accept a value of 20 % for the relative error in ice thickness. This choice is not totally arbitrary, because a closer inspection of the error field shows a minimum error in ice thickness of 18.2 %, i.e. there is indeed a plateau defined by ice thickness error values $\leq 20\,\%$, as illustrated in Fig. 5a by the medium green shading. Within the parameter space, the error in ice thickness varies much more strongly for values higher than 20 %. This plateau structure of the error field motivated us to choose 20 % as the error limit.

As mentioned in the introduction, the mass balance partition is the amount of total ice discharge compared to total precipitation. In our work, we always refer to MBP as a characteristic of the ice sheet defined in its present-day state. Its practical definition is the total ice discharge divided by the total precipitation for the simulated present-day ice sheet. In Robinson et al. (2011), the MBP was diagnosed by REMBO from simulations with prescribed observed present-day ice-sheet topography. This was done because of systematic (and regionally significant) deviations of the simulated present-day GrIS from observational data. With the ice discharge parameterization, we can now safely operate with the simulated present day ice-sheet topography for determining MBP, because of the better match between simulated and observed topography. This means that our new MBP values simulated with ice discharge parameterization slightly differ from our former approach ($c_d = 0$), and the valid MBP range (Fig. 5b) corresponds to somewhat lower values of c_m compared to those by Robinson et al. (2011). However, the MBP has large inherent uncertainty, which we derived from results of regional climate models, following Robinson et al. (2011) and yielding a range of 45 to 65 %.

From measurement of air content in the NEEM borehole samples, the NEEM community members (2013) found that the surface elevation of the source area was 130 ± 300 m lower during the Eemian than at present day, which we exploit as our third constraint. Following these findings, we assume the maximal surface elevation drop at this location during the Eemian (130 to 115 kyr BP) did not exceed 430 m (compared to present day). Accounting for the trajectory tracing results by NEEM community members (2013), we defined the deposition position of Eemian ice in the NEEM ice core at the location about 200 km upstream from the NEEM drilling site at (45° W, 76° N) (see Fig. 7d), denoted NEEMup hereafter. We use the Eemian elevation drop at the NEEM source location as additional empirical constraint.

Figure 5 illustrates that none of the constraints is redundant, because the regions of valid simulations for all three constraints intersect each other and there is plenty of space without a common crossover for every constraint. In particular, the error in ice thickness excludes low values of the ice

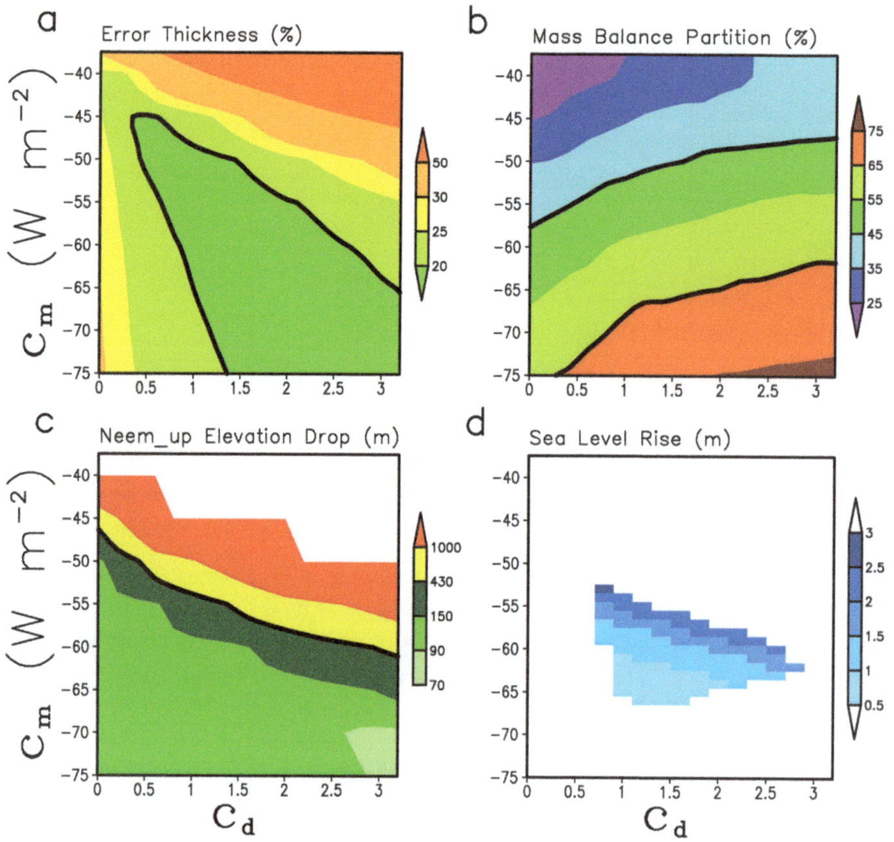

Figure 5. Estimated constraints on the parameters c_d (ice discharge) and c_m (surface melt) illustrated with our ensemble simulations together with our estimate of the GrIS contribution to Eemian sea-level rise. **(a)** Relative error in ice thickness (%). **(b)** Mass balance partition (%). **(c)** Maximum elevation reduction during the Eemian compared to present day, 200 km upstream from NEEM (m). Here, regions where no Eemian ice is simulated at the upstream position of NEEM are displayed in white. **(d)** Simulated contribution of the Greenland ice sheet to sea-level rise between the Eemian and present day under our constraints. The black lines in **(a)**–**(c)** indicate our constraints: error values < 20 % for ice thickness, a 45 to 65 % range for mass balance partition and an Eemian to present-day surface elevation reduction of < = 430 m at the upstream position of NEEM.

discharge parameter, while the mass balance partition constrains the range of the melt parameter, the upper bound of which is then further constrained by the NEEMup elevation data. While the valid region of all three constraints cover an approximately equally large part of the parameter space (Fig. 5a–c), only a relatively small subset of model parameters (Fig. 5d) is consistent with all of these constraints simultaneously.

Figure 6 shows the range of ice margins which are consistent with the different constraints shown in Fig. 5. It can be seen that the MBP constraint alone (Fig. 6a) gives a rather broad band of valid ice margins, while the NEEMup constraint (Fig. 6b) alone results in an ice margin range, which is quite comparable with that of the err(H) constraint (Fig. 6c). Finally, all three constraints together give a pronounced reduction of spread of ice margins (Fig. 6d) compared to the single-constraint cases. Simulated ice margins in the south, mid-west and north-east of Greenland compare well with observations. However, there are regions with rather strong mismatch in the south-west and in the north. Parts of this mismatch can be attributed to our model biases in precipitation. For example, REMBO simulates too much snow accumulation in the north-east and south-west of Greenland compared to the compilation by Bales et al. (2009).

Figure 5d depicts the simulated difference in GrIS volume between Eemian and present day expressed in units of global sea level equivalent. Compared to the other figure panels, we show here results from simulations with refined melt parameter spacing of 1 W m^{-2}. We enhanced the resolution of the melt parameter sampling, because the region of valid simulations appears rather elongated in the parameter space. The estimated Eemian sea-level contribution increases with increasing (less negative) c_m. This is understandable, because surface melting increases with a higher (less negative) melt parameter. Nevertheless, there is also an increase of the GrIS contribution to the Eemian sea-level highstand for increasing discharge parameter values. Obviously, there is an interplay between ice discharge and surface melt, because the ice

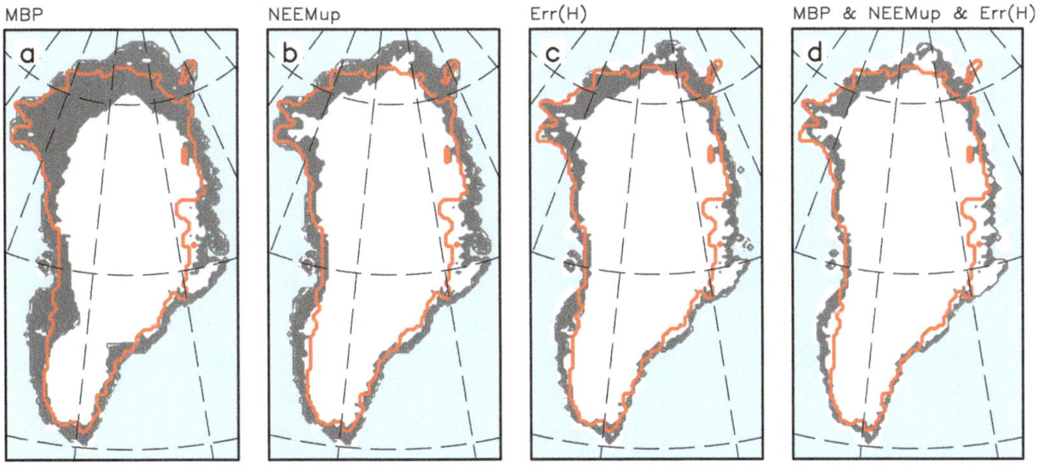

Figure 6. Simulated geographical position of present-day ice margins for simulations with the discharge parameterization (gray areas) compared to observations (red line). The gray shaded areas cover the range of simulated ice margins determined by different constraints. **(a)**–**(c)** Single constraint applies: **(a)** mass balance partition, **(b)** elevation reduction during the Eemian referenced to present day at the upstream position of NEEM, and **(c)** error in ice thickness. **(d)** All three constraints apply.

discharge removes ice from the ice sheet and brings the ice surface into lower regions of the atmosphere, where stronger surface melt can occur. Averaged over the parameter space of valid simulations, we have a contribution of the GrIS to Eemian sea-level rise (above present-day value) of 1.4 m. The minimum contribution of the GrIS sea level rise among all valid simulations is 0.6 m and the corresponding maximum is 2.5 m.

8 Eemian versus present day and GrIS stability

Figure 7 shows the simulated present-day and Eemian ice sheet distributions from model versions with high, medium and low sea-level rise contributions of the Eemian compared to present day. While all fields look rather similar for the present day, there is a considerable difference between the corresponding Eemian fields. However, the present-day surface elevations for the different valid parameters sets still show slight differences. Naturally, these differences appear mainly near the ice margin, while the interior of the ice sheet remains almost unchanged for any valid parameter set. As is often the case in such optimization problems, there is a trade-off concerning agreement with observations in certain regions (see Fig. 9a for the observed surface elevation). While the simulation with $c_d = 0.8$, $c_m = -53\ \mathrm{W\,m^{-2}}$ (Fig. 7a) better resembles the ice-free south-western region, the northern region around Petermann Glacier matches the observation less well. This situation is opposite for the simulation with $c_d = 1.2$, $c_m = -66\ \mathrm{W\,m^{-2}}$ (Fig. 7c).

For all valid parameter sets, our simulated reduction in Eemian ice volume is accompanied by a strong retreat of ice in Greenland in particular in its northern part, see Fig. 7d–f, which spans the simulated lowest and highest Eemian to present-day GrIS contribution to sea level rise. For model versions with high sensitivity to climate forcing, the GrIS splits into two parts: a small ice cap in southern Greenland and a larger ice sheet in central Greenland (Fig. 7d). For the intermediate and low sensitivity model versions, the GrIS remains in one piece (Fig. 7e and f). In all valid model versions, there is a strong retreat of ice, mainly in western and northern Greenland. Our estimates showing a strong retreat of the GrIS during the Eemian rather correspond to the simulations by Otto-Bliesner et al. (2006), while the medium to modest retreat of the Eemian GrIS was found in simulations by Helsen et al. (2013) and Quiquet et al. (2013).

Interestingly, the NEEM location almost becomes ice free at 121 kyr BP in our most sensitive model version (see Fig. 7d). Nonetheless, an ice-free NEEM position during the Eemian would not contradict the existence of Eemian ice and most probably pre-Eemian ice in the NEEM ice core at present day, as reported by NEEM community members (2013), since the Eemian ice was accumulated farther upstream of NEEM. Similar argumentation would hold for Camp Century as well.

Figure 8 shows time series of ice volume and the NEEMup surface elevation for simulations over the last two glacial cycles from previous work with the same model (Robinson et al., 2011) and our present approach, which includes the sub-grid scale discharge parameterization. At all times, the valid model versions with the discharge parameterization simulate less ice volume than that without the discharge parameterization (Fig. 8a). This has two reasons: (i) previously, the model was not tuned for agreement with present-day surface elevation (ice volume). The present-day surface mass balance partition was (and is here) regarded as the more adequate characteristic to capture the sensitivity of the GrIS to long-term climate change. (ii) In our present approach,

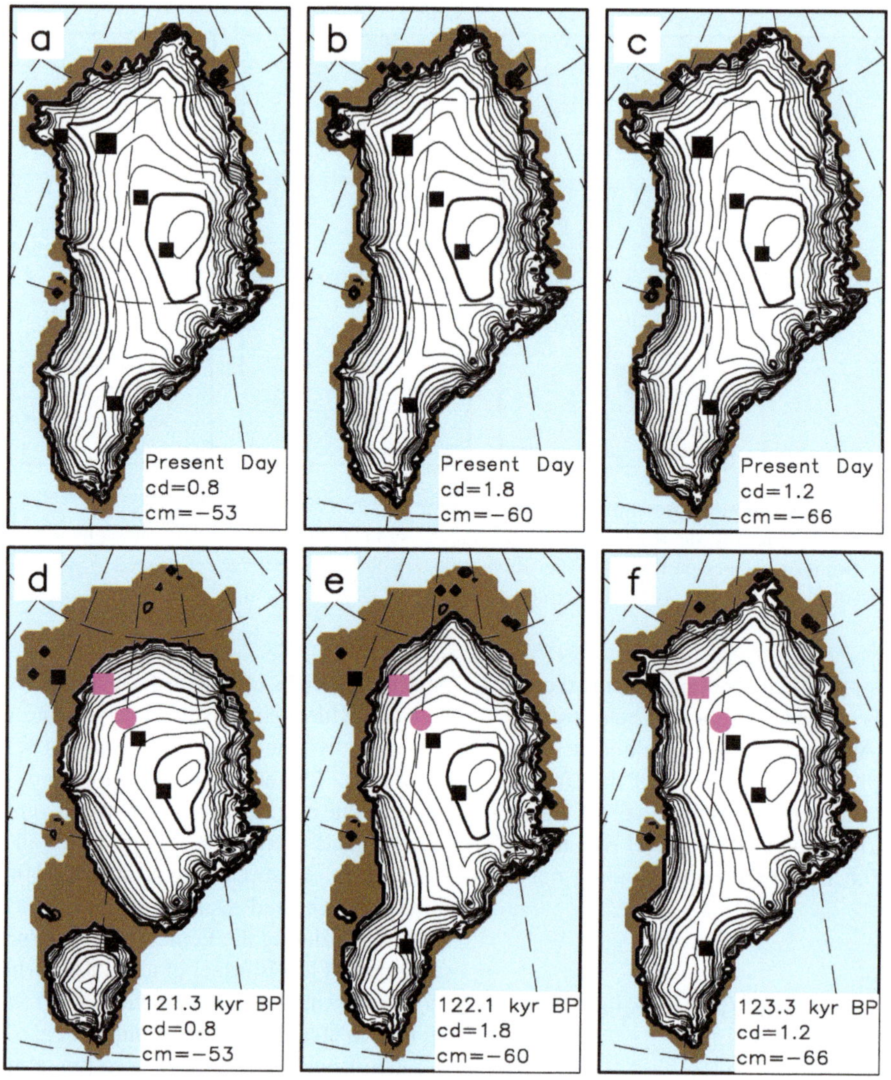

Figure 7. Present-day (**a–c**) and Eemian (**d–f**) surface topography for varying melt and discharge parameters. Simulations correspond to those giving high medium and low contributions to Eemian sea-level rise (2.5, 1.5 and 0.6 m), respectively. The Eemian snapshots correspond to times with the simulated minimum ice volume during the Eemian for the respective simulation. NEEM locations are marked in magenta (square for borehole and circle for upstream).

the inclusion of the discharge parameterization enables our rather coarse resolution model to mimic the calving of the small-scale outlet glaciers (i.e. removal in ice into the ocean by ice discharge) without an overestimation of contact regions of the ice sheet with the ocean, which leads to a smaller ice sheet.

Additionally, two extreme and unrealistic simulations, depicted by the red lines, were set up in order to demonstrate, what happens when a shape-only tuning applies in a coarse-resolution model that disregards fast sub-grid processes of small outlet glaciers. Technically, we restrict the parameter space by setting $c_d = 0$ (discharge parameterization off) and minimize the error measure err(H) and, alternatively, the weaker error measure err(V) to get the right present-day shape. The former belongs to the parameter setting $c_d = 0$, $c_m = -42\,\mathrm{W\,m^{-2}}$ and the latter to $c_d = 0$, $c_m = -40\,\mathrm{W\,m^{-2}}$. Please, note that these melt parameter values are outside the valid range of MBP as determined by Robinson et al. (2011) using observed present-day topography as well as outside the valid c_m values in MBP space of this work (Fig. 5b) using simulated present-day topography to determine MBP. Because we consider the present-day ice-sheet shape as the only constraint (for demonstration), the model without the discharge parameterization ($c_d = 0$) requires a rather high value of the melt parameter c_m to minimized shape errors (Fig. 4). As one can see in Fig. 8a, around present-day the red line corresponding to minimal err(H) is very close to the upper value of the range of the simulations with sub-grid discharge parameterization (blue shading), while the other red line (minimal err(V)) even merges with that valid range. Sim-

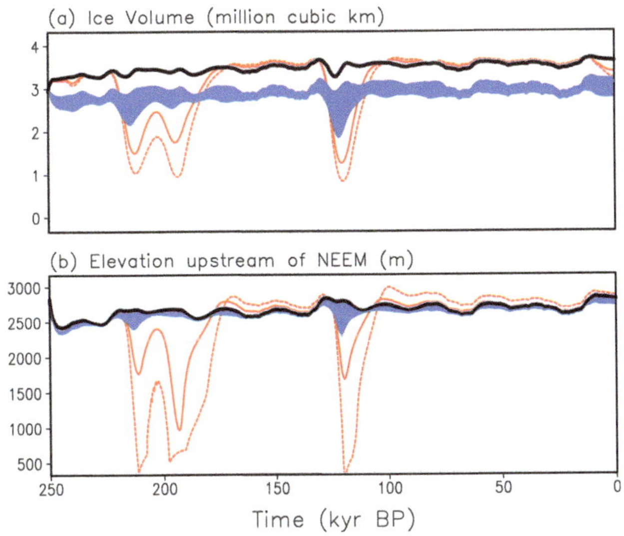

Figure 8. Time series of the simulated Greenland ice sheet evolution during the last two glacial cycles. Blue shading represents the range of valid model versions including our discharge parameterization. Black and red lines show simulations without the discharge parameterization ($c_d = 0$). Solid black lines indicate the central run of a set of optimized simulations by Robinson et al. (2011). The red lines are from simulations found via shape-only tuning of the melt parameter (see main text for explanation). In particular, a simulation with $c_m = -42\,\mathrm{W\,m^{-2}}$, found by minimizing err(H) (solid red line) and, alternatively, with $c_m = -40\,\mathrm{W\,m^{-2}}$, determined by minimizing err(V) (dashed red line). **(a)** Ice volume of the Greenland ice sheet. **(b)** Surface elevation at the NEEMup location.

ulated present-day elevations at NEEMup lie rather close to each other in different model versions. However, during the Eemian interglacial, the runs from the shape-only constraints show strong downward excursions for ice volume as well as for the NEEMup elevation (Fig. 8b). Whether such a small Eemian ice volume is still realistic might be disputable but in any case the simulated Eemian reduction in NEEMup elevation is by far larger than that estimated from the ice core (NEEM community members, 2013). The NEEMup position was even ice free during the Eemian in the simulation without the ice discharge parameterization and minimized err(V), which certainly contradicts observational data.

Moreover, the strong drop in Eemian sea-level and NEEMup elevation hints at very different stability properties of the model version without ice discharge parameterization and shape-only tuning compared to all our valid model versions which contain the sub-grid scale discharge parameterization. Even more, the models with shape-only tuning are much less stable with respect to applied positive temperature anomalies than all the model versions that are constrained using the MBP and palaeo data, whether they include discharge parameterization or not. In other words, the models with shape-only tuning of the melt parameter are less stable than both the valid model versions of our former approach without

Table 1. Temperature threshold of the stability of the Greenland ice sheet for a number of valid model parameters.

c_d	c_m [$\mathrm{W\,m^{-2}}$]	Decay threshold [°C]
0.8	−53	1.25
1.8	−60	1.75
1.2	−66	2.5

the discharge parameterization (Robinson et al., 2011, 2012) and our present ones with the discharge parameterization.

To achieve more detailed information about the stability of the GrIS, we performed an analysis based on many steady state runs as in Calov and Ganopolski (2005), but in temperature space instead of insolation space. Namely, we performed a suite of steady state simulations each 300 kyr long imposing different spatially uniform and temporarily constant surface air temperature anomalies to the lateral boundary conditions of the REMBO model for each single simulation. We use the simulated present-day GrIS as an initial condition. We sample with a temperature increment of 0.25 °C, i.e. we add a $\Delta T = 0.25\,°\mathrm{C}, 0.5\,°\mathrm{C}, \ldots, 2.75\,°\mathrm{C}$ to the present-day temperature at the lateral boundaries of the REMBO model domain and run the model for each individual ΔT into a steady state. If the GrIS decays for a certain ΔT but covers most of the Greenland land surface (see Robinson et al., 2012) for $\Delta T - 0.25\,°\mathrm{C}$, we define ΔT as the temperature threshold of GrIS decay.

We applied the procedure to three representative valid simulations with the discharge parameterization. From these simulations, we obtain thresholds of decay of the GrIS between 1.25 and 2.5 °C, depending on values of model parameters (see also Table 1). We also applied our stability analysis with the steady state method to a model version without the discharge parameterization using the shape-only constraint. The threshold estimated with this shape-only setting with err(V) minimisation ($c_d = 0$, $c_m = -40\,\mathrm{W\,m^{-2}}$) is much lower – only 0.25 °C. The higher values for GrIS decay between 1.25 and 2.5 °C from our new REMBO-SICOPOLIS version with discharge parameterization are in the range of those that we found previously without using the discharge parameterization (Robinson et al., 2012). This similarity clearly is an implication of the use of the MBP as one common constraint in both approaches. In this work, the shape and palaeo constraints are important as well, because they cover different regions in parameters space as discussed in Sect. 7. It should be noted that a complete uncertainty analysis with our new model version is planned in the future, which will likely widen the range of our estimates.

In summary, if we optimize the melt parameter in the coarse resolution model without the sub-grid scale ice discharge parameterization for only err(H) or err(V), the re-

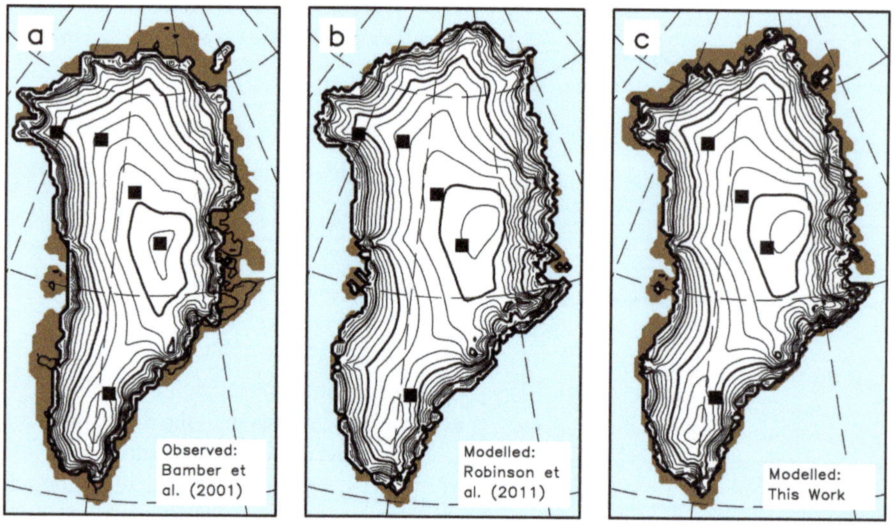

Figure 9. Present-day Greenland ice sheet topography. **(a)** Observed data compilation by Bamber et al. (2001). **(b)** Simulation after Robinson et al. (2011). **(c)** Simulation from present work, given the parameter combination with smallest error in err(H), i.e. $c_{\mathrm{d}} = 1.4$, $c_{\mathrm{m}} = -60\,\mathrm{W\,m^{-2}}$.

sulting models all overestimate surface melt and violate the MBP criterion. This leads to a strong drop of the NEEM elevation far below the one reconstructed from palaeo data, and the Greenland ice sheet becomes too sensitive to temperature anomalies. This is why Robinson et al. (2011) and Robinson et al. (2012) used the MBP criterion together with a palaeo constraint for calibration of their coarse resolution model, ensuring the correct long-term stability properties reported in this work. In our improved model with sub-grid scale discharge parameterization, we found a stability behaviour similar to that found by Robinson et al. (2012) when using the MBP and NEEMup constraints, but now – still in the coarse resolution ice-sheet model – we can additionally fulfil a strong present-day shape constraint and achieve err(H) $< 20\,\%$. We expect that all of our constraints would play a similar role in a model of the Greenland glacial system, which explicitly describes small-scale fast processes. Development of such a model is in our future plans.

One major advantage of our simple parameterization is that it applies easily for climates far away from present day – a fully explicit modelling of present-day outlet glaciers could fail for the Eemian, because many present-day outlet glaciers just vanish in the Eemian. Figure 3a and b compares simulated divergence of fast flow during present-day and the Eemian.

As stated in Sect. 3, our discharge parameterization is not intended to resolve every small individual outlet glacier; it is rather designed to capture their draining effect on the spatial and temporal average in a sub-grid scale statistical approach. Consequently, the simulated ice discharge of a certain region is the integral of the divergence of fast flow (Eq. 3) over this region. In general, our simulated present-day ice discharge (Fig. 3a) is high over regions with many observed outlet glaciers and low over regions with fewer observed outlet glaciers, see Fig. 3 in Moon and Joughin (2008) for comparison. However, in the south-western part of Greenland, the model overestimates the (sectoral/regional) ice discharge, in part due to too high simulated accumulation over that region (see Sect. 9).

Furthermore, we demonstrate that the regions of fast flow can be reduced drastically for the Eemian time period compared to the present-day state. For the Eemian, there is practically no ice discharge from regions far away from the coast. In particular, the land bridge between the large ice sheet in the north and the smaller ice cap in the south of Greenland shows diminishing ice discharge. In general, our model results suggest that during the Eemian more ice calves into the ocean from the eastern coast of Greenland than from its western coast. In particular, the Kangerlussuaq Glacier region delivers ice into the ocean during the Eemian in all our valid model versions.

9 Comparison with present-day observations and findings by others

A direct comparison of our simulated Greenland surface elevation with the observed elevation by Bamber et al. (2001) and the former approach of Robinson et al. (2011) is shown in Fig. 9. Overall, we improved agreement with observations significantly. In particular, in the simulation with the discharge parameterization several regions are now ice-free, which look very similar to reality. The remaining deficiencies are partly due to the simple discharge parameterization and the limitations in the REMBO climatology, e.g. biases in representation of precipitation as discussed earlier. In this

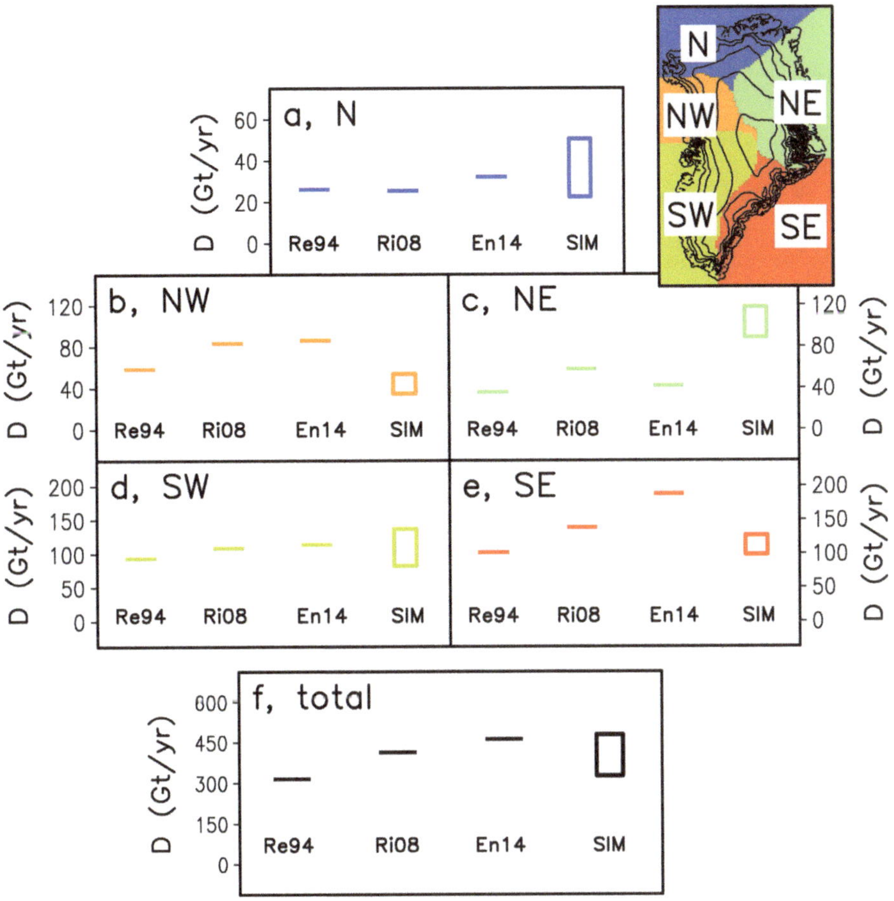

Figure 10. Simulated ice discharge (open bars) versus observations and findings by others (horizontal lines) at present day (i.e. pre-industrial conditions). The heights of the open rectangles indicate the range of our simulated discharge. Acronyms are as follows: Re94: Reeh (1994), Ri08: Rignot et al. (2008) and En14: Enderlin et al. (2014). Further on, SIM indicates our simulations. **(a)–(e)** Sectoral Greenland ice discharge in Gt yr^{-1} for the northern (N), north-western (NW), north-eastern (NE), south-western (SW), and south-eastern (SE) parts (colours of the sectors are indicated in the inset). **(f)** Total ice discharge in Gt yr^{-1}. Note that the y axes have different scales.

context, we would like to stress, that our model is a fully interactive one, where no observational data over Greenland are prescribed. This ensures that REMBO is applicable to climates far away from the present state, what is vital because the Eemian climate can deliver additional constraints for the model.

Figure 10 compares our simulated present-day sectoral and total ice discharge with findings by others. The sectors (see inset in the upper right of Fig. 10) correspond with those of Reeh (1994) and sub-divide the GrIS into northern (sector N), north-western (sector NW), north-eastern (sector NE), south-western (sector SW), and south-eastern (sector SE) parts. This subdivision is also adequate to the degree of complexity (Claussen et al., 2002) of our model in its current stage of development (a refinement of the sectors is planned for our later work).

Over the sectors N and SW, our simulated range of ice discharge compares well with previous estimates. While our simulated ice discharge range is somewhat low over sector NW, it is certainly too high over sector NE. The latter can be explained by the overestimation of our simulated present-day accumulation over sector NE by some 10 Gt yr^{-1} compared to the compilation by Bales et al. (2009). In sector SE, our results are consistent with Reeh (1994) and Rignot et al. (2008) but are significantly lower than those by Enderlin et al. (2014). The reason for the mismatch of our simulated ice discharge over sector SE with that by Enderlin et al. (2014) could be because Enderlin's data, from the year 2000, already reflect the response of the outlet glaciers to warming in the fjords due to subtropic water transported by the ocean towards Greenland (Straneo et al., 2012). The SE sector may be particularly sensitive to this change, because in this region several outlet glaciers have contact with the ocean via fjords.

Our range of valid model versions is 326–479 Gt yr^{-1} simulated total ice discharge. The lower estimate matches that by Reeh (1994), while the upper end of this range is somewhat larger than the ice discharge by Enderlin et al. (2014). The relative small total ice discharge by Reeh (1994) corresponds to the rather small accumulation estimate of the Ohmura and Reeh (1991) compilation, which Reeh used together with the assumption that the GrIS is in equilibrium to derive his discharge values. The data by Reeh (1994) can be regarded as roughly similar to pre-industrial, because it is based on accumulation, which contains several old data points, certainly before the 1990s. Meanwhile, the data by Enderlin et al. (2014) belongs to a more recent time, which corresponds to the upper value of the time interval of the reanalysis data used to prescribed the lateral boundary conditions of REMBO (1958–2001). Because by the year 2000, Greenland already started to respond to global warming, it is reasonable that the upper end of our simulated ice discharge range approximately agrees with the finding by Enderlin et al. (2014).

10 Discussion

In spite of significant improvements in the simulated GrIS topography with our discharge parameterization, for all of our simulations it was impossible to yield an error in ice thickness smaller than about 18 %. These rather large errors partly underline the limits of our ice discharge parameterization and modelling approach in general. The errors can be reduced by incorporating additional parameters, in particular such parameters which affect the interior of the ice sheet, like the basal sliding parameter (e.g. Price et al., 2011; Larour et al., 2012a). In this paper, we restricted our analysis to the parameters for surface melt and ice discharge, which rather affect the marginal regions of the ice sheet. Nonetheless, we use the error metric of the entire ice sheet here in order to keep our approach simple, and because we intend to use the metric in forthcoming work, which will include a more comprehensive approach.

We designed this parameterization as a workaround until a more comprehensive whole-Greenland glacial system model becomes available. Of course, additional improvements are possible, like introducing physically based models for individual outlet glaciers and fjords. Nevertheless, note that the relative high error in ice thickness (up to 20 %) also results from the fact that this is a stronger measure of the error in ice-sheet shape than the error in total ice area or in ice volume.

Although our model enables us to reduce the cumulative error in ice thickness from about 30 to 18 %, there is still room for further improvement. For example, higher-order models can play an important role in capturing the GrIS dynamics (Larour et al., 2012b), as inclusion of membrane stress can reduce ice volume change in short-term climate projections by up to 20 % (Fürst et al., 2013). In particular, the role of stress transfer might become important in short-term climate projections (Nick et al., 2009). Inclusion of membrane stress (e.g. Bueler and Brown, 2009) in our modelling approach together with better resolved topography (Morlighem et al., 2014) is planned for the future. In addition, a comparison of our discharge parameterization with the high-resolution and high-order models will help to improve the parameterization, because it will still be useful for large-ensemble simulations.

The term d in the thickness evolution equation (Eq. 2) represents the divergence of the ice flow associated with non-resolved ice streams/outlet glaciers, see also Appendix A. In this sense, d reduces the ice thickness over areas with fast flow. This is why we only consider total and sectoral ice discharge, which is equivalent to the spatial integral of d over areas near the ice margin where fast flow appears. Of course, our parameterization cannot explicitly simulate ice calving and we remove some mass over inland regions where in reality no calving appears. The distance parameter accounts for this, because it appears in the denominator of our parameterization and, therefore, our method removes the highest amount of mass near the ocean. One advantages of our method is that it enables us to distinguish between the mass that is removed by surface melt and by calving.

We restricted our simulations to the spatial horizontal resolution of 20 km and have not inspected a possible dependence of our ice discharge parameterization on resolution. We cannot rule out that a recalibration of the parameters will be necessary for a different resolution. For a better understanding, we plan to investigate the potential of resolution dependency in the future. At the same time, we regard such a parameterization as an important tool for use in a fixed resolution model.

While the model agrees reasonably well with observations overall, there are some significant biases in simulated ice discharge at the regional scale. For example, we have too much ice discharge in the north-eastern and too little in the north-western sector. The disagreements can be partly attributed to regional biases of simulated precipitation by REMBO and to difficulties in interpretation of the data used for comparison. When designing our constraints, we took the reduction in Eemian surface elevation upstream of the NEEM ice core from the NEEM community members (2013). In their statistics, the NEEM community members gave a one σ error for this value. In principle, one could have included the more uncertain values too by using the two σ range. Nonetheless, all of our simulations with valid parameter sets show a strong retreat of the ice in northern Greenland during Eemian times. Such a retreat strongly influences the local climate and might lead to an additional Eemian temperature rise over that region, although it is unlikely to be as strong as that reported by the NEEM community members (2013). These and other uncertainties in Eemian temperature and precipitation will be examined in future work.

11 Conclusions

We introduced a new sub-grid scale ice discharge parameterization aimed at mimicking Greenland's fast outlet glaciers in a coarse resolution ice-sheet model. Our simulated ice discharge compares reasonably well with observations and other model estimates. The ice discharge parameterization enables us to simulate an ice sheet, whose shape is in good agreement with observations and whose partition between total ice discharge and total surface melt is in good agreement with state-of-the-art regional climate models.

We used various constraints to reduce the range of valid melt and discharge parameters of the REMBO-SICOPOLIS model: a shape constraint, a constraint on the mass balance partition (Robinson et al., 2011), and a palaeo constraint on Greenland's surface elevation drop (upstream of the NEEM borehole) during the Eemian interglacial compared to present. We favoured a measure of ice thickness error at each grid point instead of just considering total Greenland area or volume, since it is a stronger measure of the quality of simulated ice-sheet shape.

The NEEM constraint proved to be a complementary constraint to the other two present-day constraints. It was the strongest constraint in controlling the upper end of the range of valid melt parameter values and thereby the highest Greenland's contribution to Eemian sea-level rise. Taken individually, this constraint was also comparable to the shape constraint in determining the range of simulated present-day GrIS margins. This demonstrates the importance of palaeoclimate information for determining the range of model parameters applicable for future prediction of the contribution of the GrIS to sea level.

We can satisfy all constraints if our sub-grid scale ice discharge parameterization is included in a coarse resolution ice sheet model in order to mimic small-scale fast processes. When using a shape-only constraints in a coarse resolution model without the parameterization of fast processes, we obtained a very unstable ice sheet – i.e. a regional temperature rise of as low as 0.25 °C was sufficient to melt the GrIS almost completely on longer time scales. When applying the MBP constraint in a coarse resolution model without the sub-grid scale ice discharge parameterization, the model has about the same stability properties as with the discharge parameterization.

The inclusion of our ice discharge parameterization along with the above-described constraints leads to similar results concerning long-term stability as Robinson et al. (2012), with a decay threshold between 1.25 and 2.5 °C. Note that although this range is consistent with previous work (Robinson et al., 2012), it does not result from an exhaustive uncertainty analysis. An updated range comparable with Robinson et al. (2012) will be the provided in future work. Finally, complying with all three constraints leads to a GrIS contribution to sea level rise during the Eemian compared to present day in the range of 0.6–2.5 m, with an average of 1.4 m. Again, this range could widen if further uncertainties were included.

Appendix A: The ice discharge parameterization as divergence of fast flow

We stated that our ice discharge parameterization locally represents the divergence of fast flow. Here we cannot give a complete proof of that, however we can present a plausibility argument.

We start with rewriting Eq. (3)

$$\mathrm{d} \propto \frac{h}{l^3}. \tag{A1}$$

If we assume that the basal topography has a small spatial gradient, then the surface gradient equals approximately the gradient of ice thickness. Furthermore, let us rename l, the distance from a point in the ice sheet to the nearest ocean grid: $l = \Delta x$. Because the ice thickness at an ocean grid point is zero in our coarse resolution model, we can write $\Delta h = h$. With that we can very crudely define the surface gradient $\Delta h / \Delta x$ at the ice margin. We can then write

$$\begin{aligned}\frac{h}{l^3} &= \frac{h}{\Delta x^3} = \left(\frac{\Delta h}{\Delta h}\right)^2 \frac{h}{\Delta x^3} = \frac{1}{\Delta x}\frac{h}{\Delta h^2}\left(\frac{\Delta h}{\Delta x}\right)^2 \\ &\approx \frac{1}{\Delta x}\frac{h}{h^2}\left(\frac{\Delta h}{\Delta x}\right)^2 = \frac{1}{\Delta x} h \frac{\left(\frac{\Delta h}{\Delta x}\right)^2}{h^2}.\end{aligned} \tag{A2}$$

Making Δx very small and using Eq. (A1) yields

$$\mathrm{d} \propto \frac{1}{\mathrm{d}x}\left(h \frac{\left(\frac{\mathrm{d}h}{\mathrm{d}x}\right)^2}{h^2}\right). \tag{A3}$$

The fraction in the latter equation is a Weertman-type sliding law

$$u_\mathrm{b}(x) \propto \frac{\left(\frac{\mathrm{d}h}{\mathrm{d}x}\right)^2}{h^2}. \tag{A4}$$

As sliding contributes by far the most to the velocity of ice streams, we can thus generalize to two dimensions

$$\mathrm{d} = \nabla \cdot \boldsymbol{q}_\mathrm{fast}, \tag{A5}$$

where $\boldsymbol{q}_\mathrm{fast} = h\,\boldsymbol{u}_\mathrm{b}$. This we interpret as the divergence of the fast flow of outlet glaciers. Of course, this only corresponds to our rather simple approach. Nonetheless, we believe that the similarity of our parameterization to the divergence of a Weertman-type sliding velocity substantiates our approach.

Acknowledgements. We would like to thank Ellyn Enderlin for providing us with ice discharge data, as well as Roger Bales and Qinghua Guo, who provided accumulation and precipitation data. We are grateful to Fuyuki Saito and two anonymous reviewers for their constructive comments.

Edited by: E. Larour

References

Applegate, P. J., Kirchner, N., Stone, E. J., Keller, K., and Greve, R.: An assessment of key model parametric uncertainties in projections of Greenland Ice Sheet behavior, The Cryosphere, 6, 589–606, doi:10.5194/tc-6-589-2012, 2012.

Bales, R. C., Guo, Q., Shen, D., McConnell, J. R., Du, G., Burkhart, J. F., Spikes, V. B., Hanna, E., and Cappelen, J.: Annual accumulation for Greenland updated using ice core data developed during 2000–2006 and analysis of daily coastal meteorological data, J. Geophys. Res., 114, D06116, doi:10.1029/2008JD011208, 2009.

Bamber, J. L., Layberry, R. L., and Gogenini, S. P.: A new ice thickness and bed data set for the Greenland ice sheet, 1. Measurement, data reduction, and errors, J. Geophys. Res., 106, 33773–33780, 2001.

Born, A. and Nisancioglu, K. H.: Melting of Northern Greenland during the last interglaciation, The Cryosphere, 6, 1239–1250, doi:10.5194/tc-6-1239-2012, 2012.

Bueler, E. and Brown, J.: Shallow shelf approximation as a "sliding law" in a thermomechanically coupled ice sheet model, J. Geophys. Res., 114, F03008, doi:10.1029/2008JF001179, 2009.

Calov, R. and Ganopolski, A.: Multistability and hysteresis in the climate-cryosphere system under orbital forcing, Geophys. Res. Lett., 32, L21717, doi:10.1029/2005GL024518, 2005.

Calov, R., Ganopolski, A., Claussen, M., Petoukhov, V., and Greve, R.: Transient simulation of the last glacial inception, Part I: glacial inception as a bifurcation of the climate system, Clim. Dynam., 24, 545–561, doi:10.1007/s00382-005-0007-6, 2005.

Claussen, M., Mysak, L. A., Weaver, A. J., Crucifix, M., Fichefet, T., Loutre, M.-F., Weber, S. L., Alcamo, J., Alexeev, V. A., Berger, A., Calov, R., Ganopolski, A., Goosse, H., Lohman, G., Lunkeit, F., Mokhov, I. I., Petoukhov, V., Stone, P., and Wang, Z.: Earth system models of intermediate complexity: Closing the gap in the spectrum of climate system models, Clim. Dynam., 18, 579–586, doi:10.1007/s00382-001-0200-1, 2002.

Enderlin, E. M., Howat, I. M., Jeong, S., Noh, M.-J., van Angelen, J. H., and van den Broeke, M. R.: An improved mass budget for the Greenland ice sheet, Geophys. Res. Lett., 41, 866–872, doi:10.1002/2013GL059010, 2014.

Fürst, J. J., Goelzer, H., and Huybrechts, P.: Effect of higher-order stress gradients on the centennial mass evolution of the Greenland ice sheet, The Cryosphere, 7, 183–199, doi:10.5194/tc-7-183-2013, 2013.

Ganopolski, A. and Calov, R.: The role of orbital forcing, carbon dioxide and regolith in 100 kyr glacial cycles, Clim. Past, 7, 1415–1425, doi:10.5194/cp-7-1415-2011, 2011.

Ganopolski, A., Calov, R., and Claussen, M.: Simulation of the last glacial cycle with a coupled climate ice-sheet model of intermediate complexity, Clim. Past, 6, 229–244, doi:10.5194/cp-6-229-2010, 2010.

Goelzer, H., Huybrechts, P., Loutre, M. F., Goosse, H., Fichefet, T., and Mouchet, A.: Impact of Greenland and Antarctic ice sheet interactions on climate sensitivity, Clim. Dynam., 37, 1005–1018, doi:10.1007/s00382-010-0885-0, 2011.

Goelzer, H., Huybrechts, P., Fürst, J. J., Nick, F. M., Andersen, M. L., Edwards, T. L., Fettweis, X., Payne, A. J., and Shannon, S.: Sensitivity of Greenland ice sheet projections to model formulations, J. Glaciol., 59, 733–549, doi:10.3189/2013JoG12J182, 2013.

Graversen, R. G., Drijfhout, S., Hazeleger, W., van de Wal, R., Bintanja, R., and Helsen, M.: Greenland's contribution to global sea-level rise by the end of the 21st century, Clim. Dynam., 37, 1427–1442, doi:10.1007/s00382-010-0918-8, 2011.

Greve, R.: A continuum-mechanical formulation for shallow polythermal ice sheets, Philos. T. Roy. Soc. Lond. A, 355, 921–974, doi:10.1098/rsta.1997.0050, 1997.

Greve, R.: On the response of the Greenland ice sheet to greenhouse climate change, Climatic Change, 46, 289–303, 2000.

Greve, R., Saito, F., and Abe-Ouchi, A.: Initial results of the SeaRISE numerical experiments with the models SICOPOLIS and IcIES for the Greenland ice sheet., Ann. Glaciol., 52, 23–30, 2011.

Hanna, E., Navarro, F. J., Pattyn, F., Domingues, C. M., Fettweis, X., Ivins, E. R., Nicholls, R. J., Ritz, C., Smith, B., Tulaczyk., S., Whitehouse, P. L., and Zwally, H. J.: Ice-sheet mass balance and climate change, Nature, 498, 51–59, doi:10.1038/nature12238, 2013.

Helsen, M. M., van de Berg, W. J., van de Wal, R. S. W., van den Broeke, M. R., and Oerlemans, J.: Coupled regional climate–ice-sheet simulation shows limited Greenland ice loss during the Eemian, Clim. Past, 9, 1773–1788, doi:10.5194/cp-9-1773-2013, 2013.

Hutter, K.: Theoretical Glaciology; Material Science of Ice and the Mechanics of Glaciers and Ice Sheets, D. Reidel Publishing Company, Dordrecht, the Netherlands, 1983.

Huybrechts, P. and de Wolde, J.: The dynamic response of the Antarctic and Greenland ice sheets to multiple-century climatic warming, J. Climate, 12, 2169–2188, doi:10.1175/1520-0442(1999)012<2169:TDROTG>2.0.CO;2, 1999.

Huybrechts, P., Letréguilly, A., and Reeh, N.: The Greenland ice sheet and greenhouse warming, Global Planet. Change, 3, 399–412, 1991.

Joughin, I., Smith, B. E., Howat, I. M., Scambos, T., and Moon, T.: Greenland flow variability from ice-sheet-wide velocity mapping, J. Glaciol., 56, 415–430, 2010.

Larour, E., Schiermeier, J., Rignot, E., Seroussi, H., Morlighem, M., and Paden, J.: Sensitivity analysis of Pine Island Glacier ice flow using ISSM and DAKOTA, J. Geophys. Res., 117, F02009, doi:10.1029/2011JF002146, 2012a.

Larour, E., Seroussi, H., Morlighem, M., and Rignot, E.: Continental scale, high order, high spatial resolution, ice sheet modeling using the Ice Sheet System Model (ISSM), J. Geophys. Res., 117, F01022, doi:10.1029/2011JF002140, 2012b.

Lipscomb, W. H., Fyke, J. G., Vizcaíno, M., Sacks, W. J., Wolfe, J., Vertenstein, M., Craig, A., Kluzek, E., and Lawrence, D. M.:

Implementation and initial evaluation of the Glimmer Community Ice Sheet Model in the Community Earth System Model, J. Climate, 26, 7352–7371, doi:10.1175/JCLI-D-12-00557.1, 2013.

Moon, T. and Joughin, I.: Changes in ice front position on Greenland's outlet glaciers from 1992 to 2007, J. Geophys. Res., 113, F02022, doi:10.1029/2007JF000927, 2008.

Moon, T., Joughin, I., Smith, B., and Howat, I.: 21st-century evolution of Greenland outlet glacier velocities, Science, 336, 576–578, doi:10.1126/science.1219985, 2012.

Morlighem, M., Rignot, E., Mouginot, J., Seroussi, H., and Larour, E.: Deeply incised submarine glacial valleys beneath the Greenland ice sheet, Nat. Geosci, 7, 418–422, doi:10.1038/ngeo2167, 2014.

NEEM community members: Eemian interglacial reconstructed from a Greenland folded ice core, Nature, 493, 489–494, doi:10.1038/nature11789, 2013.

Nick, F. M., Vieli, A., Howat, I. M., and Joughin, I.: Large-scale changes in Greenland outlet glacier dynamics triggered at the terminus, Nat. Geosci., 2, 110–114, doi:10.1038/ngeo394, 2009.

Nowicki, S., Bindschadler, R. A., Abe-Ouchi, A., Aschwanden, A., Bueler, E., Choi, H., Fastook, J., Granzow, G., Greve, R., Gutowski, G., Herzfeld, U., Jackson, C., Johnson, J., Khroulev, C., Larour, E., Levermann, A., Lipscomb, W. H., Martin, M. A., Morlighem, M., Parizek, B. R., Pollard, D., Price, S. F., Ren, D. D., Rignot, E., Saito, F., Sato, T., Seddik, H., Seroussi, H., Takahashi, K., Walker, R., and Wang, W. L.: Insights into spatial sensitivities of ice mass response to environmental change from the SeaRISE ice sheet modeling project II: Greenland, Geophys. Res. Lett., 118, 1025–1044, doi:10.1002/jgrf.20076, 2013.

Ohmura, A. and Reeh, N.: New precipitation and accumulation maps for Greenland, J. Glaciol., 37, 140–148, 1991.

Otto-Bliesner, B. L., Marshall, S. J., Overpeck, J. T., Miller, G. H., and Hu, A.: Simulating Arctic climate warmth and icefield retreat in the last interglaciation, Science, 311, 1751–1753, doi:10.1126/science.1120808, 2006.

Petoukhov, V., Ganopolski, A., Brovkin, V., Claussen, M., Eliseev, A., Kubatzki, C., and Rahmstorf, S.: CLIMBER-2: a climate system model of intermediate complexity, Part I: model description and performance for present climate, Clim. Dynam., 16, 1–17, 2000.

Price, S. F., Payne, A. J., Howat, I., and Smith, B. E.: Committed sea-level rise for the next century from Greenland ice sheet dynamics during the past decade, P. Natl. Acad. Sci. USA, 108, 8978–8983, doi:10.1073/pnas.1017313108, 2011.

Quiquet, A., Ritz, C., Punge, H. J., and Salas y Mélia, D.: Greenland ice sheet contribution to sea level rise during the last interglacial period: a modelling study driven and constrained by ice core data, Clim. Past, 9, 353–366, doi:10.5194/cp-9-353-2013, 2013.

Reeh, N.: Calving from Greenland glaciers: Observations, balance estimates of calving rates, calving laws, in: Report on the workshop on the calving rate of West Greenland glaciers in response to climate change, edited by: Reeh, N., Danish Polar Center, Copenhagen, 85–102, 1994.

Ridley, J. K., Huybrechts, P., Gregory, J. M., and Lowe, J. A.: Elimination of the Greenland ice sheet in a high CO_2 climate, J. Climate, 18, 3409–3427, 2005.

Rignot, E. and Kanagaratnam, P.: Changes in the velocity structure of the Greenland ice sheet, Science, 311, 986–990, doi:10.1126/science.1121381, 2006.

Rignot, E. and Mouginot, J.: Ice flow in Greenland for the International Polar Year 2008–2009, Geophys. Res. Lett., 39, L11501, doi:10.1029/2012GL051634, 2012.

Rignot, E., Box, J. E., Burgess, E., and Hanna, E.: Mass balance of the Greenland ice sheet from 1958 to 2007, Geophys. Res. Lett., 35, L20502, doi:10.1029/2008GL035417, 2008.

Robinson, A., Calov, R., and Ganopolski, A.: An efficient regional energy-moisture balance model for simulation of the Greenland Ice Sheet response to climate change, The Cryosphere, 4, 129–144, doi:10.5194/tc-4-129-2010, 2010.

Robinson, A., Calov, R., and Ganopolski, A.: Greenland ice sheet model parameters constrained using simulations of the Eemian Interglacial, Clim. Past, 7, 381–396, doi:10.5194/cp-7-381-2011, 2011.

Robinson, A., Calov, R., and Ganopolski, A.: Multistability and critical thresholds of the Greenland ice sheet, Nat. Clim. Change, 2, 429–432, doi:10.1038/NCLIMATE1449, 2012.

Seddik, H., Greve, R., Zwinger, T., Gillet-Chaulet, F., and Gagliardini, O.: Simulations of the Greenland ice sheet 100 years into the future with the full Stokes model Elmer/Ice, J. Glaciol., 58, 427–440, doi:10.3189/2012JoG11J177, 2012.

Shepherd, A., Ivins, E. R., Geruo, A., Barletta, V. R., Bentley, M. J., Bettadpur, S., Briggs, K. H., Bromwich, D. H., Forsberg, R., Galin, N., Horwath, M., Jacobs, S., Joughin, I., King, M. A., Lenaerts, J. T. M., Li, J. L., Ligtenberg, S. R. M., Luckman, A., Luthcke, S. B., McMillan, M., Meister, R., Milne, G., Mouginot, J., Muir, A., Nicolas, J. P., Paden, J., Payne, A. J., Pritchard, H., Rignot, E., Rott, H., Sorensen, L. S., Scambos, T. A., Scheuchl, B., Schrama, E. J. O., Smith, B., Sundal, A. V., van Angelen, J. H., van de Berg, W. J., van den Broeke, M. R., Vaughan, D. G., Velicogna, I., Wahr, J., Whitehouse, P. L., Wingham, D. J., Yi, D. H., Young, D., and Zwally, H. J.: A reconciled estimate of ice-sheet mass balance, Science, 338, 1183–1189, doi:10.1126/science.1228102, 2012.

Stone, E. J., Lunt, D. J., Rutt, I. C., and Hanna, E.: Investigating the sensitivity of numerical model simulations of the modern state of the Greenland ice-sheet and its future response to climate change, The Cryosphere, 4, 397–417, doi:10.5194/tc-4-397-2010, 2010.

Stone, E. J., Lunt, D. J., Annan, J. D., and Hargreaves, J. C.: Quantification of the Greenland ice sheet contribution to Last Interglacial sea level rise, Clim. Past, 9, 621–639, doi:10.5194/cp-9-621-2013, 2013.

Straneo, F., Sutherland, D. A., Holland, D., Gladish, C., Hamilton, G. S., Johnson, H. L., Rignot, E., Xu, Y., and Koppes, M.: Characteristics of ocean waters reaching Greenland's glaciers, Ann. Glaciol., 53, 202–210, doi:10.3189/2012AoG60A059, 2012.

Tarasov, L. and Peltier, W. R.: Greenland glacial history, borehole constraints, and Eemian extent, J. Geophys. Res., 108, 2143–2163, doi:10.1029/2001JB001731, 2003.

Uppala, S. M., Kallberg, P. W., Simmons, A. J., Andrae, U., Bechtold, V. D., Fiorino, M., Gibson, J. K., Haseler, J., Hernandez, A., Kelly, G. A., Li, X., Onogi, K., Saarinen, S., Sokka, N., Allan, R. P., Andersson, E., Arpe, K., Balmaseda, M. A., Beljaars, A. C. M., van de Berg, L., Bidlot, J., Bormann, N., Caires, S., Chevallier, F., Dethof, A., Dragosavac, M., Fisher, M., Fuentes, M., Hagemann, S., Holm, E., Hoskins, B. J., Isaksen, L., Janssen,

P. A. E. M., Jenne, R., McNally, A. P., Mahfouf, J. F., Morcrette, J. J., Rayner, N. A., Saunders, R. W., Simon, P., Sterl, A., Trenberth, K. E., Untch, A., Vasiljevic, D., Viterbo, P., and Woollen, J.: The ERA-40 re-analysis, Q. J. Roy. Meteorol. Soc., 131, 2143–2163, doi:10.1256/qj.04.176, 2005.

van de Berg, W. J., van den Broeke, M., Ettema, J., van Meijgaard, E., and Kaspar, F.: Significant contribution of insolation to Eemian melting of the Greenland ice sheet, Nat. Geosci., 4, 679–683, doi:10.1038/ngeo1245, 2011.

van den Berg, J., van de Wal, R. S. W., and Oerlemans, J.: A mass balance model for the Eurasian Ice Sheet for the last 120000 years, Global Planet. Change, 61, 194–208, doi:10.1016/j.gloplacha.2007.08.015, 2008.

van de Wal, R. S. W. and Oerlemans, J.: Modelling the short-term response of the Greenland ice-sheet to global warming, Clim. Dynam., 13, 733–744, 1997.

Vizcaíno, M., Mikolajewicz, U., Jungclaus, J., and Schurgers, G.: Climate modification by future ice sheet changes and consequences for ice sheet mass balance, Clim. Dynam., 34, 301–324, doi:10.1007/s00382-009-0591-y, 2010.

Glacier-surge mechanisms promoted by a hydro-thermodynamic feedback to summer melt

T. Dunse[1], T. Schellenberger[1], J. O. Hagen[1], A. Kääb[1], T. V. Schuler[1], and C. H. Reijmer[2]

[1]Department of Geosciences, University of Oslo, P.O. Box 1047, Blindern, 0316 Oslo, Norway
[2]Institute for Marine and Atmospheric Research Utrecht, Utrecht University, Princetonplein 5, 3584 CC Utrecht, the Netherlands

Correspondence to: T. Dunse (thorben.dunse@geo.uio.no)

Abstract. Mass loss from glaciers and ice sheets currently accounts for two-thirds of the observed global sea-level rise and has accelerated since the 1990s, coincident with strong atmospheric warming in the polar regions. Here we present continuous GPS measurements and satellite synthetic-aperture-radar-based velocity maps from Basin-3, the largest drainage basin of the Austfonna ice cap, Svalbard. Our observations demonstrate strong links between surface-melt and multiannual ice-flow acceleration. We identify a hydro-thermodynamic feedback that successively mobilizes stagnant ice regions, initially frozen to their bed, thereby facilitating fast basal motion over an expanding area. By autumn 2012, successive destabilization of the marine terminus escalated in a surge of Basin-3. The resulting iceberg discharge of $4.2 \pm 1.6\,\mathrm{Gt\,a^{-1}}$ over the period April 2012 to May 2013 triples the calving loss from the entire ice cap. With the seawater displacement by the terminus advance accounted for, the related sea-level rise contribution amounts to $7.2 \pm 2.6\,\mathrm{Gt\,a^{-1}}$. This rate matches the annual ice-mass loss from the entire Svalbard archipelago over the period 2003–2008, highlighting the importance of dynamic mass loss for glacier mass balance and sea-level rise. The active role of surface melt, i.e. external forcing, contrasts with previous views of glacier surges as purely internal dynamic instabilities. Given sustained climatic warming and rising significance of surface melt, we propose a potential impact of the hydro-thermodynamic feedback on the future stability of ice-sheet regions, namely at the presence of a cold-based marginal ice plug that restricts fast drainage of inland ice. The possibility of large-scale dynamic instabilities such as the partial disintegration of ice sheets is acknowledged but not quantified in global projections of sea-level rise.

1 Introduction

Glacier mass loss constitutes the largest contributor to global mean sea-level rise (SLR), followed by ocean thermal expansion (Church et al., 2011). Over the last two decades, both glaciers and ice caps (Gardner et al., 2013), as well as ice sheets, have lost mass at accelerating rates (AMAP, 2011; Shepherd et al., 2012). These ice mass changes coincide with atmospheric warming that causes record summer temperatures and glacier melt in the Arctic (Gardner et al., 2011) and across the Greenland Ice Sheet (Tedesco et al., 2013). In addition to mass loss by melting, glacier dynamics have the potential to significantly amplify glacier response to climate change by altering the ice discharge to the ocean. Acceleration of outlet glaciers of the Greenland Ice Sheet are generally attributed to hydraulic lubrication and marine terminus destabilization by oceanic warming and calving (Nick et al., 2009; Bartholomew et al., 2010; Moon et al., 2014). West Antarctic ice-stream acceleration is generally attributed to reduced buttressing by thinning or loss of ice shelves (Shepherd et al., 2012). Recent observations of sustained ice flow and mass loss from northeastern Greenland and West Antarctica are also reported as evidence of a marine ice-sheet instability (Khan et al., 2014; Mouginot et al., 2014). This phenomenon, also known as tidewater-glacier instability, refers to glacier speed-up due to terminus retreat into deeper water (Meier and Post, 1987; Pfeffer, 2007).

To date, glacier-dynamic feedback processes remain poorly constrained and are therefore not yet incorporated in global projections of future glacier mass loss (IPCC, 2013). SLR projections released with the IPCC Fifth Assessment Report range from 0.26 to 0.82 m over the 21st century (IPCC, 2013). The projections include ice discharge from the ice sheets into the ocean, however, based on current rates and not accounting for a future dynamic response to climate change. The IPCC acknowledges but is not able to quantify the probability of significantly higher SLR, such as that associated with disintegration of marine-based sectors of the Antarctic Ice Sheet (IPCC, 2013).

Variations in glacier flow, and hence ice discharge, encompass wide ranges of timescales and magnitudes, from diurnal velocity fluctuations to century-scale surge-type behaviour, and are mainly attributed to changes in basal drag (Clarke, 1987). Cyclic surge type behaviour is thought to arise from internal instabilities (Meier and Post, 1969), whereas seasonal velocity variations are externally controlled by surface-melt-induced acceleration through basal lubrication (Iken and Bindschadler, 1986; Schoof, 2010; Hewitt, 2013). Basal water pressure, and thus basal drag, is controlled by the rate of water supply to the glacier bed and the capacity of the subglacial drainage system to accommodate increased discharge (Schoof, 2010). Increased meltwater supply may thus accelerate but also slow down ice flow if it facilitates the establishment of a hydraulically efficient drainage system (Schoof, 2010). Several recent studies therefore suggest that hydraulic lubrication alone has a more limited effect on the future net mass balance of the Greenland Ice Sheet (Nick et al., 2013; Shannon et al., 2013) than previously anticipated (Zwally et al., 2002). However, the two recent assessments do not account for changes in the extent of the basal area subjected to hydraulic lubrication or do not quantify changes in calving loss.

The effect of surface meltwater on the thermal structure of glaciers has only recently been considered in modelling studies and was termed cryo-hydrological warming (CHW; Phillips et al., 2010, 2013). CHW includes latent heat released during refreezing of meltwater, as well as direct heat transfer between water and ice. CHW has the potential to change englacial and basal temperatures within years, whereas changes in basal thermal regime by heat conduction alone would require decades to centuries (Phillips et al., 2010). Phillips et al. (2010) regard CHW as a phenomenon specific to the ablation area and suggest that more areas of the Greenland Ice Sheet will be subjected to CHW in the case of sustained climate warming and thus upward migration of the equilibrium-line altitude (ELA). CHW has the potential to enhance ice deformation through the temperature dependency of the ice viscosity (Phillips et al., 2010), as well as basal motion, if it were to increase the extent of the basal area at the pressure-melting point (Phillips et al., 2013). It has previously been suggested that present outlet glaciers may accelerate significantly if meltwater drainage to the bed were to spread inland (Alley et al., 2008; Bartholomew et al., 2011). Phillips et al. (2013) simulated ice dynamics within the wet snow zone of western Greenland. They achieved a best fit with an observed increase in wintertime velocities if both changes in englacial and basal temperatures, and hence ice deformation and basal motion, were considered.

Cyclic surge-type behaviour is an extreme example of variations in glacier flow, and characterized by long quiescent phases (decades to centuries) with slow ice flow followed by short-lived active phases (months to years) with orders-of-magnitude increases in flow velocities (Raymond, 1987; Dowdeswell et al., 1991). Resistance to sliding during the quiescent phase leads to build-up of an upper reservoir (Clarke et al., 1984). During a surge, mass is transferred from the reservoir area to a lower receiving area, often associated with kilometre-scale advances of the terminus and greatly enhanced calving flux in the case of tidewater glaciers (Murray et al., 2003a). Glacier surges are generally regarded as internal instabilities, and not thought to be driven by external factors (Meier and Post, 1969). Nevertheless, the climatic mass balance influences the build-up of the reservoir area and thus surge occurrence and periodicity (Dowdeswell et al., 1995). In addition, external factors may drive feedback processes that lead to amplified dynamic response, pushing the glacier towards an instability threshold.

On Svalbard, glacier surges are thought to be controlled by a thermally controlled soft-bed surge mechanism (Hamilton and Dowdeswell, 1996; Jiskoot et al., 2000; Murray et al., 2000, 2003b), as previously proposed for surges of polythermal glaciers in subpolar environments (Robin, 1955; Clarke, 1976; Clarke et al., 1984). This theory concerns the slow adjustment of the basal thermal regime and driving stresses in response to geometric changes during the quiescent phase, ultimately leading to a thermal switch from cold to temperate basal conditions, permitting for fast basal motion. Basal melting sets in and then subglacial till, if present, is thawed and its shear strength reduced by rising pore-water pressure, destabilizing the glacier and initiating the surge (Clarke et al., 1984; Tulaczyk et al., 2000). Dynamic thinning and reduction in surface slope reduce the driving stress and sliding rates, eventually bringing the surge to a halt (Clarke et al., 1984).

Thermal models of glaciers generally account for heat conduction through the ice, heat transfer by ice advection, and strain heating and frictional heating. The geothermal heat flux and the annual air temperature (plus firn warming) form the lower and upper boundary conditions, respectively. Thermal effects by water flow within glaciers have previously been mentioned and considered important (e.g. Clarke, 1976), but only recently has "cryo-hydrologic warming (CHW)" been put back into focus (Phillips et al., 2010, 2013). Clarke (1976) neglected CHW in thermal models of glacier surges by assuming that water flow within cold ice is limited. However, recent studies suggest meltwater connections from the surface to the bed through cold ice of substantial thickness (e.g. Das et al., 2008; Clason et al., 2014).

Direct observations of ice dynamics during glacier surges are scarce. The evolution of several surges on Svalbard was reconstructed from archived satellite radar data only after a surge was noticed. Snapshot velocities were derived preferably for the winter season, i.e. when surface melt is absent,

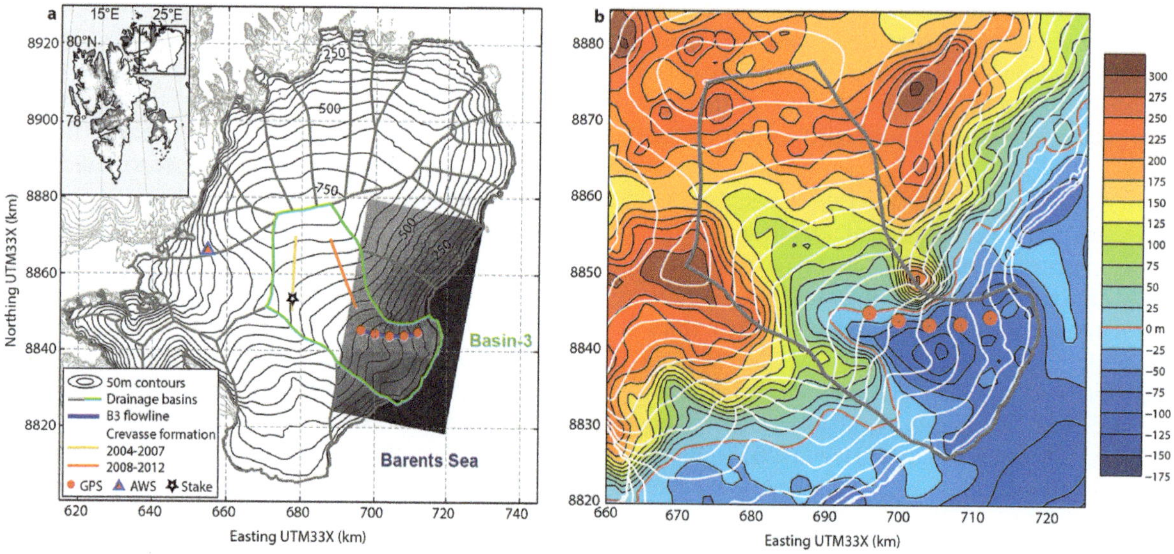

Figure 1. Surface and bedrock topography of Austfonna/Basin-3. **(a)** Surface elevation contours at 50 m interval (solid black), overlain on a TerraSAR-X backscatter image (30 April 2012). The insert provides the ice cap's location within the Svalbard archipelago. Drainage basins are outlined in solid grey; Basin-3 is outlined in solid green. The positions of five GPS receivers and one stake on Basin-3, as well as the automatic weather station (AWS), are marked. Repeat-GPR profiling revealed crevasse formation in upper reaches of Basin-3 between 2004 and 2007 and between 2008 and 2012 (Appendix D). **(b)** Bedrock contours (black, colour-filled) are at 25 m intervals, with the bedrock sea-level contour highlighted in red and 50 m surface elevation contours superimposed (white).

as liquid water deteriorates the quality of the derived motion maps using satellite interferometric radar (Luckman et al., 2002; Murray et al., 2003b, a). Changes in ice dynamics specifically during the summer melt season have therefore not been captured. Based on the available motion snapshots, Murray et al. (2003b) proposed that Svalbard glacier surges are characterized by gradual acceleration over several years, at a higher rate during the months towards the peak of the surge, and followed by multiannual, gradual slow-down.

Here, we present a 5-year record of continuous GPS measurements (May 2008 to May 2013) and satellite synthetic-aperture-radar-based velocity maps (since April 2012) from Basin-3, a marine-terminating drainage basin of the Austfonna ice cap on Svalbard. Our observations demonstrate strong links between surface-melt and ice-flow acceleration, culminating by October 2012 in a surge and drastically enhanced ice discharge. To date (January 2015), the surge continues. We propose a hydro-thermodynamic feedback mechanism to surface melt, subsequently mobilizing the ice within the reservoir area and weakening cold-based marginal ice that restricts inland ice from draining into the ocean. Finally, we discuss possible implications of the proposed mechanism for the stability of marine ice-sheet regions if exposed to significant surface melt in a warming climate.

2 Austfonna, Basin-3

At $\sim 7800\,\mathrm{km}^2$, Austfonna is the largest ice cap in the Eurasian Arctic (Fig. 1a). The ice cap consists of a main dome with an ice thickness of up to 600 m that feeds a number of drainage basins, some of which are known to have surged in the past (Dowdeswell, 1986). Austfonna has a polythermal structure. In the ice cap's interior, basal temperatures are likely at the pressure-melting point (Zagorodnov et al., 1989), while the thinner ice-cap margins are cold and frozen to the bed, except for distinct fast-flowing outlets that are dominated by fast basal motion and imbedded within the generally slowly deforming ice cap (Dowdeswell et al., 1999). The southeastern basins are to a large extent grounded below sea level, and form a continuous, non-floating calving front towards the Barents Sea (Dowdeswell et al., 2008). The regions grounded below sea level are at least partially underlain by marine sediments (Dowdeswell et al., 2008). Over 2002–2008, the climatic mass balance of Austfonna was close to zero (Moholdt et al., 2010a). Yet the ice cap was losing mass due to calving and retreat of the marine ice margin that accounted for $2.5\,\mathrm{Gt\,a}^{-1}$ or $33 \pm 5\,\%$ of the total ablation (Dowdeswell et al., 2008).

Basin-3 ($\sim 1200\,\mathrm{km}^2$ and a length of ~ 60 km) is the largest drainage basin of Austfonna and topographically constrained by a subglacial valley that extends from south of the main dome eastwards towards the Barents Sea (Fig. 1b). About one-third of the ice base is grounded below sea level down to a maximum depth of ~ 150 m within an overdeepening in the central terminus region (Fig. 1b; Dowdeswell et al., 1986; Dunse et al., 2011). A previous velocity map based on satellite radar interferometry from data acquired in the mid-1990s revealed an ice stream in the northern lower reaches of Basin-3 with flow velocities $\leq 200\,\mathrm{m\,a^{-1}}$ (Dowdeswell et al., 2008). The ice stream was topographically constrained by a subglacial mountain to the north (Isdomen) and near-stagnant ice, likely frozen to its bed, to the south. The absence of surface lineations (e.g. crevasses) as late as 1991 identified the ice stream of Basin-3 as a recent feature (Dowdeswell et al., 1999). During the 1990s, the front retreated on average by $70 \pm 10\,\mathrm{m\,a^{-1}}$, accounting for two-thirds of the calving flux of $\sim 0.4\,\mathrm{Gt\,a^{-1}}$ (Dowdeswell et al., 2008). Basin-3 is known to have surged some time prior to 1870 (Lefauconnier and Hagen, 1991), and the advancing terminus created a pronounced surge lobe nearly 25 km in width. The terminal moraine associated with the last surge lies $\sim$ 8 km from the present-day position of the calving front (Robinson and Dowdeswell, 2011). Based on lobe volume and accumulation rate, the duration of the quiescent phase of Basin-3 was estimated to last 130–140 years (Solheim, 1991), a very good estimate considering the basin's renewed surge activity since autumn 2012, which is reported on here.

3 Methods

3.1 Velocity time series

In spring 2008, five GPS receivers were deployed along the mid-1990s central flowline, 5 to 21 km upglacier from the calving front (Fig. 1; Dunse et al., 2012). We used GPS single-frequency code receivers (L1 band, C/A code only). Geographical positions were logged at hourly intervals, and every third hour for instruments installed after May 2011, at an accuracy typically better than 2 m (den Ouden et al., 2010). Filtering in the time domain was applied to reduce random errors, i.e. a 7-day running mean was applied to the daily mean position, velocities were computed, and, finally, the velocity was smoothed by applying another 7-day running mean (Dunse et al., 2012).

3.2 Velocity maps from synthetic aperture radar

Velocity maps of Basin-3, Austfonna, were produced from twenty-four 2 m resolution TerraSAR-X (TSX) scenes acquired between April 2012 and November 2013 and provided by the German Aerospace Center (DLR). Displacement fields were derived by using cross-correlation between two consecutive acquisitions (Strozzi et al., 2002) and geocoded using a DEM of Austfonna (Moholdt and Kääb, 2012). Appendix A provides more detailed information on the TSX data and processing. Calving front outlines were digitized from geocoded backscatter images (Appendix B). The calving flux of Basin-3 was derived by the ice flux through a fixed fluxgate upglacier of the calving front and the area change downglacier of that fluxgate, multiplied by an average ice thickness. The ice thickness was derived by differencing of a bedrock map (Fig. 1b; Dowdeswell, 1986; Dunse et al., 2011) and a digital elevation model (Moholdt and Kääb, 2012).

3.3 Additional data

To approximate timing and magnitude of surface-melt periods, cumulative positive-degree days (PDD) were computed from the temperature record of an automatic weather station. The station has operated since April 2004 on the western flank of Austfonna at 22°25′12″ E, 79°43′48″ N and 370 m a.s.l. (Schuler et al., 2014). Surface crevasses were identified and mapped using ground-penetrating radar (Ramac GPR; Målå Geoscience) at a centre frequency of 800 MHz, providing structural information from the glacier surface down to a depth of $\sim$ 12 m (Dunse et al., 2009, Appendix C). GPR profiles were annually repeated subsequent to 2004 and geolocated using a kinematic global navigation satellite system (GNSS; GPS and GLONASS).

4 Results

4.1 Multiannual ice-flow acceleration

Winter velocities observed by GPS in May 2008 were significantly higher than in the mid-1990s (Fig. 2). The GPS time series also reveal considerable overall acceleration, occurring in pronounced steps, each of which coincides with the summer melt period, as indicated by the temperature record of an automatic weather station (Fig. 2; Schuler et al., 2014). The 2008 summer speed-up is followed by a gradual winter deceleration. In total, 60–95% of the velocity increase during summer is reversible (Appendix D). However, during subsequent years, an increasing fraction of the summer speed-up is of lasting nature, i.e. increased summer velocities are sustained throughout the winter although surface melt had ceased.

Multiannual changes in ice dynamics are also evident from within the reservoir area, higher up on Basin-3. The annual position change of a mass-balance stake close to the ice divide (Fig. 1) revealed strong acceleration from $11\,\mathrm{m\,a^{-1}}$ between 2004 and 2007 to 28, 43 and $114\,\mathrm{m\,a^{-1}}$ for the period 2010–2013. In addition, annually repeated GPR surveys reveal first and cumulative occurrence of surface crevasses from $\sim$ 2004 onwards, between $\sim$ 2004 and 2007 along the western profile, and between 2008 and 2012 along the eastern profile (Fig. 1a; Appendix C).

4.2 Mobilization of stagnant ice regions

The drastic acceleration during autumn 2012 coincided with the expansion of the fast-flowing region across the entire basin, as revealed by a time series of ice-surface velocity maps based on intensity tracking of repeat-pass TerraSAR-X satellite radar images (Fig. 3; Appendix B). Velocities in April 2012 were up to $3\,\mathrm{m\,d^{-1}}$, 1 order of magnitude larger

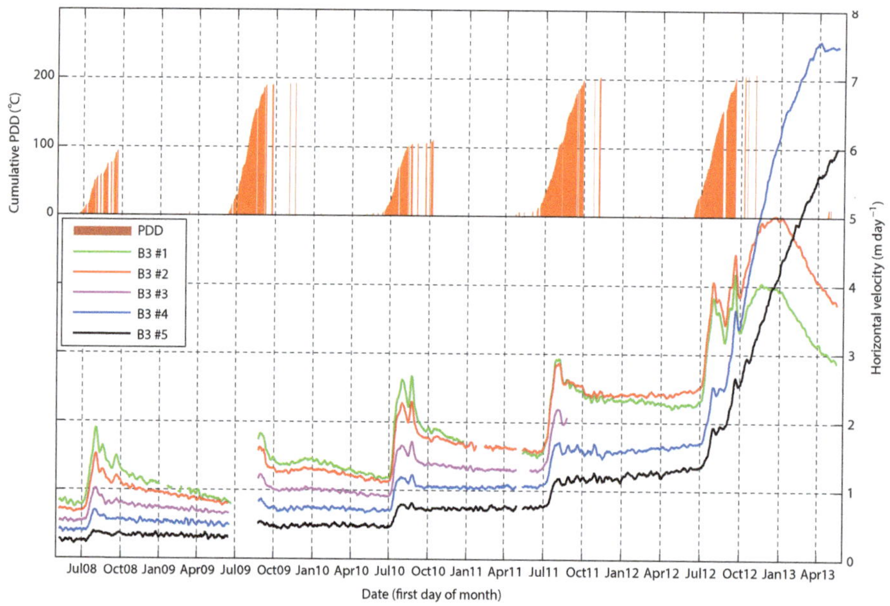

Figure 2. Flow velocities along the centreline of the fast-flow region of Basin-3, Austfonna, between May 2008 and May 2013. GPS stations are numbered from 1 at the lowest elevation to 5 at the highest (Fig. 1). Red bars (upper panel) indicate potential melt days and cumulative positive-degree days (PDD) for each summer, inferred from the temperature record of an automatic weather station.

than in the mid-1990s (Fig. 3a). In April 2012, the fast-flow region extended farther inland and had widened southward to a width of $\sim$6–8 km, as compared to $\sim$5–6 km in the mid-1990s. As a result of southward expansion, the GPS receivers were located $\sim$1 km north from the central flowline and did not capture fastest flow velocities. In contrast to the mid-1990s, the southeastern corner of Basin-3 also displayed fast motion, at velocities of up to 1 m d^{-1}. These two distinct fast-flow regions were completely separated by almost stagnant ice, notably including the calving front. Low ice velocities, <0.1 m d^{-1}, indicate the absence of considerable basal motion, and suggest frozen-bed conditions in this region. In August 2012, i.e. at the end of the summer melt season, velocities had increased significantly, up to 6 m d^{-1} for the northern and 4 m d^{-1} for the southeastern fast-flow region, along with further lateral expansion of the fast-flowing areas. Consequently, the slow-moving ice region in between had decreased in size (Fig. 3b) and disappeared by October 2012, when ice flow escalated into a surge comprising the entire width of the basin, reaching velocities >10 m d^{-1} (Fig. 3c). Velocities increased further until January 2013, reaching a maximum of 20 m d^{-1} (Fig. 3d). Between January and May 2013 the maximum velocities decreased to 15.2 m d^{-1}, while the upglacier regions continued to accelerate (Fig. A1b). By the end of 2013, fast flow of Basin-3 continued (Appendix A).

4.3 Calving flux

The TSX data allowed for calculation of the calving flux components, i.e. (i) the ice flux through a fixed fluxgate near the calving front, and (ii) the mass change of the terminus downglacier of that fluxgate, accounting for front position changes (Fig. 4; Appendix B). The observed ice flux peaked at a rate of 13.0$\pm$4.2 Gt a^{-1} in December 2012/January 2013, after which it decreased slightly (Fig. 4a). Prior to October 2012, the position of the entire calving front of Basin-3 was remarkably stable (slight retreat; Fig. 4b), indicating that the entire ice flux was balanced by iceberg calving. After November 2012, the southern and central parts of the front advanced by >1 km, reducing ice mass loss through calving by 61 %, as compared to a stable front position.

Direct conversion of calving mass loss to SLR contribution is only meaningful for a static calving front. Glacier surges are typically accompanied by significant terminus advances. An advancing terminus reduces the mass loss from the glacier; however, the submerged part of the terminus replaces sea water instantaneously, causing an instantaneous sea-level rise. We therefore distinguish between a glacier mass balance and a sea-level perspective on the calving flux (Fig. 4c; Table 1). From 19 April 2012 to 9 May 2013, calving mass loss [yearly rate] from Basin-3 accounted for 4.4 $\pm$ 1.6 Gt [4.2 $\pm$ 1.6 Gt a^{-1}], an increase by an order of magnitude compared to 1991–2008 (Dowdeswell et al., 2008), nearly tripling the calving loss from the entire Austfonna ice cap. The related sea-level rise contribution of 7.6 $\pm$ 2.7 Gt [7.2 $\pm$ 2.6 Gt a^{-1}] is as large as the total glacier mass change from the entire Svalbard archipelago for the period 2003 to 2008, estimated to be -6.6 ± 2.6 Gt a^{-1} (Moholdt et al., 2010b). Rates of sea-level rise contribution are expected to decline once the surge of Basin-3 has terminated, i.e. once the

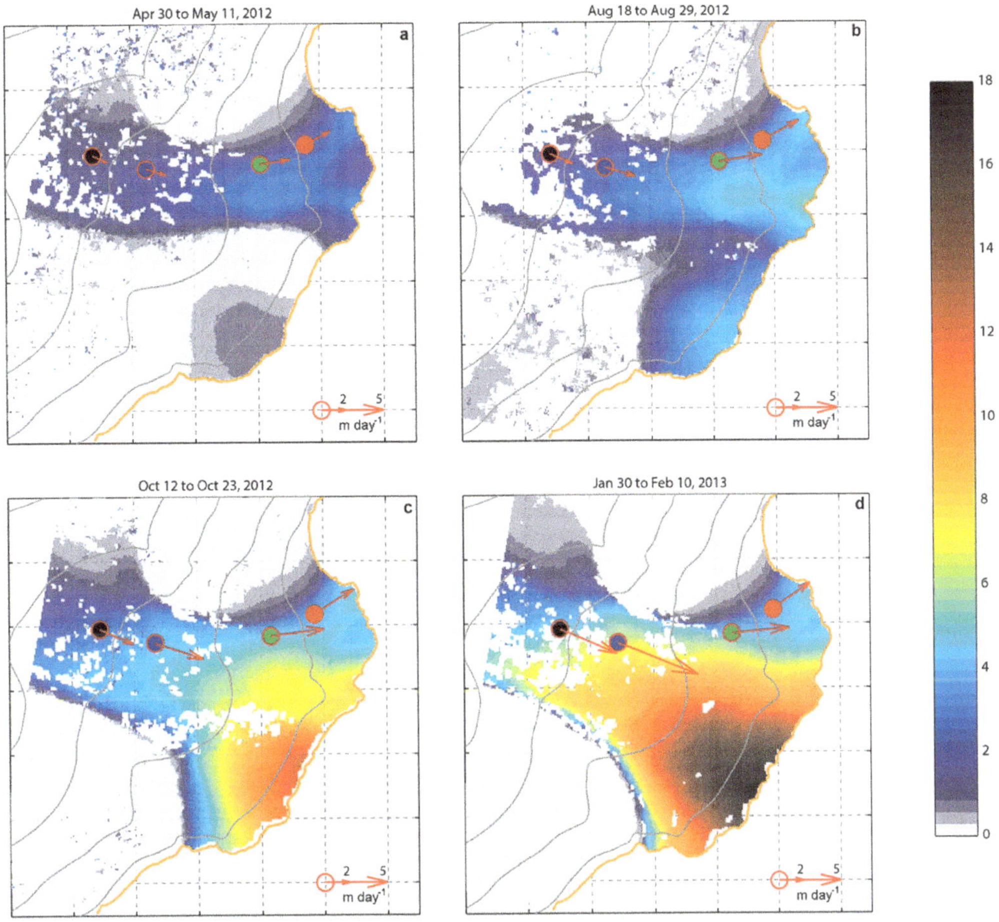

Figure 3. Surface velocity fields of Basin-3, Austfonna, derived from TerraSAR-X feature tracking: **(a)** April/May 2012, **(b)** August, **(c)** October, and **(d)** January/February 2013. Red circles represent mean position of GPS receivers over the particular repeat-pass period; fill colour according to colour coding of receivers in Fig. 2. The red arrows indicate associated GPS velocity vectors. Glacier elevation contours plotted in grey at 100 m intervals; front position at time of repeat pass in orange.

ice flux has diminished and the terminus advance has come to a halt. Nevertheless, iceberg calving and hence ice mass loss will be maintained, depending on future rates of marine-terminus retreat.

5 Discussion

The dynamic changes that have been observed at Basin-3 over the last two decades can be separated into three phases: (1) activation of a spatially confined fast-flow region in the early 1990s (Dowdeswell et al., 1999), (2) multiannual acceleration and expansion of the fast-flow region from < 2008 to 2012, and (3) active surge phase following the destabilization of the entire terminus in autumn 2012.

5.1 Phase 1: Initiation of spatially confined fast flow

Spatially confined fast flow was initiated in the northern part of lower Basin-3 in the early 1990s, and interpreted as a temporary flow instability (Dowdeswell et al., 1999). Activation of fast flow within this region could be explained by internal mechanisms, i.e. by gradual changes in basal thermal regime and driving stress, associated with long-term geometric changes during the quiescent phase (e.g. Clarke, 1976; Clarke et al., 1984). Numerical simulations of Austfonna support the concept of a thermally controlled soft-bed surge mechanism (Dunse et al., 2011), as proposed earlier for Svalbard glacier surges (Hamilton and Dowdeswell, 1996; Jiskoot et al., 2000; Murray et al., 2000, 2003b). Spatially confined fast flow since the early 1990s entailed an ice flux in excess of the balance flux (Dowdeswell et al., 1999) and thus a drawdown of the reservoir area. Extensional flow within the reservoir area is expected to form crevasses (van der Veen, 1999), as indeed observed by GPR since ~2004 (Fig. 1a; Appendix C).

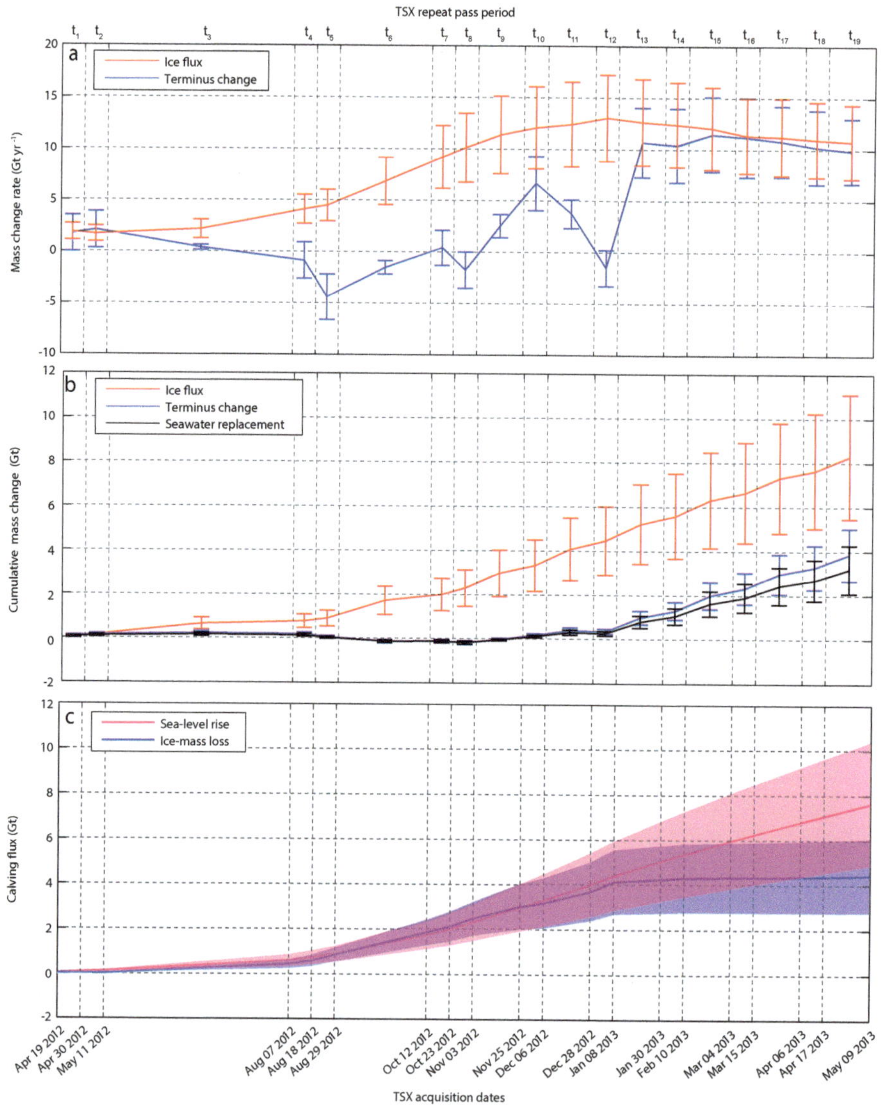

Figure 4. Calving flux from Basin-3, Austfonna, April 2012 to May 2013. Calving components are expressed in terms of the instantaneous **(a)** and cumulative mass change **(b)** and allocated to the effect on glacier mass balance and sea level **(c)** (see Sect. 4.3). Whiskers in **(a, b)** indicate uncertainty bounds calculated using propagation-of-uncertainty analysis, and shaded areas in **(c)** indicate upper and lower bounds given maximum or minimum ice thickness.

Table 1. Estimate of calving flux components and total calving flux over the TSX observation period from 19 April 2012 to 9 May 2013.

Calving flux components	(Gt)	(Gt a^{-1})
Ice flux, Q_{fg}	8.3 ± 2.8	7.8 ± 2.7
Terminus change, Q_{t}	3.8 ± 1.2	3.6 ± 1.1
Terminus-seawater displacement, Q_{tsd}	3.2 ± 1.1	3.0 ± 1.0
Total calving flux		
Mb perspective, $Q_{\mathrm{mb}} = Q_{\mathrm{fg}} - Q_{\mathrm{t}}$	4.4 ± 1.6	4.2 ± 1.5
SLR perspective, $Q_{\mathrm{sl}} = Q_{\mathrm{mb}} + Q_{\mathrm{tsd}}$	7.6 ± 3.9	7.2 ± 2.6

5.2 Phase 2: Mobilization of the reservoir – multiannual, stepwise acceleration

The multiannual acceleration of Basin-3, observed by GPS since 2008, occurred in discrete steps, each of which coincident with consecutive summer melt periods (Fig. 2). The high sensitivity and short response time (days) of glacier dynamics to melt events clearly suggest surface-melt-triggered acceleration. Short-lived acceleration during the melt season is consistent with current understanding of hydraulic lubrication (Schoof, 2010). In contrast, the multiannual acceleration of background velocities (Appendix D) cannot be explained by this mechanism and suggests a fundamental change in dynamics.

Enhanced post-summer velocities sustained throughout the winter can be explained by successive activation of previously stagnant ice regions during the previous summer melt period. Mobilization of increased ice volumes within the reservoir area leads to increased velocities and discharge. A widening of the fast-flow region itself allows for higher centreline velocities due to the increased distance from the lateral shear margin, analogous to the behaviour of Antarctic ice streams (Joughin and Alley, 2011).

Mobilization of the reservoir area is evident from the surface-crevasse formation within the upper accumulation area of Basin-3 since ~2004 (Fig. 1a; Appendix C). Crevasses are a manifestation of longitudinal extension and evidence of upglacier migration of the fast-flowing region. This is in line with the upglacier expansion of the ice stream revealed by TSX and the observed strong acceleration of the mass-balance stake within the reservoir area (Fig. 1). The first occurrence of crevasses signifies the development of potential meltwater routes to the glacier bed, subjecting an increasing region to CHW and basal hydraulic lubrication. Similarly, meltwater reaching the bed beneath the heavily crevassed shear margin of the fast-flow region efficiently weakens the basal drag that balances the lateral drag, which in turn regulates ice velocities (Joughin and Alley, 2011). CHW and increased frictional heating both act to enhance this mechanism, weakening the flow resistance exerted by the lateral shear margins and possibly cold-based ice patches acting as "sticky spots" (Alley, 1993). The interplay between CHW and the emergence of basal hydraulic lubrication over an expanding area of the ice base constitutes a hydro-thermodynamic feedback. Unconsolidated subglacial sediments underneath the marine-based ice of Basin-3 act to further enhance this feedback. Input of surface meltwater during consecutive summers progressively raises the pore-water pressure within the sediment, reducing its shear strength and favouring rapid destabilization (Clarke et al., 1984; Tulaczyk et al., 2000).

5.3 Phase 3: Destabilization of the marginal ice plug – basin-wide surge

A surge comprising the full width of Basin-3 and the subsequent advance of the terminus followed the mobilization of the remaining stagnant ice regions, notably including the calving front. Water flow may have accessed the cold-ice base through partially unfrozen subglacial sediments, as suggested in the case of Trapridge Glacier, Yukon, Canada (Clarke et al., 1984). Field measurements of thermal structure and ice flow during a surge of Trapridge Glacier do not indicate that the cold ice below the surge front acts as a thermal barrier to water flow. Clarke et al. (1984) observed very high temperature gradients in the cold basal ice below a surge front, about 10 times larger than expected from geothermal heat flux, and attributed this to water flow through an unfrozen substrate beneath the cold ice. The abrupt onset of fast flow in previously stagnant regions of Basin-3 suggests sheer tearing of the glacier from its bed. This activation was possibly related to a combination of warming of the glacier bed through CHW and mechanical stress transfer from active ice further upstream. During the fieldwork in spring 2012, large ice blocks were observed to be pushed upward above the glacier surface near the lower shear margin, an indication of strong longitudinal stresses.

While fast flow continued by the end of 2013, it is expected to slow down or come to a halt within a few years. Massive ice redistribution from the reservoir into the receiving area and towards the calving front efficiently lowers the driving stress. Eventually, temperate basal conditions underneath the dynamically thinned and decelerating terminus are expected to be no longer maintained, at which time the base of the ice will start to refreeze to its bed (Clarke et al., 1984; Dunse et al., 2011).

5.4 Hydro-thermodynamic feedback mechanism

We cannot explain our observations using standard theory of glacier surges alone (e.g. Clarke et al., 1984). The multiannual acceleration did not occur uniformly or gradually, as previously thought characteristic of Svalbard glacier surges (Murray et al., 2003b, a). Incorporating the effect of CHW over an expanding area of the glacier bed provides, we believe, a good explanation of our observations (Fig. 5). The hydro-thermodynamic feedback proposed here is not limited to CHW; it also includes the additional hydraulic lubrication effects. Ice regions undergoing a switch from cold to temperate basal conditions were previously characterized by frozen conditions, i.e. absence of water. In the initial stage of basal–hydraulic drainage system development, water input is likely to raise water pressure and enhance basal motion, in line with the established theory on basal lubrication (e.g. Schoof, 2010). In addition, thawing of underlying frozen till and rising pore-water pressure lead to enhanced sediment deformation. These processes all result in an enhancement of ice flow. Longitudinal extension opens new surface crevasses that provide englacial pathways for meltwater in subsequent summers, subjecting a larger region of the bed to surface melt. Eventually, the entire drainage basin is sufficiently destabilized, initiating the surge. Once the entire basin is moving by fast basal motion, i.e. the entire base is at pressure melting, and driving stresses are decreasing, the hydro-thermodynamic feedback no longer operates.

We would like to stress that the hydro-thermodynamic feedback proposed here does not oppose, but can be understood as an integral part of the theory of thermally controlled soft-bed glacier surges, as illustrated in Fig. 5. An increasing fraction of the glacier bed is subjected to surface melt, and basal processes conductive for surge behaviour are amplified. Basal processes include, in particular, (1) CHW that leads to a switch from cold to temperate basal conditions, permitting for fast basal motion; (2) expansion of the area

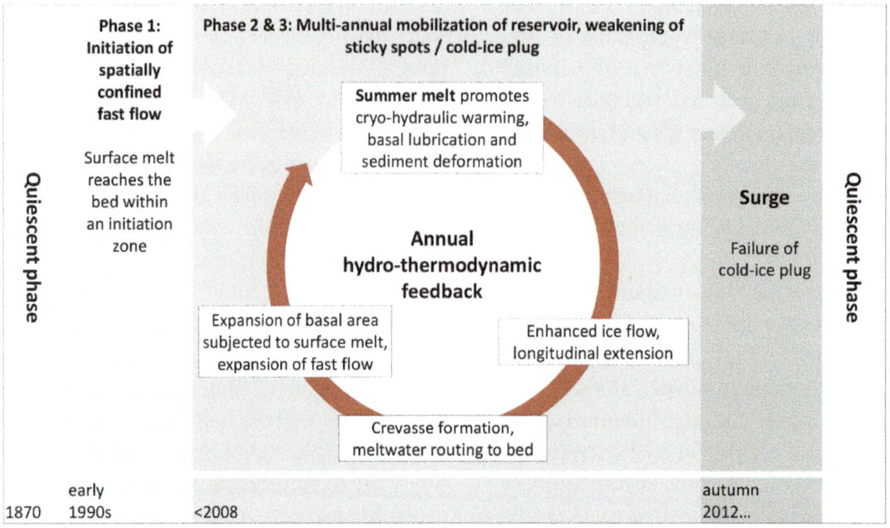

Figure 5. Schematic illustration of the proposed hydro-thermodynamic feedback to summer melt, imbedded within the surge cycle of Basin-3, Austfonna. The approximate start of each phase is indicated at the bottom. Phase 1 follows from long-term changes in glacier geometry, i.e. build-up of a reservoir, and associated changes in driving stress and basal thermal regime. The hydro-thermodynamic feedback loop operates over several years during phase 2 and 3, each loop coinciding with consecutive summer melt periods. Successive mobilization and destabilization initiates the surge. Dynamic thinning, reduction in driving stress and basal heat dissipation eventually terminate the surge.

subjected to basal lubrication; and (3) rising pore-water pressure and sediment deformation. The feedback requires an initiation zone in which surface meltwater can access the bed, such as the spatially confined fast-flow region described in phase 1. Alternatively, supra-glacial lake drainage may establish full ice-thickness meltwater pathways (Das et al., 2008). Our observations of Basin-3 suggest, on the one hand, a significant contribution of the hydro-thermodynamic feedback on the mobilization of the reservoir area and, on the other hand, weakening of the flow resistance within the receiving area.

5.5 Implications for the future stability of ice sheets

Surface melt has likely been a widespread phenomenon on Svalbard throughout most of the Holocene. The current surge of Basin-3 should therefore not be mistaken as a response to recent Arctic warming. The hydro-thermal feedback could also have played a role in previous Svalbard glacier surges. Given continued global warming, characterized by more widespread and intense occurrence of surface melt, we hypothesize that the hydro-thermodynamic feedback may gain importance in other glaciated regions, including the ice sheets.

Rapid marine ice-sheet disintegration is evident from geological records both in the Northern and Southern Hemisphere and typically associated with air temperatures similar to or warmer than those predicted for the end of the 21st century (Cook et al., 2013; Deschamps et al., 2012). Ice-rafted debris distributed across the North Atlantic ocean floor provides evidence of substantial calving associated with the rapid disintegration of the Laurentide Ice Sheet, so-called Heinrich events (Bond et al., 1992). Palaeoclimatic records suggest rates of sea-level change much greater than currently observed or projected for the 21st century, e.g. 3.5 to 5 m SLR per century (35–50 mm a^{-1}) at the end of the Last Glacial Maximum, ~ 14.5 ka ago (Deschamps et al., 2012).

Analogies between glacier surges and partial ice-sheet collapses are widespread throughout the glaciological literature (e.g. Clarke et al., 1984; Jiskoot et al., 2000). Analogous to the ice sheets, ice caps like Austfonna consist of slow-moving inland ice interspersed with faster-flowing outlet glaciers and ice streams that deliver inland ice towards the calving front. Drainage basins of Austfonna in their quiescent phase are characterized by margins frozen to their bed (Dowdeswell et al., 1999). Similarly, the coastal margins of the Antarctic Ice Sheet contain large regions of cold-based ice (Pattyn, 2010) that may currently prohibit efficient drainage of warm-based interior ice towards the ocean – e.g. in the Wilkes Basin, East Antarctica, which is known to have been dynamically active during the Pliocene warmth (Cook et al., 2013). A very recent model study confirms the potential dynamic instability of the Wilkes Basin, following the removal of a cold-ice plug (Mengel and Levermann, 2014). In that study, a retreat of the grounding line is forced by oceanic warming, thereby eliminating the cold-ice plug.

Our proposed mobilization of the reservoir area of Basin-3 is in line with Phillips et al. (2013), pointing at the potential of CHW in mobilizing more inland ice regions of the Greenland Ice Sheet, thereby increasing ice discharge through existing outlet glaciers. An obvious feature that distinguishes Basin-3 from Greenland outlet glaciers is the presence of a cold-based ice plug that, until autumn 2012, restricted drainage from the reservoir area. In the case of fast-flowing Greenland outlet glaciers, the thermo-dynamic feedback would therefore lead to gradual acceleration, as long as the ELA continues migrating upglacier, rather than leading to surge-type behaviour. In southwestern Greenland, further acceleration of Jacobshavn Isbræ followed the 2012 record summer melt season (Joughin et al., 2014). Although the authors attribute the acceleration to terminus retreat into a bedrock depression, the occurrence of pronounced summer speed-up at higher elevations may also reflect inland migration of surface melt and the effect of associated hydraulic feedbacks on glacier dynamics (Alley et al., 2008; Meierbachtol et al., 2013; Phillips et al., 2013). Similarly, the hydro-thermodynamic feedback may play a role in the sustained mass loss recently reported from northeastern Greenland and mainly attributed to sea-ice decline due to regional warming (Khan et al., 2014).

6 Summary and conclusion

For the first time, to our best knowledge, we have obtained continuous in situ velocity observations during glacier-surge initiation. Our observations from Basin-3 of Austfonna reveal details of a surge that have never been recorded before. Unlike proposed earlier for Svalbard glacier surges (Murray et al., 2003b, a), multiannual acceleration during surge initiation of Basin-3 was not gradual but occurred in discrete steps, coincident with successive summers. We propose a hydro-thermodynamic feedback mechanism triggered by surface melt reaching a growing fraction of the glacier bed. Intrusion of surface melt to the glacier bed provides an efficient heat source through CHW, facilitating a thermal switch from cold to temperate basal conditions, permitting for basal motion. Initiation of hydraulic lubrication, along with rising pore-water pressure within subglacial sediments, further enhances basal motion, eventually destabilizing the overlying ice. These processes have earlier been summarized as thermally controlled soft-bed surge mechanism (e.g. Clarke et al., 1984; Murray et al., 2003b). However, the active role of surface melt and the associated hydro-thermodynamic feedback mechanism contrast the previous understanding of glacier surges as purely internal instabilities.

The recent calving flux of Basin-3 has strong implications for the mass balance of the ice cap and its contribution to sea-level rise. From 19 April 2012 to 9 May 2013, the calving flux [yearly rate] of Basin-3 amounted to 4.4 ± 1.6 Gt [4.2 ± 1.6 Gt a^{-1}], an increase by an order of magnitude compared to 1991–2008 (Dowdeswell et al., 2008). With the terminus advance accounted for, the related sea-level rise contribution of 7.6 ± 2.7 Gt [7.2 ± 2.6 Gt a^{-1}] equals the total annual glacier mass loss from the entire Svalbard archipelago for the period 2003 to 2008, estimated to be 6.6 ± 2.6 Gt a^{-1} (Moholdt et al., 2010b).

Given continued climatic warming and increasing surface melt, we hypothesize that the hydro-thermodynamic feedback may gain significance in other glaciated areas, including the ice sheets. In light of recent record melt and rising ELA of the Greenland Ice Sheet, the proposed mechanism has the potential to lead to a long-term enhancement of outlet-glacier discharge and calving loss, as earlier proposed by Phillips et al. (2013). Our expectation contrasts with recent studies that indicate limited effects of surface-melt-induced acceleration on the future net mass balance of the Greenland Ice Sheet (Nick et al., 2013; Shannon et al., 2013). Surface melt in Antarctica is presently mainly constrained to the ice shelves (Comiso, 2000). Given strong continued warming, surface melt will increasingly occur over coastal areas of Antarctica, making the grounded ice-sheet margins vulnerable to the hydro-thermodynamic feedback.

Our study of the Austfonna ice cap highlights the importance of dynamic ice-mass loss for glacier mass balance. Current model projections of future SLR (IPCC, 2013) still do not account for the dynamic response of glaciers to continued global warming and might need to be revised upward after incorporating mechanisms such as CHW and the hydro-thermodynamic feedback.

Appendix A: TerraSAR-X velocity maps

Twenty-four TerraSAR-X (TSX) satellite synthetic aperture radar scenes of ~2 m ground resolution (TSX Stripmap mode) were acquired between April 2012 and November 2013 (Table A1), providing information on the spatial evolution of the surge (Fig. 3). For each pair of consecutive acquisitions, displacements were determined using cross-correlation of the intensity images (Strozzi et al., 2002) and 21 velocity maps produced by accounting for the respective repeat-pass period (Table A1). The size of the matching window was adjusted according to the expected maximum displacements during the repeat pass cycle: 300 × 344 pixels in range and azimuth direction in the case of 11 days or 599 × 688 pixels for 22 days or longer. Displacements were measured in discrete steps of 50 × 57 pixels in range and azimuth to achieve a resolution of ~100 × 100 m. Velocity maps were calculated from displacement maps by accounting for the time interval between the two underlying TSX images (t_1 to t_{21}, Table A1) and geocoded using a DEM of Austfonna (Moholdt and Kääb, 2012). Velocities larger than a carefully estimated maximum (Table A1) were classified as mismatches and removed. To estimate the ice flux (Appendix B), obvious erroneous velocities remaining in the vicinity of a fixed fluxgate were manually removed and the maps interpolated using inverse distance weighting to provide continuous velocity profiles along the fluxgate (Fig. A1a). A comparison between TSX and GPS velocities for each GPS station and all repeat-pass periods revealed that TSX underestimated local velocities at the GPS stations by 0.3 m d^{-1} (Fig. A2). We explain this small bias by large horizontal velocity gradients within large matching windows, which could lead to the average velocities within a matching window being typically smaller than the strictly local GPS velocity. Consequently, more stable areas produce best matches.

Table A1. TerraSAR-X acquisitions of Basin-3 and repeat-pass period and maximum velocities of inferred velocity maps.

ID of period	Repeat-pass period (days)	Start and end-date	Maximum velocity (m d^{-1})
t_1	11	19 Apr 2012–30 Apr 2012	4.5
t_2	11	30 Apr 2012–11 May 2012	4.5
t_3	88	30 Apr 2012–7 Aug 2012	4.5
t_4	11	7 Aug 2012–18 Aug 2012	6.4
t_5	11	18 Aug 2012–29 Aug 2012	6.8
t_6	44	29 Aug 2012–12 Oct 2012	8.4
t_7	11	12 Oct 2012–23 Oct 2012	13.6
t_8	11	23 Oct 2012–3 Nov 2012	13.6
t_9	22	3 Nov 2012–25 Nov 2012	15.9
t_{10}	11	25 Nov 2012–6 Dec 2012	18.2
t_{11}	22	6 Dec 2012–28 Dec 2012	18.4
t_{12}	11	28 Dec 2012–8 Jan 2013	20.0
t_{13}	22	8 Jan 2013–30 Jan 2013	18.6
t_{14}	11	30 Jan 2013–10 Feb 2013	18.6
t_{15}	22	10 Feb 2013–4 Mar 2013	17.5
t_{16}	11	4 Mar 2013–15 Mar 2013	17.3
t_{17}	22	15 Mar 2013–6 Apr 2013	16.1
t_{18}	11	6 Apr 2013–17 Apr 2013	16.4
t_{19}	22	17 Apr 2013–9 May 2013	15.2
t_{20}	11	16 Aug 2013–27 Aug 2013	15.5
t_{21}	11	12 Nov 2013–23 Nov 2013	12.5

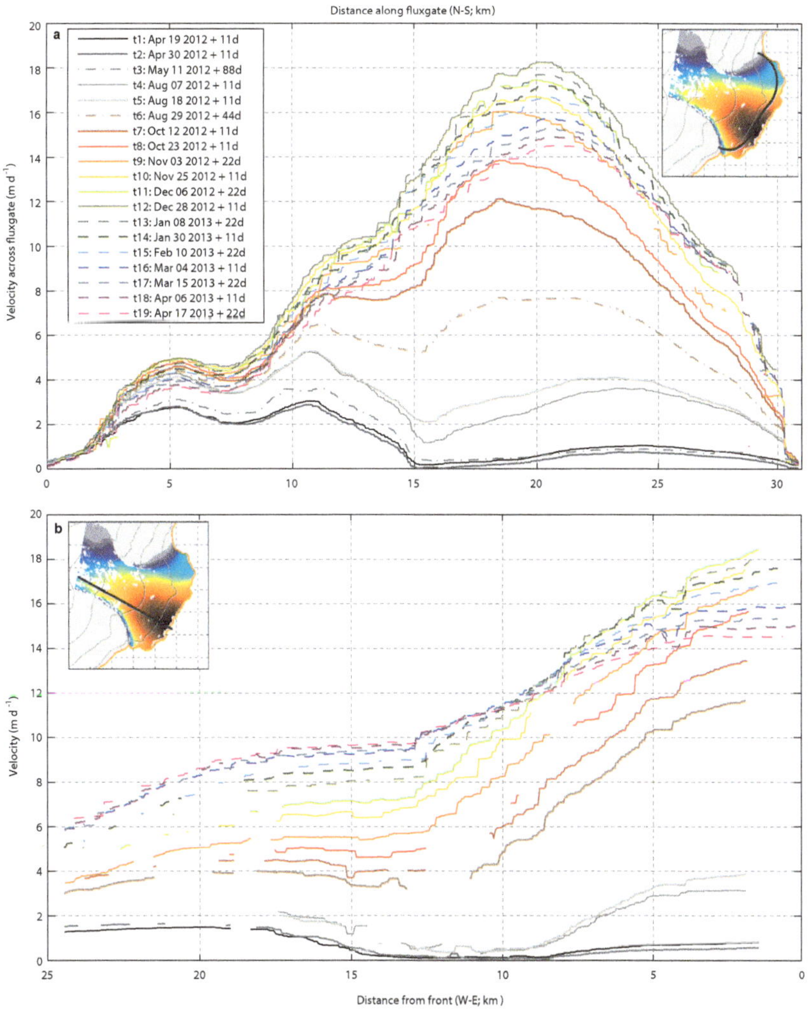

Figure A1. Evolution of ice-flow velocities across a calving fluxgate **(a)** and along a central flowline of Basin-3 **(b)**. Velocity profiles extracted from TerraSAR-X velocity maps April 2012 to May 2013 and median filtered to remove outliers. The insert shows the location of the profiles overlain onto the velocity map of t_{14}. Fluxgate velocities are used to compute the calving ice flux. Velocities for long repeat passes t_3 and t_6 have been constructed as described in B2.

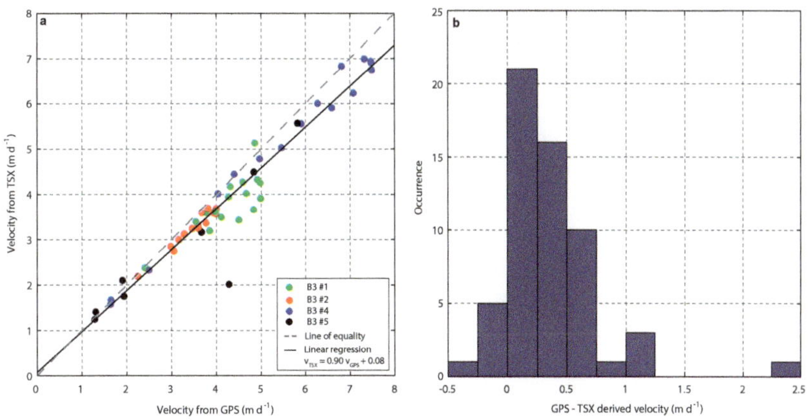

Figure A2. Comparison of TSX and GPS velocities at GPS locations and for each repeat-pass period. **(a)** Linear regression (solid black; $R^2 = 0.94$) predicts 94 % of the variance in TSX velocities ($0.14\,\mathrm{m\,d^{-1}}$). **(b)** TSX velocities typically slightly underestimate local (sub-pixel-scale) velocities measured at the GPS stations by $0.3\,\mathrm{m\,d^{-1}}$.

Appendix B: Calving flux

The calving flux, q, of Basin-3 was calculated based on TSX velocity maps and changes in the extent of the glacier by $q = q_{\rm fg} + q_{\rm t}$, with $q_{\rm fg}$ being the ice flux through a fluxgate near the calving front, and $q_{\rm t}$ the volume change of the terminus downglacier of that fluxgate due to advance or retreat of the calving front. Here, we defined a spatially fixed fluxgate approximately perpendicular to the ice flow, typically 1–3 km upglacier from the actual calving front (Fig. A1a), and where ice surface velocities could be inferred from all TSX-image pairs. The ice-flux can be written as

$$q_{\rm fg} = H_{\rm fg} \cdot w_{\rm fg} \cdot v_{\rm fg}, \tag{B1}$$

where $H_{\rm fg} = zs_{\rm fg} - zb_{\rm fg}$ is the ice thickness along the fluxgate, with $zs_{\rm fg}$ and $zb_{\rm fg}$ the surface and bedrock elevation, respectively; $w_{\rm fg}$ is the width of the fluxgate; and $v_{\rm fg}$ the velocity across the fluxgate. We assume plug flow, i.e. depth-averaged velocities equal surface velocities, because at the high flow velocities observed, ice deformation is negligible compared to basal motion (Clarke, 1987). Position changes of the calving front are addressed by changes in areal extent of the glacier downstream from that fluxgate per repeat-pass period

$$q_{\rm t} = H_{\rm t} \cdot \frac{\Delta A_{\rm t}}{\Delta t}, \tag{B2}$$

where $H_{\rm t} = zs_{\rm t} - zb_{\rm t}$ is the ice thickness at the terminus in vicinity of the calving front, where $zs_{\rm t}$ represents a typical height of the calving front and $zb_{\rm t}$ is the mean bedrock elevation for the area encompassed by the observed maximum and minimum glacier extent. $\Delta A_{\rm t}$ is the areal change of the terminus over the repeat-pass period Δt between successive TSX acquisitions. The uncertainty in the calving flux estimate is associated with uncertainties of the input variables listed in Table B1. If the variables are derived from several independent components, their uncertainties can be summed by the root of the sum of squares (RSS) of the uncertainty of the components.

B1 Ice flux

To calculate the ice flux $q_{{\rm fg},i}$ for each repeat-pass period t_i (Fig. 4a), the ice flux through 617 fluxgate segments 50 m in width was integrated, accounting for local ice thickness and velocity. Ice volume is converted to mass using an ice density $\rho_{\rm ice} = 917\,{\rm kg\,m^{-3}}$. The cumulative ice flux over the period 19 April 2012 to 9 May 2013 is derived by summing up the ice flux over 19 successive TSX repeat-pass periods, weighted by their time duration (Fig. 4b):

$$q_{\rm fg} = w_{\rm fg} \cdot \sum_{t_i=1}^{19} \sum_{n=1}^{617} H|_{{\rm fg},n} \cdot v|_{{\rm fg},i,n} \cdot \rho_{\rm ice}. \tag{B3}$$

To derive continuous velocity profiles along the fluxgates, the velocity maps were interpolated horizontally in the vicinity of the fluxgate where mismatches occurred. The velocity maps associated with long repeat-pass periods, t_3 and t_6, suffer from large gaps: only 51 and 46 % of the fluxgate segments hold sound velocity estimates. In the case of t_3, the available data align with the previous velocity profile for t_2, a factor x relation was determined, and the gaps were filled by means of piecewise polynomial interpolation so that $v(t_3) = x \cdot v(t_2)$. In the case of t_6, the velocity profile was constructed by using the mean of the previous and following repeat-pass cycle, $v(t_6) = \frac{1}{2} \cdot (v(t_5) + v(t_7))$ (dash-dotted lines in Fig. A1a). The accuracy of the resulting velocity profiles was evaluated by comparing the constructed velocity profiles and the available reliable matches, revealing a standard deviation of 0.21 and 0.25 m d^{-1} and a negligible mean offset of 0.03 and 0.04 m d^{-1} in the case of t_3 and t_6, respectively.

The uncertainty in the ice-flux term, $\Delta q_{\rm fg}$, is derived according to the law of propagation of uncertainty:

$$\Delta q_{\rm fg} = w_{\rm fg} \cdot \sum_{t_i=1}^{19} \sum_{n=1}^{617} \sigma(q_{\rm fg})_{i,n}, \tag{B4}$$

with $\left(\frac{\sigma(q_{\rm fg})}{q_{\rm fg}}\right)^2 = \left(\frac{\sigma(H_{\rm fg})}{H_{\rm fg}}\right)^2 + \left(\frac{\sigma(v_{\rm fg})}{v_{\rm fg}}\right)^2$, omitting indexes i and n.

B2 Position changes of calving front

Similarly, the calving flux term related to position changes of the calving front can be expressed by the volume change of the terminus in vicinity of the calving front, $q_{\rm t}$:

$$q_{\rm t} = H_{\rm t} \cdot \sum_{t_i=1}^{19} \frac{\Delta A_{t,i}}{\Delta t_i} \cdot \rho_{\rm ice}. \tag{B5}$$

The calving front outlines of Basin-3 were digitized from geocoded TSX intensity images. Due to layover and shadow effects along the calving front a systematic error of the order of twice the height of the calving front of 30 m is assumed (Moholdt and Kääb, 2012). The digitizing accuracy is conservatively estimated to be 4 pixels, i.e. ±8 m.

Analogous to the ice-flux term, the uncertainty in the terminus-change term, $\Delta q_{\rm t}$, is determined by the law of propagation of uncertainty:

$$\Delta q_{\rm t} = \sum_{t_i=1}^{19} \sigma(q_{\rm t})_i \cdot \rho_{\rm ice}. \tag{B6}$$

For each repeat-pass period, $\sigma(q_{\rm t})$ is determined by $\left(\frac{\sigma(q_{\rm t})}{q_{\rm t}}\right)^2 = \left(\frac{\sigma(H_{\rm t})}{H_{\rm t}}\right)^2 + \left(\frac{\sigma(\Delta A_{\rm t})}{\Delta A_{\rm t}}\right)^2$, omitting index i.

Table B1. Calving flux input variables – values, sources and uncertainties.

Variable	Value/source	Uncertainty	Explanation
zs_{fg}	40 m (constant)	±30 m	The chosen values allow for elevations from flotation height of 10 m as lower limit and mean DEM height of 67 m as upper limit. The DEM originates from prior to surge initiation. GPS data since 2008 indicate extensional flow and hence dynamic thinning.
zb_{fg}	Local bedrock map values	±30 m	Twice the accuracy in ice thickness measurement of ±15 m used to derive the bedrock map along fluxgate (Dunse et al., 2012), thereby accounting for uncertainties introduced by gridding of spatially inhomogeneous measurements.
H_{fg}	$zs_{fg} - zb_{fg}$	±42 m	RSS of errors in zs_{fg} and zb_{fg}.
zs_t	30 m (constant)	±20 m	Allows for calving front heights down to flotation and significantly larger than typical front height of 30 m (Moholdt and Kääb, 2012).
zb_t	−87 m (constant)	±30 m	Value represents the mean bedrock elevation within the observed range in front position with an uncertainty analogous to the one of zb_{fg}.
H_t	$zs_t - zb_t$	±36 m	RSS of errors in zs_t and zb_t.
v_{fg}	Local value from TSX velocity maps	±0.37 m d^{-1} (0.42 for t_3 and 0.44 for t_6)	Uncertainty based on standard deviation (SD) of TSX and GPS velocities, yielding 0.37 m d^{-1}; for the long repeat cycles t_3 and t_6, additional uncertainty is added based on a comparison of the available data (51 and 46 % coverage) and reconstructed velocity profile, resulting in a SD of 0.21 and 0.25 m d^{-1}.
Front position	2 m resolution TSX backscatter image	±8 m (4 pixels)	Digitizing error of calving front position results in uncertainty of ΔA_{cf}, determined by RSS of deviation from minimum and maximum extent of A_{cf} at times $t_{i,\mathrm{start}}$ and $t_{i,\mathrm{start}}$.

Appendix C: Crevasse formation

Ground-penetrating radar (GPR) profiling has been performed on Austfonna on an annual basis since spring 2004 (Dunse et al., 2009). The GPR was operated at a centre frequency of 800 MHz and provides information over a depth range of about ~ 12 m. The data allow for identification of individual surface crevasses. Survey profiles initially focused on the accumulation area. Since 2008, the flowline along which GPS receivers are deployed has also repeatedly been surveyed. In the accumulation area, crevasse formation took place between 2004 and 2007 (Fig. C1a and b) as well as between 2008 and 2012 (Fig. C1c and d) for the western and eastern profile, respectively. The monitored flowline towards the terminus was already heavily crevassed at the time of the first GPR survey (Fig. C1c).

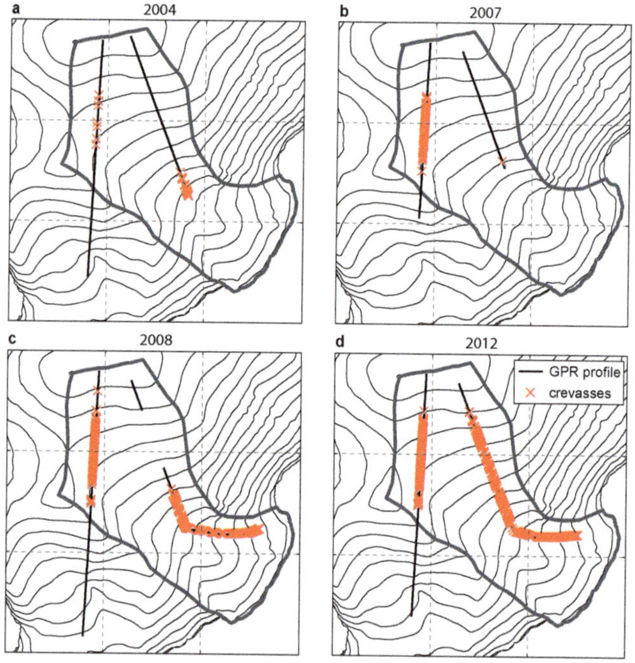

Figure C1. Crevasse formation on upper Basin-3 in 2004–2012 inferred from GPR. GPR profiles are plotted in solid black, detected crevasses are marked as red crosses. The outline of Basin-3 is shown in dark grey.

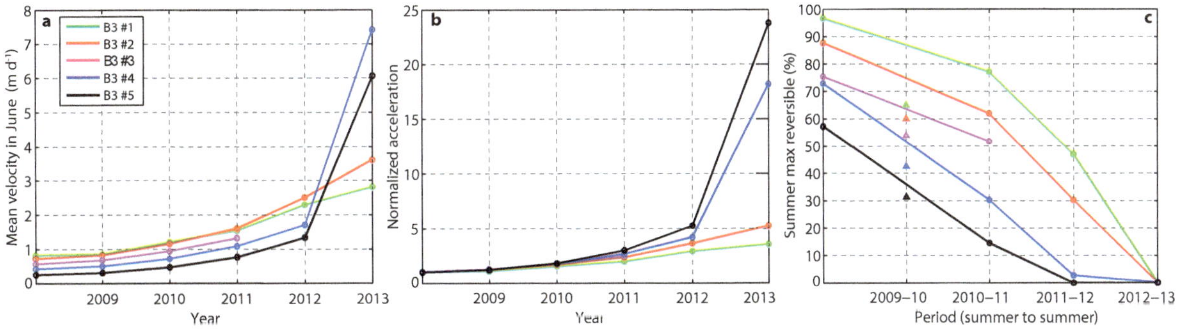

Figure D1. GPS-velocity evolution along a central flowline of Basin-3: **(a)** annual velocity minimum defined as the mean value in June 2008 to 2013 and **(b)** normalized to mean June velocity 2008, and **(c)** reversibility of summer speed-up defined as the ratio of the velocity increase from pre-summer minimum (mean June) to maximum summer velocity and subsequent velocity decrease to pre-summer velocity of the subsequent year. In 2009, summer maximum velocities were not captured due to GPS power loss. The triangles only indicate a minimum estimate for 2009–2010, based on elevated late-summer velocities measured after maintenance of GPS receivers at the end of August.

Appendix D: Background GPS velocities and summer speed-up

Our 5-year GPS record from Basin-3 shows that the annual velocity minimum typically occurs in June, just prior to the onset of summer speed-up (Dunse et al., 2012). Since 2008, the mean velocity in June has increased dramatically, with values in 2012 that were 3 to 5 times higher than in 2008 at the lowest (GPS B3 #1) and highest elevation (B3 #5), respectively (Fig. D1a and b). The strong acceleration results from flow velocities remaining above their pre-summer values after the annual summer speed-up. In particular, the reversible fraction of the summer speed-up decreases with time and diminishes in 2013, when the entire basin was surging (Fig. D1c). While the two upper GPS stations (B3 #4 and B3 #5) experience further drastic acceleration (factor of 24 and 18 increase in June 2013 compared to June 2008, respectively), the lower GPS stations (B3 #1 and B3 #2) only show moderate velocity increases, because the locations are close to the lateral shear margins of the surging basin (Fig. 3c and d).

Author contributions. J. O. Hagen, T. V. Schuler and T. Dunse set up the field experiment. T. Dunse analysed GPS and GPR data, wrote the manuscript and designed the figures. T. Schellenberger processed TSX data and produced velocity maps and front positions. A. Kääb provided TSX-data access and assisted in TSX processing. T. V. Schuler analysed AWS data. C. H. Reijmer provided GPS instruments and access to a quality-controlled data set. All co-authors assisted in data interpretation and commented on/edited the paper.

Acknowledgements. T. Dunse thanks T. Eiken and G. Moholdt for assistance in the field, W. Boot for construction of GPS receivers, and C. Nuth for discussions regarding the manuscript. A. Kääb and T. Schellenberger thank the DLR for provision of TSX data (LAN_0211). The GPSes were funded by the Netherlands Polar Programme (NPP) and the Netherlands Organisation for Scientific Research, Earth and Life Sciences section (NWO/ALW). This is contribution number 54 of the Nordic Centre of Excellence SVALI, "Stability and Variations of Arctic Land Ice", funded by the Nordic Top-level Research Initiative (TRI). This study was further supported by the European Union 7th Framework Programme through the project ice2sea, grant no. 226375; ERC grant agreement no. 320816; the Norwegian Space Centre/European Space Agency (ESA) through CryoVEx; the Research Council of Norway through CRYOMET, grant no. 214465; and RASTAR, grant no. 208013.

References

Alley, R. B.: In search of ice-stream sticky spots, J. Glaciol., 39, 447–454, 1993.

Alley, R. B., Fahnestock, M., and Joughin, I.: Climate Change. Understanding Glacier Flow in Changing Times, Science, 322, 1061–1062, doi:10.1126/science.1166366, 2008.

AMAP: Snow and Water and Ice and Permafrost in the Arctic (SWIPA): Climate Change and the Cryosphere, Arctic Monitoring and Assessment Programme (AMAP), vol. xii, Oslo, Norway, 2011.

Bartholomew, I., Nienow, P., Mair, D., Hubbard, A., King, M. A., and Sole, A.: Seasonal evolution of subglacial drainage and acceleration in a Greenland outlet glacier, Nat. Geosci., 3, 408–411, doi:10.1038/NGEO863, 2010.

Bartholomew, I., Nienow, P., Sole, A., Mair, D., Cowton, T., King, M., and Palmer, S.: Seasonal variations in Greenland Ice Sheet motion: Inland extent and behaviour at higher elevations, Earth Planet. Sci. Lett., 307, 271–278, doi:10.1016/j.epsl.2011.04.014, 2011.

Bond, G., Heinrich, H., Broecker, W., Labeyrie, L., McManus, J., Andrews, J., Huon, S., Jantschik, R., Clasen, S., Simet, C., Tedesco, K., Mieczyslawa, K., Bonani, G., and Ivy, S.: Evidence For Massive Discharges of Icebergs Into the North-atlantic Ocean During the Last Glacial Period, Nature, 360, 245–249, 1992.

Church, J. A., White, N. J., Konikow, L. F., Domingues, C. M., Cogley, J. G., Rignot, E., Gregory, J. M., van den Broeke, M. R., Monaghan, A. J., and Velicogna, I.: Revisiting the Earth's sea-level and energy budgets from 1961 to 2008, Geophys. Res. Lett., 38, L18601, doi:10.1029/2011GL048794, 2011.

Clarke, G.: Thermal regulation of glacier surging, J. Glaciol., 16, 231–250, 1976.

Clarke, G.: Fast glacier flow - ice streams and surging and and tide-water glaciers, J. Geophys. Res.-Solid, 92, 8835–8841, 1987.

Clarke, G., Collins, S., and Thompson, D.: Flow and thermal structure and and subglacial conditions of a surge-type glacier, Can. J. Earth Sci., 21, 232–240, 1984.

Clason, C. C., Mair, D. W. F., Nienow, P. W., Bartholomew, I. D., Sole, A., Palmer, S., and Schwanghart, W.: Modelling the transfer of supraglacial meltwater to the bed of Leverett Glacier, southwest Greenland, The Cryosphere Discuss., 8, 4243–4280, doi:10.5194/tcd-8-4243-2014, 2014.

Comiso, J. C.: Variability and trends in Antarctic surface temperatures from in situ and satellite infrared measurements, J. Climate, 13, 1674–1696, doi:10.1175/1520-0442(2000)013<1674:VATIAS>2.0.CO;2, 2000.

Cook, C. P., van de Flierdt, T., Williams, T., Hemming, S. R., Iwai, M., Kobayashi, M., Jimenez-Espejo, F. J., Escutia, C., Jairo Gonzalez, J., Khim, B.-K., McKay, R. M., Passchier, S., Bohaty, S. M., Riesselman, C. R., Tauxe, L., Sugisaki, S., Lopez Galindo, A., Patterson, M. O., Sangiorgi, F., Pierce, E. L., and Brinkhuis, H.: Dynamic behaviour of the East Antarctic ice sheet during Pliocene warmth, Nat. Geosci., 6, 765–769, doi:10.1038/NGEO1889, 2013.

Das, S. B., Joughin, I., Behn, M. D., Howat, I. M., King, M. A., Lizarralde, D., and Bhatia, M. P.: Fracture propagation to the base of the Greenland Ice Sheet during supraglacial lake drainage, Science, 320, 778–781, doi:10.1126/science.1153360, 2008.

den Ouden, M. A. G., Reijmer, C. H., Pohjola, V., van de Wal, R. S. W., Oerlemans, J., and Boot, W.: Stand-alone single-frequency GPS ice velocity observations on Nordenskiöldbreen, Svalbard, The Cryosphere, 4, 593–604, doi:10.5194/tc-4-593-2010, 2010.

Deschamps, P., Durand, N., Bard, E., Hamelin, B., Camoin, G., Thomas, A. L., Henderson, G. M., Okuno, J., and Yokoyama, Y.: Ice-sheet collapse and sea-level rise at the Bolling warming 14 and 600 years ago, Nature, 483, 559–564, doi:10.1038/nature10902, 2012.

Dowdeswell, J.: Drainage-basin characteristics of Nordaustlandet ice caps and Svalbard, J. Glaciol., 32, 31–38, 1986.

Dowdeswell, J., Drewry, D., Cooper, A., Gorman, M., Liestøl, O., and Orheim, O.: Digital mapping of the Nordaustlandet ice caps from airborne geophysical investigations, Ann. Glaciol., 8, 51–58, 1986.

Dowdeswell, J., Hamilton, G., and Hagen, J. O.: The duration of the active phase on surge-type glaciers: contrasts between Svalbard and other regions, J. Glaciol., 37, 388–400, 1991.

Dowdeswell, J., Hodgkins, R., Nutall, A., Hagen, J. O., and Hamilton, G.: Mass-balance Change As A Control On the Frequency and Occurrence of Glacier Surges In Svalbard and Norwegian High Arctic, Geophys. Res. Lett., 22, 2909–2912, 1995.

Dowdeswell, J., Ottesen, D., Evans, J., Cofaigh, C., and Anderson, J.: Submarine glacial landforms and rates of ice-stream collapse, Geology, 36, 819–822, 2008.

Dowdeswell, J. A., Unwin, B., Nuttall, A. M., and Wingham, D. J.: Velocity structure and flow instability and mass flux on a large Arctic ice cap from satellite radar interferometry, Earth Planet. Sci. Lett., 167, 131–140, 1999.

Dunse, T., Schuler, T. V., Hagen, J. O., Eiken, T., Brandt, O., and Høgda, K.: Recent fluctuations in the extent of the firn area of Austfonna and Svalbard and inferred from GPR, Ann. Glaciol., 50, 155–162, doi:10.3189/172756409787769780, 2009.

Dunse, T., Greve, R., Schuler, T. V., and Hagen, J. O.: Permanent fast flow versus cyclic surge behaviour: numerical simulations of the Austfonna ice cap and Svalbard, J. Glaciol., 57, 247–259, 2011.

Dunse, T., Schuler, T. V., Hagen, J. O., and Reijmer, C. H.: Seasonal speed-up of two outlet glaciers of Austfonna, Svalbard, inferred from continuous GPS measurements, The Cryosphere, 6, 453–466, doi:10.5194/tc-6-453-2012, 2012.

Gardner, A. S., Moholdt, G., Wouters, B., Wolken, G. J., Burgess, D. O., Sharp, M. J., Cogley, J. G., Braun, C., and Labine, C.: Sharply increased mass loss from glaciers and ice caps in the Canadian Arctic Archipelago, Nature, 473, 357–360, doi:10.1038/nature10089, 2011.

Gardner, A. S., Moholdt, G., Cogley, J. G., Wouters, B., Arendt, A. A., Wahr, J., Berthier, E., Hock, R., Pfeffer, W. T., Kaser, G., Ligtenberg, S. R. M., Bolch, T., Sharp, M. J., Hagen, J. O., van den Broeke, M. R., and Paul, F.: A Reconciled Estimate of Glacier Contributions to Sea Level Rise: 2003 to 2009, Science, 340, 852–857, doi:10.1126/science.1234532, 2013.

Hamilton, G. S. and Dowdeswell, J. A.: Controls on glacier surging in Svalbard, J. Glaciol., 42, 157–168, 1996.

Hewitt, I.: Seasonal changes in ice sheet motion due to melt water lubrication, Earth Planet. Sci. Lett., 371–372, 16–25, doi:10.1016/j.epsl.2013.04.022, 2013.

Iken, A. and Bindschadler, R.: Combined measurements of subglacial water-pressure and surface velocity of Findelengletscher and Switzerland – conclusions about drainage system and sliding mechanism, J. Glaciol., 32, 101–119, 1986.

IPCC: Climate Change 2013: The Physical Science Basis, in: Contribution of Working Group I to the Fifth Assessment Report of the Intergovernmental Panel on Climate Change, edited by: Stocker, T. F., Qin, D., Plattner, G.-K., Tignor, M., Allen, S. K., Boschung, J., Nauels, A., Xia, Y., Bex, V., and Midgley, P. M., Cambridge University Press, Cambridge, UK and New York, USA, 1535 pp., doi:10.1017/CBO9781107415324, 2013.

Jiskoot, H., Murray, T., and Boyle, P.: Controls on the distribution of surge-type glaciers in Svalbard, J. Glaciol., 46, 412–422, 2000.

Joughin, I. and Alley, R. B.: Stability of the West Antarctic ice sheet in a warming world, Nat. Geosci., 4, 506–513, doi:10.1038/NGEO1194, 2011.

Joughin, I., Smith, B. E., Shean, D. E., and Floricioiu, D.: Brief Communication: Further summer speedup of Jakobshavn Isbræ, The Cryosphere, 8, 209–214, doi:10.5194/tc-8-209-2014, 2014.

Khan, S. A., Kjaer, K. H., Bevis, M., Bamber, J. L., Wahr, J., Kjeldsen, K. K., Bjork, A. A., Korsgaard, N. J., Stearns, L. A., van den Broeke, M. R., Liu, L., Larsen, N. K., and Muresan, I. S.: Sustained mass loss of the northeast Greenland ice sheet triggered by regional warming, Nat. Clim. Change, 4, 292–299, doi:10.1038/NCLIMATE2161, 2014.

Lefauconnier, B. and Hagen, J. O.: Surging and Calving Glaciers in Eastern Svalbard, Norsk Polarinstitutt, Oslo, Norway, 1991.

Luckman, A., Murray, T., and Strozzi, T.: Surface flow evolution throughout a glacier surge measured by satellite radar interferometry, Geophys. Res. Lett., 29, 10-1–10-4, doi:10.1029/2001GL014570, 2002.

Meier, M. and Post, A.: What are glacier surges, Can. J. Earth Sci., 6, 807–817, 1969.

Meier, M. and Post, A.: Fast tidewater glaciers, J. Geophys. Res.-Solid, 92, 9051–9058, 1987.

Meierbachtol, T., Harper, J., and Humphrey, N.: Basal Drainage System Response to Increasing Surface Melt on the Greenland Ice Sheet, Science, 341, 777–779, doi:10.1126/science.1235905, 2013.

Mengel, M. and Levermann, A.: Ice plug prevents irreversible discharge from East Antarctica, Nat. Clim. Change, 4, 451–455, doi:10.1038/nclimate2226, 2014.

Moholdt, G. and Kääb, A.: A new DEM of the Austfonna ice cap by combining differential SAR interferometry with ICESat laser altimetry, Polar Res., 31, 18460, doi:10.3402/polar.v31i0.18460, 2012.

Moholdt, G., Hagen, J. O., Eiken, T., and Schuler, T. V.: Geometric changes and mass balance of the Austfonna ice cap, Svalbard, The Cryosphere, 4, 21–34, doi:10.5194/tc-4-21-2010, 2010a.

Moholdt, G., Nuth, C., Hagen, J. O., and Kohler, J.: Recent elevation changes of Svalbard glaciers derived from ICESat laser altimetry, Remote Sens. Environ., 114, 2756–2767, 2010b.

Moon, T., Joughin, I., Smith, B., van den Broeke, M. R., van de Berg, W. J., Noël, B., and Usher, M.: Distinct patterns of seasonal Greenland glacier velocity, Geophys. Res. Lett., 41, 7209–7216, doi:10.1002/2014GL061836, 2014.

Mouginot, J., Rignot, E., and Scheuchl, B.: Sustained increase in ice discharge from the Amundsen Sea Embayment, West Antarctica, from 1973 to 2013, Geophys. Res. Lett., 41, 1576–1584, doi:10.1002/2013GL059069, 2014.

Murray, T., Stuart, G., Miller, P., Woodward, J., Smith, A., Porter, P., and Jiskoot, H.: Glacier surge propagation by thermal evolution at the bed, J. Geophys. Res.-Sol. Ea., 105, 13491–13507, 2000.

Murray, T., Luckman, A., Strozzi, T., and Nuttall, A. M.: The initiation of glacier surging at Fridtjovbreen and Svalbard, Ann. Glaciol., 36, 110–116, doi:10.3189/172756403781816275, 2003a.

Murray, T., Strozzi, T., Luckman, A., Jiskoot, H., and Christakos, P.: Is there a single surge mechanism? Contrasts in dynamics between glacier surges in Svalbard and other regions, J. Geophys. Res.-Sol. Ea., 108, 2237, doi:10.1029/2002JB001906, 2003b.

Nick, F., Vieli, A., Howat, I., and Joughin, I.: Large-scale changes in Greenland outlet glacier dynamics triggered at the terminus, Nat. Geosci., 2, 110–114, 2009.

Nick, F. M., Vieli, A., Andersen, M. L., Joughin, I., Payne, A., Edwards, T. L., Pattyn, F., and van de Wal, R. S. W.: Future sea-level rise from Greenland/'s main outlet glaciers in a warming climate, Nature, 497, 235–238, doi:10.1038/nature12068, 2013.

Pattyn, F.: Antarctic subglacial conditions inferred from a hybrid ice sheet/ice stream model, Earth Planet. Sci. Lett., 295, 451–461, doi:10.1016/j.epsl.2010.04.025, 2010.

Pfeffer, W. T.: A simple mechanism for irreversible tidewater glacier retreat, J. Geophys. Res.-Earth, 112, F03S25, doi:10.1029/2006JF000590, 2007.

Phillips, T., Rajaram, H., and Steffen, K.: Cryo-hydrologic warming: A potential mechanism for rapid thermal response of ice sheets, Geophys. Res. Lett., 37, L20503, doi:10.1029/2010GL044397, 2010.

Phillips, T., Rajaram, H., Colgan, W., Steffen, K., and Abdalati, W.: Evaluation of cryo-hydrologic warming as an explanation for increased ice velocities in the wet snow zone and Sermeq Avannarleq and West Greenland, J. Geophys. Res.: Earth, 118, 1241–1256, doi:10.1002/jgrf.20079, 2013.

Raymond, C.: How do glaciers surge – a review, J. Geophys. Res.-Solid, 92, 9121–9134, 1987.

Robin, G. D. Q.: Ice movement and temperature distribution in glaciers and ice sheets, J. Glaciol., 2, 589–606, 1955.

Robinson, P. and Dowdeswell, J.: Submarine landforms and the behavior of a surging ice cap since the last glacial maximum: The open-marine setting of eastern Austfonna and Svalbard, Mar. Geol., 286, 82–94, doi:10.1016/j.margeo.2011.06.004, 2011.

Schoof, C.: Ice-sheet acceleration driven by melt supply variability, Nature, 468, 803–806, doi:10.1038/nature09618, 2010.

Schuler, T. V., Dunse, T., Ostby, T. I., and Hagen, J. O.: Meteorological conditions on an Arctic ice cap-8 years of automatic weather station data from Austfonna, Svalbard, Int. J. Climatol., 34, 2047–2058, doi:10.1002/joc.3821, 2014.

Shannon, S. R., Payne, A. J., Bartholomew, I. D., van den Broeke, M. R., Edwards, T. L., Fettweis, X., Gagliardini, O., Gillet-Chaulet, F., Goelzer, H., Hoffman, M. J., Huybrechts, P., Mair, D. W. F., Nienow, P. W., Perego, M., Price, S. F., Smeets, C. J. P. P., Sole, A. J., van de Wal, R. S. W., and Zwinger, T.: Enhanced basal lubrication and the contribution of the Greenland ice sheet to future sea-level rise, P. Natl. Acad. Sci. USA, 110, 14156–14161, doi:10.1073/pnas.1212647110, 2013.

Shepherd, A., Ivins, E. R., Geruo, A., Barletta, V. R., Bentley, M. J., Bettadpur, S., Briggs, K. H., Bromwich, D. H., Forsberg, R., Galin, N., Horwath, M., Jacobs, S., Joughin, I., King, M. A., Lenaerts, J. T. M., Li, J., Ligtenberg, S. R. M., Luckman, A., Luthcke, S. B., McMillan, M., Meister, R., Milne, G., Mouginot, J., Muir, A., Nicolas, J. P., Paden, J., Payne, A. J., Pritchard, H., Rignot, E., Rott, H., Sorensen, L. S., Scambos, T. A., Scheuchl, B., Schrama, E. J. O., Smith, B., Sundal, A. V., van Angelen, J. H., van de Berg, W. J., van den Broeke, M. R., Vaughan, D. G., Velicogna, I., Wahr, J., Whitehouse, P. L., Wingham, D. J., Yi, D., Young, D., and Zwally, H. J.: A Reconciled Estimate of Ice-Sheet Mass Balance, Science, 338, 1183–1189, doi:10.1126/science.1228102, 2012.

Solheim, A.: The depositional environment of surging sub-polar tidewater glaciers: a case study of the morphology and sedimentation and sediment properties in a surgeaffected marine basin outside Nordaustlandet and the Northern Barents Sea, Skrifter – Norsk Polarinstitutt, Oslo, Norway, 194, 5–97, 1991.

Strozzi, T., Luckman, A., Murray, T., Wegmüller, U., and Werner, C. L.: Glacier motion estimation using SAR offset-tracking procedures, IEEE T. Eosci. Remote, 40, 2384–2391, 2002.

Tedesco, M., Fettweis, X., Mote, T., Wahr, J., Alexander, P., Box, J. E., and Wouters, B.: Evidence and analysis of 2012 Greenland records from spaceborne observations, a regional climate model and reanalysis data, The Cryosphere, 7, 615–630, doi:10.5194/tc-7-615-2013, 2013.

Tulaczyk, S., Kamb, W. B., and Engelhardt, H. F.: Basal mechanics of Ice Stream B and West Antarctica 2. Undrained plastic bed model, J. Geophys. Res.-Sol. Ea., 105, 483–494, 2000.

van der Veen, C.: Fundamentals of Glacier Dynamics, Taylor & Francis, the Netherlands, 472 pp., 1999.

Zagorodnov, V., Sinkevich, S., and Arkhipov, S.: Ice core express-analysis for structure and thermal regime studies of Austfonna, Data of Glaciological Studies, 66, 149–158, 1989.

Zwally, H. J., Abdalati, W., Herring, T., Larson, K., Saba, J., and Steffen, K.: Surface melt-induced acceleration of Greenland ice-sheet flow, Science, 297, 218–222, 2002.

Permissions

All chapters in this book were first published in The Cryosphere (TC), by Copernicus Publications; hereby published with permission under the Creative Commons Attribution License or equivalent. Every chapter published in this book has been scrutinized by our experts. Their significance has been extensively debated. The topics covered herein carry significant findings which will fuel the growth of the discipline. They may even be implemented as practical applications or may be referred to as a beginning point for another development.

The contributors of this book come from diverse backgrounds, making this book a truly international effort. This book will bring forth new frontiers with its revolutionizing research information and detailed analysis of the nascent developments around the world.

We would like to thank all the contributing authors for lending their expertise to make the book truly unique. They have played a crucial role in the development of this book. Without their invaluable contributions this book wouldn't have been possible. They have made vital efforts to compile up to date information on the varied aspects of this subject to make this book a valuable addition to the collection of many professionals and students.

This book was conceptualized with the vision of imparting up-to-date information and advanced data in this field. To ensure the same, a matchless editorial board was set up. Every individual on the board went through rigorous rounds of assessment to prove their worth. After which they invested a large part of their time researching and compiling the most relevant data for our readers.

The editorial board has been involved in producing this book since its inception. They have spent rigorous hours researching and exploring the diverse topics which have resulted in the successful publishing of this book. They have passed on their knowledge of decades through this book. To expedite this challenging task, the publisher supported the team at every step. A small team of assistant editors was also appointed to further simplify the editing procedure and attain best results for the readers.

Apart from the editorial board, the designing team has also invested a significant amount of their time in understanding the subject and creating the most relevant covers. They scrutinized every image to scout for the most suitable representation of the subject and create an appropriate cover for the book.

The publishing team has been an ardent support to the editorial, designing and production team. Their endless efforts to recruit the best for this project, has resulted in the accomplishment of this book. They are a veteran in the field of academics and their pool of knowledge is as vast as their experience in printing. Their expertise and guidance has proved useful at every step. Their uncompromising quality standards have made this book an exceptional effort. Their encouragement from time to time has been an inspiration for everyone.

The publisher and the editorial board hope that this book will prove to be a valuable piece of knowledge for researchers, students, practitioners and scholars across the globe.

List of Contributors

A. M. Brisbourne
British Antarctic Survey, High Cross, Madingley Road, Cambridge, CB3 0ET, UK

A. M. Smith
British Antarctic Survey, High Cross, Madingley Road, Cambridge, CB3 0ET, UK

E. C. King
British Antarctic Survey, High Cross, Madingley Road, Cambridge, CB3 0ET, UK

K. W. Nicholls
British Antarctic Survey, High Cross, Madingley Road, Cambridge, CB3 0ET, UK

P. R. Holland
British Antarctic Survey, High Cross, Madingley Road, Cambridge, CB3 0ET, UK

K. Makinson
British Antarctic Survey, High Cross, Madingley Road, Cambridge, CB3 0ET, UK

B. Marzeion
Center of Climate and Cryopshere, Institute of Meteorology and Geophysics, University of Innsbruck, Austria

A.H. Jarosch
Institute of Earth Sciences, University of Iceland, Reykjavík, Iceland

J. M. Gregory
NCAS-Climate, University of Reading, Reading, and Met Office Hadley Centre, Exeter, UK

I. M. Howat
School of Earth Sciences and Byrd Polar Research Center, Ohio State University, Columbus, Ohio, USA

C. Porter
Polar Geospatial Center, University of Minnesota, St. Paul, Minnesota, USA

M. J. Noh
School of Earth Sciences and Byrd Polar Research Center, Ohio State University, Columbus, Ohio, USA

B. E. Smith
Polar Science Center, Applied Physics Lab, University of Washington, Seattle, Washington, USA

S. Jeong
School of Earth Sciences and Byrd Polar Research Center, Ohio State University, Columbus, Ohio, USA

T. L. Edwards
Department of Geographical Sciences, University of Bristol, Bristol BS8 1SS, UK

X. Fettweis
Department of Geography, University of Liege, Laboratory of Climatology (Bat B11), Allée du 6 Août, 2, 4000 Liège, Belgium

O. Gagliardini
Laboratoire de Glaciologie et Géophysique de l'Environnement, UJF - Grenoble 1/CNRS, 54, rue Molière BP 96, 38402 Saint-Martin-d'Hères Cedex, France
Institut Universitaire de France, Paris, France

F. Gillet-Chaulet
Laboratoire de Glaciologie et Géophysique de l'Environnement, UJF - Grenoble 1/CNRS, 54, rue Molière BP 96, 38402 Saint-Martin-d'Hères Cedex, France

H. Goelzer
Earth System Sciences & Departement Geografie, Vrije Universiteit Brussel, Pleinlaan 2, 1050 Brussel, Belgium

J. M. Gregory
NCAS-Climate, Department of Meteorology, University of Reading, Reading, UK
Met Office Hadley Centre, Exeter, UK

M. Hoffman
Fluid Dynamics and Solid Mechanics Group, Los Alamos National Laboratory, T3 MS B216, Los Alamos, NM 87545, USA

P. Huybrechts
Earth System Sciences & Departement Geografie, Vrije Universiteit Brussel, Pleinlaan 2, 1050 Brussel, Belgium

A. J. Payne
Department of Geographical Sciences, University of Bristol, Bristol BS8 1SS, UK

M. Perego
Department of Scientific Computing, Florida State University, 400 Dirac Science Library, Tallahassee, FL 32306, USA

S. Price
Fluid Dynamics and Solid Mechanics Group, Los Alamos National Laboratory, T3 MS B216, Los Alamos, NM 87545, USA

A. Quiquet
Laboratoire de Glaciologie et Géophysique de l'Environnement, UJF - Grenoble 1/CNRS, 54, rue Molière BP 96, 38402 Saint-Martin-d'Hères Cedex, France

C. Ritz
Laboratoire de Glaciologie et Géophysique de l'Environnement, UJF - Grenoble 1/CNRS, 54, rue Molière BP 96, 38402 Saint-Martin-d'Hères Cedex, France

S. A. Gregory
Thayer School of Engineering, Dartmouth College, Hanover, NH 03766, USA

M. R. Albert
Thayer School of Engineering, Dartmouth College, Hanover, NH 03766, USA

I. Baker
Thayer School of Engineering, Dartmouth College, Hanover, NH 03766, USA

C. G. Schmitt
National Center for Atmospheric Research, Boulder, Colorado, USA
American Climber Science Program, Eldora, Colorado, USA

J. D. All
Department of Geography and Geology, Western Kentucky University, Bowling Green, Kentucky, USA
American Climber Science Program, Eldora, Colorado, USA

J. P. Schwarz
Chemical Sciences Division, Earth System Research Laboratory, National Oceanic and Atmospheric Administration, Boulder, Colorado, USA
Cooperative Institute for Research in Environmental Sciences, University of Colorado, Boulder, Colorado, USA

W. P. Arnott
Department of Physics, University of Nevada, Reno, Nevada, USA

R. J. Cole
Institute of Arctic and Alpine Research, University of Colorado, Boulder, Colorado, USA
Department of Natural Resources and Environmental Management, University of Hawaii, Manoa, Hawaii, USA
American Climber Science Program, Eldora, Colorado, USA

E. Lapham
American Climber Science Program, Eldora, Colorado, USA

A. Celestian
Department of Geography and Geology, Western Kentucky University, Bowling Green, Kentucky, USA

N. Steiner
The City College of New York at the City University of New York, New York, USA

M. Tedesco
The City College of New York at the City University of New York, New York, USA

B. de Fleurian
CNRS, LGGE UMR5183, 38041 Grenoble, France
Univ. Grenoble Alpes, LGGE, 38041 Grenoble, France

O. Gagliardini
CNRS, LGGE UMR5183, 38041 Grenoble, France
Univ. Grenoble Alpes, LGGE, 38041 Grenoble, France
Institut Universitaire de France, Paris, France

T. Zwinger
CSC - IT Center for Science Ltd, Espoo, Finland

G. Durand
CNRS, LGGE UMR5183, 38041 Grenoble, France
Univ. Grenoble Alpes, LGGE, 38041 Grenoble, France

E. Le Meur
CNRS, LGGE UMR5183, 38041 Grenoble, France
Univ. Grenoble Alpes, LGGE, 38041 Grenoble, France

D. Mair
Department of Geography and Environment, University of Aberdeen, Aberdeen, UK

P. Råback
CSC - IT Center for Science Ltd, Espoo, Finland

J. C. Ryan
Centre for Glaciology, Institute of Geography and Earth Sciences, Aberystwyth University, Aberystwyth, SY233DB, UK

A. L. Hubbard
Department of Geology, University of Tromsø, 9037 Tromsø, Norway

J. E. Box
Geological Survey of Denmark and Greenland, Copenhagen, Denmark

J. Todd
Scott Polar Research Institute, University of Cambridge, Cambridge, UK

P. Christoffersen
Scott Polar Research Institute, University of Cambridge, Cambridge, UK

J. R. Carr
Centre for Glaciology, Institute of Geography and Earth Sciences, Aberystwyth University, Aberystwyth, SY233DB, UK

T. O. Holt
Centre for Glaciology, Institute of Geography and Earth Sciences, Aberystwyth University, Aberystwyth, SY233DB, UK

N. Snooke
Department of Computer Science, Aberystwyth University, Aberystwyth, SY23 3DB, UK

C. J. Legleiter
Department of Geography, University of Wyoming, Dept. 3371, 1000 E. University Ave., Laramie, WY 82071, USA

M. Tedesco
Department of Earth and Atmospheric Sciences, The City College of New York, Marshak MR-106, 160 Convent Avenue, New York, NY 10031, USA

L. C. Smith
Department of Geography, University of California Los Angeles, Box 951524, Los Angeles, CA 90095-1524, USA

A. E. Behar
Jet Propulsion Laboratory, 4800 Oak Grove Drive, Pasadena, CA 91109-8099, USA

B. T. Overstreet
Department of Geography, University of Wyoming, Dept. 3371, 1000 E. University Ave., Laramie, WY 82071, USA

M. J. Beedle
Geography Program, University of Northern British Columbia, Prince George, BC, Canada

B. Menounos
Geography Program, University of Northern British Columbia, Prince George, BC, Canada

R. Wheate
Geography Program, University of Northern British Columbia, Prince George, BC, Canada

C. C. Clason
Department of Physical Geography and Quaternary Geology, Stockholm University, 106 91 Stockholm, Sweden

D. W. F. Mair
Geography and Environment, University of Aberdeen, Aberdeen, AB24 3UF, UK

P. W. Nienow
School of Geosciences, University of Edinburgh, Edinburgh, EH8 9XP, UK

I. D. Bartholomew
School of Geosciences, University of Edinburgh, Edinburgh, EH8 9XP, UK

A. Sole
Department of Geography, University of Sheffield, Sheffield, S10 2TN, UK

S. Palmer
Geography, College of Life and Environmental Sciences, University of Exeter, Exeter, EX4 4RJ, UK

W. Schwanghart
Institute of Earth and Environmental Science, University of Potsdam, 14476 Potsdam-Golm, Germany

A. A.W. Fitzpatrick
Institute of Geography and Earth Sciences, Aberystwyth University, Aberystwyth, UK

A. L. Hubbard
Institute of Geography and Earth Sciences, Aberystwyth University, Aberystwyth, UK

J. E. Box
Geological Survey of Denmark and Greenland (GEUS), Copenhagen, Denmark

D. J. Quincey
School of Geography, University of Leeds, Leeds, LS2 9JT, UK

D. van As
Geological Survey of Denmark and Greenland (GEUS), Copenhagen, Denmark

A. P. B. Mikkelsen
Department of Geography and Geology, University of Copenhagen, Denmark

S. H. Doyle
Institute of Geography and Earth Sciences, Aberystwyth University, Aberystwyth, UK

C. F. Dow
Glaciology Group, College of Science, Swansea University, Swansea, SA2 8PP, UK

B. Hasholt
Department of Geography and Geology, University of Copenhagen, Denmark

G. A. Jones
Institute of Geography and Earth Sciences, Aberystwyth University, Aberystwyth, UK

A. Legchenko
IRD, UJF-Grenoble 1/CNRS/G-INP, LTHE UMR5564, BP 53, Grenoble Cedex 9, 38041, France

C. Vincent
Laboratoire de Glaciologie et Géophysique de l'Environnement and CNRS - LGGE, 38041 Grenoble Cedex 9, France

J. M. Baltassat
BRGM, BP 6009, 45060, Orléans Cedex 2, France

J. F. Girard
BRGM, BP 6009, 45060, Orléans Cedex 2, France

E. Thibert
IRSTEA, UR ETGR, Erosion torrentielle neige et avalanches, 2 rue de la Papeterie-BP 76, 38402 Saint Martin d'Hères, France
Université Grenoble Alpes, 38041 Grenoble, France

O. Gagliardini
Laboratoire de Glaciologie et Géophysique de l'Environnement and CNRS - LGGE, 38041 Grenoble Cedex 9, France
Institut Universitaire de France, IUF, Paris, France

M. Descloitres
IRD, UJF-Grenoble 1/CNRS/G-INP, LTHE UMR5564, BP 53, Grenoble Cedex 9, 38041, France

A. Gilbert
Laboratoire de Glaciologie et Géophysique de l'Environnement and CNRS - LGGE, 38041 Grenoble Cedex 9, France

S. Garambois
Institut des Sciences de la terre (ISTerre), CNRS/ Université Joseph Fourier, Grenoble, France

A. Chevalier
IRD, UJF-Grenoble 1/CNRS/G-INP, LTHE UMR5564, BP 53, Grenoble Cedex 9, 38041, France

H. Guyard
IRD, UJF-Grenoble 1/CNRS/G-INP, LTHE UMR5564, BP 53, Grenoble Cedex 9, 38041, France

M. P. Lüthi
Versuchsanstalt für Wasserbau, Hydrologie und Glaziologie (VAW), ETH Zurich, 8093 Zurich, Switzerland

C. Ryser
Versuchsanstalt für Wasserbau, Hydrologie und Glaziologie (VAW), ETH Zurich, 8093 Zurich, Switzerland

L. C. Andrews
Institute for Geophysics, The University of Texas at Austin, Austin, Texas, 78758, USA
Dept. of Geological Sciences, The University of Texas at Austin, Austin, Texas, 78713, USA

G. A. Catania
Institute for Geophysics, The University of Texas at Austin, Austin, Texas, 78758, USA
Dept. of Geological Sciences, The University of Texas at Austin, Austin, Texas, 78713, USA

M. Funk
Versuchsanstalt für Wasserbau, Hydrologie und Glaziologie (VAW), ETH Zurich, 8093 Zurich, Switzerland

R. L. Hawley
Dept. of Earth Sciences, Dartmouth College, Hanover, New Hampshire, 03755, USA

M. J. Hoffman
Fluid Dynamics and Solid Mechanics Group, Los Alamos National Laboratory, Los Alamos, New Mexico, 87545, USA

T. A. Neumann
NASA Goddard Space Flight Center, Code 615, Greenbelt, Maryland, 20770, USA
Geographical Institute, University of Zurich, 8057 Zurich, Switzerland

T. Kleiner
Alfred Wegener Institute, Helmholtz Centre for Polar and Marine Research, Am Alten Hafen 26, 27568 Bremerhaven, Germany

M. Rückamp
Alfred Wegener Institute, Helmholtz Centre for Polar and Marine Research, Am Alten Hafen 26, 27568 Bremerhaven, Germany

J. H. Bondzio
Alfred Wegener Institute, Helmholtz Centre for Polar and Marine Research, Am Alten Hafen 26, 27568 Bremerhaven, Germany

A.Humbert
Alfred Wegener Institute, Helmholtz Centre for Polar and Marine Research, Am Alten Hafen 26,

27568 Bremerhaven, Germany

R. Calov
Potsdam Institute for Climate Impact Research, Potsdam, Germany

A.Robinson
Universidad Complutense Madrid, 28040 Madrid, Spain
Instituto de Geociencias, UCM-CSIC, 28040 Madrid, Spain
Potsdam Institute for Climate Impact Research, Potsdam, Germany

M. Perrette
Potsdam Institute for Climate Impact Research, Potsdam, Germany

A.Ganopolski
Potsdam Institute for Climate Impact Research, Potsdam, Germany

T. Dunse
Department of Geosciences, University of Oslo, P.O. Box 1047, Blindern, 0316 Oslo, Norway

T. Schellenberger
Department of Geosciences, University of Oslo, P.O. Box 1047, Blindern, 0316 Oslo, Norway

J. O. Hagen
Department of Geosciences, University of Oslo, P.O. Box 1047, Blindern, 0316 Oslo, Norway

A. Kääb
Department of Geosciences, University of Oslo, P.O. Box 1047, Blindern, 0316 Oslo, Norway

T. V. Schuler
Department of Geosciences, University of Oslo, P.O. Box 1047, Blindern, 0316 Oslo, Norway

C. H. Reijmer
Institute for Marine and Atmospheric Research Utrecht, Utrecht University, Princetonplein 5, 3584 CC Utrecht, the Netherlands